Lecture Notes in Computer Science   16578

Founding Editors

Gerhard Goos
Juris Hartmanis

Editorial Board Members

Elisa Bertino, *Purdue University, West Lafayette, IN, USA*
Wen Gao, *Peking University, Beijing, China*
Bernhard Steffen, *TU Dortmund University, Dortmund, Germany*
Moti Yung, *Columbia University, New York, NY, USA*

The series Lecture Notes in Computer Science (LNCS), including its subseries Lecture Notes in Artificial Intelligence (LNAI) and Lecture Notes in Bioinformatics (LNBI), has established itself as a medium for the publication of new developments in computer science and information technology research, teaching, and education.

LNCS enjoys close cooperation with the computer science R & D community, the series counts many renowned academics among its volume editors and paper authors, and collaborates with prestigious societies. Its mission is to serve this international community by providing an invaluable service, mainly focused on the publication of conference and workshop proceedings and postproceedings. LNCS commenced publication in 1973.

Marie-Pierre Béal · Pascal Caron

Editors

Developments in Language Theory

30th International Conference, DLT 2026
Rouen, France, June 30 – July 3, 2026
Proceedings

 Springer

Editors
Marie-Pierre Béal
LIGM Université Gustave Eiffel
Champs-sur-Marne, France

Pascal Caron
Université de Rouen Normandie
Saint-Étienne du Rouvray, France

ISSN 0302-9743   ISSN 1611-3349 (electronic)
Lecture Notes in Computer Science
ISBN 978-3-032-28403-7   ISBN 978-3-032-28404-4 (eBook)
https://doi.org/10.1007/978-3-032-28404-4

Preface

The 30th International Conference on Developments in Language Theory (DLT 2026) was co-organized by the three computer science laboratories of Normandy, including the Groupe de Recherche Rouennais en Informatique Fondamentale (GR^{2}IF), the Groupe de recherche en informatique, image, automatique et instrumentation de Caen (GREYC), and the Laboratoire d'Informatique, de Traitement de l'Information et des Systèmes (LITIS) of Université de Rouen Normandie. The conference took place from June 30 to July 3, 2026.

The DLT (Developments in Language Theory) conference series brings together researchers in the field of formal language theory and its applications, fostering the exchange and dissemination of results in these areas.

This volume of Lecture Notes in Computer Science contains the scientific papers presented at DLT 2026. The volume also includes extended abstracts or articles of the three invited talks presented by Pamela Fleischmann, Beatrice Palano, and Jean-Eric Pin, who we wish to warmly thank.

The 22 regular papers were selected from 49 submissions covering various fields in formal language theory. Each paper was reviewed by three Program Committee members with the assistance of external referees and thoroughly discussed by the Program Committee.

Authors from the following countries submitted papers: Belgium, Canada, Czech Republic, Finland, France, Germany, Italy, Japan, Portugal, Russia, Slovakia, Switzerland, United Kingdom, United Arab Emirates, and United States of America.

We wish to thank everybody who contributed to the success of this conference: the authors for submitting their carefully prepared manuscripts, the Program Committee members and external referees for their valuable judgment of the submitted manuscripts, and the invited speakers for their excellent presentations of topics related to the theme of the conference. Last but not least, we would like to express our sincere thanks to the local organizers Nicolas Bedon, Julien Clément, Julien David, Solène Guérin, Jean-Gabriel Luque, Florent Nicart, and Bruno Patrou.

April 2026

Marie-Pierre Béal
Pascal Caron

Organization

Organization Committee

Nicolas Bedon	Université de Rouen Normandie, France
Pascal Caron (Chair)	Université de Rouen Normandie, France
Julien Clément	Université de Caen Normandie, France
Julien David	Université de Caen Normandie, France
Jean-Gabriel Luque	Université de Rouen Normandie, France
Ludovic Mignot	Université de Rouen Normandie, France
Florent Nicart	Université de Rouen Normandie, France
Bruno Patrou	Université de Rouen Normandie, France

Program Committee Chairs

Marie-Pierre Béal	Université Gustave Eiffel, France
Pascal Caron	Université de Rouen Normandie, France

Steering Committee

Marie-Pierre Béal	Université Gustave Eiffel, France
Volker Diekert	Universität Stuttgart, Germany
Dora Giammarresi	Università di Roma "Tor Vergata", Italy
Yo-Sub Han	Yonsei University, South Korea
Natasha Jonoska	University of South Florida, USA
Martin Kutrib	Universität Giessen, Germany
Ian McQuillan	University of Saskatchewan, Canada
Giovanni Pighizzini (Chair)	Università degli Studi di Milano, Italy
Michel Rigo	Université de Liège, Belgium
Kai Salomaa	Queen's University, Canada
Shinnosuke Seki	University of Electro-Communications, Japan
Mikhail Volkov	Ural Federal University, Russia

Program Committee

Marie-Pierre Béal	Université Gustave Eiffel, France
Véronique Bruyère	Université de Mons, Belgium
Pascal Caron	Université de Rouen Normandie, France
Émilie Charlier	Université de Liège, Belgium
Laura Ciobanu	Heriot-Watt University, United Kingdom
Frank Drewes	Umeå University, Sweden
Szilárd Zsolt Fazekas	Akita University, Japan
Dora Giammarresi	University of Roma Tor Vergata, Italy
Yo-Sub Han	Yonsei University, South Korea
Jarkko Kari	University of Turku, Finland
Christof Löding	RWTH Aachen University, Germany
Ian McQuillan	University of Saskatchewan, Canada
ThomasPlace	Université de Bordeaux, France
Marinella Sciortino	University of Palermo, Italy
Marek Szykuła	University of Wrocław, Poland

Additional Reviewers

Duncan Adamson	Shibashis Guha
Aistis Atminas	Vesa Halava
Bernard Boigelot	Benjamin Hellouin de Menibus
Tobias Brockmeyer	David Hyland
Arnaud Carayol	Yusuke Inoue
Olivier Carton	Kamil Khadiev
Giuseppa Castiglione	Sungmin Kim
Thomas Colcombet	Jakub Konieczny
Maxime Crochemore	Savinien Kreczman
Flavio D'Alessandro	Denis Kuperberg
Alessandro De Luca	Chris Köcher
Henk Don	Arnaud Lefebvre
Ruiwen Dong	Julien Leroy
Jean-Philippe Dubernard	Alex Levine
Kévin Dubrulle	Brennan Lockinger
Jonas Ellert	Ali Lotfi
Henning Fernau	Jean-Gabriel Luque
Emmanuel Filiot	Florin Manea
Islam Foniqi	Benjamin Monmege
Anna Frid	Etienne Moutot
Estéban Gabory	András Murvai
Christophe Grandmont	Yuto Nakashima

Alexander Okhotin
Bruno Patrou
Maciej Piróg
Aditya Prakash
Luca Prigioniero
Narad Rampersad
Priscilla Raucci
Antoine Renard
Antonio Restivo
Tina Ringleb
Giuseppe Romana
Andrew Ryzhikov
Aleksi Saarela
Pierluigi San Pietro
A. C. Cem Say
Shinnosuke Seki
Carla Selmi

Jeffrey Shallit
Arseny Shur
Ryoma Sin'ya
Michał Skrzypczak
Timo Specht
Wolfgang Steiner
Géraud Sénizergues
Guillaume Theyssier
Alexander Thumm
Pascal Weil
Matthias Wendlandt
Markus Whiteland
Max Wiedenhöft
Maximilian Winkler
Sarah Winter
Abuzer Yakaryilmaz

Contents

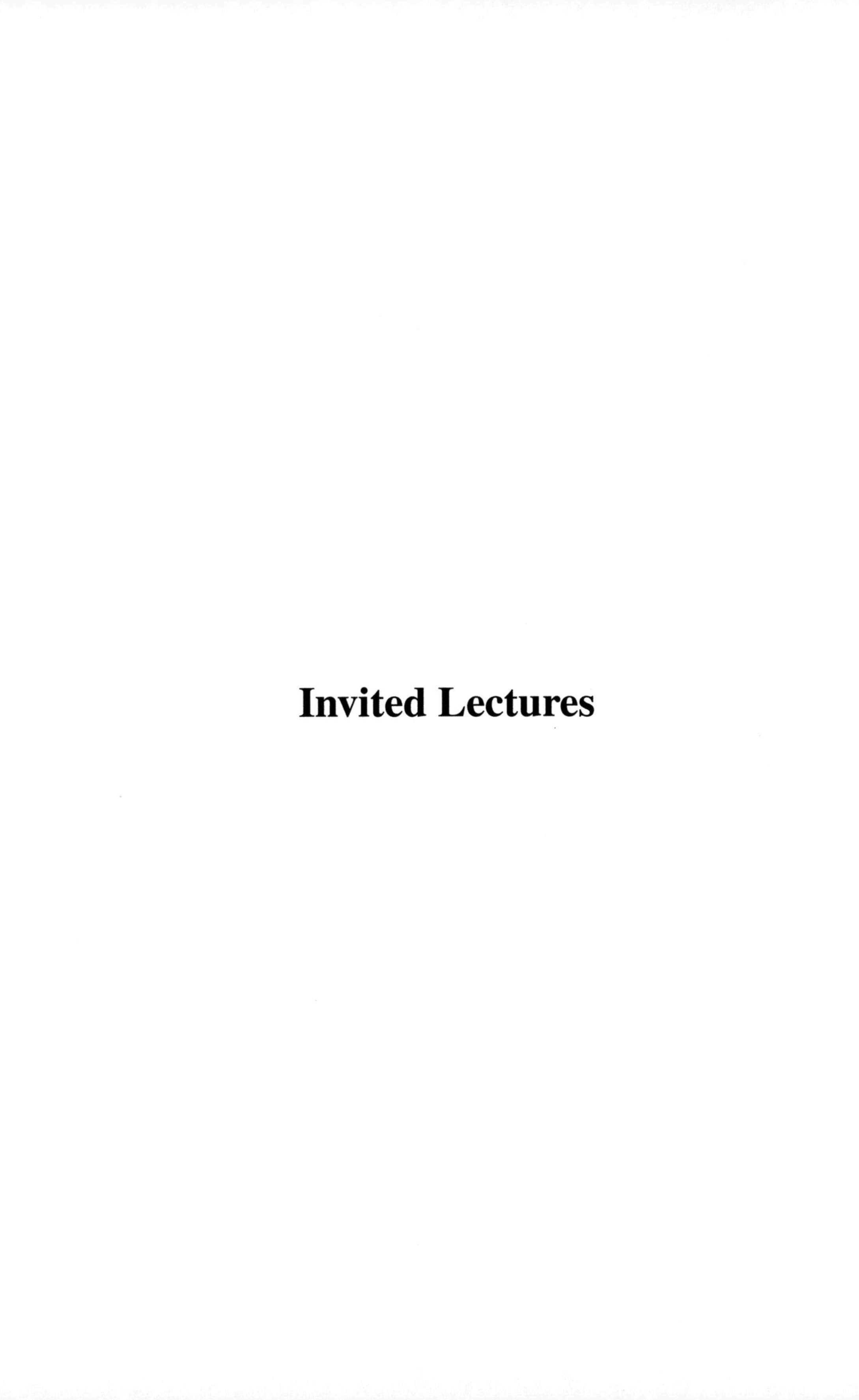

Invited Lectures

On Some Decision Problems on Quantum Automata

Flavio D'Alessandro[1], Carlo Mereghetti[2], Beatrice Palano[2(✉)],
and Paolo Papi[1]

[1] Dipartimento di Matematica "G. Castelnuovo", Sapienza Università di Roma, 00185
Roma, Italy
{flavio.dalessandro,paolo.papi}@uniroma1.it
[2] Dipartimento di Informatica "G. Degli Antoni", Università degli Studi di Milano,
20133 Milano, Italy
{carlo.mereghetti,beatrice.palano}@unimi.it

Abstract. This work continues the investigation on *measure-once quantum finite automata* by topological arguments, building on the work by Bertoni et al. [6,10]. We study conditions ensuring the decidability of *intersection* and *classification problems* for quantum finite automata, with respect to families of context-free languages and some variants.

Keywords: Quantum automata · Context-free languages · Algebraic groups · Decidability

1 Introduction

Quantum finite automata (QFAs) represent a theoretical model for a quantum computer with finite memory. While we can hardly expect to see a full-featured quantum computer in the near future, small quantum components, theoretically modeled by QFAs, seem to be promising from a physical implementation viewpoint (see, e.g., [17,28]). Very roughly speaking, a QFA is obtained by imposing the quantum paradigm—superposition, unitary evolution, observation—to a classical finite state automaton. The state of the QFA is described by a linear combination of classical basis states, called superposition. The QFA steps from a superposition to the next one by a unitary (reversible) evolution. Superpositions can transfer the complexity of a computation from a large number of sequential steps to a large number of coherently superposed classical basis states (this phenomenon is sometimes referred to as quantum parallelism). Along its computation, the QFA can be "observed", that is, some features, called observables, can be measured. From measuring an observable, an outcome is seen with a certain

The first and the fourth author acknowledge their membership to the National Group for Algebraic and Geometric Structures, and their Applications (GNSAGA–INdAM). The second and the third author acknowledge their membership to the National Group for Scientific Computing (GNCS–INdAM).

M.-P. Béal and P. Caron (Eds.): DLT 2026, LNCS 16578, pp. 3–17, 2026.
https://doi.org/10.1007/978-3-032-28404-4_1

probability and the current superposition irreversibly "collapses", with the same probability, to a particular superposition coherent with the observed outcome.

Variants of QFAs have been introduced (see, e.g., [8,29,38]), basically differing on the observation policy. Here, we focus on the original and simplest model of a QFA, namely, the *measure-once* QFA [4,5,16,30]. For this type of quantum device, an observation takes place only once, at the end of input processing. This observation yields the probability of seeing an accepting state, and hence, of accepting the input string. Being the only model under consideration in this paper, for the sake of conciseness, we will be simply writing QFA in the sequel, understanding the designation "measure-once".

A well consolidated trend in the literature has been pointing out strength and weakness of QFAs. From a computational point of view, they are strictly less powerful than classical models, in that they recognize a proper subclass of regular languages, namely, the group languages [4,5,16,30,35,40]. On the other hand, they can greatly outperform their classical counterparts when size, measured by the number of basis states, is at stake [9,11,12,27,41].

The decidability of several properties for QFAs has been deeply investigated (see, e.g. [6,10,14,15,20,24,25,29]). Here, we tackle some decision problems for QFAs, by using a topological approach first introduced in [15] and further exploited in, e.g., [2,6,10]. This approach is strictly related to some computational issues on matrices. Precisely, a crucial step in decision algorithm design amounts to effectively computing the Zariski closure of a linear group over $\mathbb{Q}$.

This topic has been object of a recent intensive research, which we would like to quickly overview. For the case of finitely generated groups of matrices, an algorithm has been first developed in [20], by Derksen, Jeandel, and Koiran. In 2022, following an approach alternative to [20], Nosan, Pouly, Schmitz, Shirmohammadi, and Worrell, have devised in [32] a bound on the degree of the polynomials defining the Zariski closure of a matrix set, thus obtaining an algorithm returning the closure in elementary time. In 2023, Hrushovski, Ouaknine, Pouly, and Worrell, have investigated the effective computation of polynomial invariants for affine programs [23]. Therein, an algorithm is exhibited which, given a finite set of square matrices of the same size with coefficients in $\mathbb{Q}$, computes the Zariski closure of the semigroup generated by the set. These techniques have been recently considered in [26], for computing the Zariski closure of matrix sets representing context-free languages in the family 1-VASS.

It should be stressed that the decidability issues for matrix semigroups turns out to be a challenging problem. Even for small dimension, a well-known result by Paterson shows that the problems of freeness and membership are both undecidable for 3×3 matrix semigroups over $\mathbb{Z}$ [18]. In contrast, in 2017, by following an approach developed by Choffrut and Karhumaki in [18], a result of Potapov and Semukhin shows that membership is decidable, for non-singular 2×2 integer matrix semigroups [37].

Let us now get back to the subject matter of this paper. To properly frame the line of research and our contribution, we quickly lay down some formalism. To our purpose, a QFA with n basis states on input alphabet Σ, is a triple

$Q = (s, \varphi, P)$, where $s \in \mathbb{R}^n$ is the row-vector of unit Euclidean norm containing the amplitudes of the initial superposition of basis states, $P \in \{0, 1\}^{n \times n}$ is the projector onto the accepting subspace, and $\varphi : \Sigma^* \to O_n$ is the morphism into the group $O_n \subset \mathbb{R}^{n \times n}$ of *orthogonal* matrices, generated by the orthogonal evolution steps $\{\varphi(\sigma)\}_{\sigma \in \Sigma}$. The probability that Q accepts an input string $w \in \Sigma^*$ is computed as $f_Q(w) = \|s\, \varphi(w)\, P\|^2$, so that the language accepted by Q with cut-point $\lambda \in \mathbb{R}$ is defined to be the set $L_{Q,\lambda} = \{w \in \Sigma^* : f_Q(w) > \lambda\}$. The cut-point is said to be isolated whenever $\delta \in \mathbb{R}_+$ exists, such that $|f_Q(w) - \lambda| \geq \delta$ for every $w \in \Sigma^*$. Since we are mainly interested in algorithmically verifiable properties, we require QFAs to be effectively provided as input to our deciding algorithms. Thus, we consider *rational* QFAs, i.e., in which all the components and cut-points are rationals.

In [15,20], it is first emphasized the relevance of the topological structure of the *set* $\varphi(\Sigma^*) = \{\varphi(w) : w \in \Sigma^*\} \subset O_n$ *of the dynamics of* Q, when deciding properties for QFAs. In particular, the authors exploit the fact, proved in [33, Ch. 3, Sec. 4, Thm. 5], that the Euclidean closure $Cl(\varphi(\Sigma^*))$ coincides with the Zariski closure $\overline{\varphi(\Sigma^*)}$ over $\mathbb{R}$ or, stated in other words, that $Cl(\varphi(\Sigma^*))$ is algebraic over $\mathbb{R}$, i.e., is the vanishing set of a polynomial $p : \mathbb{R}^{n \times n} \to \mathbb{R}$. In addition, for rational QFAs, an algorithm is provided, effectively returning such p. As a consequence, several decision problems on rational QFAs are turned into deciding the truth value of some formulae in the first-order logic over $\mathbb{R}$. This latter task is well known to be algorithmically solvable by the Tarski-Seidenberg quantifier elimination [1,39]. It is worth stresseing that the obtained decidability of some problems on QFAs sharply contrasts the undecidability of the same problems on classical probabilistic finite automata [7,34].

Intersection. The topological approach is further pursued in [2,6], where the decision problem (L, Q) INTERSECTION is settled: given a language L from a fixed effective family of languages, and a rational QFA Q with cut-point λ, decide whether or not $L \cap L_{Q,\lambda} = \emptyset$ holds. This problem actually generalizes some decision problems considered in [15,20]. In this scenario, the key issue under investigation now becomes the topological structure of the *set* $\varphi(L)$ *of the dynamics of* Q *on* L. The authors prove that (L, Q) INTERSECTION is decidable for those languages yielding an effective semialgebraic $Cl(\varphi(L))$. A set is (effective) semialgebraic whenever it can be defined by (algorithmically definable) polynomial inequalities. The class of semialgebraic sets contains that of algebraic sets. As in the case of algebraic sets, semialgebricity enables to turn decision problems into deciding first order formulae over $\mathbb{R}$. Context-free languages are shown to yield semialgebraic $Cl(\varphi(L))$, while effectiveness is proved for several language families, such as, linear context-free, bounded semilinear, and monoidal languages. As a consequence, (L, Q) INTERSECTION turns out to be decidable, for L coming from such families.

Classification. Decidability issues related to quantum classification are tackled in [10], again from a topological viewpoint. A *k-quantum classifier* is a k-tuple of QFAs on input alphabet Σ, satisfying: (i) each QFA accepts at least one word in Σ^*, and (ii) every word in Σ^* is accepted by at most one QFA. The intended meaning is that all the QFAs in the team compete for accepting a word, but at most one of them succeeds. A classifier is said to be *complete* if, in addition, every word in Σ^* is accepted by exactly one among the k QFAs. The authors study the two problems (COMPLETE) k-Q.CLASSIFIER of deciding whether a k-tuple of rational QFAs is a (complete) k-quantum classifier. It is worth noticing that (COMPLETE) 1-Q.CLASSIFIER is the non-emptiness (universality) problem for QFAs, which is proved to be (un)decidable in [15,20]. From this point of view, the authors extend these two results to the general case $k > 1$.

1.1 Our Contribution

In this paper, we build and expand on these results in the literature, again adopting topological tools. We obtain the following improvements:

Intersection. A context-free language L is said to be of finite index $k \in \mathbb{N}$ whenever it can be generated by a context-free grammar in which, for every word in L, there exists a derivation exhibiting at most k nonterminals in its sentential forms. The family of *finite index context-free languages* possesses several interesting characterizations [3,22], among which we focus on that stating that L can be generated by a grammar $G = \mathcal{G}_1 \circ \cdots \circ \mathcal{G}_k$, obtained as a composition of k families of linear grammars. This characterization enables us to express the closure $\mathbf{Cl}(\varphi(L))$ of the dynamics of a rational QFA $\mathcal{Q}$ on L as a suitable combination of the *monoids of cycles* M_A, for every variable A of G. Roughly speaking, any monoid M_A corresponds (up to a technical detail) to the matrix image, according to φ, of the language of cycles in G associated with the variable A [2,6]. In particular, owing to a result in [32, Thm. 11] stating the finiteness of ascending chains of linear algebraic groups over the field $\mathbb{Q}$, and to the structuring in terms of cycle monoids mentioned above, we prove that $(L, \mathcal{Q})$ INTERSECTION is decidable for finite index context-free language and rational QFAs, thus extending the results provided in [2].

Classification. First of all, we define "relativized" versions of the quantum classification decision problems. More precisely, we introduce the decision problems (COMPLETE) $[L]\,k$-Q.CLASSIFIER, which now require to check whether or not the (complete) quantum classifications take place on the language $L \subseteq \Sigma^*$, instead of the whole Σ^*. As for the problem $(L, \mathcal{Q})$ INTERSECTION, we first prove that $[L]\,k$-Q.CLASSIFIER is decidable for languages L for which $\mathbf{Cl}(\varphi(L))$ is effective semialgebraic where, roughly speaking, φ here represents the direct sum of the dynamics of the k rational QFAs to be tested. This enables us the obtain the decidability of $[L]\,k$-Q.CLASSIFIER, for L being a simple matrix linear language or finite index context-free language.

Concerning COMPLETE $[L]\,k$-Q.CLASSIFIER, as above recalled, we already have that COMPLETE $[\Sigma^*]\,k$-Q.CLASSIFIER is undecidable. Therefore, we find it interesting, to consider the version of the problem, in which we test whether or not k-tuples of rational QFAs perform quantum classification by working with *isolated cut-points*. We prove that this "isolated cut-point" variant of the problem COMPLETE $[L]\,k$-Q.CLASSIFIER turns out to be decidable for L being a regular language. In addition, to provide more generality, we introduce the decision problem SUPER-COMPLETE $[L]$ k-Q.CLASSIFIER, in which the input k-tuple of rational QFAs is allowed to classify over any super-language of L. Again, by topological arguments, we prove that the isolated cut-point version of the problem SUPER-COMPLETE $[L]$ k-Q.CLASSIFIER is decidable for L being a simple matrix linear language or finite index context-free language.

Section 2 collects basics on topology, algebraic, and semialgebraic sets. In Sect. 3, we review introductory notions on context-free languages, with a focus on the family of finite index context-free languages. Several characterization of this family are recalled, particularly the one involving grammar compositions. In Sect. 4, we highlight the main ingredients of the topological approach in assessing problem decidability. In Sect. 5, we present our original contribution on quantum intersection and classification decidability.

Due to page limit, some background material and proofs are omitted. In particular, we refer the reader to, e.g., [8, 13, 38], for standard notions, notations, and results on **quantum finite automata** (QFAs).

2   Algebraic and Semialgebraic Sets

We recall the definition of algebraic set over the field of real numbers [1, 21, 33]:

Definition 1. *A subset $\mathcal{A} \subseteq \mathbb{R}^n$ is* algebraic *(over the field of real numbers), if and only if it is the zero set of an arbitrary subset $\mathcal{P}$ of polynomials in $\mathbb{R}[x_1, \ldots, x_n]$, i.e., $\mathcal{A} = \{v \in \mathbb{R}^n : p(v) = 0,\ for\ every\ p \in \mathcal{P}\}$.*

Note that by Hilbert's basis theorem (see, e.g., [21, Ch. 1, Sec. 1, Thm. 1]), one may assume that the set $\mathcal{P}$ is finite. Even more, since we are dealing with algebraic sets over $\mathbb{R}$, then $\mathcal{P}$ can be reduced to a singleton since the condition in Definition 1 can be equivalently replaced by the single condition $\sum_{p \in \mathcal{P}} p(x)^2 = 0$. The family of algebraic sets is closed under intersections and finite unions.

In the *Zariski topology* on $\mathbb{R}^n$, a set $\mathcal{Z} \subseteq \mathbb{R}^n$ is an *open* whenever its complement $\mathbb{R}^n \setminus \mathcal{Z}$ is algebraic. Thus, in this topology, *closed* subsets are algebraic. The *Zariski closure* of any given set $\mathcal{X} \subseteq \mathbb{R}^n$, i.e., the smallest closed set containing $\mathcal{X}$, will be denoted by $\overline{\mathcal{X}}$ (see, e.g., [21, Ch. 1]).

We now apply these definitions to square matrices:

Definition 2. *A set $\mathcal{A} \subseteq \mathbb{R}^{n \times n}$ of matrices is* algebraic *if and only if, by considering each matrix in $\mathcal{A}$ as an n^2-dimensional vector, the resulting subset of $\mathbb{R}^{n^2}$ is algebraic.*

Given a subset E of matrices, we denote by $\mathbf{Cl}(E)$ its closure under the topology induced by the *Euclidean norm*. The following useful result holds:

Proposition 1. [6, Cor. 2] *The Euclidean topological closure of the product of two sets of matrices contained in a compact subspace coincides with the product of the topological closures of the two sets of matrices.*

2.1 Algebraic Groups

Given a subset E of a group, we denote by $\langle E \rangle$ and E^* the subgroup and the submonoid generated by E, respectively. We recall that, for an arbitrary set E of orthogonal matrices, $\mathbf{Cl}(E^*) = \mathbf{Cl}(\langle E \rangle)$ holds true. In particular, $\mathbf{Cl}(E^*)$ is a group with respect to the same product operation defined in E^* (see, e.g., [15], [6, Thm. 6]). The next theorem shows that the Euclidean topological closure of a monoid of orthogonal matrices is algebraic.

Theorem 1. [33, Ch. 3, Sec. 4, Thm. 5], [6, Thm. 7] *Let E be a set of $n \times n$ orthogonal matrices. Then, $\mathbf{Cl}(\langle E \rangle)$ is a subgroup and it is the zero set of all polynomials $p(x_{1,1}, \ldots, x_{n,n})$ satisfying the condition $p(I) = 0$ and $p(eX) = p(X)$ for every $e \in E$. Furthermore, if the matrices in E have rational entries, the above condition can be restricted to polynomials having coefficients in $\mathbb{Q}$.*

Let $\mathbb{K}$ be a field. We denote by $\mathrm{GL}_n(\mathbb{K})$ the general linear group of $n \times n$ matrices with entries in $\mathbb{K}$. In [20], Derksen *et al.* have provided an algorithm for computing the Zariski closure of finitely generated linear subgroups of $\mathrm{GL}_n(\mathbb{K})$, for several families of fields $\mathbb{K}$. In particular, the following holds:

Theorem 2. [20, Thm. 9] *Let $E \subseteq \mathrm{GL}_n(\mathbb{Q})$ be a finite set of invertible matrices. Then, the Zariski closure $\overline{\langle E \rangle}$ is effectively computable.*

Any non-empty algebraic subset $\mathcal{Z} \subseteq \mathbb{R}^n$ is said to be *irreducible* if and only if $\mathcal{Z}$ cannot be written as the union of two algebraic subsets of $\mathcal{Z}$, distinct from $\emptyset$ and $\mathcal{Z}$ itself. Let $\mathcal{V}, \mathcal{W} \subseteq \mathbb{R}^n$ be algebraic sets and suppose that $\mathcal{W}$ is irreducible. If $\mathcal{V} \subset \mathcal{W}$, then we have that $\dim(\mathcal{V}) < \dim(\mathcal{W})$, where $\dim(\mathcal{V})$ denotes the dimension of $\mathcal{V}$ as a variety (see, e.g., [21, Def. 1.12.25]). This implies that the length of every ascending chain of irreducible algebraic subsets of $\mathrm{GL}_n(\mathbb{R})$ is bounded above by $\dim(\mathrm{GL}_n(\mathbb{R})) = n^2$ [21, Prop. 1.2.20]. A striking result by Nosan *et al.* shows that this property of termination for strictly ascending chains of algebraic sets holds true for linear algebraic groups over the field $\mathbb{Q}$. Precisely:

Theorem 3. [32, Thm. 11] *Let $n \in \mathbb{N}$ and let $G_i = \overline{\langle S_i \rangle}$, for $S_i \subseteq \mathrm{GL}_n(\mathbb{Q})$ with $1 \leq i \leq \ell$, be such that $G_1 \subset G_2 \subset \cdots \subset G_\ell$. Then, $\ell \leq \exp^4(poly(n))$.*

Notice that Theorem 3 fails on linear groups over $\mathbb{C}$ since the groups of 2^k-roots of unity $G_k = \{\xi^m I : 0 \leq m \leq 2^k - 1\}$, where ξ is a primitive 2^k-th root of unity, form a nested infinite chain $G_1 \subset G_2 \subset \cdots$ of subgroups of $\mathrm{GL}_n(\mathbb{C})$.

2.2 Semialgebraic Sets

Algebraic sets are not closed under complement and projection. The following more general class of sets enjoys extra closure properties and is therefore more robust (see, e.g., [1] for an in-depth exposition):

Definition 3. *A subset $\mathcal{A} \subseteq \mathbb{R}^n$ is* semialgebraic *(over the field of real numbers), if and only if it satisfies one of the following two equivalent conditions:*

(i) $\mathcal{A}$ is the set of vectors satisfying a finite Boolean combination of predicates of the form $p(x_1, \ldots, x_n) > 0$, for polynomials $p \in \mathbb{R}[x_1, \ldots, x_n]$.

(ii) $\mathcal{A}$ is first-order definable within the theory of the structure whose domain are the reals and whose predicates are of the form $p(x_1, \ldots, x_n) > 0$ and $p(x_1, \ldots, x_n) = 0$, for polynomials $p \in \mathbb{R}[x_1, \ldots, x_n]$.

Similarly to Definition 2, a set $\mathcal{A} \subseteq \mathbb{R}^{n \times n}$ of matrices is *semialgebraic* if and only if, by considering matrices in $\mathcal{A}$ as n^2-dimensional vectors, the resulting set is semialgebraic in $\mathbb{R}^{n^2}$. Closure properties of algebraic and semialgebraic sets of matrices are discussed in [6, Sec. 2.4].

The *direct sum* of $n \times n$ square matrices $M_1, M_2, \ldots, M_k$ is the $kn \times kn$ square block matrix

$$M_1 \oplus M_2 \oplus \cdots \oplus M_k = \begin{pmatrix} M_1 & 0 & \cdots & 0 \\ 0 & M_2 & \cdots & 0 \\ \vdots & \vdots & \ddots & \vdots \\ 0 & 0 & 0 & M_k \end{pmatrix}. \tag{1}$$

Clearly, the direct sum of orthogonal matrices yields an orthogonal matrix. The set of matrices obtained by direct-summing orthogonal matrices form a subgroup of $kn \times kn$ orthogonal matrices.

Effectiveness. Given the algorithimic investigation we are going to pursue, we will adopt the following terminology. A set $\mathcal{A}$ of matrices will be called *effective algebraic* (resp., *effective semialgebraic*) if the polynomials (resp., the formula) defining $\mathcal{A}$ can be algorithmically computed from the input of the problem, namely, an effectively defined language L and a *rational* QFA $\mathcal{Q}$, i.e., a QFA with rational components.

3   Context-Free Languages

We recall some basics on context-free languages (see, e.g., [3]). A *context-free grammar* is a quadruple $G = (V, \Sigma, P, S)$ where Σ is the alphabet of *terminal symbols*, V is the set of *variables* (or *nonterminal symbols*), P is the set of *productions* (or *rules*), and S is the *axiom* (or *start symbol*) of the grammar. A word from Σ^* is called *terminal*, while a word from $(V \cup \Sigma)^*$ is called *sentential form*. As usual, nonterminal symbols are denoted by uppercase letters $A, B, \ldots$ A production of G writes as $A \to \alpha$, with α being a sentential form. For $\ell \in \mathbb{N}$, the ℓ-*step derivation relation* of G is denoted by $\overset{\ell}{\underset{G}{\Rightarrow}}$; we let $\overset{*}{\underset{G}{\Rightarrow}} = \bigcup_{\ell \in \mathbb{N}} \overset{\ell}{\underset{G}{\Rightarrow}}$ denote the *derivation relation* of G. We simply write $\overset{\ell}{\Rightarrow}$ and $\overset{*}{\Rightarrow}$, whenever G is understood. The *language generated by G* is defined to be the set of terminal words $L(G) = \{w \in \Sigma^* : S \overset{*}{\underset{G}{\Rightarrow}} w\}$. A *context-free language* is any language generated by a context-free grammar.

A language is *linear* whenever it can be generated by a linear grammar, i.e., by a context-free grammar in which the right part of each production contains at most one variable. The family of linear languages is denoted by LIN. A context-free language is said to be *of index* $k \in \mathbb{N}$ whenever it can be generated by a context-free grammar in which, for every word in the language, there exists a derivation passing through sentential forms each displaying at most k occurrences of variables. The family of context-free languages of index k will be denoted by $\mathrm{Ind}(k)$, while with $\mathrm{IND}_{\mathrm{FIN}}$ we denote the family of context-free languages *of finite index*, i.e., $\mathrm{IND}_{\mathrm{FIN}} = \bigcup_{k \in \mathbb{N}} \mathrm{Ind}(k)$.

We now overview a characterization by Ginsburg and Spanier [22] (see also Nivat [31]) of the family $\mathrm{IND}_{\mathrm{FIN}}$, that will be used in Sect. 5.1. To this aim, we recall that, given the alphabets Σ_1, Σ_2 and a family $\mathcal{F}$ of languages on Σ_2, an $\mathcal{F}$-*substitution* is a morphism $\theta : \Sigma_1^* \to \Pi(\Sigma_2^*)$ from the free monoid Σ_1^* into the multiplicative monoid $\Pi(\Sigma_2^*)$ of subsets of the free monoid Σ_2^* satisfying $\theta(x) \in \mathcal{F}$ for every $x \in \Sigma_1$. Let us recursively define the family $\{\mathrm{Qrt}(k)\}_{k \in \mathbb{N}}$ of *quasi-rational languages of rank k* (also called *bounded derivation languages of rank k*) as:

$$\mathrm{Qrt}(k) = \begin{cases} \mathrm{LIN} & \text{if } k = 1 \\ \mathrm{LIN} \circ \mathrm{Qrt}(k-1) & \text{if } k > 1, \end{cases} \tag{2}$$

where, for $k > 1$, the operation $\mathrm{LIN} \circ \mathrm{Qrt}(k-1)$ returns the family of the languages obtained as the images of linear languages *via* $\mathrm{Qrt}(k-1)$-substitutions, i.e., $\mathrm{Qrt}(k) = \{\theta(L) : L \subseteq \Sigma^* \text{ belongs to LIN and } \theta(\sigma) \in \mathrm{Qrt}(k-1) \text{ for } \sigma \in \Sigma\}$. The family $\mathrm{Qrt}(k)$ can equivalently be formulated in terms of *grammar composition* (we refer the reader to [3, Sec. II.2], for details on grammar composition).

The following theorem collects some characterizations for index-bounded languages:

Theorem 4. *[22, Thm 4.2], [3, Sec. VII.5, Thm. 5.2] Let L be a context-free language and $k \geq 1$. Then, $L \in \mathrm{Ind}(k)$ if and only if $L \in \mathrm{Qrt}(k)$. In turn, $L \in \mathrm{Qrt}(k)$ if and only if L can be generated by a grammar $\mathcal{G}_1 \circ \cdots \circ \mathcal{G}_k$ obtained as a composition of k families of linear grammars.*

4 Approaching Decision Problems on QFAs

We start our decidability investigations by over viewing some ideas and results on the *intersection problem*. This enables us either to pinpoint some facts that will be relevant in the sequel, and go through some theoretical tools and machinery we will be using.

Let us begin by formally stating the intersection problem:

$(L, \mathcal{Q})$ INTERSECTION
> INSTANCE: A language $L \subseteq \Sigma^*$ from a fixed effective family $\mathcal{L}$ of languages, a rational QFA $\mathcal{Q}$ on input alphabet Σ, with rational cut-point λ.
> QUESTION: Is $L \cap L_{\mathcal{Q},\lambda} = \emptyset$?

Recall that, for a language $L \subseteq \Sigma^*$ and a QFA $\mathcal{Q} = (s, \varphi, P)$ with n basis states on input alphabet Σ, the set of the dynamics of $\mathcal{Q}$ on L is the set of orthogonal matrices $\varphi(L) = \{\varphi(w) : w \in L\} \subseteq O_n$. The following result from [2,6] states the decidability of $(L, \mathcal{Q})$ INTERSECTION greatly depends on the topological structure of $\varphi(L)$:

Proposition 2. *Let $L \subseteq \Sigma^*$ be a language and $\mathcal{Q} = (s, \varphi, P)$ be a rational* QFA *on input alphabet Σ, such that $\mathbf{Cl}(\varphi(L))$ is effective semialgebraic. Then, $(L, \mathcal{Q})$* INTERSECTION *is decidable.*

Thus, by Proposition 2, showing the effective semialgebricity of $\mathbf{Cl}(\varphi(L))$ turns out to be a key point to asses the decidability of $(L, \mathcal{Q})$ INTERSECTION. To this aim, let us focus on the *context-free* case, and let the language L be generated by the context-free grammar $G = (V, \Sigma, P, S)$. We get that the semi-algebraicity of $\mathbf{Cl}(\varphi(L))$ and its effective computation reduce to those of an algebro-combinatorial object associated with every variable $A \in V$, namely, its *monoid of cycles*. Let us state its definition. A derivation of the form $A \underset{G}{\overset{*}{\Rightarrow}} uAv$, with $u, v \in \Sigma^*$, is called *cycle of A*. The *set of cycles of A* is defined as

$$C_A(G) = \{(u, v) \in \Sigma^* \times \Sigma^* : A \underset{G}{\overset{*}{\Rightarrow}} uAv\}. \tag{3}$$

Correspondingly, the *monoid of cycles of A* is the set of orthogonal matrices

$$M_A(G) = \{\varphi(u) \oplus \varphi(v)^T : A \underset{G}{\overset{*}{\Rightarrow}} uAv\} \tag{4}$$

$$= \{\varphi(u) \oplus \varphi(v)^T : (u, v) \in C_A(G)\},$$

where $\varphi(v)^T$ is the transpose of the matrix $\varphi(v)$ and $\oplus$ denotes the direct sum of matrices, as displayed in (1). When no ambiguity arises, $C_A(G)$ and $M_A(G)$ will be simply denoted by C_A and M_A, respectively. The following lemma states algebraic and topological properties of M_A and its Eucledean closure:

Lemma 1. [2, Lm. 3] *M_A is a monoid and $\mathbf{Cl}(M_A)$ is an algebraic group.*

The following proposition states that the (effective) algebricity of the closures $\mathbf{Cl}(M_A)$ yields the (effective) semialgebricity of $\mathbf{Cl}(\varphi(L))$, for our context-free language L and QFA $\mathcal{Q}$:

Proposition 3. [2, Prop. 9] *Let the context-free language L be generated by the context-free grammar $G = (V, \Sigma, P, S)$, and let the* QFA *$\mathcal{Q} = (s, \varphi, P)$. Then, $\mathbf{Cl}(\varphi(L))$ is semialgebraic. Moreover, if $\mathbf{Cl}(M_A)$ turns out to be effective algebraic for every $A \in V$, then $\mathbf{Cl}(\varphi(L))$ is effective semialgebraic.*

Proposition 3 enables us to emphasize interesting classes of languages for which $(L, \mathcal{Q})$ INTERSECTION becomes decidable, as in the following

Lemma 2. [6, Prop. 22] *Let $G = (V, \Sigma, P, S)$ be a context-free linear grammar, and let the rational* QFA *$\mathcal{Q} = (s, \varphi, P)$. Then, $\mathbf{Cl}(M_A)$ is effective algebraic for every $A \in V$, and hence $(L, \mathcal{Q})$* INTERSECTION *is decidable for L being a* linear *context-free language.*

In the next section, we improve Lemma 2, by showing the decidability of $(L, \mathcal{Q})$ INTERSECTION on a wider class of languages. Moreover, we will apply the techniques so far discussed, to deciding certain quantum classification problems.

5   Main Results

We are now ready to put our topology-based approach to work, in decidability investigations on QFAs. We start by expanding on the intersection problem, and then switch to quantum classifications.

5.1   On Quantum Intersection Problems

Theorem 5. $(L, \mathcal{Q})$ INTERSECTION *is decidable for L being a* finite index *context-free language.*

In order to prove Theorem 5, we need to show some preliminary results. So, given the finite index context-free language $L \in \mathrm{IND}_{\mathrm{FIN}}$, according to Theorem 4 we have that $L \in \mathrm{Ind}(k)$, for some $k \geq 1$, so that L is generated by a grammar which is a composition $G = \mathcal{G}_1 \circ \cdots \circ \mathcal{G}_k$ of k families of linear grammars.

From here on, we continue our reasoning by assuming $k = 2$, i.e., $G = \mathcal{G}_1 \circ \mathcal{G}_2$, since the general case $k \geq 3$ can be treated with a similar argument.

Let the context-free linear grammar $\mathcal{G}_1 = (V_1, \Sigma_1, P_1, S)$ and let $\mathcal{G}_2 = \{G_\sigma : \sigma \in \Sigma_1\}$ be the family of context-free linear grammars where, for every $\sigma \in \Sigma_1$, we have the grammar $G_\sigma = (V_\sigma, \Sigma_2, P_\sigma, S_\sigma)$. We assume that the variable sets $V_1, \{V_\sigma\}_{\sigma \in \Sigma_1}$ are mutually disjoint, and let $\Sigma = \Sigma_2$. We are now going to provide a structuring of the cycles of G with respect to the variables A of the grammar $\mathcal{G}_1$. We recall by (3), that $C_A = \{(u, v) \in \Sigma^* \times \Sigma^* : A \overset{*}{\Rightarrow} uAv\}$ is the set of cycles of A while, by (4), the corresponding monoid writes as $M_A = \{\varphi(u) \oplus \varphi(v)^T : (u, v) \in C_A\}$.

Take $(u, v) \in C_A$ and let $\delta := (A \overset{*}{\Rightarrow} uAv)$ be a corresponding cycle. Given the compositional structure of G, we may rearrange the productions of δ in such a way that δ rewrites as

$$A \underset{\mathcal{G}_1}{\overset{*}{\Rightarrow}} u_1 \cdots u_h A v_h \cdots v_1 \underset{\mathcal{G}_2}{\overset{*}{\Rightarrow}} w_1 \cdots w_h A \omega_h \cdots \omega_1, \tag{5}$$

where, for every $i = 1, \ldots, h$, we have

$$u_i \underset{\mathcal{G}_2}{\overset{*}{\Rightarrow}} w_i \quad \text{and} \quad v_i \underset{\mathcal{G}_2}{\overset{*}{\Rightarrow}} \omega_i, \quad \text{with } u_i, v_i \in \Sigma_1^* \text{ and } w_i, \omega_i \in \Sigma^*. \tag{6}$$

Let $u_i \underset{\mathcal{G}_2}{\overset{*}{\Rightarrow}} w_i$ be any of the h many derivations pointed out in (6), and let $u_i = \sigma_1 \cdots \sigma_{n_i} \in \Sigma_1^*$. Then, for every $j = 1, \ldots, n_i$, there exists $\ell_j \in \mathbb{N}$ such that

$$\sigma_j \underset{\mathcal{G}_2}{\overset{\ell_j}{\Rightarrow}} w_i^{(j)}, \quad \text{with } w_i^{(j)} \in \Sigma^*, \tag{7}$$

is a derivation in the grammar $G_{\sigma_j} \in \mathcal{G}_2$, with $w_i = w_i^{(1)} \cdots w_i^{(n_i)}$. Similarly, for $i = 1, \ldots h$, one gets $v_i \underset{\mathcal{G}_2}{\overset{*}{\Rightarrow}} \omega_i$ with $v_i = \tau_1 \cdots \tau_{n_i} \in \Sigma_1^*$, so that, for $j = 1, \ldots, n_i$ and some $m_j \in \mathbb{N}$, we have $\tau_j \underset{\mathcal{G}_2}{\overset{m_j}{\Rightarrow}} \omega_i^{(j)}$, with $\omega_i^{(j)} \in \Sigma^*$ and $\omega_i = \omega_i^{(1)} \cdots \omega_i^{(n_i)}$.

Now, let $\gamma \in \mathbb{N}$, and define C_A^γ as the set of all pairs $(u, v) \in \Sigma^* \times \Sigma^*$ for whycle δ in the form (5,6), where the lengths ℓ_j, m_j of every derivation as in (7) and in the corresponding derivation from τ_j do not exceed γ. With C_A^γ, we associate the set of matrices $M_A^\gamma = \{\varphi(u) \oplus \varphi(v)^T : (u, v) \in C_A^\gamma\}$. Algebraic and topological properties of these sets and their Euclidean closure are displayed in the following

Lemma 3. *The following properties hold true:*

(i) $M_A^\gamma \subseteq M_A^{\gamma+1}$, *for every* $\gamma \in \mathbb{N}$,
(ii) $M_A = \bigcup_{\gamma \in \mathbb{N}} M_A^\gamma$,
(iii) $\mathbf{Cl}(M_A^\gamma)$ *is effective algebraic, for every* $\gamma \in \mathbb{N}$.

We are now able to complete the proof of Theorem 5:

Proof of Theorem 5: Let the language $L \in \text{IND}_{\text{FIN}}$, and let $\mathcal{Q} = (s, \varphi, P)$ be a rational QFA. As declared after the statement of Theorem 5 above, we prove the case $L = L(G)$ with $G = \mathcal{G}_1 \circ \mathcal{G}_2$, since the general case $k \geq 3$ can be treated with a similar argument.

By Proposition 2, we need to prove that $\mathbf{Cl}(\varphi(L))$ is effective semialgebraic. To this aim, in the light of Proposition 3, the key point is to show that $\mathbf{Cl}(M_A)$ is effective algebraic, for every variable A of G. Since $V_1, \{V_\sigma\}_{\sigma \in \Sigma_1}$ are mutually disjoint by construction, then either $A \in \bigcup_{\sigma \in \Sigma_1} V_\sigma$ or $A \in V_1$ holds true. In the former case, A is a variable of some linear grammar G_σ from the family $\mathcal{G}_2$, and the claim follows by applying Lemma 2 to G_σ. Instead, assume $A \in V_1$ and recall, by Lemma 1, that $\mathbf{Cl}(M_A) = \overline{M_A}$ and $\mathbf{Cl}(M_A^\gamma) = \overline{M_A^\gamma}$ for any $\gamma \in \mathbb{N}$. So, from Lemma 3*(ii)* and basic facts in topology, one gets

$$\mathbf{Cl}(M_A) \;=\; \overline{\bigcup_{\gamma \in \mathbb{N}} M_A^\gamma} \;=\; \overline{\bigcup_{\gamma \in \mathbb{N}} \overline{M_A^\gamma}}. \tag{8}$$

Now, recall that $\overline{M_A^\gamma} = \langle M_A^\gamma \rangle$ is an algebraic group, for any $\gamma \in \mathbb{N}$. Hence, from Lemma 3*(i)*, one has that $\{\overline{M_A^\gamma}\}_{\gamma \in \mathbb{N}}$ is an ascending chain of linear algebraic subgroups of $\text{GL}_{2n}(\mathbb{Q})$ and so, by Theorem 3, we have

$$\bigcup_{\gamma \in \mathbb{N}} \overline{M_A^\gamma} = \bigcup_{\gamma=1}^{m} \overline{M_A^\gamma}, \tag{9}$$

for a positive (computable) constant $m = c(n)$, depending on n only (n being the number of basis states of $\mathcal{Q}$). Clearly, from (9) we get $\bigcup_{\gamma \in \mathbb{N}} \overline{M_A^\gamma} = \bigcup_{\gamma=1}^{m} \overline{M_A^\gamma}$ which, together with (8), yields

$$\mathbf{Cl}(M_A) = \bigcup_{\gamma=1}^{m} \overline{M_A^\gamma}. \tag{10}$$

To conclude our proof, it is enough to observe that $\overline{M_A^\gamma} = \mathbf{Cl}(M_A^\gamma)$ is effective algebraic by Lemma 3*(iii)*, for every $\gamma = 1, \ldots, m$. $\square$

5.2   On Quantum Classification Problems

We apply our topological approach to the investigation of certain decision problems on string classification by QFAs. First of all, let us formally state the notion of a quantum classifier:

Definition 4. *Let $L \subseteq \Sigma^*$ be a language from a fixed family $\mathcal{L}$ of languages. For any given $k \geq 2$, a $[L]$ k-Q.CLASSIFIER is a k-tuple $(\mathcal{Q}_1, \ldots, \mathcal{Q}_k)$ of QFAs on input alphabet Σ, with cut-points $(\lambda_1, \ldots, \lambda_k)$, satisfying:*

(i) $L \cap L_{\mathcal{Q}_i, \lambda_i} \neq \emptyset$, for every $i = 1, \ldots, k$,
(ii) $L \cap L_{\mathcal{Q}_i, \lambda_i} \cap L_{\mathcal{Q}_j, \lambda_j} = \emptyset$, for every $1 \leq i \neq j \leq k$.

Moreover, the $[L]$ k-Q.CLASSIFIER is said to be complete *(resp.,* super-complete*) whenever*

(iii) $L = \bigcup_{i=1}^k L_{\mathcal{Q}_i, \lambda_i}$ (resp., $L \subseteq \bigcup_{i=1}^k L_{\mathcal{Q}_i, \lambda_i}$).

We point out that the notions of *quantum classifier* and *complete quantum classifier*, introduced and initially investigated in [10], come out as special cases from Definition 4, by considering the case $[\Sigma^*]$ k-Q.CLASSIFIER.

Two natural decision problems on quantum classifiers can be formally stated as follows:

$[L]$ k-Q.CLASSIFIER
> INSTANCE: A language $L \subseteq \Sigma^*$ from a fixed effective family $\mathcal{L}$ of languages, a k-tuple $(\mathcal{Q}_1, \ldots, \mathcal{Q}_k)$ of rational QFAs on input alphabet Σ, with rational cut-points $(\lambda_1, \ldots, \lambda_k)$.
> QUESTION: Is $(\mathcal{Q}_1, \ldots, \mathcal{Q}_k)$ a $[L]$ k-Q.CLASSIFIER?

(SUPER-)COMPLETE $[L]$ k-Q.CLASSIFIER
> INSTANCE: A language $L \subseteq \Sigma^*$ from a fixed effective family $\mathcal{L}$ of languages, a k-tuple $(\mathcal{Q}_1, \ldots, \mathcal{Q}_k)$ of rational QFAs on input alphabet Σ, with rational cut-points $(\lambda_1, \ldots, \lambda_k)$.
> QUESTION: Is $(\mathcal{Q}_1, \ldots, \mathcal{Q}_k)$ a (super-)complete $[L]$ k-Q.CLASSIFIER?

Let us start with the first of these decision problems. For the sake of exposition clarity, we focus on the case $[L]$ 2-Q.CLASSIFIER: scaling to an arbitrary k should come straightforwardly. So, let $(\mathcal{Q}_1 = (s_1, \varphi_1, P_1), \mathcal{Q}_2 = (s_2, \varphi_2, P_2))$ be the input pair of rational QFAs on input alphabet Σ, with rational cut-points (λ_1, λ_2), and n_1 and n_2 basis states. By (??), for $i = 1, 2$, the corresponding accepted languages write as $L_{\mathcal{Q}_i, \lambda_i} = \{w \in \Sigma^* : f_{\mathcal{Q}_i}(w) > \lambda_i\}$. Now, we let $n = n_1 + n_2$, and define the morphism $\varphi : \Sigma^* \to O_n$, generated by mapping every $\sigma \in \Sigma$ as

$$\varphi(\sigma) = \varphi_1(\sigma) \oplus \varphi_2(\sigma) = \begin{pmatrix} \varphi_1(\sigma) & 0 \\ 0 & \varphi_2(\sigma) \end{pmatrix}. \tag{11}$$

In turn, for $i = 1, 2$, we let $g_i : O_{n_1} \oplus O_{n_2} \to [0, \infty)$ be the map defined as

$$g_i(X) = g_i \left(\begin{pmatrix} X_1 & 0 \\ 0 & X_2 \end{pmatrix} \right) = \| s_i X_i P_i \|^2. \tag{12}$$

We are going to tackle the decidability of property *(ii)* in Definition 4:

Proposition 4. *Let $L \subseteq \Sigma^*$ be a language for which $\mathbf{Cl}(\varphi(L))$ is effective semi-algebraic, φ being the morphism generated by (11). Then, it is decidable whether or not $L \cap L_{\mathcal{Q}_1, \lambda_1} \cap L_{\mathcal{Q}_2, \lambda_2} \neq \emptyset$.*

Proof. Let the formula $\Phi \equiv (\forall\, X \in \mathbf{Cl}(\varphi(L)),\ g_1(X) \leq \lambda_1 \ \vee\ g_2(X) \leq \lambda_2)$, where g_i is the map defined in (12). Since, by hypothesis, $\mathbf{Cl}(\varphi(L))$ is effective semialgebraic, then Φ is a first order formula in the arithmetics of real numbers, and can be constructed from (11) and (12).

We have that Φ holds true if and only if $L \cap L_{\mathcal{Q}_1, \lambda_1} \cap L_{\mathcal{Q}_2, \lambda_2} = \emptyset$. In fact, we first observe, by (12), that if $X = \varphi(w)$ for some $w \in \Sigma^*$, then we have $f_{\mathcal{Q}_i}(w) = g_i(X)$, for $i = 1, 2$. In addition, since $\varphi(L) \subseteq \mathbf{Cl}(\varphi(L))$, the necessity follows. Let us prove sufficiency. The condition $L \cap L_{\mathcal{Q}_1, \lambda_1} \cap L_{\mathcal{Q}_2, \lambda_2} = \emptyset$ implies that $\|s_1\, \varphi(w)\, P_1\|^2 \leq \lambda_1 \ \vee\ \|s_2\, \varphi(w)\, P_2\|^2 \leq \lambda_2$ holds true for every $w \in L$, i.e.,

$$\forall\, X \in \varphi(L),\ \ g_1(X) \leq \lambda_1 \ \vee\ g_2(X) \leq \lambda_2. \tag{13}$$

By the continuity of the map g_i, for $i = 1, 2$, one has that the validity of (13) extends to $\mathbf{Cl}(\varphi(L))$, as encompassed in Φ.

Therefore, we are left to decide whether or not Φ holds true. This can be achieved by applying the Tarski-Seidenberg elimination result. $\qquad\square$

Proposition 4 enables out to single out some meaningful language families for which quantum classification can be decided:

Theorem 6. *Let L be any given language from the family of the simple matrix linear languages (see [19, Ch. 1, Sec. 5]) or finite index context-free languages. Then, $[L]$ k-Q.CLASSIFIER is decidable.*

Proof. For the two considered language families, we have to ensure the decidability of both property *(i)* and *(ii)* in Definition 4.

For simple matrix linear languages, the decidability of property *(i)* and the effective semialgebricity of $\mathbf{Cl}(\varphi(L))$ comes from [2, Prop. 10]. Therefore, Proposition 4, suitably extended to arbitrary k-tuples of rational QFAs, yields the decidability of the corresponding quantum classification.

For finite index context-free languages, the decidability of properties *(i)* and *(ii)* comes from Theorem 5 and the k-extended version of Proposition 4. $\qquad\square$

Concerning the decidability of COMPLETE $[L]$ k-Q.CLASSIFIER, we stress that already the problem COMPLETE $[\Sigma^*]$ k-Q.CLASSIFIER is proved to be undecidable in [10, Thm. 6]. However, in the following theorem, we point out decidable instances of COMPLETE $[L]$ k-Q.CLASSIFIER, for quantum classification taking place by QFAs working with *isolated cut-points*:

Theorem 7. *Suppose rational QFAs working with* isolated *rational cut-points are considered. Then:*

(i) SUPER-COMPLETE $[L]$ k-Q.CLASSIFIER *is decidable for L being any given language from the family of simple matrix linear languages or finite index context-free languages.*

(ii) COMPLETE $[L]$ k-Q.CLASSIFIER *is decidable for L being any given regular language.*

References

1. Basu, S., Pollack, R., Roy, M.-F.: Algorithms in Real Algebraic Geometry. Springer, Berlin (2003)
2. Benso, A., D'Alessandro, F., Papi, P.: On the intersection problem for quantum finite automata. Th. Comp. Sci. **1053**, 115454 (2025)
3. Berstel, J.: Transductions and Context-Free Languages. Teubner, Stuttgart (1979)
4. Bertoni, A., Carpentieri, M.: Regular languages accepted by quantum automata. Inf. Comp. **165**, 174–182 (2001)
5. Bertoni, A., Carpentieri, M.: Analogies and differences between quantum and stochastic automata. Th. Comp. Sci. **262**, 69–81 (2001)
6. Bertoni, A., Choffrut, C., D'Alessandro, F.: On the decidability of the intersection problem for quantum automata and context-free languages. Int. J. Found. Comp. Sci. **25**, 1065–1081 (2014)
7. Bertoni, A., Mauri, G., Torelli, M.: Some recursively unsolvable problems relating to isolated cutpoints in probabilistic automata. In: Salomaa, A., Steinby, M. (eds.) ICALP 1977. LNCS, vol. 52, pp. 87–94. Springer, Heidelberg (1977). https://doi.org/10.1007/3-540-08342-1_7
8. Bertoni, A., Mereghetti, C., Palano, B.: Quantum computing: 1-way quantum automata. In: Ésik, Z., Fülöp, Z. (eds.) DLT 2003. LNCS, vol. 2710, pp. 1–20. Springer, Heidelberg (2003). https://doi.org/10.1007/3-540-45007-6_1
9. Bertoni, A., Mereghetti, C., Palano, B.: Golomb rulers and difference sets for succinct quantum automata. Int. J. Found. Comp. Sci. **14**, 871–888 (2003)
10. Bertoni, A., Mereghetti, C., Palano, B.: Some formal tools for analyzing quantum automata. Th. Comp. Sci. **356**, 14–25 (2006)
11. Bianchi, M.P., Mereghetti, C., Palano, B.: Complexity of promise problems on classical and quantum automata. In: Calude, C.S., Freivalds, R., Kazuo, I. (eds.) Computing with New Resources. LNCS, vol. 8808, pp. 161–175. Springer, Cham (2014). https://doi.org/10.1007/978-3-319-13350-8_12
12. Bianchi, M.P., Mereghetti, C., Palano, B.: Size lower bounds for quantum automata. Th. Comp. Sci. **551**, 102–115 (2014)
13. Bianchi, M.P., Mereghetti, C., Palano, B.: Quantum finite automata: advances on Bertoni's ideas. Th. Comp. Sci. **664**, 39–53 (2017)
14. Bianchi, M.P., Palano, B.: Behaviours of unary quantum automata. Fund. Inf. **104**, 1–15 (2010)
15. Blondel, V.D., Jeandel, E., Koiran, P., Portier, N.: Decidable and undecidable problems about quantum automata. SIAM J. Comput. **34**, 1464–1473 (2005)
16. Brodsky, A., Pippenger, N.: Characterizations of 1-way quantum finite automata. SIAM J. Comput. **34**, 1456–1478 (2001)
17. Candeloro, A., Mereghetti, C., Palano, B., Cialdi, S., Paris, M.G.A., Olivares, S.: An enhanced photonic quantum finite automaton. App. Sci. **11**, 8768 (2021)
18. Choffrut, C., Karhumaki, J.: Some decision problems on integer matrices. RAIRO-Theor. Inf. Appl. **39**(1), 125–131 (2005)
19. Dassow, J., Paun, G.: Regulated Rewriting in Formal Language Theory. EATCS Monographs in Theoretical Computer Science 18. Springer (1989)
20. Derksen, H., Jeandel, E., Koiran, P.: Quantum automata and algebraic groups. J. Symb. Comp. **39**, 357–371 (2005)
21. Geck, M.: An Introduction to Algebraic Geometry and Algebraic Groups. Oxford University Press, UK (2003)

22. Ginsburg, S., Spanier, E.H.: Derivation-bounded languages. J. Comp. Sys. Sci. **2**, 228–250 (1968)
23. Hrushovski, E., Ouaknine, J., Pouly, A., Worrell, J.: On strongest algebraic program invariants. J. ACM **70**, 1–22 (2023)
24. Jeandel, E.: Indécidabilité sur les automates quantiques. Master thesis, ENS Lyon (2002)
25. Li, L., Qiu, D.: Determining the equivalence for one-way quantum finite automata. Th. Comp. Sci. **403**, 42–51 (2008)
26. El Manssour, R.A., Naraghi, M., Shirmohammadi, M., Worrell, J.: Algebraic closure of matrix sets recognized by 1-VASS. arXiv:2507.09373v2 (2025)
27. Mereghetti, C., Palano, B.: Quantum automata for some multiperiodic languages. Th. Comp. Sci. **387**, 177–186 (2007)
28. Mereghetti, C., et al.: Photonic realization of a quantum finite automaton. Phys. Rev. Res. **2**, 890130 (2020)
29. Mereghetti, C., Palano, B., Raucci, P.: Unary quantum finite state automata with control language. App. Sci. **14**, 1490 (2024)
30. Moore, C., Crutchfield, J.: Quantum automata and quantum grammars. Th. Comp. Sci. **237**, 275–306 (2000)
31. Nivat, M.: Transductions des langages de Chomsky. Ann. Inst. Fourier (Grenoble) **18**, 339–455 (1968)
32. Nosan, K., Pouly, A., Schmitz, S., Shirmohammadi, M., Worrell, J.: On the computation of the Zariski closure of finitely generated groups of matrices. In: Proc. ISSAC 2022, Int. Symp. Symbolic and Algebraic Computation, pp. 129–138. ACM (2022)
33. Onishchik, A., Vinberg, E.: Lie Groups and Algebraic Groups. Springer, Berlin (1990)
34. Paz, A.: Introduction to Probabilistic Automata. Academic Press, New York (1971)
35. Pin, J.E.: Varieties of Formal Languages. North Oxford Academic, Oxford (1986)
36. Rabin, M.O.: Probabilistic automata. Inf. Contr. **6**, 230–245 (1963)
37. Potapov, I., Semukhin, P.: Decidability of the membership problem for 2×2 integer matrices. In: Proc. SODA 2017, 28-th Annual ACM-SIAM Symposium on Discrete Algorithms, Barcelona, Spain, pp. 170–186 (2017)
38. Qiu, D., Li, L., Mateus, P., Gruska, J.: Quantum finite automata. In: Handbook of Finite State Based Models and Applications, pp. 113–144. CRC Press (2016)
39. Tarski, A.: A decision method for elementary algebra and geometry. In: Quantifier Elimination and Cylindrical Algebraic Decomposition, pp. 156–172. Springer, Vienna (1998)
40. Thierrin, G.: Permutation automata. Math. Sys. Th. **2**, 83–90 (1968)
41. Xiao, L., Qiu, D.: State complexity of one-way quantum finite automata together with classical states. J. Comp. Sys. Sci. **154**, 103659 (2025)

Scattered Factor Universality - A Survey

Pamela Fleischmann[✉]

Kiel University, Kiel, Germany
`fpa@informatik.uni-kiel.de`

Abstract. Since the seminal work of Imre Simon in the 1970s, *scattered factors* (also known as subsequences or scattered subwords), a lot of research time has been invested in this notion. A scattered factor is a non-necessarily consecutive part of a word but in correct order, e.g., `mai` and `rade` are scattered factors of `normandie` but `monde` is not since the `o` occurs before the `m`.

Within these last 50 years, the main question by Simon about the index of the nowadays called Simon congruence relation is still open: two words are called k-Simon congruent if they have exactly the same scattered factors up to length k. For instance, the words `abab` and `abba` are 2-Simon-congruent but not 3-Simon-congruent. Thus, the research in the field around scattered factors has broadened in order to find an angle to tackle Simon's original problem.

1 Introduction

A word u is a scattered factor of a word w, ($u \in \mathrm{ScatFact}(w)$) if there exist possibly empty words $v_1, \ldots, v_{n+1}$ such that $w = v_1 u[1] v_2 u[2] \cdots v_n u[n] v_{n+1}$ where $u[i]$ denotes the ith letter of u. The research on scattered factors started with the PhD thesis of Imre Simon in the 1970s with the fundamental work about piecewise-testable languages [44]. Simon introduced the relationship, nowadays known as Simon's congruence, in which two words are said to be related if they have the same scattered factors up to a given length k. One approach to get closer to determine the index of this congruence relation is the notion of m-nearly k-universal words [8, 11, 15, 16, 42]: a word is called m-*nearly k-universal* if exactly m words of length k are not scattered factors of it (cf. the notion of k-richness [25–27]). By this approach, the general problem is partitioned into problems parametrised by m. Note that with this notion the perspective is shifted from the appearing scattered factors to also the absent like `monde` in `normandie` (cf. [3, 16, 17, 34, 45]). Thus, as scattered factors serve as a complexity measure for a word's absent or existing information, the relation between words and their scattered factors is studied in combinatorics on words, stringology, language and automata theory (see [37, Chapter 6, Subwords] for a profound overview on the topic). Recently, Simon's congruence got attention in language theory [1, 2, 14, 29, 30] and pattern matching [18, 31]. Moreover, scattered factors have been studied in combinatorics on words [34, 40], algorithms and stringology [6, 7, 10, 21, 38, 46], logics [9, 23, 35, 36], as well as language and automata theory [2, 25,

M.-P. Béal and P. Caron (Eds.): DLT 2026, LNCS 16578, pp. 18–24, 2026.
https://doi.org/10.1007/978-3-032-28404-4_2

28, 43, 44, 48]. From a more applicative point of view, scattered factors are used in bioinformatics [12, 47], formal software verification [23, 48] and database theory [5, 22, 32, 33, 41]. Further, scattered factors are also used to model corrupted data in the context of the theoretical problem of reconstruction [13, 19, 39]. In 2025, the research has been extended to *jumbled scattered factors*, which means that not all letters have to be in the correct order. For instance, monde is a 1-jumbled scattered factor of normandie since the longest scattered factor of monde, namely onde, which is also a scattered factor of normandie has length four and thus one letter is not in order [20].

2   Main Definitions

For the basic definitions of combinatorics on words see [37]. In the following Σ is an alphabet of cardinality σ.

Definition 1. *A word $u \in \Sigma^*$ is called a* scattered factor *of $w \in \Sigma^*$ if there exist $v_1, \ldots, v_{n+1} \in \Sigma^*$ such that $w = v_1 u[1] v_2 u[2] \cdots v_{|u|} u[|u|] v_{|u|+1}$ holds. Set* $\mathrm{ScatFact}_k(w) = \{u \in \mathrm{ScatFact}(w) \mid |u| = k\}$ *for $k \in [|w|]$ and* $\mathrm{ScatFact}(w) = \bigcup_{k \in [|w|]} \mathrm{ScatFact}_k(w)$.

Definition 2. *A word $w \in \Sigma^*$ is called m-nearly k-universal if $|\mathrm{ScatFact}_k(w)| = \sigma^k - m$. If $m = 0$, the words are called k-universal, and if $m = 1$, they are called nearly k-universal. The largest $k \in \mathbb{N}_0$, such that a word is k-universal, is called the* universality index *$\iota(w)$. A word is called k-circular universal if a conjugate of w is k-universal; the largest k such that w is k-circular universal is the* circular universality index *$\zeta(w)$.*

The word ababcc is 1-universal, 2-nearly 2-universal (ca, cb are absent), and 2-circular universal (the conjugate abccab is 2-universal). One of the main tools for the investigation of a words' scattered factors is the *arch factorisation* which was introduced by Hébrard [24]. Clearly, the number of arches is the universality index of the word.

Definition 3. *For a word $w \in \Sigma^*$ the* arch factorisation *is given by $w = \mathrm{ar}_1(w) \cdots \mathrm{ar}_k(w)\, \mathrm{r}(w)$ for $k \in \mathbb{N}_0$ with*
  1. *$\iota(\mathrm{ar}_i(w)) = 1$ for all $i \in [k]$ and $\mathrm{alph}(\mathrm{r}(w)) \subsetneq \Sigma$,*
  2. *$\mathrm{ar}_i(w)[|\mathrm{ar}_i(w)|] \notin \mathrm{alph}(\mathrm{ar}_i(w)[1..|\mathrm{ar}_i(w)| - 1])$ for all $i \in [k]$.*

The words $\mathrm{ar}_i(w)$ are called arches *of w and $\mathrm{r}(w)$ is the* rest *of w. If we have $\mathrm{r}(w) = \varepsilon$, we call w perfect k-universal ($\mathrm{PUniv}_{\Sigma,k}$). The* modus *$\mathrm{m}(w)$ is given by $\mathrm{ar}_1(w)[|\mathrm{ar}_1(w)|] \cdots \mathrm{ar}_k(w)[|\mathrm{ar}_k(w)|]$, i.e., it is the word containing the unique last letters of each arch.*

The α-β-factorisation is based on the arch factorisations of w and w^R.

Definition 4. *Let $\bar{\mathrm{ar}}_i(w) := (\mathrm{ar}_{\iota(w)-i+1}(w^R))^R$ the ith reverse arch, let $\bar{\mathrm{r}}(w) := (\mathrm{r}(w^R))^R$ the* reverse rest, *and define the* reverse modus *$\bar{\mathrm{m}}(w)$ as $\mathrm{m}(w^R)^R$. For $w \in \Sigma^*$ define w's α-β-factorization by $w =: \alpha_0 \beta_1 \alpha_1 \cdots \alpha_{\iota(w)-1} \beta_{\iota(w)} \alpha_{\iota(w)}$ with $\mathrm{ar}_i(w) = \alpha_{i-1}\beta_i$ and $\bar{\mathrm{ar}}_i(w) = \beta_i \alpha_i$ for all $i \in [\iota(w)]$, $\bar{\mathrm{r}}(w) = \alpha_0$, as well as $\mathrm{r}(w) = \alpha_{\iota(w)}$.*

We conclude this section with the notion of jumbled scattered factors. Here, $p(u)$ denotes the Parikh vector of u and thus, we assume a fixed order on Σ.

Definition 5. *The word $u \in \Sigma^*$ is called a* scattered factor *of $w \in \Sigma^*$ with ℓ* jumbles *(or u is ℓ-jumbled in w) for $\ell \in [|u|-1]_0$ if $p(u) \le p(w)$ and there exists $v \in \mathrm{ScatFact}_{|u|-\ell}(u)$ such that $v \in \mathrm{ScatFact}(w)$ holds. The set of scattered factors of w with ℓ-jumbles is denoted by $\mathrm{JScatFact}(w, \ell)$. We define the* jumble index *of u w.r.t. w by $\delta_w(u) = \min\{\ell \in \mathbb{N}_0 \mid u \in \mathrm{JScatFact}(w, \ell)\}$. Let $\mathrm{SJScatFact}(w, \ell) = \{u \in \mathrm{JScatFact}(w, \ell) \mid \delta_w(u) = \ell\}$.*

For $w = \mathtt{normandie}$, we have $\mathtt{monde} \in \mathrm{JScatFact}(w, 1)$ and $\mathtt{roman}$ is 2-jumbled in w.

3 Scattered Factors

In this section, we present the results regarding scattered factor universality since its introduction in 2020. The first observation is clear having the arch factorization in mind: one can determine $\iota(w)$ in $O(|w|)$.

Theorem 6 ([8]). *A word $w \in \Sigma^*$ is k-universal for some $k \in \mathbb{N}_0$ iff*
    1. $\mathrm{ScatFact}_k(w^n) = \mathrm{ScatFact}_k(w^{n+1})$ for an $n \in \mathbb{N}$ and $\iota(w^n) \ge kn$, iff
    2. $\mathrm{ScatFact}_k(w) = \mathrm{ScatFact}_k(ww^R)$.

Theorem 7 ([8]). *Let $w \in \Sigma^*$. If $\iota(w) = k$ and $\zeta(w) = k+1$ then $\iota(w^s) = sk + s - 1$, for all $s \in \mathbb{N}$. If $w \in \{\mathtt{a}, \mathtt{b}\}^*$ with $\iota(w) = k$ and $s \in \mathbb{N}$ then $\iota(w^s) = sk + s - 1$ if $\zeta(w) = k+1$ and sk otherwise.*

The other direction of Theorem 7 does not hold for arbitrary alphabets: Consider the 2-universal non-binary word $w = \mathtt{babccaabc}$. We have that w^2 is 5-universal but w is not 3-circular universal. The following theorem presents an equivalence for $\zeta(w) = k+1$ based on the function $\iota_w(s) = \iota(w^s)$ and its growth $\nabla \iota_w(s) = \iota_w(s) - \iota_w(s-1)$ with $\iota_w(0) = 0$.

Theorem 8 ([15]). *For $w \in \Sigma^*$, the following statements are equivalent:*
    1. $\nabla \iota_w(s) = k+1$ for all $2 \le s \le \sigma$,
    2. $\nabla \iota_w(s) = k+1$ for all $s \in \mathbb{N}_{\ge 2}$,
    3. $\zeta(w) = k+1$.

We move now to m-nearly k-universal words. In the case of nearly k-universal words, we obtain that for each $u \in \Sigma^k$ exists exactly one congruence class.

Theorem 9 ([16]). *For $w \in \Sigma^*$, the following statements are equivalent*
    1. w is nearly k-universal,
    2. $\iota(w) = k - 1$, and for all $\hat{k}, \tilde{k} \in \mathbb{N}_0$ with $\hat{k} + \tilde{k} + 1 = k$ there exist $u \in \mathrm{PUniv}_{\Sigma, \hat{k}}$, $v \in \mathrm{PUniv}_{\Sigma, \tilde{k}}$, and $x \in \Sigma^+$ with $|\mathrm{alph}(x)| = \sigma - 1$ such that $w = uxv^R$.

Remark 10 ([8,16]). *For $m = 0$ and $m = \sigma^k$, we have exactly one congruence class each, for $m = \sigma^k - 1$ we have σ^k, and for $m = \sigma^k - 2$ we have $2\binom{\sigma}{2}k$ classes.*

Theorem 11 ([16]). *A word $w \in \Sigma^*$ is 2-nearly k-universal iff $\iota(w) = k - 1$ and there exists $\ell \in [k]$ such that $|\Sigma \setminus \operatorname{alph}(\alpha_\ell)| = 2$ and $|\Sigma \setminus \operatorname{alph}(\alpha_i)| = 1$ holds for all $i \in [k] \setminus \{\ell\}$.*

In the binary alphabet one can characterize the singleton congruence classes.

Theorem 12 ([17]). *Let $w \in \{a, b\}^*$. Then $|[w]_{\sim_k}| < \infty$ and in particular $|[w]_{\sim_k}| = 1$ iff $\iota(w) < k$ and $|\alpha_i| < k - \iota(w)$ for all $i \in [\iota(w)]_0$.*

The following results investigate the set of words which remain if one deletes a scattered factor from a word, e.g., $C(\mathsf{banana}, \mathsf{ana}) = \{\mathsf{bna}, \mathsf{ban}\}$.

Theorem 13 ([4]). *Given $w \in \Sigma^*$ and $u \in \operatorname{ScatFact}(w)$ we can compute the set $C(w, u)$ in time $O(|w||u|\binom{w}{u})$ and space $O(|u|\binom{w}{u})$.*

Theorem 14 ([4]). *Given a set $S \subseteq \Sigma^m$ and $w \in \Sigma^n$, we can determine if there exists a word u such that $C(w, u) = S$ in $O(n^2\binom{n}{m})$ time and $O(n\binom{n}{m})$ space.*

Theorem 15 ([4]). *Given a set of words $V = \{v_1, v_2, \ldots, v_r\} \subseteq \Sigma^\ell$ for some $n, r, \ell \in \mathbb{N}$ and a single word $u \in \Sigma^{n-\ell}$ we can determine if there exists some word $w \in \Sigma^n$ such that $C(w, u) = V$ in $O(rn^{r+1})$ time and further, if such a word exists, we can construct it in $O(rn^{r+1})$ time.*

The last two results are related to the universality in the jumbled context. Notice that extending Simon's congruence in a natural way to jumbled scattered factors ($\operatorname{JScatFact}_{\leq k}(v, \ell) = \operatorname{JScatFact}_{\leq k}(w, \ell)$) is on costs of the congruence condition: we only have an equivalence relation.

Proposition 16 ([20]). *For all $w \in \Sigma^*$ there exist $\ell \leq k - \iota(w)$ and $k \in I$, such that $\bigcup_{\ell' \in [0,\ell]} \operatorname{JScatFact}_k(w, \ell') = \Sigma^k$.*

Proposition 17 ([20]). *Let $w \in \Sigma^*$, $k \in I$ and $\operatorname{JScatFact}_k(w, \ell) \subset \Sigma^k$. Then, we have $\operatorname{JScatFact}_k(w, \ell + 1) \setminus \operatorname{JScatFact}_k(w, \ell) \neq \emptyset$.*

4 Conclusion

In this work, the main results regarding scattered factor universality from its beginning in 2020 till now are given. The main idea of this notion and the associated notion of m-nearly k-universality is to get closer to the determination of the index of Simon's congruence. In some cases, we were able to characterise these subclasses including their cardinality. Since studying the scattered factors of a word and its absent scattered factors is tightly related, the new research complement scattered factor set aims to combine the information: what remains if we delete a scattered factor. Lastly, two results about jumbled scattered factors, i.e. scattered factors where not all letters have to occur in the correct order, w.r.t. universality were presented.

Acknowledgement. I would like to thank my various co-authors on this topic: D. Adamson, L. Barker, J.D. Day, S.B. Germann, K. Harwardt, L. Haschke, J. Höfer, A. Huch, M. Kammholz, S. Kim, T. Koß, M. Kosche, F. Manea, A. Mayrock, D. Nowotka, P. Sarnighausen-Cahn, S. Siemer, M. Wiedenhöft.

References

1. Adamson, D., Fleischmann, P., Huch, A., Koß, T., Manea, F.: k-universality of regular languages revisited. Front. Algorithmics, 16–32 (2025)
2. Adamson, D., Fleischmann, P., Huch, A., Koß, T., Manea, F., Nowotka, D.: k-Universality of regular languages. In: ISAAC 2023. LIPIcs, vol. 283, pp. 4:1–4:21. Schloss Dagstuhl - Leibniz-Zentrum für Informatik (2023)
3. Adamson, D., Fleischmann, P., Huch, A., Manea, F., Sarnighausen-Cahn, P., Wiedenhöft, M.: Tight bounds for the number of absent subsequences **16106**, 15–29 (2025)
4. Adamson, D., Fleischmann, P., Huch, A.: (Sets of) complement scattered factors. CoRR, 2603.20790v1 (2026)
5. Artikis, A., Margara, A., Ugarte, M., Vansummeren, S., Weidlich, M.: Complex event recognition languages: tutorial. In: DEBS, pp. 7–10 (2017)
6. Baeza-Yates, R.A.: Searching subsequences. TCS **78**(2), 363–376 (1991)
7. Bannai, H., Tomohiro, I., Kociumaka, T., Köppl, D., Puglisi, S.J.: Computing longest Lyndon subsequences and longest common Lyndon subsequences. Algorithmica **86**(3), 735–756 (2024)
8. Barker, L., Fleischmann, P., Harwardt, K., Manea, F., Nowotka, D.: Scattered factor-universality of words. In: Jonoska, N., Savchuk, D. (eds.) DLT 2020. LNCS, vol. 12086, pp. 14–28. Springer, Cham (2020). https://doi.org/10.1007/978-3-030-48516-0_2
9. Baumann, P., Ganardi, M., Thinniyam, R.S., Zetzsche, G.: Existential definability over the subword ordering. Log. Methods Comput. Sci. **19**(4) (2023)
10. Bringmann, K., Künnemann, M.: Multivariate fine-grained complexity of longest common subsequence. In: SODA 2018, pp. 1216–1235. SIAM (2018)
11. Day, J.D., Fleischmann, P., Kosche, M., Koß, T., Manea, F., Siemer, S.: The edit distance to k-subsequence universality. In: STACS 2021. LIPIcs, vol. 187, pp. 25:1–25:19. Schloss Dagstuhl - Leibniz-Zentrum für Informatik (2021)
12. Do, D.T., Le, T.T.Q., Le, N.-Q.-K.: Using deep neural networks and biological subwords to detect protein s-sulfenylation sites. Briefings Bioinform. **22**(3) (2021)
13. Dress, A.W.M., Erdős, P.L.: Reconstructing words from subwords in linear time. Ann. Comb. **8**, 457–462 (2005)
14. Fazekas, S.Z., Koß, T., Manea, F., Mercas, R., Specht, T.: Subsequence matching and analysis problems for formal languages. In: ISAAC 2024. LIPIcs, vol. 322, pp. 28:1–28:23. Schloss Dagstuhl - Leibniz-Zentrum für Informatik (2024)
15. Fleischmann, P., Germann, S.B., Nowotka, D.: Scattered factor universality-the power of the remainder. preprint arXiv:2104.09063 (published at RuFiDim) (2021)
16. Fleischmann, P., Haschke, L., Höfer, J., Huch, A., Mayrock, A., Nowotka, D.: Nearly k-universal words - investigating a part of Simon's congruence. TCS **974**, 114113 (2023)
17. Fleischmann, P., Höfer, J., Huch, A., Nowotka, D.: α-β-factorization and the binary case of Simon's congruence. In: FCT 2023. LNCS, vol. 14292, pp. 190–204. Springer (2023)

18. Fleischmann, P., et al.: Matching patterns with variables under Simon's congruence. In: RP 2023. LNCS, vol. 14235, pp. 155–170. Springer (2023)
19. Fleischmann, P., Lejeune, M., Manea, F., Nowotka, D., Rigo, M.: Reconstructing words from right-bounded-block words. Int. J. Found. Comput. Sci. **32**(6), 619–640 (2021)
20. Fleischmann, P., Huch, A., Kammholz, M., Koß, T.: Jumbled scattered factors. In: Ko, S.-K., Manea, F. (eds.) Developments in Language Theory - 29th International Conference, DLT 2025, Seoul, South Korea, 19–22 August 2025, Proceedings. LNCS, vol. 16036, pp. 261–277. Springer (2025)
21. Forrester, P.J., Mays, A.: Finite size corrections relating to distributions of the length of longest increasing subsequences. Adv. Appl. Math. **145**, 102482 (2023)
22. Frochaux, A., Kleest-Meißner, S.: Puzzling over subsequence-query extensions: disjunction and generalised gaps. In: AMW 2023. CEUR Workshop Proceedings, vol. 3409 (2023)
23. Halfon, S., Schnoebelen, P., Zetzsche, G.: Decidability, complexity, and expressiveness of first-order logic over the subword ordering. In: LICS 2017, pp. 1–12. IEEE Computer Society (2017)
24. Hébrard, J.-J.: An algorithm for distinguishing efficiently bit-strings by their subsequences. TCS **82**(1), 35–49 (1991)
25. Karandikar, P., Kufleitner, M., Schnoebelen, P.: On the index of Simon's congruence for piecewise testability. Inf. Process. Lett. **115**(4), 515–519 (2015)
26. Karandikar, P., Schnoebelen, P.: The height of piecewise-testable languages with applications in logical complexity. In: Proc. CSL. LIPIcs, vol. 62, pp. 37:1–37:22 (2016)
27. Karandikar, P., Schnoebelen, P.: The height of piecewise-testable languages and the complexity of the logic of subwords. LICS **15**(2) (2019)
28. Karandikar, P., Schnoebelen, P.: The height of piecewise-testable languages and the complexity of the logic of subwords. Log. Methods Comput. Sci. **15**(2) (2019)
29. Kim, S., Han, Y.-S., Ko, S.-K., Salomaa, K.: On Simon's congruence closure of a string. TCS **972**, 114078 (2023)
30. Kim, S., Han, Y.-S., Ko, S.-K., Salomaa, K.: On the Simon's congruence neighborhood of languages. In: DLT. LNCS, vol. 13911, pp. 168–181. Springer (2023)
31. Kim, S., Ko, S.-K., Han, Y.-S.: Simon's congruence pattern matching. TCS **994**, 114478 (2024)
32. Kleest-Meißner, S., Sattler, R., Schmid, M.L., Schweikardt, N., Weidlich, M.: Discovering event queries from traces: laying foundations for subsequence-queries with wildcards and gap-size constraints. In: ICDT. LIPIcs, vol. 220, pp. 18:1–18:21 (2022)
33. Kleest-Meißner, S., Sattler, R., Schmid, M.L., Schweikardt, N., Weidlich, M.: Discovering multi-dimensional subsequence queries from traces - from theory to practice. In: BTW. LNI, vol. P-331, pp. 511–533 (2023)
34. Kosche, M., Koß, T., Manea, F., Siemer, S.: Absent subsequences in words. Fundam. Informaticae **189**(3–4), 199–240 (2022)
35. Kuske, D.: The subtrace order and counting first-order logic. In: Fernau, H. (ed.) CSR 2020. LNCS, vol. 12159, pp. 289–302. Springer, Cham (2020). https://doi.org/10.1007/978-3-030-50026-9_21
36. Kuske, D., Zetzsche, G.: Languages ordered by the subword order. In: Bojańczyk, M., Simpson, A. (eds.) FoSSaCS 2019. LNCS, vol. 11425, pp. 348–364. Springer, Cham (2019). https://doi.org/10.1007/978-3-030-17127-8_20
37. Lothaire, M.: Combinatorics on Words. Cambridge Mathematical Library. Cambridge University Press (1997)

38. Maier, D.: The complexity of some problems on subsequences and supersequences. J. ACM **25**(2), 322–336 (1978)
39. Manuch, J.: Characterization of a word by its subwords. In: DLT 1999, pp. 210–219. World Scientific (1999)
40. Mateescu, A., Salomaa, A., Yu, S.: Subword histories and Parikh matrices. J. Comput. Syst. Sci. **68**(1), 1–21 (2004)
41. Sattler, R., Kleest-Meißner, S., Lange, S., Schmid, M.L., Schweikardt, N., Weidlich, M.: DISCES: systematic discovery of event stream queries. Proc. ACM Manag. Data **3**(1) (2025)
42. Schnoebelen, P., Veron, J.: On arch factorization and subword universality for words and compressed words. In: WORDS 2023. LNCS, vol. 13899, pp. 274–287. Springer (2023)
43. Simon, I.: Hierarchies of events with dot-depth one, University of Waterloo. Ph.D. thesis (1972)
44. Simon, I.: Piecewise testable events. In: Brakhage, H. (ed.) GI-Fachtagung 1975. LNCS, vol. 33, pp. 214–222. Springer, Heidelberg (1975). https://doi.org/10.1007/3-540-07407-4_23
45. Tronícek, Z.: On problems related to absent subsequences. In: COCOA 2023. LNCS, vol. 14462, pp. 351–363. Springer (2023)
46. Wagner, R.A., Fischer, M.J.: The string-to-string correction problem. J. ACM **21**(1), 168–173 (1974)
47. Wang, C., Cho, K., Gu, J.: Neural machine translation with byte-level subwords. In: EAAI 2020, pp. 9154–9160. AAAI Press (2020)
48. Zetzsche, G.: The complexity of downward closure comparisons. In: ICALP 2016. LIPIcs, vol. 55, pp. 123:1–123:14. Schloss Dagstuhl - Leibniz-Zentrum für Informatik (2016)

Seventy Years of Algebraic, Logical, and Topological Methods for Regular Languages

Jean-Éric Pin[✉]

Paris, France
Jean-Eric.Pin@irif.fr

Abstract. This article presents the evolution of ideas from algebra, topology, and model theory used for the study of regular languages over the past seventy years.

Keywords: Regular languages · Syntactic monoid · Ordered syntactic monoids · equations · profinite · Model theory · Duality · operations on languages

In this lecture, I will present a historical overview of the algebraic and topological tools used to study regular languages. Further details and references can be found in one of the survey articles [13, 40, 43, 44, 47, 49–53, 56, 66, 67, 86], in the books [1, 22, 68, 81], or in the Chaps. 1, 6, 7, 9, 14, 16, 17, 22, 27, 29 of the Handbook of Automata Theory [54, 55].

1 Early Results (1956-1976)

The *syntactic congruence* of a language was first introduced in 1956 by Schützenberger in a paper [73] that also founded the theory of variable-length codes. The *syntactic congruence* of a language L of A^* is the equivalence $\sim_L$ on A^* defined by $u \sim_L v$ if and only if, for all $x, y \in A^*$,

$$xuy \in L \iff xvy \in L \tag{1}$$

The *syntactic monoid* of L is the quotient monoid $A^*/\!\sim_L$. Rabin and Scott proved in 1959 that a language is regular if and only if its syntactic monoid is finite, which opened up the possibility to study regular languages through properties of their syntactic monoid. Several major results have confirmed the validity of this approach.

Theorem 1.1. (Schützenberger 1965 [74]**)** *A regular language is star-free if and only if its syntactic monoid is aperiodic.*

© The Author(s), under exclusive license to Springer Nature Switzerland AG 2026
M.-P. Béal and P. Caron (Eds.): DLT 2026, LNCS 16578, pp. 25–40, 2026.
https://doi.org/10.1007/978-3-032-28404-4_3

Theorem 1.2. (Brzozowski-Simon 1973 [15]**, McNaughton 1974** [35]**)** *A regular language of A^+ is locally testable if and only if its syntactic semigroup is locally idempotent and commutative.*

Theorem 1.3. (Simon 1975 [78]**)** *A regular language is piecewise testable if and only if the $\mathcal{J}$-classes of its syntactic monoid are trivial.*

Theorem 1.4. (Schützenberger 1976 [76]**)** *A regular language is unambiguous star-free if and only if the regular $\mathcal{J}$-classes of its syntactic monoid are idempotent semigroups.*

In 1976, Eilenberg proposed a common framework for these results. A *variety of finite monoids* is a class of finite monoids closed under taking submonoids, quotients and finite direct products. A *variety of languages* is a class of regular languages closed under Boolean operations, left and right residuals and inverses of morphisms between free monoids.

Theorem 1.5. (Eilenberg 1976 [22]**)** *There is a bijection between varieties of finite monoids and varieties of languages.*

For instance, rational languages correspond to the variety of all finite monoids, star-free languages correspond to the variety **A** of finite aperiodic monoids, and piecewise testable languages correspond to the variety **J** of finite $\mathcal{J}$-trivial monoids.

To cover the case of locally testable languages, Eilenberg also needed to consider varieties of finite semigroups and $+$-*varieties of languages*, in which languages never contain the empty word. Again, there is a bijection between varieties of finite semigroups and $+$-varieties of languages.

From there, various results made it possible to describe the varieties of languages corresponding to several varieties of finite monoids: commutative monoids, idempotent monoids [60, 77], $\mathcal{R}$-trivial or $\mathcal{L}$-trivial monoids [14, 22, 37], monoids whose groups are commutative [33, 75], nilpotent [22] or solvable [79], monoids with commuting idempotents [34], $\mathcal{J}$-trivial monoids with commuting idempotents [5], monoids whose regular $\mathcal{J}$-classes are rectangular bands [76], block groups [41], etc. For finite groups, the varieties of p-groups and nilpotent groups [22, 83], solvable groups [79] and supersolvable groups [19].

Curiously, the variety of finite groups does not appear in this list, because this would require describing the languages whose syntactic monoid is a given simple group.

2 Equations

Equations have long been used in mathematics to concisely describe mathematical objects. Regarding monoids, the story truly began in 1935 with Birkhoff's theorem on equational classes [11]. However, Birkhoff's theorem does not apply to varieties of *finite* monoids. To cover this case, two solutions were proposed. The first uses ultimate equations and the second uses profinite equations.

2.1   Ultimate Equations

Let u, v be two words over a countable alphabet A. A monoid M *satisfies the equation* $u = v$ if $\varphi(u) = \varphi(v)$ for every monoid morphism $\varphi : A^* \to M$. Let now $(u_n, v_n)_{n \geqslant 0}$ be a sequence of pairs of words of A^*. A monoid *ultimately satisfies the equations* $u_n = v_n$ if it satisfies these equations for n sufficiently large. In 1976, Eilenberg and Schützenberger [23] proved the following result:

Theorem 2.1. *A class of finite monoids is a variety of finite monoids if and only if it can be ultimately defined by a sequence of equations.*

For instance, a finite monoid is aperiodic if and only if it ultimately satisfies $x^{n+1} = x^n$. A finite monoid is a group if and only if it ultimately satisfies $x^{n!} = 1$.

2.2   Profinite Equations

The idea underlying profinite equations is to replace "ultimate equations" by "equations satisfied in the limit". Such limits are obtained via a metric on A^* defined as follows.

A finite monoid M *separates* two words u and v of A^* if there is a monoid morphism $\varphi : A^* \to M$ such that $\varphi(u) \neq \varphi(v)$. Let us set

$$d(u, v) = 2^{-r(u,v)}$$

where $r(u, v)$ is the minimum size of a finite monoid that separates u and v, with the usual conventions $\min \emptyset = +\infty$ and $2^{-\infty} = 0$. This defines a *metric d* on A^*, with the following properties:

(1) Every monoid morphism from (A^*, d) to a finite monoid, equipped with the discrete metric, is uniformly continuous.
(2) Monoid morphisms between free monoids are uniformly continuous.
(3) The product on A^* is uniformly continuous.

The completion of A^* for this metric, denoted by $\widehat{A^*}$, is the set of *profinite words* over A. The product of words can be extended to profinite words, making $\widehat{A^*}$ a compact topological monoid, called the *free profinite monoid* on A. Furthermore, each monoid morphism φ from A^* to a finite monoid M extends in a unique way to a uniformly continuous morphism from $\widehat{\varphi} : \widehat{A^*} \to M$.

It is not so easy to give examples of profinite words which are not words, but here is one. In a compact monoid, the smallest closed subsemigroup containing a given element x has a unique idempotent, denoted x^ω. In a finite monoid or in a free profinite monoid, one can show that x^ω is the limit of the Cauchy sequence $x^{n!}$.

Let u and v be two profinite words on the alphabet A. A finite monoid M *satisfies the profinite equation* $u = v$ if, for each monoid morphism $\varphi : A^* \to M$, one has $\widehat{\varphi}(u) = \widehat{\varphi}(v)$. Based on an early paper of Banaschewski [6], Reiterman [70] and Banaschewski [7] proved the following result.

Theorem 2.2. *A class of monoids is a variety of finite monoids if and only if it can be defined by a (possibly infinite) set of profinite equations.*

By combining Theorems 1.5 and 2.2, one gets the following corollary.

Corollary 2.3. *Every variety of languages can be defined by a set of profinite equations.*

2.3 Extensions of the Equational Approach

Although Corollary 2.3 has proven to be a powerful tool for the study of classes of regular languages, it has some weaknesses. First, the distinction between varieties and +-varieties seems somewhat artificial. Next, many natural families of regular languages do not form a variety. Fortunately, several improvements have made it possible to correct these shortcomings.

The first improvement relies on a definition already given in Schützenberger's seminal 1956 article. The *syntactic preorder* of a language L of A^* is the relation $\leqslant_L$ defined on A^* by[1] $u \leqslant_L v$ if and only if, for all $x, y \in A^*$,

$$xuy \in L \implies xvy \in L \tag{2}$$

The syntactic preorder induces a partial order on the syntactic monoid of L. The resulting ordered monoid is called the *syntactic ordered monoid* of L. For instance, the syntactic ordered monoid of the language $\{a, aba\}$ is the monoid $M = \{1, a, b, ab, ba, aba, 0\}$ presented by the relations $a^2 = b^2 = bab = 0$ and its syntactic order is defined by $0 < ab < 1$, $0 < ba < 1$, $0 < aba < a$ and $0 < b$.

A *positive variety of languages* is a class of regular languages closed under finite intersection and finite union, left and right residuals and inverses of morphisms between free monoids. The main advantage of positive varieties is the ability to study classes of languages that are not closed by complement. Theorem 1.5 extends as follows [42].

Theorem 2.4. *There is a bijection between varieties of finite ordered monoids and positive varieties of languages.*

The more general notion of *conjunctive varieties* was introduced in [32], but to my knowledge, it has not yet found convincing applications.

It is proved in [64] that varieties of finite ordered monoids can be defined by a set of profinite inequations of the form $u \leqslant v$. Consequently, Corollary 2.3 extends as follows:

Corollary 2.5. *Every positive variety of languages can be defined by a set of profinite inequations.*

---

[1] In previous articles, between 1995 and 2010, I unfortunately used the reverse order.

The next step dates from 2002, but it is simpler to return to it after discussing the next one, which concerns lattices of languages. A *lattice of languages* over an alphabet A is a set of languages over A closed under finite union and finite intersection, and which contains A^* and $\emptyset^2$.

Let L be a regular language on the alphabet A and let $\overline{L}$ be its topological closure in the free profinite monoid over A. Let u and v be profinite words on A. We say that L *satisfies the profinite inequation* $u \to v$ if

$$u \in \overline{L} \implies v \in \overline{L} \tag{3}$$

If η is the syntactic morphism of L, it is equivalent to state that

$$\widehat{\eta}(u) \in \eta(L) \implies \widehat{\eta}(v) \in \eta(L) \tag{4}$$

In 2008, Gehrke, Grigorieff and the author [27] proved the following result.

Theorem 2.6. *A set of regular languages over the alphabet A is a lattice of languages if and only if it can be defined by a set of equations of the form $u \to v$, where u and v are profinite words over A.*

To recover Corollary 2.5 from Theorem 2.6, we need the following result.

Proposition 2.7. *Let L be a regular language over A satisfying the profinite equation $u \to v$.*

(1) The complement of L satisfies the profinite equation $v \to u$.
(2) For each $x, y \in A^*$, the residual $x^{-1}Ly^{-1}$ satisfies the profinite equation $xuy \to xvy$.
(3) Let $\varphi : B^* \to A^*$ be a monoid morphism. Then the language $\varphi^{-1}(L)$ satisfies the profinite equation $\widehat{\varphi}(u) \to \widehat{\varphi}(v)$.

An intermediate case was already studied independently by Straubing [82] and Ésik [25] in 2002. Let $\mathcal{C}$ be a composition-closed class of morphisms between free monoids. A *positive $\mathcal{C}$-variety of languages* is defined like a positive variety except that it is only closed under *inverses of $\mathcal{C}$-morphisms*.

Examples of such classes $\mathcal{C}$ include the classes of length-preserving, length-decreasing, length-increasing, length-multiplying and also of course the class of all morphisms. A morphism between free monoids is *length-multiplying* if there is an integer k such that the image of each letter is a word of length k.

Let us say that a language L satisfies the *profinite $\mathcal{C}$-inequation* $u \leqslant v$ if, for every monoid $\mathcal{C}$-morphism $\varphi : B^* \to A^*$, L satisfies $\widehat{\varphi}(u) \to \widehat{\varphi}(v)$.

Corollary 2.8. *Any positive $\mathcal{C}$-variety of languages can be defined by a set of profinite $\mathcal{C}$-inequations.*

---

[2] respectively, the union and the intersection of an empty family of languages...

When $\mathcal{C}$ is the class of length-increasing morphisms, one essentially recovers +-varieties. Let us give two other instances of this result when $\mathcal{C}$ is the class ot length-multiplying morphisms. Let us call *modular simple* a language of the form

$$(A^d)^* a_1 (A^d)^* a_2 (A^d)^* \cdots a_k (A^d)^*$$

where $d > 0$, $k \geqslant 0$ and $a_1, a_2, \ldots, a_k \in A$. Finite union of modular simple languages form a positive $\mathcal{C}$-variety, defined by the $\mathcal{C}$-identities $1 \leqslant x^{\omega-1}y$ and $1 \leqslant yx^{\omega-1}$ [21]. If η is the syntactic morphism of L, it is equivalent to say that, if x and y are words *of the same length*, then $1 \leqslant \widehat{\eta}(x^{\omega-1}y)$ and $1 \leqslant \widehat{\eta}(yx^{\omega-1})$.

The second example is related to Boolean circuits. Recall that AC^0 is the set of unbounded fan-in, polynomial size, constant-depth Boolean circuits. The regular languages recognised by a circuit in AC^0 have been characterized in [9,82]. They form a $\mathcal{C}$-variety defined by the $\mathcal{C}$-equation $(x^{\omega-1}y)^\omega = (x^{\omega-1}y)^{\omega+1}$.

3   Operations on Languages

A general tool for studying operations on languages has been proposed in [58,59]. See also [53] for further examples. The possibility of extending operations on languages to classes of languages arose shortly after the variety theorem was published. In particular, a natural question is to determine the *closure* of a class of languages by a given operation.

3.1   Renaming

The first results in this direction concerned the *length preserving morphisms*, also called *renamings* by Esik. The set $\mathcal{P}(M)$ of subsets of a monoid M is a monoid, called the *power monoid* of M, under the multiplication of subsets defined by

$$XY = \{xy \mid x \in X \text{ and } y \in Y\}$$

For a variety $\mathbf{V}$ of finite monoids, let $\mathbf{PV}$ denote the variety of finite monoids generated by the monoids of the form $\mathcal{P}(M)$, where $M \in \mathbf{V}$. Reutenauer [71] and Straubing [80] independently proved the following result:

Theorem 3.1. *A variety of languages is closed under renaming if and only if the corresponding variety of finite monoids is a fixed point of the operator $\mathbf{P}$.*

This result was later extended to the ordered case by defining an appropriate operator $\mathbf{P}^{\downarrow}$ on varieties of finite ordered monoids. Let $\mathcal{V}$ be a positive variety of languages and let $\mathbf{V}$ be its corresponding variety of finite ordered monoids. Let also $\Lambda\mathcal{V}$ be the closure of $\mathcal{V}$ under renaming. The following result was proved by Polak [69, Theorem 4.2], Cano [18] and Cano-Pin [17, Proposition 6.3] and improved in [16][3]

---

[3] Unfortunately, as indicated earlier, the opposite order was used in this paper, so that upper sets were used instead of lower sets.

Theorem 3.2. *The variety of finite ordered monoids corresponding to* $\Lambda\mathcal{V}$ *is* $\mathbf{P}^{\downarrow}\mathbf{V}$.

On the algebraic side, there is a huge literature on the operators $\mathbf{P}$, whose ultimate goal would be the complete classification of the varieties $\mathbf{PV}$. See [1,2] for an overview of the known results. A detailed study of the operator $\mathbf{P}^{\downarrow}$ can also be found in [4].

3.2 Shuffle Product

In 1978, Perrot characterised all commutative varieties of languages closed under shuffle. This result was supplemented twenty years later by Esik and Simon [24].

Theorem 3.3. *The unique non-commutative variety of languages closed under shuffle is the variety of all regular languages.*

Thus the variety of commutative languages is the unique maximal proper variety of languages closed under shuffle. The counterpart for positive varieties was given by Cano and the author [17].

Theorem 3.4. *There is a largest proper positive variety closed under shuffle. It is the largest positive variety not containing the language* $(ab)^*$. *It is also the largest proper positive variety closed under renaming.*

This positive variety is decidable, as there is an effective characterization of the corresponding variety of finite ordered monoids. See [56] for further results.

3.3 Concatenation Product

Concatenation product and its variants are probably the most studied operations on regular languages. The article [49] presents most of the results known before 2011. A detailed account of recent results can be found in [66,67].

A language L of A^* is a *marked product* of the languages $L_0, L_1, \ldots, L_n$ if $L = L_0 a_1 L_1 \cdots a_n L_n$ for some letters $a_1, \ldots, a_n$ of A. A marked product $L = L_0 a_1 L_1 \cdots a_n L_n$ is said to be *unambiguous* if every word of L admits a unique decomposition of the form $u = u_0 a_1 u_1 \cdots a_n u_n$ with $u_0 \in L_0, \ldots, u_n \in L_n$. For instance, the marked product $\{a, c\}^* a \{1\} b \{b, c\}^*$ is unambiguous. Other variants include deterministic and modular products.

Let $\mathcal{L}$ be a lattice of languages. The *polynomial closure* of $\mathcal{L}$ is the set of languages that are finite unions of marked products of languages of $\mathcal{L}$. It is denoted $\mathrm{Pol}\,\mathcal{L}$. Similarly, the *unambiguous polynomial closure* $\mathrm{UPol}\,\mathcal{L}$ of $\mathcal{L}$ is the set of languages that are finite unions of unambiguous marked products of languages of $\mathcal{L}$. It is denoted $\mathrm{UPol}\,\mathcal{L}$. Finally, let co-$\mathrm{Pol}\,\mathcal{L}$ denote the set of complements of languages in $\mathrm{Pol}\,\mathcal{L}$ and let $\mathscr{B}\mathrm{Pol}\,\mathcal{L}$ be the Boolean closure of $\mathrm{Pol}\,\mathcal{L}$.

An algebraic characterization of the polynomial closure of a variety of languages was first given in [62,65]. It was extended to positive varieties in [46] and to lattices of regular languages in [12].

Theorem 3.5. *If $\mathcal{L}$ is a lattice of regular languages closed under residuals, then $\mathrm{Pol}\,\mathcal{L}$ is defined by the equations of the form $x^\omega y x^\omega \leqslant x^\omega$, where x, y are profinite words such that the equations $x = x^2$ and $y \leqslant x$ are satisfied by $\mathcal{L}$.*

Straubing gave an algebraic description of the varieties of languages closed under marked products [80] in terms of Mal'cev products of monoids. Combined with a result of [63], it leads to the following statement.

Theorem 3.6. *Let $\mathbf{V}$ be a variety of monoids and let $\mathcal{V}$ be the corresponding variety of languages. Then $\mathcal{V}$ is closed under marked products if and only if, for each profinite word x such that the equation $x = x^2$ is satisfied by $\mathcal{V}$, the equation $x^{\omega+1} = x^\omega$ is also satisfied by $\mathcal{V}$.*

This result was later extended to $\mathcal{C}$-varieties in [20, Theorem 4.1].

4   Concatenation Hierarchies

Further details on the results presented in this section can be found in [51] for results up to 2017 and in the recent book of Place and Zeitoun [68] for the most recent ones.

Concatenation hierarchies are defined by alternating Boolean operations and polynomial operations. For historical reasons, they are indexed by half-integers.

In the Straubing-Thérien hierarchy, the level 0 is the trivial variety of languages, defined by $\mathcal{V}_0(A) = \{\emptyset, A^*\}$. For each $n \geqslant 0$, the level $n + 1/2$ is the polynomial closure of the level n, and the level $n + 1$ is the Boolean closure of the level $n + 1/2$. It is pictured in the diagram below.

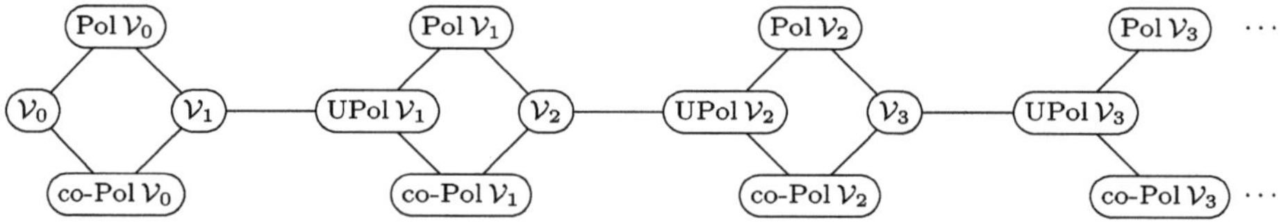

In this diagram, the central row consists of varieties of languages and the two other rows of positive varieties. It follows that there is a corresponding diagram with varieties of finite (ordered) monoids. For instance, $\mathbf{V}_1 = \mathbf{J}$ and $\mathbf{V}_2 = \mathbf{PJ}$. To date, the decidability of the Straubing-Thérien hierarchy remains a fascinating open problem. The decidability of the varieties $\mathcal{V}_n$ has been proved for $n \leqslant 3$, each step being a major advance, and is open for $n > 3$.

Other concatenation hierarchies are equally interesting. For the Brzozowski (or dot-depth) hierarchy, level zero consists of four languages: the empty one, the singleton consisting of the empty word, and their complements. Another prominent hierarchy is the group hierarchy whose basis consists of all languages recognized by a finite group.

Unfortunately, it is not possible to summarize here the impressive results obtained by Place and Zeitoun, but I recommend reading the introduction of [66] or the book [68] for a complete overview. Many arguments come from algebra and model theory, but I would like to emphasize the crucial role played by the natural notion of *separability*. Besides its intrinsic interest, it is related to that of a *pointlike set* in semigroup theory and has a natural interpretation in the profinite topology [3]. Another great collaboration between these three areas!

5 Logic

The connections with logic have been studied since the beginning of automata theory. In particular, the work of Büchi, Elgot and Trakhtenbrot in the 1960 s showed that monadic second-order on words captures the class of regular languages. The case of first-order logic was not addressed until ten years later by McNaughton and Papert [36].

Theorem 5.1. *A language is star-free if and only if it is first-order definable.*

In 1982, Thomas [85] refined this result by showing that the Straubing-Thérien hierarchy corresponds, level by level, to the quantifier alternation hierarchy of first-order formulas, depicted in the diagram below. This was a pleasant surprise, as these two hierarchies had been defined completely independently.

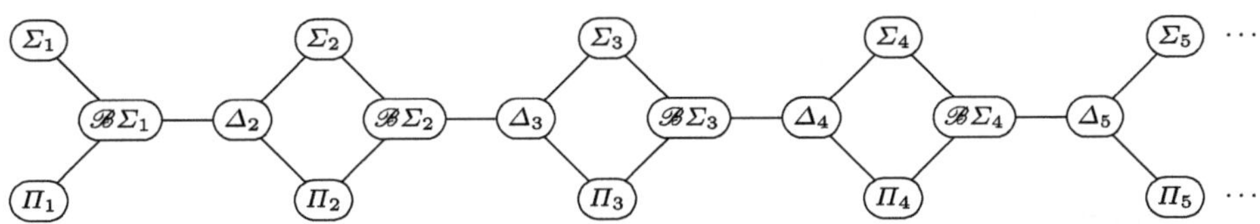

Here is another remarkable result [26, 84].

Theorem 5.2. *A language is unambiguous star-free if and only if it is definable in the two variables fragment of first-order logic.*

Many more fragments of logic have been studied, notably temporal logic. See [45] and the books [68, 81] for more examples. And as this 2026 article [8] shows, work in this field remains at the forefront of current research.

6 An Example of Success Story

Let me give a typical example of the fruitful collaboration between automata theory, algebra, topology and model theory, which took place during the 1990 s An automaton is *reversible* if every letter induces a partial one-to-one map from the set of states into itself. However, no restrictions are imposed on the sets of initial and final states. The question arose of describing the languages recognized by a reversible automaton.

As early as 1979, Reutenauer [71] had observed that these automata were related to the pro-group topology on a free group. This topology, studied by M. Hall in the fifties, is the analogue for the free group of the one we defined for the free monoid. Hall [29] proved that every finitely generated subgroup of the free group is closed for this topology.

With Reutenauer [57], we had conjectured more generally that every product of finitely generated subgroups of a free group is closed. This conjecture was proven in 1993 by Ribes and Zalesskii [72], but the proof is quite difficult. Since then, other proofs have been proposed, including one via model theory, due to Herwig and Lascar [31]. A consequence of this theorem is that the trace on A^* of the closure of a regular language for the pro-group topology is itself a regular language, and there is an algorithm to compute it.

But this is not the end of the story. In the eighties, Rhodes circulated a conjecture that was considered at the time to be one of the most important open problems in the theory of finite semigroups. This conjecture proposed an algorithm for calculating the *group kernel* of a finite semigroup and had far-reaching consequences. The conjecture was finally proved by Ash in 1991. But in the 90's, I found a surprising connection with the pro-group topology, which, through the result of Ribes and Zalesskii mentioned earlier, gave another proof of the Rhodes conjecture. When the dust settled, several conjectures had been resolved, notably a simple description of the variety **PG** of powergroups, a result that appears to be completely unrelated. An account of this exciting saga can be found in the articles [30, 41].

Coming back to the reversible automata, the following corollary was proved [38, 39, 48]. Let L be a regular language and let M be its syntactic ordered monoid.

Theorem 6.1. *The following conditions are equivalent:*

(1) L recognized by a reversible automaton,
(2) L is closed in the pro-group topology and the idempotents of M commute,
(3) the idempotents of M commute and, for each idempotent e of M, $e \leqslant 1$.

Many questions remain open, particularly regarding extending the previous results to certain varieties of finite groups. For example, there is an algorithm for computing the closure of a regular language in the pro-p topology (defined by p-groups), but in the case of solvable groups, the question is still open.

7   Open Problems

I already mentioned the numerous open problems about the concatenation hierarchies. Let me mention two other open problems. The first one is (part of) the famous *star-height problem*—see [52] for more details.

Problem 1. Is there a regular language of star height two?

Theorem 1.1 gives an algebraic characterization of the languages of star-height 0 and one may wonder whether such a characterization is possible for the languages of star-height one. It is proved in [61] that the languages of star-height $\leqslant 1$ are closed under Boolean operations, quotients and inverse of length-decreasing morphisms. According to Corollary 2.8, it would suffice to find a single nontrivial length-decreasing equation satisfied by these languages to prove the existence of a language of star-height two. Unfortunately, no such equation is known and we still have no example of a language of star height two.

The second problem was proposed by Antonio Restivo. Partial results and related questions are presented in [10].

Problem 2. What is the smallest class of languages containing the singletons and closed under finite Boolean operations, product and shuffle

Let us call *intermixed* the languages of this class. It is known that intermixed languages form a proper subclass of the class of regular languages. Furthermore, they are closed under residuals and under inverses of length-decreasing morphisms, but not under inverses of morphisms. In view of Corollary 2.8, this opens the door for a potential algebraic characterization, still to be found.

8   Conclusion

Automata and semigroups have mutually enriched each other. First, some now standard notions in semigroup theory were originally inspired or motivated by results on regular languages. For instance, the notion of a *derived category* of a semigroup morphism originates in the proof by Brzozowski and Simon of Theorem 1.2. Conversely, several concepts from semigroup theory have proven very useful for the study of regular languages. For example, the *wreath product principle*, initially introduced in 1979 by Straubing [79], is a major tool that has been generalized several times since.

Work continues to extend known results on varieties of languages: some have been extended to positive varieties, others to $\mathcal{C}$-varieties, and still others to lattices of languages. But there is still much to be done!

Furthermore, algebraic, topological and logical approaches are not limited to words. They have also been successfully applied to infinite words, words indexed by ordinals or linear orders, and trees. Again, extending all known results on words to these other structures is a vast research program.

Finally, it is possible to extend Theorem 2.6 beyond regular languages. The price to pay is to go from profinite equations to another type of equations, the *procompact* equations. Here the key space is βA^*, the *Stone-Čech compactification* of A^*, usually defined as the set of ultrafilters on the discrete space A^*. But it is also the closure of the image of A^* in the product space $\prod K$, where the product runs over all maps from A^* into a compact Hausdorff space K whose underlying set is $\mathcal{P}(\mathcal{P}(A^*))$. That is the reason why I propose to call the elements of βA^* *procompact words*, by analogy with profinite words.

It was proved in [28] that a set of languages of A^* is a lattice of languages if and only if it can be defined by a set of equations of the form $u \to v$, where u and v are procompact words over A. We only know of a few examples of such equations for specific classes of languages, and this result remains difficult to apply at present.

References

1. Almeida, J.: Finite semigroups and universal algebra. World Scientific Publishing Co., Inc (1994). Translated from the 1992 Portuguese original and revised by the author

2. Almeida, J.: Power semigroups: results and problems. In: M. Ito and C. Nehaniv (eds.)Algebraic Engineering. Proceedings of the First International Conference on Semigroups and Algebraic Engineering Held in Aizu, Japan, 24–28 March 1997, pp. 399–415. World Scientific, Singapore (1999)

3. Almeida, J.: Some algorithmic problems for pseudovarieties, vol. 54, pp. 531–552 (1999). Automata and formal languages, VIII (Salgótarján 1996)

4. Almeida, J., Cano, A., Klíma, O., Pin, J.É.: Fixed points of the lower set operator. Internat. J. Algebra Comput. 25(1–2), 259–292 (2015)

5. Ash, C.J., Hall, T.E., Pin, J.É.: On the varieties of languages associated with some varieties of finite monoids with commuting idempotents. Inform. and Comput. 86(1), 32–42 (1990). https://doi.org/10.1016/0890-5401(90)90024-C

6. Banaschewski, B.: On profinite universal algebras. In: General topology and its relations to modern analysis and algebra, III (Proc. Third Prague Topological Sympos., 1971), pp. 51–62. Academia [Publishing House of the Czechoslovak Academy of Sciences], Prague (1972)

7. Banaschewski, B.: The Birkhoff theorem for varieties of finite algebras. Algebra Universalis 17(3), 360–368 (1983)

8. Barloy, C., Tschirbs, F., Vortmeier, N., Zeume, T.: Algebraic characterizations of classes of regular languages in DynFO. In: 43rd International Symposium on Theoretical Aspects of Computer Science (STACS 2026), LIPIcs. Leibniz Int. Proc. Inform., vol. 364, pp. Art. No. 9, 19. Schloss Dagstuhl. Leibniz-Zent. Inform., Wadern (2026). https://doi.org/10.4230/lipics.stacs.2026.9

9. Barrington, D.A.M., Compton, K., Straubing, H., Thérien, D.: Regular languages in NC^1. J. Comput. System Sci. 44(3), 478–499 (1992)

10. Berstel, J., Boasson, L., Carton, O., Pin, J.É., Restivo, A.: The expressive power of the shuffle product. Inf. Comput. 208, 1258–1272 (2010)

11. Birkhoff, G.: On the structure of abstract algebras. Proc. Cambridge Phil. Soc. 31, 433–454 (1935)

12. Branco, M.J., Pin, J.É.: Equations for the polynomial closure of a lattice of regular languages. In: Albers, S., Thomas, W. (eds.) ICALP 2009, Part II. LNCS. vol. 5556, pp. 115–126. Springer, Berlin (2009)

13. Brzozowski, J.A.: Developments in the theory of regular languages, pp. 29–40. IFIP Congress (1980)

14. Brzozowski, J.A., Fich, F.E.: Languages of $\mathcal{R}$-trivial monoids. J. Comput. Syst. Sci. 20(1), 32–49 (1980)

15. Brzozowski, J.A., Simon, I.: Characterizations of locally testable events. Discrete Math. 4, 243–271 (1973)

16. Cano, A., Pin, J.É.: Upper set monoids and length preserving morphisms. J. Pure Appli. Algebra 216, 1178–1183 (2012)

17. Cano Gómez, A., Pin, J.É.: Shuffle on positive varieties of languages. Theoret. Comp. Sci. 312, 433–461 (2004)

18. Cano Gómez, A.: Semigroupes ordonnés et opérations sur les langages rationnels, Ph.D. thesis. Université Paris 7 and Departamento de Sistemas Informáticos y Computación, Universidad Politécnica de Valencia (2003)

19. Carton, O., Pin, J.É., Soler-Escrivà, X.: Languages recognized by finite supersoluble groups. J. Autom. Lang. Comb. **14**(2), 149–161 (2009)

20. Chaubard, L., Pin, J.É., Straubing, H.: Actions, wreath products of $\mathcal{C}$-varieties and concatenation product. Theor. Comput. Sci. **356**, 73–89 (2006)

21. Chaubard, L., Pin, J.É., Straubing, H.: First order formulas with modular predicates. In: 21st Annual IEEE Symposium on Logic in Computer Science (LICS 2006), pp. 211–220. IEEE (2006)

22. Eilenberg, S.: Automata, languages, and machines. Vol. B. Academic Press [Harcourt Brace Jovanovich Publishers], New York (1976), with two chapters ("Depth decomposition theorem" and "Complexity of semigroups and morphisms") by Bret Tilson, Pure and Applied Mathematics, Vol. 59

23. Eilenberg, S., Schützenberger, M.P.: On pseudovarieties. Advances in Math. **19**(3), 413–418 (1976). https://doi.org/10.1016/0001-8708(76)90029-3

24. Ésik, Z., Simon, I.: Modeling literal morphisms by shuffle. Semigroup Forum **56**(2), 225–227 (1998)

25. Ésik, Z.: Extended temporal logic on finite words and wreath products of monoids with distinguished generators. In: Ito, M.E.A. (ed.) DLT 2002, Kyoto, Japan. pp. 43–58. No. 2450 in Lecture Notes in Comput. Sci., Springer, Berlin (2002)

26. Etessami, K., Vardi, M.Y., Wilke, T.: First-order logic with two variables and unary temporal logic. In: LICS 1997, Warsaw, vol. 179, pp. 279–295 (2002). https://doi.org/10.1006/inco.2001.2953

27. Gehrke, M., Grigorieff, S., Pin, J.É.: Duality and equational theory of regular languages. In: Aceto, L., Damgård, I., Goldberg, L.A., Halldórsson, M.M., Ingólfsdóttir, A., Walukiewicz, I. (eds.) ICALP 2008. LNCS, vol. 5126, pp. 246–257. Springer, Heidelberg (2008). https://doi.org/10.1007/978-3-540-70583-3_21

28. Gehrke, M., Grigorieff, S., Pin, J.É.: A topological approach to recognition. In: Abramsky, S., Gavoille, C., Kirchner, C., Meyer auf der Heide, F., Spirakis, P.G. (eds.) ICALP 2010. LNCS, vol. 6199, pp. 151–162. Springer, Heidelberg (2010). https://doi.org/10.1007/978-3-642-14162-1_13

29. Hall, M.: A topology for free groups and related groups. Ann. Math. **52**, 127–139 (1950)

30. Henckell, K., Margolis, S.W., Pin, J.É., Rhodes, J.: Ash's type II theorem, profinite topology and Mal'cev products I. Internat. J. Algebra Comput. **1**(4), 411–436 (1991)

31. Herwig, B., Lascar, D.: Extending partial automorphisms and the profinite topology on free groups. Trans. Amer. Math. Soc. **352**(5), 1985–2021 (2000)

32. Klíma, O., Polák, L.: On varieties of meet automata. Theoret. Comput. Sci. **407**(1–3), 278–289 (2008)

33. Lallement, G.: Semigroups and Combinatorial Applications, Wiley and Sons (1979)

34. Margolis, S.W., Pin, J.É.: Products of group languages. In: Budach, L. (ed.) FCT 1985. LNCS, vol. 199, pp. 285–299. Springer, Heidelberg (1985). https://doi.org/10.1007/BFb0028813

35. McNaughton, R.: Algebraic decision procedures for local testability. Math. Systems Theory **8**(1), 60–76 (1974)

36. McNaughton, R., Papert, S.: Counter-free automata. The M.I.T. Press, Cambridge, Mass.-London (1971): with an appendix by William Henneman. M.I.T, Research Monograph, No. 65

37. Pin, J.É.: Varieties of formal languages. Plenum Publishing Corp., New York (1986), with a preface by M.-P. Schutzenberger, Translated from the French by A. Howie

38. Pin, J.É.: On the languages accepted by finite reversible automata. In: Ottmann, T. (ed.) ICALP 1987. LNCS, vol. 267, pp. 237–249. Springer, Heidelberg (1987). https://doi.org/10.1007/3-540-18088-5_19

39. Pin, J.É.: On reversible automata. In: Simon, I. (ed.) LATIN 1992. LNCS, vol. 583, pp. 401–416. Springer, Heidelberg (1992). https://doi.org/10.1007/BFb0023844

40. Pin, J.É.: Finite semigroups and recognizable languages: an introduction. In: Semigroups, Formal Languages And Groups (York, 1993), pp. 1–32. Kluwer Acad. Publ., Dordrecht (1995)

41. Pin, J.É.: PG = BG, a success story. In: Fountain, J. (ed.) Semigroups, Formal Languages and Groups (York, 1993), NATO Adv. Sci. Inst. Ser. C Math. Phys. Sci., vol. 466, pp. 33–47. Kluwer Acad. Publ., Dordrecht (1995)

42. Pin, J.É.: A variety theorem without complementation. Russian Mathematics (Izvestija vuzov.Matematika) **39**, 80–90 (1995)

43. Pin, J.É.: Logic, semigroups and automata on words. Ann. Math. Artif. Intell. **16**, 343–384 (1996)

44. Pin, J.É.: Syntactic semigroups. In: Rozenberg, G., Salomaa, A. (eds.) Handbook of formal languages, vol. 1, chap. 10, pp. 679–746. Springer (1997)

45. Pin, J.É.: Logic on words. In: G. Păun, G.R., Salomaa, A. (eds.) Current Trends in Theoretical Computer Science, Entering the 21st Century, pp. 254–273. Word Scientific (2001)

46. Pin, J.É.: Algebraic tools for the concatenation product. Theor. Comput. Sci. **292**, 317–342 (2003)

47. Pin, J.É.: Profinite methods in automata theory. In: Albers, S. (ed.) 26th International Symposium on Theoretical Aspects of Computer Science (STACS 2009). pp. 31–50. Internationales Begegnungs- Und Forschungszentrum fur Informatik (IBFI). Schloss Dagstuhl, Germany, Dagstuhl, Germany (2009)

48. Pin, J.É.: Automates réversibles: combinatoire, algèbre et topologie. In: Bayart, F., Charpentier, É. (eds.) Leçons de mathématiques d'aujourd'hui, vol. 4. Sel Fer, Paris: Cassini (2010)

49. Pin, J.É.: Theme and variations on the concatenation product. In: Winkler, F. (ed.) CAI 2011. LNCS, vol. 6742, pp. 44–64. Springer, Heidelberg (2011). https://doi.org/10.1007/978-3-642-21493-6_3

50. Pin, J.É.: Equational descriptions of languages. Int. J. Found. Comput. S. **23**, 1227–1240 (2012)

51. Pin, J.É.: The dot-depth hierarchy, 45 years later. In: Konstantinidis, S., Moreira, N., Reis, R., Jeffrey, S. (eds.) The Role of Theory in Computer Science - Essays Dedicated to Janusz Brzozowski, pp. 177–202. Word Scientific (2017)

52. Pin, J.É.: Open problems about regular languages, 35 years later. In: Konstantinidis, S., Moreira, N., Reis, R., Jeffrey, S. (eds.) The Role of Theory in Computer Science, Essays dedicated to Janusz Brzozowski, pp. 153–176. Word Scientific (2017)

53. Pin, J.É.: How to prove that a language is regular or star-free? In: Leporati, A., Martín-Vide, C., Shapira, D., Zandron, C. (eds.) LATA 2020. LNCS, vol. 12038, pp. 68–88. Springer, Cham (2020). https://doi.org/10.1007/978-3-030-40608-0_5

54. Pin, J.É. (ed.): Handbook of automata theory. Vol. I. Theoretical foundations. EMS Press, Berlin (2021). https://doi.org/10.4171/Automata

55. Pin, J.É. (ed.): Handbook of automata theory. Vol. II. Automata in mathematics and selected applications. EMS Press, Berlin (2021). https://doi.org/10.4171/Automata

56. Pin, J.É.: Shuffle product of regular languages: results and open problems. In: Poulakis, D., Rahonis, G. (eds.) CAI 2022. LNCS, vol. 13706, pp. 26–39. Springer (2022). https://doi.org/10.1007/978-3-031-19685-0_3

57. Pin, J.É., Reutenauer, C.: A conjecture on the Hall topology for the free group. Bull. London Math. Soc. **23**, 356–362 (1991)

58. Pin, J.É., Sakarovitch, J.: Some operations and transductions that preserve rationality. In: Cremers, A.B., Kriegel, H.-P. (eds.) GI-TCS 1983. LNCS, vol. 145, pp. 277–288. Springer, Heidelberg (1982). https://doi.org/10.1007/BFb0036488

59. Pin, J.É., Sakarovitch, J.: Une application de la représentation matricielle des transductions. Theor. Comput. Sci. **35**, 271–293 (1985)

60. Pin, J.É., Straubing, H., Thérien, D.: Small varieties of finite semigroups and extensions. J. Austral. Math. Soc. **37**, 269–281 (1984)

61. Pin, J.É., Straubing, H., Thérien, D.: Some results on the generalized star-height problem. Inf. Comput. **101**, 219–250 (1992)

62. Pin, J.É., Weil, P.: Polynomial closure and unambiguous product. In: 22th ICALP, pp. 348–359, vol. 944. LNCS. Springer, Berlin (1995)

63. Pin, J.É., Weil, P.: Profinite semigroups, Mal'cev products and identities. J. of Algebra **182**, 604–626 (1996)

64. Pin, J.É., Weil, P.: A Reiterman theorem for pseudovarieties of finite first-order structures. Algebra Universalis **35**, 577–595 (1996)

65. Pin, J.É., Weil, P.: Polynomial closure and unambiguous product. Theory Comput. Syst. **30**, 1–39 (1997)

66. Place, T., Zeitoun, M.: Dot-depth three, return of the $\mathcal{J}$-class. In: Proceedings of the 39th Annual ACM/IEEE Symposium on Logic in Computer Science, p. 15. ACM, New York (2024). https://doi.org/10.1145/3661814.3662082

67. Place, T., Zeitoun, M.: Closing star-free closure. ACM Trans. Comput. Log. **26**(3), Art. 15, 62 (2025)

68. Place, T., Zeitoun, M.: Classes of regular languages and their decision problems (2026). https://hal.science/hal-05570589, to appear

69. Polák, L.: Operators on classes of regular languages. In: Gomes, G.S.M., Pin, J.É., Silva, P. (eds.) Semigroups, Algorithms, Automata and Languages, pp. 407–422. World Scientific (2002)

70. Reiterman, J.: The Birkhoff theorem for finite algebras. Algebra Universalis **14**(1), 1–10 (1982)

71. Reutenauer, C.: Sur les variétés de langages et de monoïdes. In: Weihrauch, K. (ed.) GI-TCS 1979. LNCS, vol. 67, pp. 260–265. Springer, Heidelberg (1979). https://doi.org/10.1007/3-540-09118-1_27

72. Ribes, L., Zalesskii, P.A.: On the profinite topology on a free group. Bull. London Math. Soc. **25**(1), 37–43 (1993). https://doi.org/10.1112/blms/25.1.37

73. Schützenberger, M.P.: Une théorie algébrique du codage. Séminaire Dubreil. Algèbre et théorie des nombres **9**, 1–24 (1955–1956). http://eudml.org/doc/111094

74. Schützenberger, M.P.: On finite monoids having only trivial subgroups. Inf. Control **8**, 190–194 (1965)

75. Schützenberger, M.P.: Sur les monoïdes finis dont les groupes sont commutatifs. Rev. Française Automat. Informat. Recherche Opérationnelle Sér. Rouge 8(R-1), 55–61 (1974)

76. Schützenberger, M.P.: Sur le produit de concaténation non ambigu. Semigroup Forum **13**(1), 47–75 (1976/77)

77. Sezinando, H.: The varieties of languages corresponding to the varieties of finite band monoids. Semigroup Forum **44**(3), 283–305 (1992)

78. Simon, I.: Piecewise testable events. In: Brakhage, H. (ed.) GI-Fachtagung 1975. LNCS, vol. 33, pp. 214–222. Springer, Heidelberg (1975). https://doi.org/10.1007/3-540-07407-4_23

79. Straubing, H.: Families of recognizable sets corresponding to certain varieties of finite monoids. J. Pure Appl. Algebra **15**(3), 305–318 (1979)
80. Straubing, H.: Recognizable sets and power sets of finite semigroups. Semigroup Forum **18**, 331–340 (1979)
81. Straubing, H.: Finite automata, formal logic, and circuit complexity. Birkhäuser Boston Inc., Boston, MA (1994)
82. Straubing, H.: On logical descriptions of regular languages. In: Rajsbaum, S. (ed.) LATIN 2002. LNCS, vol. 2286, pp. 528–538. Springer, Heidelberg (2002). https://doi.org/10.1007/3-540-45995-2_46
83. Thérien, D.: Classification of finite monoids: the language approach. Theoret. Comput. Sci. **14**, 195–208 (1981)
84. Thérien, D., Wilke, T.: Over words, two variables are as powerful as one quantifier alternation. In: STOC 1998, Dallas, TX, pp. 234–240. ACM, New York (1999)
85. Thomas, W.: Classifying regular events in symbolic logic. J. Comput. System Sci. **25**(3), 360–376 (1982)
86. Yu, S.: Regular languages. In: Rozenberg, G., Salomaa, A. (eds.) Handbook of Language Theory, vol. 1, chap. 2, pp. 679–746. Springer (1997)

Conference Papers

Thue–Morse Series and k-ary Partitions

Mehdi Golafshan and Michel Rigo[(✉)][iD]

Department of Mathematics, University of Liège, Allée de la découverte 12 (B37),
4000 Liège, Belgium
{mgolafshan,m.rigo}@uliege.be

Abstract. We show that the product of the generating function of k-ary partitions with the Thue–Morse series whose coefficients are kth roots of unity defines a k-regular sequence $(g_k(n))_{n\geq 0}$, given explicitly as a product over the digits of the base-k expansions. We study this sequence and highlight similarities with classical congruence results for k-ary partitions. We provide an elementary proof that the unrestricted k-ary partition function is not k-regular. Finally, we study the inverse of the Thue–Morse series, derive explicit recurrence relations for its coefficients, and show that this inverse cannot be k-regular.

Keywords: Thue–Morse series · k-ary partition function · k-regularity · automatic sequences · Mahler equations · inverse formal power series · digit expansions · partition polynomials

1   Introduction

Automatic and regular sequences are a central object of study in combinatorics on words [1,2,4,6,8,9]. The Thue–Morse word $01101001\cdots$ appears across a remarkable range of contexts, from number theory [11] and combinatorics [15, 19,20] to physics, aperiodic order, and even economic modeling. To give the reader an idea, the survey paper [3] has been cited in more than 170 documents in the `zbMath` database!

Merca [18] recently exploited the fact that the generating function of binary partitions is the inverse of the Thue–Morse series with coefficients in $\{-1,1\}$. Equivalently, the convolution of the first n terms of the binary Thue–Morse sequence over $\{-1,1\}$ and the binary partition function is equal to 0 for $n > 1$. Such an identity raises a natural question: *Does a similar phenomenon occur for k-ary partitions and generalized ThueMorse sequences when $k \geq 3$?* Let us now introduce the main concepts.

The *order* of a power series p in $\mathbb{C}[\![z]\!]$, denoted by $\mathrm{ord}(\mathsf{p})$, is the least n such that the coefficient $[z^n]\mathsf{p} \neq 0$. Let $(\mathsf{p}_n(z))_{n\geq 0}$ be a sequence of power series such that $\mathrm{ord}(\mathsf{p}_n) \to \infty$ when $n \to \infty$. Then the infinite product

$$\prod_{n\geq 0}(1 + \mathsf{p}_n(z))$$

Supported by the FNRS Research grant T.196.23 (PDR).

is well-defined in the ring of formal series $\mathbb{C}[\![z]\!]$. The coefficient of z^d in the partial products depends only on factors with $\mathrm{ord}(\mathsf{p}_n) \leq d$ and thus, there are only finitely many such factors.

Let F_n be the nth Fibonacci number with the convention $F_0 = F_1 = 1$. In the theory of partitions of integers, it is natural to encounter expressions such as

$$\mathsf{c}(z) = \prod_{n \geq 1} \frac{1}{1 - z^n} = \sum_{n \geq 0} c(n)\, z^n \quad \text{or,} \quad \mathsf{f}(z) = \prod_{n \geq 0} \frac{1}{1 - z^{F_n}} = \sum_{n \geq 0} f(n)\, z^n.$$

The first (resp., second) one is the generating function of the number of partitions of a non-negative integer n as a sum of integers[1] (resp., Fibonacci numbers[2]) —each can be used any number of times, possibly zero— and the order of the terms does not matter.

Remark 1. We use the convention $F_0 = F_1 = 1$, so the product for $\mathsf{f}(z)$ contains the factor $(1 - z)^{-2}$. Equivalently, $f(n)$ counts Fibonacci partitions in which the part 1 occurs in two distinguishable copies (coming from F_0 and F_1). This convention is consistent with the example $f(2) = 4$ below.

As an example, $c(3) = 3$ and $f(2) = 4$ because

$$3 = 3 \cdot 1 = 1 \cdot 1 + 1 \cdot 2 = 0 \cdot 1 + 0 \cdot 2 + 1 \cdot 3 \quad \text{and} \quad 2 = 2 \cdot F_0 = 1 \cdot F_0 + 1 \cdot F_1 = 2 \cdot F_1 = F_2.$$

It is worth mentioning that the sequence $(f(n))_{n \geq 0}$ is Fibonacci-regular [4, p. 452]. This is a natural generalization of the concept of a k-regular sequence recalled below where one considers Fibonacci numeration system instead of integer base numeration system.

In this article, we will not distinguish between a power series and the sequence of its coefficients. Thus, by abuse of language, we say that the sequence $(f(n))_{n \geq 0}$ or the series $\mathsf{f}(z)$ is regular. Let R be a ring and let $k \geq 2$ be an integer. Recall that a sequence $(a_n)_{n \geq 0}$ with values in R is called *k-regular* if the R-module generated by its k-kernel

$$\mathcal{K}_k(a) = \left\{ (a_{k^\ell n + r})_{n \geq 0} \mid \ell \geq 0,\ 0 \leq r < k^\ell \right\}$$

is finitely generated. Equivalently, it can be produced by a finite linear representation over R depending on the base-k expansion of n. We refer the reader to [2,4,7]. Regular sequences form a natural generalization of automatic sequences. They arise naturally in combinatorics on words, number theory, and the study of formal power series, where many arithmetically defined sequences exhibit regular behavior with respect to their base-k expansions. This concept has been generalized to other numeration systems [1,8,9].

---

[1] A000041: $1, 1, 2, 3, 5, 7, 11, 15, 22, 30, 42, 56, 77, \ldots$.
[2] A007000: $1, 2, 4, 7, 11, 17, 25, 35, 49, 66, 88, 115, 148, \ldots$.

In [2, Example 18], it is shown that the sequence $(c_{\leq j}(n))_{n \geq 0}$ that counts the number of partitions of n as a sum of integers $\leq j$ is k-regular, for all $j \geq 1$ and all $k \geq 2$,

$$c_{\leq j}(z) = \prod_{\ell=1}^{j} \frac{1}{1-z^{\ell}}. \tag{1}$$

This follows from [2, Theorem 3.3] because $c_{\leq j}(z)$ is a rational function whose poles are roots of unity. For the z-adic topology, $c_{\leq j}(z) \to c(z)$ when $j \to \infty$. However, the sequence $(c(n))_{n \geq 0}$ counting *unrestricted* partitions is not k-regular because it grows too quickly (see [4, Theorem 16.3.1]).

Let $k \geq 2$. Motivated by the recent paper [18], we are interested in the following type of partition. A *k-ary partition* of the non-negative integer n is a way of writing n as a sum of powers of k, where each power can be used any number of times (possibly zero), and the order of the terms does not matter. It is well-known [5] and it follows from the uniqueness of base-k expansions that

$$b_k(z) = \sum_{n \geq 0} b_k(n)z^n = \prod_{r \geq 0} \left(1 + z^{k^r} + z^{2k^r} + \cdots\right) = \prod_{r \geq 0} \frac{1}{1-z^{k^r}}. \tag{2}$$

For instance, $b_3(10) = 5$ because

$$10 = 10 \cdot 3^0 = 7 \cdot 3^0 + 1 \cdot 3^1 = 4 \cdot 3^0 + 2 \cdot 3^1 = 1 \cdot 3^0 + 3 \cdot 3^1 = 1 \cdot 3^0 + 1 \cdot 3^2.$$

and $b_3(z)$ encodes the sequence A062051 in the OEIS.

$$1 + z + z^2 + 2z^3 + 2z^4 + 2z^5 + 3z^6 + 3z^7 + 3z^8 + 5z^9 + 5z^{10} + 5z^{11} + \cdots.$$

Mahler, and later de Bruijn were already interested in estimating $b_k(n)$ [10, 17].

Having the Fibonacci sequence and the series f, which is Fibonacci-regular, on the one side, and a base-k numeration system on the other, we seek analogues of Fibonacci-regularity phenomena in base-k systems. It is therefore natural to ask whether the sequence $(b_k(n))_{n \geq 0}$ is k-regular.

Similarly to (1), the sequence counting restricted k-ary partitions in which only powers k^r with $0 \leq r \leq j$ are allowed forms an ℓ-regular sequence, for all $j \geq 1$ and all $\ell \geq 2$.

1.1 Some Results on k-ary Partitions

Let us recall some basic results. See, for instance, [5].

Lemma 1. *Let $k \geq 2$. For all $n \geq 0$, we have*

$$b_k(kn) = b_k(kn+1) = \cdots = b_k(kn+k-1)$$

and $b_k(kn) = b_k(kn-1) + b_k(n)$, for all $n \geq 1$. Therefore, $(1-z)\, b_k(z) = b_k(z^k)$.

Proof. Based on (2), we have

$$(1 - z)\mathsf{b}_k(z) = \prod_{r \geq 1} \frac{1}{1 - z^{k^r}} = \prod_{r \geq 0} \frac{1}{1 - (z^k)^{k^r}} = \mathsf{b}_k(z^k).$$

Comparing coefficients of z^m gives

$$b_k(m) - b_k(m - 1) = \begin{cases} b_k(m/k), & \text{if } k \mid m; \\ 0, & \text{if } k \nmid m; \end{cases}$$

(with the convention $b_k(-1) = 0$). For $m = kn + r$ with $1 \leq r \leq k - 1$ this yields $b_k(kn + r) = b_k(kn + r - 1)$, hence $b_k(kn) = b_k(kn + 1) = \cdots = b_k(kn + k - 1)$. For $m = kn$ with $n \geq 1$ this yields $b_k(kn) = b_k(kn - 1) + b_k(n)$. $\square$

Andrews et al. showed that $b_k(kn) \equiv 0$ modulo k if the base-k expansion of n contains the digit $k - 1$. Their main result is the following [5].

Theorem 1. *If the base-k expansion of n is $d_\ell \cdots d_0$ (with $0 \leq d_i < k$), then* $b_k(kn) \equiv (d_\ell + 1) \cdots (d_0 + 1) \bmod k$.

Lemma 1 is a standard result showing that the k-ary partition function is k-Mahler. A power series $\mathsf{q} \in \mathbb{C}[\![z]\!]$ is *k-Mahler* if it satisfies an equation of the form

$$\sum_{i=0}^{d} p_i(z)\mathsf{q}(z^{k^i}) = 0$$

where $d \geq 1$, $p_0, \ldots, p_d$ are polynomials in $\mathbb{C}[z]$ and $p_0(z)p_d(z) \neq 0$.

Becker proved that all k-regular power series are k-Mahler [6]. Conversely, if a power series $\mathsf{q} \in \overline{\mathbb{Q}}[\![z]\!]$ satisfies

$$\mathsf{q}(z) = \sum_{i=1}^{d} p_i(z)\,\mathsf{q}(z^{k^i})$$

for some positive d and polynomials $p_1, \ldots, p_d$ in $\overline{\mathbb{Q}}[z]$, not all zero, then it is k-regular. However, for the Mahler relation given by Lemma 1, $\mathsf{b}_k(z)$ has a non-constant polynomial coefficient. So Becker's result does not directly apply.

1.2   Formal Series for Thue–Morse Words

Let $k, m \geq 2$. Let $\omega_m = \exp(2i\pi/m)$. As a consequence of Lemma 1, the convolution of $(b_k(n))_{0 \leq n < k\ell}$ —i.e., prefixes whose length is a multiple of k— with consecutive powers of ω_k does not provide any information:

$$\sum_{n=0}^{k\ell-1} b_k(n)\, \omega_k^{k\ell-1-n} = \sum_{j=0}^{\ell-1}\sum_{r=0}^{k-1} b_k(kj + r)\, \omega_k^{-1-kj-r} = \sum_{j=0}^{\ell-1} b_k(kj) \sum_{r=0}^{k-1} \omega_k^{-1-r} = 0.$$

It always vanishes, and hence cannot distinguish prefixes.

As the value of $b_k(n)$ depends on the base-k digits of n, it is thus relevant to consider the generalized Thue–Morse word $\mathbf{t}_{k,m} = (t_{k,m}(n))_{n \geq 0}$ over the alphabet $\{0, \ldots, m-1\}$ where $t_{k,m}(n)$ is the base-k sum-of-digits $\sigma_k(n)$ of n reduced modulo m, i.e.,

$$t_{k,m}(n) = \sigma_k(n) \bmod m. \tag{3}$$

See, for instance, [19, 20].

It is a standard technique, see [11], to consider the formal series in $\mathbb{C}[\![z]\!]$

$$\mathbf{t}_{k,m}(z) := \prod_{r \geq 0} \left(\sum_{d=0}^{k-1} \omega_m^d \, z^{d \cdot k^r} \right) = \sum_{n \geq 0} \omega_m^{t_{k,m}(n)} \, z^n.$$

The trick is based on the uniqueness of base-k expansions. Indeed, if the base-k expansion of n is $d_\ell \cdots d_0$, then the coefficient of z^n is obtained from

$$\prod_{i=0}^{\ell} \omega_m^{d_i} \, z^{d_i \cdot k^i} = \omega_m^{\sigma_k(n)} z^n$$

and by Euclidean division, the sum-of-digits $\sigma_k(n)$ is $q \cdot m + t_{k,m}(n)$ for some $q \geq 0$ (with $0 \leq t_{k,m}(n) < m$). Thus the series $\mathbf{t}_{k,m}(z)$ encodes the word $\mathbf{t}_{k,m}$. As an example, the first terms of $\mathbf{t}_{3,3}(z)$ are

$$1 + \omega_3 \, z + \omega_3^2 \, z^2 + \omega_3 \, z^3 + \omega_3^2 \, z^4 + z^5 + \omega_3^2 \, z^6 + z^7 + \omega_3 \, z^8 + \cdots .$$

In what follows, we fix $k = m$ and, for simplicity, write ω and $\mathbf{t}_k$ instead of ω_k and $\mathbf{t}_{k,k}$.

1.3   Our Contributions

The starting point of this note is to consider variations on Merca's paper [18]: we multiply the k-ary partition generating function $\mathbf{b}_k$ by the series $\mathbf{t}_k$, obtaining (by Theorem 2) an identity of the form

$$\mathbf{t}_k(z).\mathbf{b}_k(z) = \sum_{n \geq 0} g_k(n) \, z^n. \tag{4}$$

Such a convolution is a standard way to measure how "digit-structured" the sequence $(b_k(n))_{n \geq 0}$ may be, the ideas being similar to the discrete Fourier transform. In another direction, compositional inverses are discussed in [14].

In the case $k = 2$ and the classical Thue–Morse sequence over $\{0, 1\}$, it is known that $\mathbf{t}_2(z)$ is the inverse of $\mathbf{b}_2(z)$ in $\mathbb{C}[\![z]\!]$, i.e., $g_2(0) = 1$ and $g_2(n) = 0$ for all $n > 0$ [18]. In this paper, we observe that this fails for $k \geq 3$ and we quantify the deviation via explicit digit-products. The function $g_k : \mathbb{N} \to \mathbb{C}$ is non-zero infinitely often and k-regular (see Proposition 1, where we explain that it is digit-wise multiplicative). For $k = 3$, the sequence g_3 takes only finitely many values and is therefore 3-automatic; a deterministic finite automaton with output (DFAO) with 7 states is provided in Fig. 1b . For $k \geq 3$, we briefly discuss its summatory function $N \mapsto \sum_{n < N} g_k(n)$ using classical procedures.

In Sect. 3, using elementary arguments on the growth of $b_k(n)$, we show that $\mathsf{b}_k(z)$ is not k-regular (Theorem 3). Since g_k is k-regular, as a consequence of Theorem 2, the inverse of $\mathsf{t}_k(z)$ cannot be k-regular. We nevertheless describe the inverse of $\mathsf{t}_k(z)$ in $\mathbb{C}[\![z]\!]$: In Sect. 3.2, we obtain recurrence relations for its coefficients. Finally, we reconsider Ulas–Żmija partition polynomials [21] in the context of this article and extend the function $g_k : \mathbb{N} \to \mathbb{C}$ to $n \mapsto g_{k,n}(t) \in \mathbb{C}[t]$, where $g_{k,n}(1) = g_k(n)$, opening the way to new developments.

2   The Product of the Two Series

We need a useful description of $g_k(n)$ given in (4). First, consider the product of two factors, in general form, from the two series $\mathsf{t}_k(z)$ and $\mathsf{b}_k(z)$. Let us start with some classical cyclotomic manipulations.

Lemma 2. *Let $k \geq 2$ and $\omega = \exp(2i\pi/k)$. We have*

$$\frac{1}{1-z} \sum_{i=0}^{k-1} \omega^i z^i = -\omega^{k-1} \prod_{j=1}^{k-2} (z - \omega^j) =: P_k(z)$$

which is thus a polynomial of degree $k - 2$. In particular, we have $P_2(z) = 1$.

As an example, we have

$$P_3(z) = -\omega_3^2\, z + 1 \quad \text{and} \quad P_4(z) = iz^2 + (1+i)z + 1.$$

Proof. The numerator in the l.h.s. evaluated at $z = 1$ is the sum of the kth roots of unity and thus vanishes. Hence $1 - z$ divides the numerator whose leading term has coefficient ω^{k-1} which is non zero. So the quotient has a leading term $-\omega^{k-1}z^{k-2}$.

The numerator is also a geometric sum, so we can write

$$\frac{1}{1-z} \sum_{i=0}^{k-1} \omega^i z^i = \frac{1 - z^k}{(1-z)(1-\omega z)}. \tag{5}$$

The numerator of the r.h.s. vanishes for all k-th roots of unity. The denominator of the r.h.s. vanishes for $z = 1$ and $z = \omega^{k-1}$. The conclusion follows from the fundamental theorem of algebra. $\qquad\square$

Lemma 3. *Let $k \geq 2$ and $\omega = \exp(2i\pi/k)$. The coefficients c_j of the polynomial $P_k(z) = c_{k-2}^{(k)} z^{k-2} + \cdots + c_0^{(k)}$ (depicted in Fig. 1a) are all non-zero. More precisely, for $0 \leq j \leq k - 2$, we have*

$$c_j^{(k)} = \frac{1 - \omega^{j+1}}{1 - \omega} \quad \text{and} \quad 1 \leq \left| c_j^{(k)} \right| \leq \frac{1}{\sin(\pi/k)}.$$

Proof. Using (5), we get

$$-\omega P_k(z) = \prod_{j=1}^{k-2}(z - \omega^j) = \frac{z^k - 1}{(z-1)(z-\omega^{-1})}.$$

Hence

$$-\omega(z - \omega^{-1})\sum_{j=0}^{k-2} c_j^{(k)}\, z^j = \frac{z^k - 1}{z - 1} = \sum_{i=0}^{k-1} z^i.$$

We then compare the coefficients, $-\omega\, c_{k-2}^{(k)} = 1$, for $1 \le j \le k - 2$, $c_j^{(k)} = 1 + \omega\, c_{j-1}^{(k)}$ and $c_0^{(k)} = 1$. The result follows from the recurrence relation. $\qquad\square$

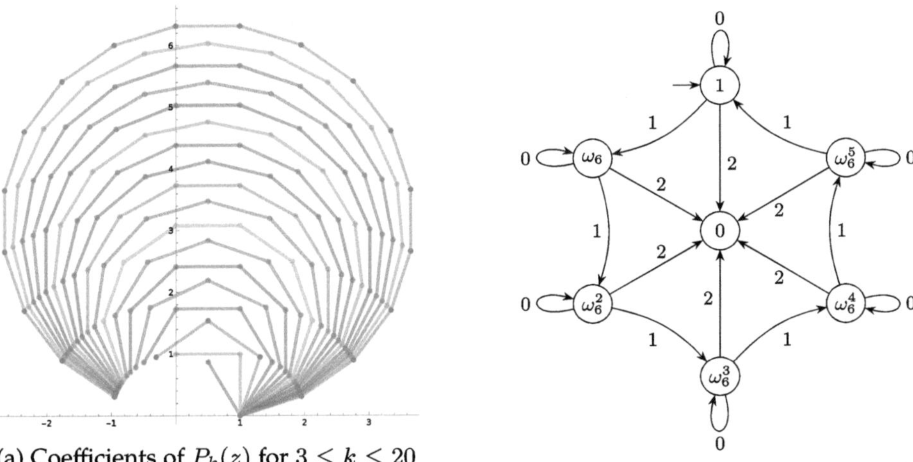

(a) Coefficients of $P_k(z)$ for $3 \le k \le 20$ in the complex plane.

(b) A DFAO for $g_3(n)$ in base 3.

Fig. 1. The coefficients of P_k (left) and a DFAO for g_3 (right).

As a consequence of Lemma 2, we have

$$\mathsf{t}_k(z).\mathsf{b}_k(z) = \prod_{r\ge 0} P_k(z^{k^r}) = \sum_{n\ge 0} g_k(n)\, z^n.$$

Compared with the case $k = 2$, studied in [18] where the r.h.s. is equal to the constant 1, the situation is a bit more involved. We have $g_k(n) = 0$ if and only if $k - 1$ appears in the base-k expansion of $n \ge 0$. Indeed, inspecting the form of

$$P_k(z^{k^r}) = \sum_{i=0}^{k-2} c_i^{(k)}\, z^{i\cdot k^r},$$

there is no term of the form $z^{(k-1)\cdot k^r}$. One may notice the similarity with Theorem 1.

If the base-k expansion of n is of the form $d_\ell \cdots d_0 \in \{0, \ldots, k-2\}^*$, then

$$g_k(n) = \prod_{j=0}^{\ell} c_{d_j}^{(k)} \tag{6}$$

and from Lemma 3, all factors have modulus at least one. We state this observation.

Proposition 1. *The function $g_k : \mathbb{N} \to \mathbb{C}$ is k-regular.*

Proof. There is a 1-dimensional linear representation associating with each digit $d \neq k-1$ the complex number $c_d^{(k)}$ and associating $k-1$ with 0. $\square$

Remark 2. In the case $k = 3$, g_3 is 3-automatic since its values lie in $\{0\} \cup \{\omega_6^i \mid 1 \le i \le 5\}$. Since $c_1^{(3)} = -\omega_3^2 = \omega_6$, we get the DFAO generating $g_3(n)$ when fed with base-3 expansion of n depicted in Fig. 1b. Since the coefficients of $P_3(z)$ are $c_0^{(3)} = 1$ and $c_1^{(3)} = \omega_6$, we get the loops with labels 0 and reading a digit 1 corresponds to a multiplication by ω_6.

Theorem 2. *Let $k \ge 2$. The following relations hold, for all $n \ge 0$,*

$$\sum_{j=0}^{n} \omega^{t_k(j)}\, b_k(n-j) = g_k(n)$$

where $\omega = \exp(2i\pi/k)$, $(t_k(j))_{j\ge 0} = (t_{k,k}(j))_{j\ge 0}$ is the k-letter Thue–Morse word defined in (3) and g_k is the k-regular function defined in (6).

Proof. For all $n \ge 0$, the coefficient of z^n in $\mathsf{t}_k \cdot \mathsf{b}_k$ is

$$\sum_{j=0}^{n} [z^j]\mathsf{t}_k.[z^{n-j}]\mathsf{b}_k = \sum_{j=0}^{n} \omega^{t_k(j)}\, b_k(n-j) = g_k(n). \tag{7}$$

$\square$

So (7) gives a triangular system that allows one to compute the values of the k-ary partition function $b_k(0), \ldots, b_k(n)$ from the Thue–Morse word $(t_k(j))_{j\ge 0}$ over $\{0, \ldots, k-1\}$ and the values $g_k(0), \ldots, g_k(n)$ of the k-regular function g. As an example, for $k = 3$ and $\omega = \exp(2i\pi/3)$:

$$\begin{aligned}
b_3(0) = g_3(0) = 1 & \quad\Rightarrow b_3(0) = 1 \\
b_3(1) + \omega\, b_3(0) = g_3(1) = -\omega^2 & \quad\Rightarrow b_3(1) = 1 \\
b_3(2) + \omega\, b_3(1) + \omega^2\, b_3(0) = g_3(2) = 0 & \quad\Rightarrow b_3(2) = 1 \\
b_3(3) + \omega\, b_3(2) + \omega^2\, b_3(1) + \omega\, b_3(0) = g_3(3) = -\omega^2 & \quad\Rightarrow b_3(3) = 2
\end{aligned}$$

$$\vdots$$

Conversely, from $b_k(0), \ldots, b_k(n)$ and $g_k(0), \ldots, g_k(n)$, we can recover the prefix of length $n+1$ of the Thue–Morse word $(t_k(j))_{j \geq 0}$. Again, for $k = 3$, we get an unconventional way to recover Thue–Morse prefixes from k-ary partition data:

$$
\begin{aligned}
\omega^{t_0} \cdot 1 &= g_3(0) = 1 & &\Rightarrow t_0 = 0 \\
\omega^{t_0} \cdot 1 + \omega^{t_1} \cdot 1 &= g_3(1) = -\omega^2 & &\Rightarrow t_1 = 1 \\
\omega^{t_0} \cdot 1 + \omega^{t_1} \cdot 1 + \omega^{t_2} \cdot 1 &= g_3(2) = 0 & &\Rightarrow t_2 = 2 \\
\omega^{t_0} \cdot 2 + \omega^{t_1} \cdot 1 + \omega^{t_2} \cdot 1 + \omega^{t_3} \cdot 1 &= g_3(3) = -\omega^2 & &\Rightarrow t_3 = 1
\end{aligned}
$$

$$\vdots$$

Consider the summatory function $S_k(N) := \sum_{n=0}^{N-1} g_k(n)$. Its asymptotic behavior follows directly from classical results, see [4, Theorem 3.5.1] or [12, 16],

$$
S_k(N) = N^{\log_k |A_k|} \, \Phi(\log_k N) + \mathcal{O}(N^\beta)
$$

where Φ is a 1-periodic Hölder continuous function, $\beta < \log_k |A_k|$ and

$$
A_k := \sum_{r=0}^{k-2} c_r^{(k)} \quad \text{and} \quad |A_k| = \frac{k}{2\sin(\pi/k)}.
$$

One has to sum the matrices of the linear representation, which is straightforward in this case. The reader may consult [13] to get precise asymptotic expansions on a larger family of series. We have $\log_3 |A_3| = 1/2$, $\log_4 |A_4| = 3/4$, $\log_5 |A_5| \simeq 0.899$, $\log_6 |A_6| = 1$ and $1 < \log_k |A_k| < 2$ for all $k \geq 7$.

We briefly highlight an easy computation to get an intuition about Fig. 2. Using the digit-wise definition of g_k, first observe that

$$
\sum_{r=0}^{k-1} g_k(kn+r) = \sum_{r=0}^{k-2} c_r^{(k)} g_k(n) = A_k \, g_k(n).
$$

Note that $S_k(k) = \sum_{r=0}^{k-2} c_r^{(k)} = A_k$. Hence

$$
S_k(kN) = \sum_{n=0}^{N-1} \sum_{r=0}^{k-1} g_k(kn+r) = \sum_{n=0}^{N-1} A_k \, g_k(n) = A_k \, S_k(N)
$$

and for all ℓ, along the corresponding subsequence, we have $|S_k(k^\ell)| = |A_k|^\ell$.

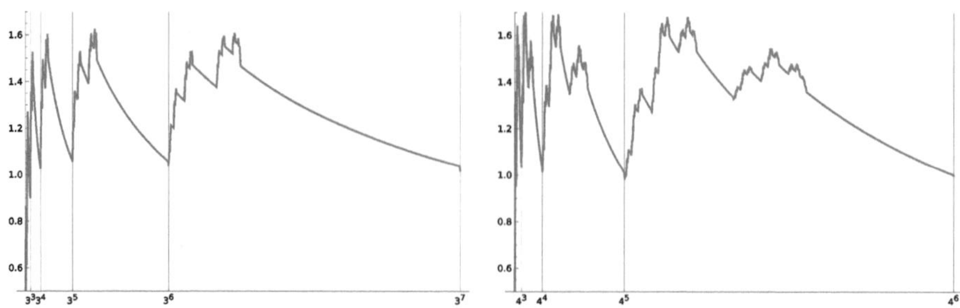

Fig. 2. Summatory functions $S_3(N)/\sqrt{N}$ and $S_4(N)/N^{3/4}$.

3   Going Further

3.1   (Non)-regularity

We have already mentioned in the introduction that the number of restricted k-partitions where appear only powers of k bounded by k^j, for some fixed j, form a ℓ-regular sequence, for all $j \geq 1$ and all $\ell \geq 2$.

$$\mathsf{b}_{k,\leq j}(z) := \prod_{r=0}^{j} \frac{1}{1 - z^{k^r}} = \sum_{n\geq 0} b_{k,\leq j}(n)\, z^n.$$

For instance, for $k = 2$ and $j = 3$, we found the sequence A008643 in the OEIS for the number of partitions of n into parts 1, 2, 4 and 8. The sequence A008642 ($k = 2$, $j = 2$) seems to have several combinatorial interpretations.

However, the unrestricted sequence $(b_k(n))_{n\geq 0}$ grows too quickly to be k-regular (see [4, Theorem. 16.3.1]).

Theorem 3. *The series* $\mathsf{b}_k(z)$ *is not k-regular.*

Proof. From Lemma 1, we have, for all $m \geq 0$ and $r \in \{0,\ldots,k-1\}$,

$$b_k(km + r) = \sum_{j=0}^{m} b_k(j). \tag{8}$$

For each $\ell \geq 0$, we consider the sequence $(s_\ell(n))_{n\geq 0}$ defined by

$$s_\ell(n) := b_k(k^\ell n),$$

i.e., it is a particular sequence of the k-kernel. Using (8), we have

$$s_{\ell+1}(n) = b_k(k^{\ell+1}n) = \sum_{j=0}^{k^\ell n} b_k(j) \geq \sum_{m=0}^{n} b_k(k^\ell m) = \sum_{m=0}^{n} s_\ell(m).$$

Roughly speaking, each time we multiply the argument by k (to extract an element from the k-kernel), we get (at least) a discrete integral of the previous element.

We prove by induction on ℓ that the sequence satisfies $s_\ell(n) \geq \binom{n+\ell}{\ell}$ and thus $s_\ell(n)$ grows at least like n^ℓ as $n \to \infty$. The base case is clear: $s_0(n) = b_k(n) \geq 1$. Assuming inequality holds for ℓ, we have

$$s_{\ell+1}(n) \geq \sum_{m=0}^{n} s_\ell(m) \geq \sum_{m=0}^{n} \binom{m+\ell}{\ell} = \binom{n+\ell+1}{\ell+1}$$

where the last equality is the classical hockey-stick identity (also called Christmas stocking theorem).

However, if a sequence $(r(n))_{n\geq 0}$ is k-regular, then there exist constants C, D such that $|r(n)| \leq Cn^D$, for all $n \geq 1$. Moreover, all elements of the finitely generated module spanned by the k-kernel share this property.

But we have shown that $s_\ell(n)$ grows at least as n^ℓ, contradicting the above property of k-regular sequences. $\qquad\square$

3.2   Inverse of the k-Automatic Series $\mathsf{t}_k(z)$

Since the constant term of $\mathsf{t}_k(z)$ is 1, the series is invertible in $\mathbb{C}[\![z]\!]$. Let $\mathsf{u}_k(z) = \sum_{n\geq 0} u_n^{(k)} z^n$ denote the inverse of $\mathsf{t}_k(z)$. Since k is fixed, we simply write u_n instead of $u_n^{(k)}$ for the sake of notation. Note that $\mathsf{u}_k(z)$ cannot be k-regular: if it were the case, then $\mathsf{b}_k(z) = \mathsf{u}_k(z) \cdot \mathsf{g}_k(z)$ would be k-regular: convolution of k-regular sequences is k-regular [2].

From the definition of t_k, it is straightforward to see that

$$\mathsf{t}_k(z) = \left(\sum_{d=0}^{k-1} \omega^d z^d \right) \mathsf{t}_k(z^k).$$

Hence, we obtain the following Mahler-type functional equations:

$$(1 - \omega z)\, \mathsf{t}_k(z) = (1 - z^k)\, \mathsf{t}_k(z^k)$$

and

$$(1 - z^k)\, \mathsf{u}_k(z) = (1 - \omega z)\, \mathsf{u}_k(z^k).$$

From this, for $n \geq k$, we get

$$u_n = u_{n-k} + \begin{cases} u_{n/k}, & \text{if } n \equiv 0 \pmod{k}; \\ -\omega\, u_{(n-1)/k}, & \text{if } n \equiv 1 \pmod{k}; \\ 0, & \text{otherwise}; \end{cases}$$

with initial values $(u_0, u_1, \ldots, u_{k-1}) = (1, -\omega, 0, \ldots, 0)$. These recurrence relations directly yield, for all $n \geq 0$ and $r \in \{2, \ldots, k-1\}$ when $k \geq 3$,

$$u_{kn} = \sum_{j=0}^{n} u_j, \quad u_{kn+1} = -\omega \sum_{j=0}^{n} u_j, \quad u_{kn+r} = 0.$$

3.3 Partition Polynomials

In [21], *binary partition polynomials* are considered. We consider their base-k analogue, a family of polynomials in t, defined by

$$\mathsf{b}_k(t, z) := \prod_{r \geq 0} \frac{1}{1 - tz^{k^r}} = \sum_{n \geq 0} b_{k,n}(t)\, z^n.$$

The interest is that $b_{k,n}(t) = \sum_{j=0}^n a_{k,n,j}\, t^j$, where $a_{k,n,j}$ counts partitions of n with exactly j parts. So $b_{k,n}(t)$ records the number of parts (powers of k, counted with multiplicity), not only the total number of partitions. It is known that $(1 - t\,z)\,\mathsf{b}_k(t, z) = \mathsf{b}_k(t, z^k)$. Two paths can therefore be followed.

We multiply, as in the first part of the paper, the above series in $\mathbb{C}[t][\![z]\!]$ by $\mathsf{t}_k(z)$

$$\mathsf{t}_k(z).\mathsf{b}_k(t, z) = \sum_{n \geq 0} g_{k,n}(t)\, z^n \tag{9}$$

and we can therefore study this sequence $(g_{k,n}(t))_{n \geq 0}$ of polynomials.

n	$g_{2,n}(t)$	n	$g_{2,n}(t)$
0	1	5	$t^5 - t^3 - t + 1$
1	$t - 1$	6	$t^6 - t^4 + t^3 - t^2 - t + 1$
2	$t^2 - 1$	7	$t^7 - t^5 + t^4 - 2t^3 - t^2 + 3t - 1$
3	$t^3 - 2t + 1$	8	$t^8 - t^6 + t^5 - t^4 - t^3 + 2t^2 - 1$
4	$t^4 - t^2 + t - 1$	9	$t^9 - t^7 + t^6 - t^5 - 2t^4 + 2t^3 + t^2 - 2t + 1$

For $k = 2$, $g_{2,0}(t) = 1$ and

$$g_{2,2n}(t) = t\, g_{2,2n-1}(t) + g_{2,n}(t) \text{ and } g_{2,2n+1}(t) = t\, g_{2,2n}(t) - g_{2,n}(t).$$

Observe that, from these relations, the sequence $(g_{2,n}(0))_{n \geq 0}$ is the Thue–Morse word over $\{-1, 1\}$.

Note that in [21], this is not the series $\mathsf{t}_2(z)$ whose coefficient of z^n is $(-1)^{\sigma_2(n)}$ which is considered, but a series whose coefficient of z^n is $(-t)^{\sigma_2(n)}$. So an alternative generalization is to consider the series

$$\mathsf{t}_k(t, z) := \prod_{r \geq 0} \left(\sum_{d=0}^{k-1} (t\,\omega_k)^d\, z^{d \cdot k^r} \right) = \sum_{n \geq 0} (t\,\omega_k)^{\sigma_k(n)} z^n.$$

In this context, we recover that $\mathsf{t}_2(t, z).\mathsf{b}_2(t, z) = 1$ [21, Thm. 2.1]. As observed in the first part of the paper, we obtain a new relation for $k \geq 3$:

$$\mathsf{t}_k(t, z).\mathsf{b}_k(t, z) = \sum_{n \geq 0} g_k(n)\, t^{\sigma_k(n)}\, z^n$$

where g_k is the k-regular function defined in (6). So, the result is quite similar to Theorem 2.

It is therefore more interesting to study the relation (9). For a uniform presentation, we set $g_2(n) = \delta_{n,0}$. The polynomials $g_{k,n}(t)$ satisfy recurrences mirroring those of $b_{k,n}(t)$. Due to space limitation, the proof is not given here.

Proposition 2. *Let $k \geq 2$ and let $\omega_k = \exp(2i\pi/k)$. Then $g_{k,0}(t) = 1$ and*

$$g_{k,kn}(t) = t\, g_{k,kn-1}(t) + g_{k,n}(t) \qquad\qquad (n \geq 1),$$
$$g_{k,kn+r}(t) = t\, g_{k,kn+r-1}(t) + \omega_k^r\, g_{k,n}(t) \qquad\qquad (n \geq 0,\ 1 \leq r \leq k-1).$$

In particular, for all $n \geq 0$, we have $g_{k,n}(1) = g_k(n)$, i.e., the sum of coefficients of the polynomial $g_{k,n}(t)$ is equal to $g_k(n)$.

Acknowledgments. We thank the anonymous reviewers. Their comments helped us to improve the presentation of this article.

References

1. Allouche, J.-P., Scheicher, K., Tichy, R.F.: Regular maps in generalized number systems. Math. Slovaca **50**, 41–58 (2000)
2. Allouche, J.-P., Shallit, J.: The ring of k-regular sequences. Theoret. Comput. Sci. **98**, 163–197 (1992)
3. Allouche, JP., Shallit, J.: The Ubiquitous Prouhet-Thue-Morse Sequence. In: Ding, C., Helleseth, T., Niederreiter, H. (eds) Sequences and their Applications. Discrete Mathematics and Theoretical Computer Science. Springer, London (1999). https:// doi.org/10.1007/978-1-4471-0551-0_1
4. Allouche, J.-P., Shallit, J.: Automatic Sequences. Theory, Applications, Generalizations. Cambridge Univ. Press (2003)
5. Andrews, G.E., Fraenkel, A.S., Sellers, J.A.: Characterizing the number of m-ary partitions modulo m. Amer. Math. Monthly **122**, 880–885 (2015)
6. Becker, P.-G.: k-regular power series and Mahler-type functional equations. J. Number Theory **49**, 269–286 (1994)
7. Berstel, J., Reutenauer, C.: Noncommutative Rational Series with Applications, volume 137 of Encycl. Math. Appl. Cambridge Univ. Press (2011)
8. É. Charlier, C. Cisternino, and M. Stipulanti. Robustness of Pisot-regular sequences. Adv. in Appl. Math. **125**, Art. 102151 (2021)
9. Charlier, É., Cisternino, C., Stipulanti, M.: Regular sequences and synchronized sequences in abstract numeration systems. Eur. J. Combin. **101**, Art. ID 103475 (2022)
10. de Bruijn, N.G.: On Mahler's partition problem. Proc. Nederl. Akad. Wetensch. **51**, 659–669 (1948)
11. Drmota, M., Stoll, T.: Newman's phenomenon for generalized Thue-Morse sequences. Discrete Math. **308**, 1191–1208 (2008)
12. Dumas, P.: Asymptotic expansions for linear homogeneous divide-and-conquer recurrences: algebraic and analytic approaches collated. Theoret. Comput. Sci. **548**, 25–53 (2014)
13. Dumas, P., Flajolet, P.: Asymptotique des récurrences mahlériennes: le cas cyclotomique. J. Théor. Nombres Bordeaux **8**, 1–30 (1996)
14. Gawron, M., Ulas, M.: On formal inverse of the Prouhet-Thue-Morse sequence. Discrete Math. **339**(5), 1459–1470 (2016)
15. Golafshan, M., Rigo, M., Whiteland, M.A.: Computing the k-binomial complexity of generalized Thue-Morse words. J. Combin. Theory Ser. A, 220, Paper No. 106152 (2026)

16. Heuberger, C., Krenn, D.: Asymptotic analysis of regular sequences. Algorithmica **82**, 429–508 (2020)
17. Mahler, K.: On a special functional equation. J. London Math. Soc. **15**, 115–123 (1940)
18. Merca, M.: Binary partitions and Thue-Morse sequence. Bol. Soc. Mat. Mex. **32**, Paper No. 11 (2026)
19. Starosta, S.: Generalized Thue-Morse words and palindromic richness. Kybernetika **48**, 361–370 (2012)
20. Tromp, J., Shallit, J.: Subword complexity of a generalized Thue-Morse word. Inform. Process. Lett. **54**, 313–316 (1995)
21. Ulas, M., Żmija, B.: On arithmetic properties of binary partition polynomials. Adv. in Appl. Math. **110**, 153–179 (2019)

Self-assembly of Strings and Languages Revisited: Efficiently Deciding Closure Under Self-assembly

Katalin Anna Lázár[1] , Florin Manea[2] , Stefan Siemer[2] ,
and Timo Specht[2(✉)]

[1] ELTE Eötvös Loránd University, 1117 Budapest, Hungary
`lazarkati@inf.elte.hu`
[2] University of Göttingen, 37077 Göttingen, Germany
`{florin.manea,stefan.siemer,timo.specht}@cs.uni-goettingen.de`

Abstract. The self-assembly operation $\mathcal{S}$, also called overlap-assembly, is a bio-inspired formal string operation that combines two strings xy and yz, with a nonempty common overlap y, to derive a new string xyz. When applied to a pair of strings u and v, this operation yields a set of strings, with one string derived for each nonempty suffix of u that is a prefix of v. The operation $\mathcal{S}$ can be naturally extended to languages: for a formal language L, $\mathcal{S}(L)$ is the set of all strings obtainable by applying the operation $\mathcal{S}$ to pairs of strings from L. We analyze the complexity of deciding whether a given regular language L is closed under the self-assembly operation $\mathcal{S}$: we present an efficient algorithm for finite languages and then provide tight upper and lower bounds for the complexity of this problem for regular languages defined by deterministic or nondeterministic finite automata. The problem is undecidable for context-free languages.

1 Introduction

The string operation known as self-assembly or overlap assembly, was introduced in [8], to formally define a framework that enables shorter strings (i.e., simpler objects) to aggregate into longer strings (i.e., larger structures) and formalizes in a string-based framework the general principles of self-assembly, one of the most important operations for nano-scale engineering. The seminal work [8], as well as some subsequent papers [10,11], thoroughly discuss the biology-rooted motivations behind this formal string operation. Interested readers are referred to these papers for more information.

In more detail, the string self-assembly operation $\mathcal{S}$ is a string operation that, given two strings $u_1 = xy$ and $u_2 = yz$, sharing a nonempty overlap y, produces the string xyz. Clearly, when multiple overlaps (suffixes of u_1 that are also prefixes of u_2) exist, each yields a distinct result. The resulting set of strings is denoted by $\mathcal{S}(u_1, u_2)$. Note that this operation is not commutative, i.e., $\mathcal{S}(u_1, u_2)$ may differ from $\mathcal{S}(u_2, u_1)$. The string operation $\mathcal{S}$ is naturally extended

M.-P. Béal and P. Caron (Eds.): DLT 2026, LNCS 16578, pp. 57–71, 2026.
https://doi.org/10.1007/978-3-032-28404-4_5

to languages. Given a language L, the set $\mathcal{S}(L)$ is the set of all strings obtained by applying the self-assembly operation, $\mathcal{S}$, to pairs of strings from L; that is, $\mathcal{S}(L) = \cup_{x,y \in L}\mathcal{S}(x,y)$. Here, we consider the transitive closure $\mathcal{S}^+$, originally introduced in [8], and focus on the closure of languages under this operation.

The seminal work [8] initiated the formal study of the generative power of self-assembly from a language theoretic viewpoint. The focus was on understanding the formal language theoretic properties (e.g., position in the Chomsky hierarchy, closure properties) of the class of languages obtained by applying (iterated) self-assembly to languages from the classes on the various levels of the Chomsky hierarchy. For instance, it is shown that (iterated) self-assembly preserves regularity. The paper also established the first decidability results for two basic questions: deciding whether a language is closed under assembly, and, for languages that are closed, computing a minimal generating set (w.r.t. inclusion, called the *minimal base*).

Subsequent papers have refined and extended these results (see, e.g., [6,7, 10,11]). In [10], the closure properties for a range of language families defined via automata- and complexity-theoretic characterizations have been examined and the complexity of several associated decision problems has been analyzed. In [11], the potential use of iterated overlap assembly in constructing combinatorial DNA libraries has been explored. Since the $\mathcal{S}-$operation preserves regularity, in [6] the state complexity of languages of $\mathcal{S}(L)$, if L is regular, has been explored.

Several related string operations have also been defined and studied from a formal-language perspective, including string blending [4,12], superposition [5], various hairpin-based string operations, such as completion [17,21,24,27], reduction [23], and lengthening [22]. From an applied perspective, parallel overlap assembly has been used in gene-shuffling methodologies [28].

In this paper, we address from an algorithmic perspective one of the two problems mentioned above, introduced in [8]: given a language L, decide whether it is closed under self-assembly, i.e., whether $L = \mathcal{S}^+(L)$. It is not hard to see that this problem is equivalent to testing whether $L = \mathcal{S}(L)$. Although it has been shown that this problem is decidable for regular languages and undecidable for context-free, context-sensitive, or recursively enumerable languages, the exact complexity of this problem for regular languages has not been thoroughly investigated in the literature. We aim to fill this gap.

First, we consider the case when L is a finite language given as list of k strings, of total length N, and show that we can decide whether $L = \mathcal{S}(L)$ in time $\mathcal{O}(N + \min\{k^2p, kN\})$, where $p \leq k$ is the number of distinct lengths of the input strings. In particular, our algorithm works in optimal time $\mathcal{O}(N)$ if $k \leq N^{1/3}$.

For a regular language given by a deterministic finite automaton (DFA) with n states, we show that deciding whether $L = \mathcal{S}(L)$ is equivalent to testing the emptiness of the intersection of two regular languages given by DFAs. This yields an $\mathcal{O}(n^2\sigma)$-time algorithm, where n is the number of states of the DFA accepting L and σ the size of the input alphabet. Moreover, one can obtain, via a reduction from the Orthogonal vectors problem [29], fine-grained lower bounds, showing that the existence of a solution running in $\mathcal{O}((n^2\sigma)^{1-\varepsilon})$ time would contradict the Strong Exponential Time Hypothesis [29]. A slight extension shows that a combinatorial algorithm running in $\mathcal{O}(n^2\sigma^{1-\varepsilon})$ time would imply a breakthrough combinatorial algorithm for Boolean matrix multiplication. When the input language is given by non-deterministic finite automaton with ε-transitions or by a regular expression, the problem becomes PSPACE-complete.

The paper is structured as follows. First, we introduce the necessary definitions and formulate the closure problem for self-assembly. Then, we establish several preliminary properties. Subsequently, we present our main results in three parts: finite languages; regular languages specified by deterministic finite automata; and regular languages given by nondeterministic finite automata with ε-transitions.

2 Preliminaries

Let $\mathbb{N} = \{1, 2, \ldots\}$ denote the natural numbers and $\mathbb{N}_0 = \mathbb{N} \cup \{0\}$. For $n \in \mathbb{N}$, we write $[n] = \{1, \ldots, n\}$, and for all $i, n \in \mathbb{N}_0$ with $i \leq n$ we define $[i : n] = \{i, i+1, \ldots, n\}$. Let Σ be a finite set of terminal symbols (*letters*). A *string* w is a finite sequence of letters. The total number of letters in w is referred to as its *length*, and is denoted by $|w|$. A string of length 0 is called an *empty string*, and is denoted by ε. The set of all finite strings over an alphabet Σ, denoted by Σ^*, forms a free monoid $(\Sigma, \cdot, \varepsilon)$ under concatenation, with ε as the identity element. The set of all nonempty strings over Σ is $\Sigma^+ = \Sigma^* \setminus \{\varepsilon\}$. Any subset $L \subseteq \Sigma^*$ is called a *(formal) language*. Let Σ^n denote all strings in Σ^* of length $n \in \mathbb{N}_0$. Given a string w and indices $1 \leq i \leq j \leq |w|$, let $w[i]$ be the ith symbol of w, and let $w[i : j] = w[i] \cdots w[j]$ be the *substring* of w from position i to j. A substring with $i = 1$ is a *prefix*, while a substring with $j = |w|$ is a *suffix* of w.

Definition 1 ([8]). *Let* $u, v \in \Sigma^*$. *The (self-)assembly operation* $\mathcal{S}(u, v)$ *is defined as* $\mathcal{S}(u, v) = \{u'wv' \in \Sigma^+ \mid u = u'w, v = wv', w \neq \varepsilon\}$. *If either* u *or* v *is empty, then* $\mathcal{S}(u, v) = \emptyset$. *For two languages* $L_1, L_2 \subseteq \Sigma^*$, *we define* $\mathcal{S}(L_1, L_2) = \bigcup_{w_1 \in L_1, w_2 \in L_2} \mathcal{S}(w_1, w_2)$. *For* $L \subseteq \Sigma^+$ *and* $n \geq 1$, *let* $\mathcal{S}^n(L) = \mathcal{S}(L, \mathcal{S}^{n-1}(L))$, *where* $\mathcal{S}^0(L) = L$. *The iterated assembly of* L *is* $\mathcal{S}^+(L) = \bigcup_{n \geq 1} \mathcal{S}^n(L)$.

If we define $\mathcal{S}^*(L) = \bigcup_{n \geq 0} \mathcal{S}^n(L)$, then $\mathcal{S}^*(L) \setminus \mathcal{S}^+(L) \subseteq \{\varepsilon\}$. Indeed, if $\varepsilon \in L$, the computation of $\mathcal{S}(L)$ excludes only ε, since this string cannot be produced by any self-assembly operation. A language $L \subseteq \Sigma^*$ is *closed under iterated assembly* if $L = \mathcal{S}^*(L)$; when $L \subseteq \Sigma^+$, it suffices to require that $L = \mathcal{S}^+(L)$. Due to the negligible distinction between $\mathcal{S}^+(L)$ and $\mathcal{S}^*(L)$, we henceforth focus on $\mathcal{S}^+(L)$ and languages L that do not contain the empty word ε.

We briefly recall standard definitions from regular language theory, as presented in [16]. An ε-NFA is a 5-tuple $A = (Q, \Sigma, q_0, F, \delta)$, where Q is a finite set of states, Σ is the input alphabet, $\delta : Q \times (\Sigma \cup \{\varepsilon\}) \to 2^Q$ is the transition function, $q_0 \in Q$ is the initial state, and $F \subseteq Q$ is the set of accepting states. We denote the language recognized by A as $L(A)$, and its size, measured as the number of transitions, as $|A|$. A DFA is an ε-NFA, with the restriction that $\delta : Q \times \Sigma \to Q$. Both ε-NFAs and DFAs recognize exactly the class of regular languages. Every ε-NFA can be determinized, and the determinization of an ε-NFA may result in an exponential increase in size. Regular languages can also be specified by regular expressions. Any regular expression r can be converted, in time $O(|r|)$, into an ε-NFA A, with $|A| = O(|r|)$, that accepts the language described by r.

Computational Model. We work in the standard unit cost word RAM model with word size w, where each memory word stores $w \geq \log n$ bits. All usual arithmetic/bitwise operations on memory words, as well as indirect addressing, are assumed to run in $\mathcal{O}(1)$ time [14]. We approach a series of decision problems whose inputs are strings and finite automata. We assume that the input strings and automata are over an integer alphabet, i.e., $\Sigma = \{1, \ldots, \sigma\}$. Moreover, we assume that the states of each automaton $A = (Q, \Sigma, q_0, F, \delta)$ are also given as integers $Q = \{0, 1, \ldots, t\}$, where $q_0 = 0$ (we will still denote them symbolically, e.g., with $q_0, \ldots, q_t$ to avoid confusions). Finally, for a decision problem $\mathcal{P}$ and an input $\mathcal{I}$ for this problem, we denote by $\mathcal{P}(\mathcal{I})$ the correct output for problem $\mathcal{P}$ on input $\mathcal{I}$.

Data Structures. For a string w of length n, over a *polynomial alphabet* $\Sigma = \{1, \ldots, \sigma\}$, with $\sigma \in \mathcal{O}(n^c)$ for some positive constant $c \in \mathcal{O}(1)$, we can construct the suffix array SA_w in $\mathcal{O}(n)$ time (see [18]). This array is defined as $SA_w[i] = j$ if and only if $w[j : n]$ is the i-th suffix of w in lexicographic order. Given the suffix array, the longest common extension data structures (for short, LCE) can be constructed in $\mathcal{O}(n)$ time allowing us to return, for all $i, j \in [n]$, the length $\text{LCE}(i, j)$ of the longest common prefix of $w[i : n]$ and $w[j : n]$ (see [19]). These data structures are particularly useful for testing if two substrings $w[i : i + \ell]$ and $w[j : j + \ell]$ are equal or if a substring $w[i : i + \ell]$ is a prefix of $w[j : j + \ell']$ (in both cases, by checking if $\text{LCE}(i, j) \geq \ell + 1$). Furthermore, as shown in [3], we can construct data structures that allow us to retrieve in constant time, for any substring $w[i : j]$ of w, the range $SA[a : b]$ of the suffix array. For every $\ell \in [a : b]$, if $SA[\ell] = h$ then the prefix of length $j - i + 1$ of $w[h : n]$ is equal to $w[i : j]$. For the rest of the paper, the aforementioned data structures are referred to as the *stringology toolbox*. When needed, we will simply say that we build the stringology toolbox for a given string, meaning that all these data structures are constructed for the respective string. Given a string w as above, and a decomposition of $w = w_1 \cdots w_\ell$ in $\ell \geq 1$ strings, it is folklore that we can compute the stringology toolbox for each of the strings w_i, with $i \in [\ell]$ in $\mathcal{O}(n)$ time overall.

For a string w, with $|w| = n$, the *border array* $\pi[\cdot]$ of size n can be computed in linear time (see [20]): for each i, the value $\pi[i]$ is the length of the longest prefix of $w[1 : i]$ that is also a suffix of $w[1 : i]$ (i.e., the longest border of $w[1 : i]$). Moreover, for a set of strings $W = \{w_1, \ldots, w_k\}$, with $\sum_{i=1,k} |w_i| = N$, over a polynomial alphabet, we can construct in $\mathcal{O}(N + k^2)$ time a data structure that allows us to retrieve in $\mathcal{O}(1)$ time, for all $i, j \in [k]$, the length $\text{LO}(i, j)$ of the longest string that is both a suffix of w_i and a prefix of w_j (i.e., the longest overlap of w_i and w_j). This follows from the construction of [15], combined with the linear time suffix tree algorithm of [13].

Finally, radix sort is a non-comparison-based sorting algorithm that lexicographically orders a sequence of n keys $(x_1, \ldots, x_n)$, where each key is a k-tuple over $[1 : m]$, in $\mathcal{O}((n+m)k)$ time [1]. If m is polynomial in n (i.e., upper bounded by $n^{\mathcal{O}(1)}$), then radix sort can be implemented in $\mathcal{O}(nk)$ time.

3 The Closure Problem

We define the Self-Assembly Closure Problem (SAC) as follows:

SAC – Self-Assembly Closure Problem
Input: Language $L \subseteq \Sigma^+$.
Output: $\mathrm{SAC}(L) = \mathrm{YES}$ if $L = \mathcal{S}^+(L)$; $\mathrm{SAC}(L) = \mathrm{NO}$ otherwise.

We consider this problem in two settings: when L is finite and when L is regular. In the finite case, L is given as an enumeration of strings $L = \{w_1, \ldots, w_k\}$. Let $N = \sum_{i=1,k} |w_i|$ and $p = |\{|w_i| \mid i \in [k]\}|$ (i.e., p is the number of distinct string lengths in L (clearly, $p \leq k$)). As is standard in stringology, we assume that L is over a *polynomial alphabet* $\Sigma = \{1, \ldots, \sigma\}$ with $\sigma \in \mathcal{O}(N^c)$ for some constant $c > 0$. In this case, the input size is N. In the regular case, L is given as either a DFA or as an ε-NFA accepting it. In both cases, every state is assumed to be involved in at least one transition. If L is regular, then the input is an automaton $A = (Q, \Sigma, q_0, F, \delta)$. The size of the input is the number of transitions of the automaton A. From now on, we denote by n the number of states in Q, i.e., $n = |Q|$, and define the polynomial alphabet as $\Sigma = \{1, \ldots, \sigma\}$.

The following lemma captures a key observation: to decide if a language L is closed under $\mathcal{S}$ it is sufficient to check whether $L = \mathcal{S}(L)$ holds.

Lemma 1. *A language L is closed under $\mathcal{S}$ if and only if $L = \mathcal{S}(L)$.*

Proof. If L is closed under $\mathcal{S}$, then $L = \mathcal{S}^+(L)$, which implies that $S(L) \subseteq L$. However, by definition $L \subseteq S(L)$ (since $\varepsilon \notin L$), and consequently, $L = \mathcal{S}(L)$. Conversely, the equality $L = \mathcal{S}(L)$ ensures that repeated applications of the operator do not change the generated language: $\mathcal{S}^2(L) = \mathcal{S}(\mathcal{S}(L)) = \mathcal{S}(L)$, and inductively $\mathcal{S}^i(L) = L$ for all $i \geq 1$. Hence, $\mathcal{S}^+(L) = \bigcup_{n \geq 1} \mathcal{S}^n(L)$. It follows that deciding closure under $\mathcal{S}$ requires examining only a single iteration. $\square$

According to Lemma 1, for a language L, $\mathrm{SAC}(L) = \mathrm{YES}$ if and only if $\mathcal{S}(L) = L$.

3.1 Finite Languages

Here, we focus on efficiently deciding whether a finite set of strings is closed under $\mathcal{S}$. Before presenting our solution to this problem, we should emphasize that this question is of broader interest. For instance, analyzing how strings overlap and how such overlaps enable their combination, seems to be a central task related to solving the shortest superstring problem by finding the shortest Hamiltonian path in a graph that encodes the overlaps between the strings, called the overlap graph (see [26] and the references therein). Intuitively, our goal is to construct a string that is not in L but that nonetheless possesses an overlapping prefix and suffix of L to test whether a finite language L is closed under $\mathcal{S}$. This stringology task seems of independent theoretical interest to us.

Theorem 1. *Given a finite language $L = \{w_1, \ldots, w_k\}$, we can compute $\mathtt{SAC}(L)$ in $\mathcal{O}(N + \min\{k^2 p, kN\})$ time.*

Proof. General approach. Let $X = \{|w_i| \mid i \in [k]\}$, and recall that $|X| = p$. Our algorithm computes all triples (ℓ, j, i), where $\ell \in [N]$, $j, i \in [k]$, such that a string $u_{\ell,j,i}$ of length $\ell < |w_i| + |w_j|$ with w_j as a suffix and w_i as a prefix, exists. Note that if such a string exists, then it is unique. We refer to such a triple as a *query*. If a query (ℓ, j, i) satisfies $\ell \notin X$, then $\mathtt{SAC}(L) = \text{NO}$, since the string $u_{\ell,j,i}$ has length $\ell \notin X$, and is not contained in L, but belongs to $\mathcal{S}(w_i, w_j)$. Similarly, $\mathtt{SAC}(L) = \text{NO}$ if there is no string in L, of length ℓ, that has w_i as prefix or w_j as suffix. Based on these arguments, we can assume that we do not receive any queries (ℓ, j, i) that lead to the conclusion that $\mathtt{SAC}(L) = \text{NO}$. In this case, it can be shown that there are $\mathcal{O}(\min\{k^2 p, kN\})$ such queries, and they can be computed in time $\mathcal{O}(N + \min\{k^2 p, kN\})$.

Our algorithm checks, for all such queries, if the corresponding strings $u_{\ell,j,i}$ are in L (without explicitly constructing $u_{\ell,j,i}$). If these queries and checks are batched (i.e., they are not performed independently each time a new triple is computed but rather grouped according to their first two components, ℓ and j, and then processed based on this grouping), all the answers can be computed in $\mathcal{O}(N + \min\{k^2 p, kN\})$ time. We can now introduce the details of our algorithm. *Preprocessing.* Firstly, we radix sort, in $\mathcal{O}(N + k)$ time, the strings $w_1, \ldots, w_k$ w.r.t. their lengths. We assume from now on that $|w_i| \leq |w_{i+1}|$ for all $i \in [k-1]$ (but the order of strings of the same length is not important) and that $\ell_1 < \ell_2 < \ldots < \ell_p$ is an enumeration of X. During the sorting, we store for each $j \in [k]$ the value $len(j)$ such that $|w_j| = \ell_{len(j)}$ and, for each $\ell \in X$, the values i_ℓ and j_ℓ such that $i_\ell \leq j_\ell$ and $|w_i| = \ell$ for all $i \in [i_\ell : j_\ell]$ and $|w_i| \neq \ell$ for all $i \notin [i_\ell : j_\ell]$. Finally, we compute in $\mathcal{O}(N)$ time an array T, with N elements, where $T[\ell_t] = t$, for $t \in [p]$, and $T[\ell] = 0$, if $\ell \notin X$; using T we can, e.g., test whether $\ell \in X$.

For every $\ell \in X$ with $t = T[\ell]$, we define $U_t = \#w_{i_\ell}\#w_{i_\ell+1}\cdots\#w_{j_\ell}\#$, and set $U = U_1 U_2 \cdots U_p$. Since $|U| \in \mathcal{O}(N)$, we can construct in $\mathcal{O}(N)$ time the *stringology toolbox* for U and all of the strings $U_1, \ldots, U_p$. This allows us to check in $\mathcal{O}(1)$ time if w_i, for any $i \in [k]$, occurs at some position t in U or at some position t' in $U_{T[|w_i|]}$. We store, for each $i \in [k]$, the position a_i where $\#w_i\#$ occurs in U and the position b_i where it occurs in $U_{T[|w_i|]}$. All the preprocessing described above can be done in $\mathcal{O}(N)$ time.

We now compute in $\mathcal{O}(k^2)$ time the 2-dimensional, $k \times k$ arrays $P[\cdot, \cdot]$ and $S[\cdot, \cdot]$, where $P[i, j] = 1$ (resp., $S[i, j] = 1$) if and only if w_i is a prefix (resp., suffix) of w_j, and $P[i, j] = 0$ (resp., $S[i, j] = 0$) otherwise. Using P and S, we compute, within the same time complexity, the 2-dimensional, $k \times p$ arrays $PL[\cdot, \cdot]$ and $SL[\cdot, \cdot]$, whose elements are lists, with $PL[i, t] = \{x \in [k] \mid \#w_i$

occurs at position b_x in $U_t\}$ (resp., $SL[i,t] = \{x \in [k] \mid$ an occurrence of $w_i\#$ ends at position $b_x + |w_x| - 1$ in $U_t\})$, and $PL[i,t] = \emptyset$ if $\#w_i$ does not occur in U_t (resp., $SL[i,t] = \emptyset$ if $w_i\#$ does not occur in U_t). That is, $PL[i,t] \neq \emptyset$ (resp., $SL[i,t] \neq \emptyset$) if and only if w_i is a prefix (resp., suffix) of some string w_j of length ℓ_t. We can compute the arrays P and PL as follows (a similar approach works for S and SL): we initialize all elements of P with 0 and all elements of PL with $\emptyset$, and, then, for every pair $i, j \in [k]$, we test whether w_i occurs in U at position a_j (i.e., w_i is a prefix of w_j) and set $P[i,j] = 1$ accordingly; if $P[i,j] = 1$, we insert j in list $PL[i, T[len(j)]]$.

Further, we compute in $\mathcal{O}(N+k^2)$ time data structures allowing us to retrieve $LO(i,j)$ for all $i, j \in [k]$, and we construct, in $\mathcal{O}(N)$ overall time, the border arrays $\pi_j[\cdot]$ of the strings w_j, for all $j \in [k]$.

Computing the Queries. We compute the set Q of queries, initialized as $\emptyset$. Recall that $(\ell, j, i) \in Q$ if and only if there exists a string $u_{\ell,j,i}$ of length $\ell < |w_i| + |w_j|$ that has w_j as a suffix and w_i as a prefix and, such L has a string with prefix w_i, and a (possibly different) string with suffix w_j.

We consider all pairs $i, j \in [k]$. For such a pair, we set $q = LO(i,j)$. While $q > 0$, we perform the following three steps: we compute $\ell = |w_i| + |w_j| - q$; if $\ell \in X$, $PL[i, T[\ell]] \neq \emptyset$, and $SL[j, T[\ell]] \neq \emptyset$ simultaneously hold, we add the query (ℓ, j, i) to Q; otherwise, we return $\texttt{SAC}(L) = \text{NO}$; we set $q \leftarrow \pi_j[q]$ and iterate.

We verify that our approach is correct. For each $i, j \in [k]$ we identify all lengths ℓ for which a string $u_{\ell,j,i}$ as above exists. This clearly reduces to finding all prefixes of w_j that are also suffixes of w_i (i.e., overlaps of w_i and w_j), and computing the length ℓ of the string obtained by applying $\mathcal{S}$ to w_i and w_j using that overlap (that is, of the string $u_{\ell,j,i}$; note that we do not compute this string, only its length). The length of the longest overlap of w_i and w_j is given by $LO(i,j)$. The other overlaps are the nontrivial borders of $w_j[1 : LO(i,j)]$, therefore, (following, e.g., [20]) their lengths are the sequence of numbers $\pi_j^{(g)}[LO(i,j)]$, for all $g \geq 1$ such that $\pi_j^{(g)}[LO(i,j)] > 0$ (where $\pi_j^{(g)}[x] = \pi_j[\pi_j^{(g-1)}[x]]$, for $g > 1$, and $\pi_j^{(1)}[x] = \pi_j[x]$). Our algorithm simply iterates through these numbers and constructs the corresponding queries, stopping after all nonempty borders of $w_j[1 : LO(i,j)]$ have been considered, or as soon as some ℓ is obtained for which $\ell \notin X$ or there is no string of length ℓ in L that has w_i as a prefix or there is no string of length ℓ in L that has w_j as suffix (in all these cases $u_{\ell,j,i}$ is not contained in L, thus we can conclude that L is not closed under $\mathcal{S}$). The correctness of our approach is now immediate.

We now focus on the complexity of this step. In each iteration of the while loop, the length of the considered overlap strictly decreases w.r.t. the previously considered one (so the length of the string $u_{\ell,j,i}$ strictly increases). We iterate until we either check all strictly positive lengths of borders of $w_j[1 : LO(i,j)]$ or reach a case when we can conclude that $u_{\ell,j,i} \notin L$, as explained above. Thus, the total number of iterations of the loop, for some pair $i, j \in [k]$, is upper bounded by $\min\{p, |w_i|\}$, and the time spent in each iteration is $\mathcal{O}(1)$. Hence, the total complexity of computing the set of queries Q (as well as $|Q|$) is $\mathcal{O}(\min\{k^2p, kN\})$.

Using the Queries. We pause the presentation of the algorithm to explain how we plan to use these queries. For each query $(\ell, j, i) \in Q$, we will check whether the corresponding string $u_{\ell,j,i}$ is (not) in L. This is equivalent to checking whether U_t contains or not a substring $\#w_g\#$ with $\#w_i$ as a prefix and $w_j\#$ as a suffix, where $t = T[\ell]$. In our algorithm, we look for an index $g \in SL[j, t]$ (i.e., $\#w_g\#$ ends with $w_j\#$) such that the position r where b_g occurs in the suffix array of U_t is in the range $[a : b]$ of the respective array, containing the suffixes starting with $\#w_i$. Note that we know that $PL[i, t] \neq \emptyset$ (as $(\ell, j, i) \in Q$), and consequently, that $\#w_i = U_t[b_c : b_c + |w_i|]$, where c is an element of $PL[i, t]$. Hence, we can retrieve the range $[a : b]$ in $\mathcal{O}(1)$ time, using the data structures from the stringology toolbox constructed for U_ℓ.

In conclusion, $L \neq \mathcal{S}(L)$ if and only if a query $(\ell, j, i) \in Q$ is found (called an *empty query* in the followings) for which there is no index $g \in SL[j, T[\ell]]$, such that the position r where b_g occurs in the suffix array of $U_{T[\ell]}$ is in the range $[a : b]$ of the suffixes starting with $\#w_i$.

Encoding and Sorting the Queries. Returning to the algorithm, we encode each query $(\ell, j, i) \in Q$, with $t = T[\ell]$, as the tuple $(t, j, a_{t,i}, b_{t,i})$, where $[a_{t,i} : b_{t,i}]$ is the range of the suffix array of U_t containing the suffixes starting with $\#w_i$. This encoding is done in $\mathcal{O}(|Q|) \subseteq \mathcal{O}(\min\{k^2 p, kN\})$ time, because the range of the suffix array of U_t containing the suffixes starting with $\#w_i$ can be retrieved in $\mathcal{O}(1)$ using the stringology toolbox for $U_{T[\ell]}$, as explained above.

Once this encoding is done, we radix sort the set of tuples $(t, j, a_{t,i}, b_{t,i})$, where $(\ell, j, i) \in Q$ with $T[\ell] = t$, in two ways (and obtain, as such, two sorted lists). In the first list $\mathcal{L}_1$, we sort the tuples in ascending order w.r.t., in order of their significance, the first, second, third, and fourth component, while in the second list $\mathcal{L}_2$, we sort in ascending order w.r.t., in order of their significance, the first, second, fourth, and third component. Clearly, this sorting takes $\mathcal{O}(N + \min\{k^2 p, kN\})$ time. Now, in $\mathcal{O}(|Q|) \subseteq \mathcal{O}(\min\{k^2 p, kN\})$, we compute for every t and j the sorted lists $\mathcal{L}_{1,t,j}$ and $\mathcal{L}_{2,t,j}$ as the sublists of $\mathcal{L}_1$ and $\mathcal{L}_2$, respectively, which contain the tuples whose first two components are t and j; we assume that each tuple from $\mathcal{L}_{1,t,j}$ has a pointer to its copy from $\mathcal{L}_{2,t,j}$. Note that the number of elements in $\mathcal{L}_{1,t,j}$ (which equal to the number of elements of $\mathcal{L}_{2,t,j}$) is at most k.

Encoding the Input Set. Each $g \in SL[j, t]$ is encoded as the triple (t, j, r), where r is the position in the suffix array of U_t where b_g occurs (i.e., the position of the suffix array corresponding to the suffix starting with $\#w_g\#$). These triples can be computed in $\mathcal{O}(k^2)$ time, during the computation of the array $SL[\cdot, \cdot]$. Once the encoding is complete, we radix sort the resulting set of triples in ascending order w.r.t. the first component and break ties in ascending order w.r.t. the second component (note that there are no two distinct triples with identical first two components). Let $\mathcal{P}$ be the resulting sorted list of triples. Let $\mathcal{P}_{t,j}$ be the (sorted) sublist of $\mathcal{P}$ containing the pairs which have t and j as the first and second component, respectively. Note that the number of elements in $\mathcal{P}_{t,j}$ is at most k. The time needed for this step is $\mathcal{O}(N + k^2)$.

Using the Queries, Revisited. We pause the presentation of the algorithm again to recall our goal: to find an empty query, that is, a triple $(\ell, j, i) \in Q$ for which there is no index $g \in SL[j, T[\ell]]$, such that the position r where b_g occurs in the suffix array of $U_{T[\ell]}$ is in the range $[a : b]$ of the suffixes starting with $\#w_i$. After applying the encoding, our goal is to find a tuple (t, j, a, b), encoding a query $(\ell, j, i) \in Q$, for which there is no triple $(t, j, r) \in \mathcal{P}$ such that $r \in [a : b]$. In other words, we want to see if there exists a tuple $(t, j, a, b) \in \mathcal{L}_{1,t,j}$ such that there is no element $(t, j, r) \in \mathcal{P}_{t,j}$ with $r \in [a : b]$.

Finding an Empty Query. We now return to presenting the algorithm. To detect if there exists a tuple $(t, j, a, b) \in \mathcal{L}_{1,t,j}$, such that there is no element $(t, j, r) \in \mathcal{P}_{t,j}$ with $r \in [a : b]$, we apply an algorithm similar to merging these two sorted lists. We use the auxiliary list $\mathcal{L}_{2,t,j}$ to keep track of the tuples (t, j, a, b) for which we have not yet found some element $(t, j, r) \in \mathcal{P}_{t,j}$ with $r \in [a : b]$. As hinted, a crucial property at this point is that the lists $\mathcal{L}_{1,t,j}$, $\mathcal{L}_{2,t,j}$, and $\mathcal{P}_{t,j}$ are sorted.

We fix some $t \in [p]$ and $j \in [k]$ and maintain a set $\mathcal{R}$, initially empty. We repeat the following steps until all three lists $\mathcal{L}_{1,t,j}$, $\mathcal{L}_{2,t,j}$, and $\mathcal{P}_{t,j}$ are empty. Let (t, j, a, b) be the first element of $\mathcal{L}_{1,t,j}$, (t, j, a', b') be the first element of $\mathcal{L}_{2,t,j}$, and (t, j, r) be the first element of $\mathcal{P}_{t,j}$. If $b' < \min\{a, r\}$, then return $\mathrm{SAC}(L) = \mathrm{NO}$. Otherwise, if $a \leq r$, insert (t, j, a, b) in $\mathcal{R}$, preserving the pointer from this element to its copy from $\mathcal{L}_{2,t,j}$, remove it from $\mathcal{L}_{1,t,j}$, and iterate. If $r < a$, remove all elements from $\mathcal{R}$, delete the corresponding tuples from $\mathcal{L}_{2,t,j}$ (by using the pointers from the elements of $\mathcal{R}$ to their copies $\mathcal{L}_{2,t,j}$).

If all $t \in [p]$, $j \in [k]$ are processed without returning NO, then $\mathrm{SAC}(L) = \mathrm{YES}$.

The correctness of the algorithm follows from a standard sweeping-line argument over the interval $[1 : N]$. The event points consist of the left margins of ranges $[a : b]$ from elements $(t, j, a, b) \in \mathcal{L}_{1,t,j}$, the right margins of ranges $[a' : b']$ from elements $(t, j, a', b') \in \mathcal{L}_{1,t,j}$, and the points r from elements $(t, j, r) \in \mathcal{P}_{t,j}$. These three sets can be traversed in increasing order. When the sweep first encounters a left margin, then a new range $[a : b]$ is opened and stored in $\mathcal{R}$ (while being removed from $\mathcal{L}_{1,t,j}$). If the first event is a point r, then we can conclude that all open ranges are nonempty and remove them from the set $\mathcal{R}$ and the list $\mathcal{L}_{2,t,j}$ (the corresponding queries cannot be empty). When the sweep instead encounters a right margin, it means that the respective interval has been opened and stored in $\mathcal{R}$, and no point from $\mathcal{P}_{t,j}$ can be found in it. We have now reached the end of this interval and need to close it - therefore, the respective range (and the corresponding query) is empty, and we can return NO.

The complexity of the final step of our algorithm is proportional to the total number of elements in the sets $\mathcal{L}_{1,t,j}$, $\mathcal{L}_{2,t,j}$, and $\mathcal{P}_{t,j}$, over all $t \in [p]$ and $j \in [k]$. This adds up to $\mathcal{O}(\min\{k^2 p, kN\})$.

Conclusion. In summary, our algorithm runs in $\mathcal{O}(N + \min\{k^2 p, kN\})$ time, and decides the existence of some $\ell \in [N]$ and $i, j \in [k]$, such that the string $u_{\ell, j, i}$ of length $|\ell| < |w_j| + |w_i|$ that has $w_j \in L$ as a suffix and $w_i \in L$ as a prefix, is contained in $\mathcal{S}(L)$ but not in L. If such a triple (ℓ, i, j) exists, the algorithm correctly returns $\mathrm{SAC}(L) = \mathrm{NO}$, and $\mathrm{SAC}(L) = \mathrm{YES}$, otherwise. $\qquad\square$

The proof of Theorem 1 can be canonically adapted to compute all strings of L that can be obtained by applying $\mathcal{S}$ on shorter strings of the same language, and, as such, compute the *minimal base* of this language w.r.t. self-assembly.

Whether $\mathrm{SAC}(L)$ can be computed faster remains an open problem.

3.2   Regular Languages

Regular Languages Given as DFA. Recall that, in this case, SAC is assumed to have as input a DFA $A = (Q, \Sigma, q_0, F, \delta)$ with $|Q| = n$ and $|\Sigma| = \sigma$; the size of the input for SAC is, therefore, $\mathcal{O}(n\sigma)$. We also assume that $B, E, \#, \$$ are pairwise distinct letters, not contained in Σ.

We first recall the Intersection Non-Emptiness Problem for DFAs.

DFA-INE – Intersection Non-Emptiness Problem for DFAs

Input: Given two deterministic finite automata $A_1 = (\Sigma, Q_1, q_0^1, \delta_1, F_1)$ and $A_2 = (\Sigma, Q_2, q_0^2, \delta_2, F_2)$, where $|Q_1| = n_1$ and $|Q_2| = n_2$.

Output: DFA-INE(A_1, A_2) = YES if $L(A_1) \cap L(A_2) \neq \emptyset$; DFA-INE$(A_1, A_2)$ = NO otherwise.

For $|Q_1| = n_1$ and $|Q_2| = n_2$, the input size for DFA-INE is $\mathcal{O}((n_1 + n_2)\sigma)$.

As the main result of this section, we show the inter-reducibility of SAC and the intersection non-emptiness problem for two DFA DFA-INE. This means that we can reduce SAC to DFA-INE, and vice-versa, in linear time w.r.t. the input size of the respective problems. To show the inter-reducibility of SAC and DFA-INE we will define two functions f, g, whose domain and codomain are the class of DFAs over Σ. For a DFA A, the input for SAC, $f(A)$ is defined such that $L(f(A)) = \{w_1 \$ w_2 \# w_3 \mid w_1 w_2 \in L(A), w_1 w_2 w_3 \notin L(A)\}$, and $g(A)$ is defined such that $L(g(A)) = \Sigma^* \$ \{u \# v \mid uv \in L(A)\}$. We can show that the language $L(A)$ is not closed under $\mathcal{S}$ (i.e., $\mathrm{SAC}(A) = \mathrm{NO}$) if and only if the intersection of $L(f(A))$ and $L(g(A))$ is nonempty (so DFA-INE$(f(A), g(A))$ = YES); further both $f(A)$ and $g(A)$ can be constructed in $\mathcal{O}(n\sigma)$. Later in this section, we will also show the converse: for two DFAs A_1 and A_2, inputs for DFA-INE, we can construct a DFA $h(A_1, A_2)$, in time $\mathcal{O}((n_1 + n_2)\sigma)$, such that DFA-INE$(A_1, A_2)$ = YES if and only if $\mathrm{SAC}(h(A_1, A_2)) = \mathrm{NO}$. This is formalized in the next lemmas.

Lemma 2. *Given DFA A, input for SAC, we can construct in $O(n\sigma)$ time DFA $f(A)$, where $L(f(A)) = \{w_1 \$ w_2 \# w_3 \mid w_1 w_2 \in L(A), w_2 \neq \varepsilon, w_1 w_2 w_3 \notin L(A)\}$ and $f(A)$ has $\mathcal{O}(n)$ states.*

Lemma 3. *Given DFA A, input for SAC, we can construct in $O(n\sigma)$ time DFA $g(A)$, where $L(g(A)) = \Sigma^* \$ \{u \# v \mid u \neq \varepsilon, uv \in L(A)\}$ and $g(A)$ has $\mathcal{O}(n)$ states.*

Finally, using the results of Lemma 2 and Lemma 3 we obtain the following.

Lemma 4. $\mathrm{SAC}(A) = \mathit{NO}$ *if and only if* DFA-INE$(f(A), g(A)) = \mathit{YES}$.

Proof. We need to show that $L(A)$ is not closed under $\mathcal{S}$ if and only if the intersection of $L(f(A))$ and $L(g(A))$ is nonempty. Recall that $L(f(A)) = \{w_1\$w_2\#w_3 \mid w_1w_2 \in L(A), w_1w_2w_3 \notin L(A)\}$ and $L(g(A)) = \Sigma^*\${u\#v \mid uv \in L(A)\}$.

Now, assume that $L(f(A)) \cap L(g(A)) \neq \emptyset$. Then, there exists w such that $w \in L(f(A))$ and $w \in L(g(A))$. Thus, $w = w_1\$w_2\#w_3$, where $w_1w_2 \in L(A)$ and $w_1w_2w_3 \notin L(A)$, and $w \in \Sigma^*\$u\#v$ with $uv \in L(A)$. Since each letter $\$$ and $\#$ only occurs once in w, we can infer that $u = w_2$ and $v = w_3$. Furthermore, since w_2 is a nonempty suffix of w_1w_2 and a nonempty prefix w_2w_3, we obtain $w_1w_2w_3 \in \mathcal{S}(w_1w_2, w_2w_3)$ and $w_1w_2w_3 \notin L(A)$; hence $\mathcal{S}(L(A)) \neq L(A)$.

For the converse, assume that $\mathcal{S}(L(A)) \neq L(A)$. Thus, there exists $w \in \mathcal{S}(L(A))$ with $w \notin L(A)$. Since $w \in \mathcal{S}(L(A))$ there are strings $w', w'' \in L(A)$ such that $w \in \mathcal{S}(w', w'')$. Because $\mathcal{S}(w', w'')$ is not empty, we can write $w' = w_1w_2$ and $w'' = w_2w_3$ for some nonempty w_2. Note that $w = w_1w_2w_3 \notin L(A)$ but $w' = w_1w_2 \in L(A)$ and $w'' = w_2w_3 \in L(A)$. Therefore, $w_1\$w_2\#w_3 \in L(f(A))$ and $w_1\$w_2\#w_3 \in L(g(A))$. Hence, it follows that $L(f(A)) \cap L(g(A)) \neq \emptyset$. $\square$

We can also show the converse of Lemma 4: for two DFAs A_1 and A_2, inputs for DFA-INE, we can construct a DFA $h(A_1, A_2)$, in time $\mathcal{O}((n_1+n_2)\sigma)$, such that DFA-INE$(A_1, A_2) = $ YES if and only if SAC$(h(A_1, A_2)) = $ NO. We define the DFAs A_1' and A_2' such that $L(A_1') = \{B\#w\$ \mid w \in L(A_1)\}$ and $L(A_2') = \{\#w\$E \mid w \in L(A_2)\}$ and a DFA $A = h(A_1, A_2)$ which accepts $L(A) = L(A_1') \cup L(A_2')$ and show that $L(A)$ is not closed under $\mathcal{S}$ if and only if DFA-INE$(A_1, A_2) = $ YES.

Lemma 5. *For DFAs A_1, A_2, inputs for* DFA-INE, *we can construct a DFA $h(A_1, A_2)$ in $\mathcal{O}((n_1 + n_2)\sigma)$ time such that $L(h(A_1, A_2)) = \{B\#w\$ \mid w \in L(A_1)\} \cup \{\#w\$E \mid w \in L(A_2)\}$.*

We may now use the results of Lemma 5 to obtain the following.

Lemma 6. DFA-INE$(A_1, A_2) = $ *YES if and only if* SAC$(L(h(A_1, A_2))) = $ *NO.*

Proof. We need to show that $L(A_1) \cap L(A_2) \neq \emptyset$ if and only if $L(h(A_1, A_2))$ is not closed under the operation $\mathcal{S}$. Recall that $L(h(A_1, A_2)) = L_1' \cup L_2'$, where $L_1' = \{B\#w\$ \mid w \in L(A_1)\}$ and $L_2' = \{\#w\$E \mid w \in L(A_2)\}$. Let $L = L(h(A_1, A_2))$.

First, suppose that $L(A_1) \cap L(A_2) \neq \emptyset$. Then there exists a string $w \in L(A_1) \cap L(A_2)$. We define $w_1 = B\#w\$$ and $w_2 = \#w\$E$. Clearly, we have $w_1, w_2 \in L$. Let $x = \#w\$$, so that $w_1 = Bx$ and $w_2 = xE$. Applying $\mathcal{S}$ to w_1 and w_2 yields $w_3 = BxE \in \mathcal{S}(w_1, w_2)$. We claim that $w_3 \notin L$. Indeed, by construction, every string in L that starts with B must end with $\$$, while every string in L that ends with E must start with $\#$. Neither condition holds for w_3. Consequently, $w_3 \in \mathcal{S}(L)$ but $w_3 \notin L$, which implies $\mathcal{S}(L) \neq L$. Therefore, L is not closed under $\mathcal{S}$.

For the reverse direction, assume that $\mathcal{S}(L) \neq L$. Then there exists a string $w \in \mathcal{S}(L)$ such that $w \notin L$. Since $w \in \mathcal{S}(L)$, there exist strings $w_1, w_2 \in L$ with $w \in \mathcal{S}(w_1, w_2)$. We first argue that $w_1 \in L_1'$ and $w_2 \in L_2'$. It is easy to see that if both w_1 and w_2 are in L_1' (respectively, in L_2') then $\mathcal{S}(w_1, w_2)$ is nonempty if and only if it is trivial (i.e., $w_1 = w_2$ and $\mathcal{S}(w_1, w_2) = \{w_1\}$). Moreover, if $w_1 \in L_2'$ and $w_2 \in L_1'$, then $\mathcal{S}(w_1, w_2) = \emptyset$, as w_1 ends with E and E does not occur inside w_2. So, the only possibility that remains is that $w_1 \in L_1'$ and $w_2 \in L_2'$. Thus, we can write $w_1 = B\#x\$$ and $w_2 = \#y\$E$ for some strings $x, y \in \Sigma^*$. Now, in order to have that $\mathcal{S}(w_1, w_2) \neq \emptyset$, as $\#$ does not occur neither in x nor in y, is to have that $x = y$ and the overlap of w_1 and w_2 equals $\#x\$$. By construction of L_1' and L_2', it follows that $x \in L(A_1)$ and $x \in L(A_2)$. Hence, $L(A_1) \cap L(A_2) \neq \emptyset$.

$\square$

The results of this section provide a good understanding of the complexity of SAC for regular input languages, given as DFAs. This is stated in the next remark.

Remark 1. Based on Lemmas 4 and 6, we obtain that every upper and, respectively, lower bound for DFA-INE applies to SAC as well. We provide an overview of what is known about the problem DFA-INE.

Firstly, DFA-INE can be solved in $\mathcal{O}(n_1 n_2 \sigma)$, by applying the standard product automaton construction from the literature [16]. Hence, SAC, when its input is a regular languages given as DFA, can be solved in $\mathcal{O}(n^2 \sigma)$ time.

On the other hand, one can show, by a reduction from the Orthogonal Vectors Problem [29], that an algorithm solving DFA-INE (on the word RAM model) in $\mathcal{O}((n_1 n_2 \sigma)^{1-\varepsilon})$ time, for some $\varepsilon > 0$, would falsify the Strong Exponential Time Hypothesis (for short, SETH). This complements the result of [31], where it was shown that there is no algorithm, running on a multi-tape deterministic Turing machine, solving DFA-INE in $\mathcal{O}((n_1 n_2 \sigma)^{1-\varepsilon})$ for some $\varepsilon > 0$ (recall also the result of [25], where a similar *conditional* lower bound was shown, for the respective computational model). In conclusion, this also means that the existence of a word RAM algorithm solving SAC in $\mathcal{O}((n^2 \sigma)^{1-\varepsilon})$ time would falsify SETH.

Finally, the above lower bounds for DFA-INE do not exclude algorithms running in $\mathcal{O}(n_1 n_2 \sigma^{1-\varepsilon})$ for DFA-INE, so it remains open whether the standard algorithm solving DFA-INE in $\mathcal{O}(n_1 n_2 \sigma)$ is (conditionally) optimal, when the alphabet is not constant. Using a reduction from the triangle detection in dense graphs problem, it can be shown that the existence of a *combinatorial algorithm* solving DFA-INE in $\mathcal{O}(n_1 n_2 \sigma^{1-\varepsilon})$ time would lead to a truly subcubic combinatorial algorithm for the triangle detection problem and, as such, for computing the Boolean Matrix Multiplication. It has been conjectured that such combinatorial algorithms do not exist [30]. Recall that combinatorial algorithms are those whose intermediate steps admit a natural combinatorial interpretation of the underlying problem [9]. A more relaxed characterization in [2] defines them simply as algorithms that avoid oracle calls to fast ring matrix multiplication. The lower bound mentioned above for combinatorial algorithms does not exclude faster algorithms solving DFA-INE based on, e.g., fast matrix multiplication. The same things can be said about SAC: the existence of a *combinatorial algorithm* solving SAC in $\mathcal{O}(n^2 \sigma^{1-\varepsilon})$ time would lead to a truly subcubic algorithm for computing the Boolean Matrix Multiplication (BMM, for short).

In conclusion, we can state the following theorem.

Theorem 2. SAC *can be solved in* $\mathcal{O}(n^2\sigma)$ *time, for input DFA A with n states and alphabet of size σ. The existence of an algorithm solving* SAC *in* $\mathcal{O}((n^2\sigma)^{1-\varepsilon})$ *time would falsify SETH, and the existence of a* combinatorial algorithm *solving* SAC *in* $\mathcal{O}(n^2\sigma^{1-\varepsilon})$ *time would lead to a truly subcubic algorithm for BMM.*

Regular Languages Given as NFA. In this case, SAC has as input an ε-NFA $A = (Q, \Sigma, q_0, F, \delta)$ with $|Q| = n$ and $|\Sigma| = \sigma$; the size of the input is the number of transitions of A. As before, we assume that $\varepsilon \notin L(A)$. By a reduction from the universality problem for NFAs we obtain the following lower bound for SAC.

Lemma 7. SAC *is PSPACE-hard, for regular input languages, given as ε-NFAs.*

We complement the result of the previous lemma with an upper bound.

Lemma 8. SAC(A) *can be computed in polynomial space.*

The next theorem follows now from Lemmas 7 and 8.

Theorem 3. SAC *is PSPACE-complete, for regular languages, given as ε-NFAs.*

The results of this section provide a detailed understanding of SAC when the input is a regular language. Note that this problem becomes undecidable for the class of context-free languages, given, e.g., as context-free grammars [8], as well as for larger classes of the Chomsky hierarchy.

References

1. Aho, A.V., Hopcroft, J.E., Ullman, J.D.: The Design and Analysis of Computer Algorithms, Addison-Wesley (1974)
2. Bansal, N., Williams, R.: Regularity lemmas and combinatorial algorithms. Theory Comput. **8**(1), 69–94 (2012)
3. Belazzougui, D., Kosolobov, D., Puglisi, S.J., Raman, R.: Weighted ancestors in suffix trees revisited. In: 32nd Annual Symposium on Combinatorial Pattern Matching (CPM 2021), vol. 191, pp. 8:1–8:15 (2021)
4. Bellamoli, F., Franco, G., Kari, L., Lampis, S., Ng, T., Wang, Z.: Conjugate word blending: formal model and experimental implementation by XPCR. Nat. Comput. **20**(4), 647–658 (2021)
5. Bottoni, P., Labella, A., Manca, V., Mitrana, V.: Superposition based on Watson-Crick-like complementarity. Theory Comput. Syst. **39**(4), 503–524 (2006)
6. Brzozowski, J.A., Kari, L., Li, B., Szykuła, M.: State complexity of overlap assembly. Int. J. Found. Comput. Sci. **31**(8), 1113–1132 (2020)
7. Csuhaj-Varjú, E., Lázár, K.A.: String assembly in networks of evolutionary processors. J. Autom. Lang. Comb. **21**(1–2), 41–54 (2016)
8. Csuhaj-Varjú, E., Petre, I., Vaszil, G.: Self-assembly of strings and languages. Theor. Comput. Sci. **374**(1), 74–81 (2007)
9. Das, D., Koucký, M., Saks, M.E.: Lower bounds for combinatorial algorithms for Boolean Matrix Multiplication. In: STACS. LIPIcs, vol. 96, pp. 23:1–23:14 (2018)
10. Enaganti, S.K., Ibarra, O.H., Kari, L., Kopecki, S.: Further remarks on DNA overlap assembly. Inf. Comput. **253**(1), 143–154 (2017)

11. Enaganti, S.K., Ibarra, O.H., Kari, L., Kopecki, S.: On the overlap assembly of strings and languages. Nat. Comput. 1–11 (2017)
12. Enaganti, S.K., Kari, L., Ng, T., Wang, Z.: Word blending in formal languages. Fundam. Inform. **171**(1–4), 151–173 (2020)
13. Farach, M.: Optimal suffix tree construction with large alphabets. In: 38th Annual Symposium on Foundations of Computer Science, FOCS 1997, Miami Beach, Florida, USA, October 19–22, 1997. pp. 137–143. IEEE Computer Society (1997)
14. Fredman, M.L., Willard, D.E.: BLASTING through the information theoretic barrier with FUSION TREES. In: Ortiz, H. (ed.) Proceedings of the 22nd Annual ACM Symposium on Theory of Computing, May 13–17, 1990, Baltimore, Maryland, USA, pp. 1–7. ACM (1990)
15. Gusfield, D., Landau, G.M., Schieber, B.: An efficient algorithm for the all pairs suffix-prefix problem. Inf. Process. Lett. **41**(4), 181–185 (1992)
16. Hopcroft, J.E., Ullman, J.D.: Introduction to Automata Theory, Languages and Computation. Addison-Wesley (1979)
17. Ito, M., Leupold, P., Manea, F., Mitrana, V.: Bounded hairpin completion. Inf. Comput. **209**(1–2), 471–485 (2011)
18. Kärkkäinen, J., Sanders, P., Burkhardt, S.: Linear work suffix array construction. J. ACM **53**(6), 918–936 (2006)
19. Kasai, T., Lee, G., Arimura, H., Arikawa, S., Park, K.: Linear-time longest-common-prefix computation in suffix arrays and its applications. In: Amir, A. (ed.) CPM 2001. LNCS, vol. 2089, pp. 181–192. Springer, Heidelberg (2001). https://doi.org/10.1007/3-540-48194-X_17
20. Knuth, D.E., Morris, J.H., Jr., Pratt, V.R.: Fast pattern matching in strings. SIAM J. Comput. **6**(2), 323–350 (1977)
21. Manea, F., Martín-Vide, C., Mitrana, V.: On some algorithmic problems regarding the hairpin completion. Discrete Appl. Math. **157**(9), 2143–2152 (2009)
22. Manea, F., Martín-Vide, C., Mitrana, V.: Hairpin lengthening: language theoretic and algorithmic results. J. Log. Comput. **25**(4), 987–1009 (2015)
23. Manea, F., Mitrana, V., Yokomori, T.: Two complementary operations inspired by the DNA hairpin formation: completion and reduction. Theor. Comput. Sci. **410**(4–5), 417–425 (2009)
24. Manea, F., Mitrana, V., Yokomori, T.: Some remarks on the hairpin completion. Int. J. Found. Comput. Sci. **21**(5), 859–872 (2010)
25. de Oliveira Oliveira, M., Wehar, M.: On the fine grained complexity of finite automata non-emptiness of intersection. In: Jonoska, N., Savchuk, D. (eds.) DLT 2020. LNCS, vol. 12086, pp. 69–82. Springer, Cham (2020). https://doi.org/10.1007/978-3-030-48516-0_6
26. Park, S., Park, S.G., Cazaux, B., Park, K., Rivals, E.: A linear time algorithm for constructing hierarchical overlap graphs. In: 32nd Annual Symposium on Combinatorial Pattern Matching (CPM 2021). Leibniz International Proceedings in Informatics (LIPIcs), vol. 191, pp. 22:1–22:9 (2021)
27. Păun, G., Rozenberg, G., Yokomori, T.: Hairpin languages. Int. J. Found. Comput. Sci. **12**, 837–847 (2001)
28. Stemmer, W.P.: DNA shuffling by random fragmentation and reassembly: in vitro recombination for molecular evolution. Proc. Natl. Acad. Sci. **91**(22), 10747–10751 (1994)
29. Vassilevska Williams, V.: On some fine-grained questions in algorithms and complexity. In: Proceedings of the International Congress of Mathematicians (ICM 2018), pp. 3447–3487. World Scientific (2018)

30. Vassilevska Williams, V., Williams, R.: Subcubic equivalences between path, matrix, and triangle problems. J. ACM **65**(5) (2018)
31. Wehar, M.: Unconditional time and space complexity lower bounds for intersection non-emptiness (2025). https://arxiv.org/abs/2512.00297

Reversible Weighted Automata over Finite Rings and Monoids with Commuting Idempotents

Peter Kostolányi$^{(\boxtimes)}$ and Andrej Ravinger

Department of Computer Science, Comenius University in Bratislava,
Mlynská dolina, 842 48 Bratislava, Slovakia
`{kostolanyi,andrej.ravinger}@fmph.uniba.sk`

Abstract. Reversible weighted automata are introduced and considered in a specific setting where the weights are taken from a nontrivial locally finite commutative ring such as a finite field. It is shown that the supports of series realised by such automata are precisely the rational languages such that the idempotents in their syntactic monoids commute. In particular, this is true for reversible weighted automata over the finite field $\mathbb{F}_2$, where the realised series can be directly identified with such languages. A new automata-theoretic characterisation is thus obtained for the variety of rational languages corresponding to the pseudovariety of finite monoids **ECom**, which also forms the Boolean closure of the reversible languages in the sense of J.-É. Pin. The problem of determining whether a rational series over a locally finite commutative ring can be realised by a reversible weighted automaton is decidable as a consequence.

Keywords: Weighted automaton · Reversible automaton · Finite field · Variety of languages · Decidability

1 Introduction

This article embarks upon the study of *reversibility* in the context of *weighted automata*. More precisely, we focus here on a very special case of weighted automata over *finite commutative rings*, including in particular the two-element Galois field $\mathbb{F}_2$. Strong connections to language theory turn out to arise.

The study of the concept of reversibility in computing goes back to the seminal work of R. Landauer [34]: according to his fundamental thermodynamical principle, any loss of information that takes place during a computation necessarily leads to some minimal amount of heat dissipation. This observation led C. H. Bennett [11] to consider logical reversibility of computations, and in particular the *reversible Turing machines*, in which any configuration has at most

The work was supported by the grant VEGA 1/0288/26. The second author was supported by the grant UK/1299/2026.

one preceding and at most one successor configuration. Similar ideas gradually inspired the development of an entire field of reversible computing.

Starting by D. Angluin [3] defining – in the context of language inference – what is now usually called *bideterministic finite automata* [43,51–53], many different variants of *reversible finite automata* have been introduced and extensively studied throughout the literature [2,18,19,21–23,35,43,45,46], while some generalisations of the usual concepts of reversibility in finite automata have been considered as well [5,6,17,20]. What all these models have in common is that unlike in the case of reversible Turing machines, reversibility turns out to be a real restriction for finite automata and leads to a loss in their expressivity. Hence the study of the corresponding classes of languages has been a substantial part of the aforementioned efforts.

The notion of a *reversible finite automaton* considered in this article is the one of J.-É. Pin [43] – in other words, a finite automaton is *reversible* if its transition relation is both deterministic and codeterministic; however, the automaton is allowed to have more than one initial as well as more than one final state. Thus, using the terminology usual in weighted automata theory, we may say that a reversible finite automaton is a nondeterministic finite automaton that is finitely sequential [8,31,33,39–41] (i.e., "deterministic" with possibly more than one initial state) and its transpose is finitely sequential as well. This definition is not only very natural, but also leads to a well-behaved class of *reversible languages*, which forms a *positive variety* [43]; the corresponding pseudovariety of ordered monoids can be captured by the pseudoinequalities $x^\omega y^\omega = y^\omega x^\omega$ and $x^\omega \leq 1$, which also give rise to a natural characterisation of reversible languages in terms of forbidden configurations in their minimal deterministic automata [27,28,43]. Reversible finite automata with at most one initial and at most one final state are precisely the *bideterministic finite automata* [43,51–53].

In this article, we lift the notion of reversibility in the sense of J.-É. Pin [43] to the setting of *weighted finite automata* – i.e., finite automata with transitions carrying weights taken from some algebra such as a semiring, encompassing a necessary shift from recognising languages towards realisation of formal *power series* in several noncommuting variables [12,15,16,47,48]. While the very origins of weighted automata theory go back already to the seminal article of M.-P. Schützenberger [49], the field has experienced a wave of renewed interest during the last decades, and still represents a very active area of study [15,16]. It is thus relatively surprising that reversible weighted automata have attracted almost no attention so far, the only exception being the recent research on bideterministic weighted automata [29,32].

The study of reversible weighted automata would not only be natural in view of the previous research on reversibility in automata theory, but it can also be motivated by the study of certain *decision problems* for weighted automata. The *determinisability* problem, in which one asks about the existence of a deterministic – or sequential [36] – equivalent of a given weighted automaton, attracted significant attention in recent years [9,10,24,30]. While the problem was proved to be decidable in important settings such as over fields [9] and some complexity bounds have been obtained as well [10,24], one still has no efficient algorithms for deciding the determinisability problem for sufficiently general classes of weighted automata. On the other hand, it has been shown that one can do better when deciding the existence of a *bideterministic* equivalent for a given weighted automaton [29,32]. This indicates that it might make sense to also look at some other restrictions of weighted automata related to determinism, and in particular at the *reversible weighted automata*, which represent both a natural generalisation of bideterministic weighted automata, as well as a class of finitely sequential automata that is in general incomparable with the deterministic weighted automata when it comes to expressivity.

The aim of this article is to initiate a systematic research on reversible weighted automata, while we mostly focus here on a very special case, in which the weights are from a *locally finite commutative ring*. This includes the case of weighted automata and formal power series over *finite fields*, and in particular over the *two-element field* $\mathbb{F}_2$. The latter setting gives rise to a natural way of describing languages, as any rational series over $\mathbb{F}_2$ is at the same time a characteristic series of some rational language – in fact, the series over $\mathbb{F}_2$ correspond to what has been studied as *formal languages over* GF(2) by E. Bakinova et al. [7,38]; the idea of describing languages using weighted automata over fields also appears in the study of *image-binary automata* of S. Kiefer and s C. Widdershoven [25,26]. In a similar spirit, a weighted automaton over a *finite* or *locally finite* (semi)ring gives rise to a *rational* language by taking the support of its behaviour, i.e., the language of all words with a nonzero coefficient in the realised series.

We show that regardless of a *nontrivial locally finite commutative ring* R considered, the languages described by the *reversible weighted automata* over R in this way always form the same class, namely the variety of languages corresponding to the pseudovariety **ECom** of all finite monoids with commuting idempotents. This is at the same time the Boolean closure of the positive variety of all reversible languages [43], i.e., the variety generated by such languages.

Note that the result applies in particular to reversible weighted automata over $\mathbb{F}_2$ – the languages described by such automata are precisely the languages from the aforementioned variety. This means that interpreting a reversible finite automaton as a reversible weighted automaton over $\mathbb{F}_2$ leads to an increase in the expressive power of the model and to better closure properties of the corresponding class of languages. At the same time, a new automata-theoretic characterisation of the variety of languages corresponding to **ECom** is obtained.

We also characterise the *series* realised by the reversible weighted automata over nontrivial *finite* commutative rings. This characterisation implies, together with effective decidability of membership of a monoid in **ECom**, the existence of an algorithm for deciding whether a weighted automaton over an effective *locally finite* commutative ring R admits a reversible equivalent over R or not.

2   Preliminaries

We write $\mathbb{N}$, $\mathbb{Z}$, and $\mathbb{Q}$ for the sets of all *nonnegative* integers, integers, and rational numbers, $\mathbb{B}$ for the Boolean domain $\mathbb{B} = \{0, 1\}$, and $[n] = \{1, \ldots, n\}$ for all $n \in \mathbb{N}$. Alphabets are finite and nonempty, ε denotes the empty word.

A *semiring* is a quintuple $(S, +, \cdot, 0, 1)$, or simply S, such that $(S, +, 0)$ is a commutative monoid, $(S, \cdot, 1)$ is a monoid, $\cdot$ distributes over $+$ from both sides, and 0 is a zero in $(S, \cdot, 1)$. A *subsemiring* of $(S, +, \cdot, 0, 1)$ is a semiring $(T, +_T, \cdot_T, 0, 1)$, where $T \subseteq S$ and $+_T, \cdot_T$ are the restrictions of $+$ and $\cdot$ to T. The *subsemiring of S generated by* $G \subseteq S$ is the smallest subsemiring $\langle G \rangle$ of S such that $G \subseteq \langle G \rangle$. A semiring $(S, +, \cdot, 0, 1)$ is *commutative* when $\cdot$ is, *finite* when S is, *finitely generated* when $S = \langle G \rangle$ for some finite $G \subseteq S$, and *locally finite* when all finitely generated subsemirings of S are finite. A semiring $(S, +, \cdot, 0, 1)$ is *nontrivial* if S contains at least two elements, that is, if $0 \neq 1$. An important example of a semiring is the *Boolean semiring* $(\mathbb{B}, \vee, \wedge, 0, 1)$.

A *ring* (with unity) is a semiring $(R, +, \cdot, 0, 1)$ such that $(R, +, 0)$ is an abelian group. A *field* is a nontrivial commutative ring $(\mathbb{F}, +, \cdot, 0, 1)$ such that $(\mathbb{F} \backslash \{0\}, \cdot, 1)$ is an abelian group. The finite field with two elements is denoted by $\mathbb{F}_2$.

The *characteristic* of a ring R is, in case it exists, the smallest $n \in \mathbb{N} \setminus \{0\}$ such that $\sum_{k=1}^{n} 1 = 0$ in R – or zero when there is no such n.

The reader can consult $[12, 15, 16, 47, 48]$ for the basics on formal power series in several noncommuting variables and weighted automata. A *formal power series* over a semiring S and alphabet Σ is a mapping $r\colon \Sigma^* \to S$; the value of r upon $w \in \Sigma^*$ is denoted by (r, w) and called the *coefficient* of r at w, while one writes $r = \sum_{w \in \Sigma^*} (r, w)\, w$. The set of all series over S and Σ is denoted by $S\langle\!\langle \Sigma^* \rangle\!\rangle$. Given $r, s \in S\langle\!\langle \Sigma^* \rangle\!\rangle$, we define $r + s$ by $(r + s, w) = (r, w) + (s, w)$ and $r \cdot s$ by $(r \cdot s, w) = \sum_{u, v \in \Sigma^*, uv = w} (r, u)(s, v)$ for all $w \in \Sigma^*$. Each $a \in S$ is identified with $r_a \in S\langle\!\langle \Sigma^* \rangle\!\rangle$ such that $(r_a, \varepsilon) = a$ and $(r_a, w) = 0$ for all $w \in \Sigma^+$; the left or right multiplication by a thus corresponds to left or right *scalar multiplication*. Similarly, each $w \in \Sigma^*$ is identified with $r_w \in S\langle\!\langle \Sigma^* \rangle\!\rangle$ such that $(r_w, w) = 1$ and $(r_w, x) = 0$ for all $x \in \Sigma^* \setminus \{w\}$.

The *support* of $r \in S\langle\!\langle \Sigma^* \rangle\!\rangle$ is a language $\mathrm{supp}(r) = \{w \in \Sigma^* \mid (r, w) \neq 0\}$, and the *characteristic series* $\underline{L} \in S\langle\!\langle \Sigma^* \rangle\!\rangle$ of a language $L \subseteq \Sigma^*$ over a nontrivial semiring S is given by $(\underline{L}, w) = 1$ for $w \in L$ and $(\underline{L}, w) = 0$ for $w \in \Sigma^* \setminus L$.

A *nondeterministic finite automaton* over an alphabet Σ is a quadruple $\mathcal{A} = (Q, \to, I, F)$, where Q is a finite set of states, $\to \subseteq Q \times \Sigma \times Q$ is a transition relation, and $I, F \subseteq Q$ are the sets of initial and final states, respectively. Given $p, q \in Q$ and $w \in \Sigma^*$ such that $w = a_1 \ldots a_n$ for some $n \in \mathbb{N}$ and $a_1, \ldots, a_n \in \Sigma$, we write $p \xrightarrow{w} q$ if there are states $p_0, \ldots, p_n \in Q$ such that $p_0 = p$, $p_n = q$, and $(p_{k-1}, a_k, p_k) \in \to$ for $k = 1, \ldots, n$. The *language* described by $\mathcal{A}$ is given by $\|\mathcal{A}\| = \{w \in \Sigma^* \mid \exists p \in I \; \exists q \in F : p \xrightarrow{w} q\}$. The automaton $\mathcal{A}$ is *deterministic* if $|I| = 1$ and $q = q'$ whenever $p \xrightarrow{a} q$ and $p \xrightarrow{a} q'$ for some $p, q, q' \in Q$ and $a \in \Sigma$.

A *deterministic finite automaton* with a complete transition function over Σ can also be written as $\mathcal{A} = (Q, \cdot, i, F)$, where $\cdot\colon Q \times \Sigma^* \to Q$ is a right action of Σ^* on Q, where $i \in Q$ is the initial state, and $F \subseteq Q$ is a set of final states. The *language* recognised by $\mathcal{A}$ is given by $\|\mathcal{A}\| = \{w \in \Sigma^* \mid i \cdot w \in F\}$.

A *weighted automaton* $\mathcal{A} = (Q, \sigma, \iota, \tau)$ over a semiring S and alphabet Σ is given by a finite set of states Q, a transition weighting function $\sigma \colon Q \times \Sigma \times Q \to S$, and functions $\iota \colon Q \to S$, $\tau \colon Q \to S$ assigning initial and final weights to states. A *run* of $\mathcal{A}$ is a word $\gamma = q_0 a_1 q_1 a_2 q_2 \ldots q_{t-1} a_t q_t$ with $t \in \mathbb{N}$, $q_0, \ldots, q_t \in Q$, and $a_1, \ldots, a_t \in \Sigma$ such that $\sigma(q_{k-1}, a_k, q_k) \neq 0$ for $k = 1, \ldots, t$; we say that γ is a run on $w = a_1 \ldots a_t$ leading from q_0 to q_t. Given any such run γ, we set $\sigma(\gamma) = \sigma(q_0, a_1, q_1) \ldots \sigma(q_{t-1}, a_t, q_t)$ and $\overline{\sigma}(\gamma) = \iota(q_0)\sigma(\gamma)\tau(q_t)$. Let $\mathcal{R}(\mathcal{A}, w)$ be the set of all runs of $\mathcal{A}$ on $w \in \Sigma^*$. The *series* $\|\mathcal{A}\| \in S\langle\!\langle \Sigma^* \rangle\!\rangle$ *realised by* $\mathcal{A}$ is then given by $(\|\mathcal{A}\|, w) = \sum_{\gamma \in \mathcal{R}(\mathcal{A}, w)} \overline{\sigma}(\gamma)$ for all $w \in \Sigma^*$. A series $r \in S\langle\!\langle \Sigma^* \rangle\!\rangle$ is *rational* over S if $r = \|\mathcal{A}\|$ for some weighted automaton $\mathcal{A}$ over S.

Every $\mathcal{A} = (Q, \sigma, \iota, \tau)$ over S and Σ such that $Q = [n]$ for some $n \in \mathbb{N}$ (which can always be assumed) determines a *linear representation* $\mathcal{P}_\mathcal{A} = (n, \mathbf{i}, \mu, \mathbf{f})$, where $\mathbf{i} = (\iota(1), \ldots, \iota(n))$ is a row vector of initial weights, μ is a homomorphism from Σ^* to the monoid $S^{n \times n}$ of all $n \times n$ matrices over S with matrix multiplication such that $\mu(a) = (\sigma(i, a, j))_{n \times n}$ for all $a \in \Sigma$, and $\mathbf{f} = (\tau(1), \ldots, \tau(n))^T$ is a column vector of final weights. One then has $(\|\mathcal{A}\|, w) = \mathbf{i}\mu(w)\mathbf{f}$ for all $w \in \Sigma^*$.

Let $\mathcal{A}_k = (Q_k, \sigma_k, \iota_k, \tau_k)$ be weighted automata over S and Σ for $k = 1, \ldots, n$, where $n \in \mathbb{N}$. The *disjoint union* of $\mathcal{A}_1, \ldots, \mathcal{A}_n$ is an automaton $\mathcal{A} = (Q, \sigma, \iota, \tau)$, where $Q = \bigcup_{k=1}^n (Q_k \times \{k\})$, $\sigma((p, k), a, (q, k)) = \sigma_k(p, a, q)$, $\iota(p, k) = \iota_k(p)$, and $\tau(q, k) = \tau_k(q)$ for $k = 1, \ldots, n$ and all $p, q \in Q_k$ and $a \in \Sigma$; moreover, $\sigma((p, k), a, (q, \ell)) = 0$ for all $k, \ell \in [n]$ such that $k \neq \ell$, and all $p \in Q_k$, $q \in Q_\ell$, and $a \in \Sigma$. Clearly $\|\mathcal{A}\| = \|\mathcal{A}_1\| + \ldots + \|\mathcal{A}_n\|$.

We also use some elementary concepts from *algebraic language theory*; we refer the reader to [44,50] for the basic theory of *varieties of languages*, and to [1] for the theory of *pseudovarieties of finite monoids*.

3   Reversible Weighted Automata

We now define the *reversible weighted automata* over a semiring S by generalising the notion of reversible finite automata as understood by J.-É. Pin [43].

Definition 1. Let S be a semiring and Σ an alphabet. A weighted automaton $\mathcal{A} = (Q, \sigma, \iota, \tau)$ over S and Σ is *reversible* if the following two conditions are satisfied for all $p, p', q, q' \in Q$ and $a \in \Sigma$:

(i) If $\sigma(p, a, q)$ and $\sigma(p, a, q')$ are both nonzero, then $q = q'$;
(ii) If $\sigma(p, a, q)$ and $\sigma(p', a, q)$ are both nonzero, then $p = p'$.

The condition *(i)* says that the transitions of $\mathcal{A}$ are deterministic; as there still can be multiple initial states, this is equivalent to saying that $\mathcal{A}$ is *finitely sequential* [8,31,33,39–41]. The condition *(ii)* says the same for the transpose of $\mathcal{A}$. If in addition there is at most one state $p \in Q$ with $\iota(p) \neq 0$ and at most one state $q \in Q$ with $\tau(q) \neq 0$, the automaton $\mathcal{A}$ is called *bideterministic* [29,32].

We call a series over a semiring S and over an alphabet Σ *reversible* over S if it is realised by some reversible weighted automaton over S and Σ. The set of all reversible series over S and Σ is denoted by $\mathrm{RevS}(S, \Sigma)$. Moreover, by a reversible *language over S and Σ*, we understand a support of some reversible series over S and Σ; we write $\mathrm{RevL}(S, \Sigma) = \{\mathrm{supp}(r) \mid r \in \mathrm{RevS}(S, \Sigma)\}$ for the set of all such languages, and $\mathrm{RevL}(S)$ for the class of all reversible languages over S and any alphabet. Observe that the reversible languages over the Boolean semiring $\mathbb{B}$ are precisely the usual reversible languages in the sense of J.-É. Pin [43]. We thus denote the positive variety of all such languages by $\mathrm{RevL}(\mathbb{B})$.

Example 2. We mostly consider a very special class of reversible weighted automata over *finite* – or slightly more generally, *locally finite* – *commutative rings* in this article. What can be readily observed right now is that given any nontrivial locally finite ring R, the class $\mathrm{RevL}(R)$ of all reversible languages over R contains at least some languages that are not reversible in the usual sense, i.e., which do not belong to $\mathrm{RevL}(\mathbb{B})$. Indeed, let R be of characteristic $n \geq 2$. Then the reversible weighted automaton $\mathcal{A}$ in Fig. 1 clearly realises the series

$$\|\mathcal{A}\| = \sum_{t \in \mathbb{N} \setminus \{0\}} a^t.$$

As a consequence, $\mathrm{supp}(\|\mathcal{A}\|) = a^+$ is a reversible language over R, so that $a^+ \in \mathrm{RevL}(R, \{a\})$. On the other hand, $a^+ \notin \mathrm{RevL}(\mathbb{B}, \{a\})$, as it does not satisfy the characterisation of reversible languages from [43]; more precisely, the ordered syntactic monoid of a^+ does not satisfy the pseudoinequality $x^\omega \leq 1$.

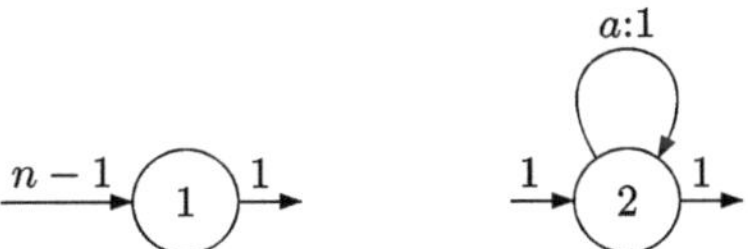

Fig. 1. A reversible weighted automaton $\mathcal{A}$ over a nontrivial locally finite ring R of characteristic $n \geq 2$ and over a unary alphabet $\Sigma = \{a\}$.

Given the observation from the previous example, it seems to be worthwhile to take a closer look at the properties of the classes of reversible series and languages introduced above. At least a few observations can be made at the abstract level of semirings. Given any semiring S and alphabet Σ, we denote by $\mathrm{1RevS}(S, \Sigma)$ the set of all series realised by reversible weighted automata $\mathcal{A} = (Q, \sigma, \iota, \tau)$ over S and Σ with *precisely one initial state*, i.e., a state q satisfying $\iota(q) \neq 0$.[1] We also write $\mathrm{1RevL}(S, \Sigma) = \{\mathrm{supp}(r) \mid r \in \mathrm{1RevS}(S, \Sigma)\}$. We then easily obtain the following characterisation of reversible series over S.

---

[1] Note that such automata form a weighted generalisation of reversible deterministic finite automata in the sense of M. Holzer, S. Jakobi, and M. Kutrib [21,22].

Proposition 3. *Let S be a semiring and Σ an alphabet. Then $\mathrm{RevS}(S, \Sigma)$ consists of precisely all finite sums of series from $1\mathrm{RevS}(S, \Sigma)$.*

Proof. Let $\mathcal{A} = (Q, \sigma, \iota, \tau)$ be a reversible weighted automaton over S and Σ. The automaton $\mathcal{A}_q = (Q, \sigma, \iota_q, \tau)$, with $\iota_q(q) = \iota(q)$ and $\iota_q(p) = 0$ for all $p \in Q \setminus \{q\}$, is then reversible with precisely one initial state for each $q \in Q$ such that $\iota(q) \neq 0$; as a result, $\|\mathcal{A}_q\| \in 1\mathrm{RevS}(S, \Sigma)$. Thus

$$\|\mathcal{A}\| = \sum_{\substack{q \in Q \\ \iota(q) \neq 0}} \|\mathcal{A}_q\|$$

is a finite sum of series from $1\mathrm{RevS}(S, \Sigma)$. Conversely, given $k \in \mathbb{N}$ and series $r_1, \ldots, r_k \in 1\mathrm{RevS}(S, \Sigma)$, each of the series r_j for $j = 1, \ldots, k$ is realised by some reversible weighted automaton $\mathcal{A}_j$ over S and Σ. The disjoint union of these automata is then clearly reversible as well, hence $r_1 + \ldots + r_k \in \mathrm{RevS}(S, \Sigma)$. $\square$

Proposition 4. *The set $\mathrm{RevS}(S, \Sigma)$ is closed under addition and under – both left and right – scalar multiplication for every semiring S and alphabet Σ.*

Proof. Closure under addition follows directly by Proposition 3. For left or right scalar multiplication by $\alpha \in S$, it is clearly sufficient to multiply all initial or final weights of a reversible weighted automaton by α. $\square$

Any finite automaton can be turned into a weighted automaton over a nontrivial semiring S by assigning the weight 1 to all its transitions, initial states, and final states. It is standard that for *deterministic* finite automata recognising some language L, the resulting automaton realises the characteristic series $\underline{L}$ of L over S. The following proposition records this observation specifically for the reversible automata with one initial state.

Proposition 5. *Let S be a nontrivial semiring, $L \in 1\mathrm{RevL}(\mathbb{B}, \Sigma)$, and $\underline{L}$ be the characteristic series of L over S. Then $\underline{L} \in 1\mathrm{RevS}(S, \Sigma)$. As a consequence, $L \in 1\mathrm{RevL}(S, \Sigma)$ and $1\mathrm{RevL}(\mathbb{B}, \Sigma) \subseteq 1\mathrm{RevL}(S, \Sigma)$.*

Proof. As $L \in 1\mathrm{RevL}(\mathbb{B}, \Sigma)$, the language L is surely recognised by some reversible finite automaton $\mathcal{A} = (Q, \rightarrow, I, F)$ with precisely one initial state $i \in I$. Let $\mathcal{A}' = (Q, \sigma, \iota, \tau)$ be a weighted automaton such that for all $p, q \in Q$ and $a \in \Sigma$, one has $\sigma(p, a, q) = 1$ if $p \xrightarrow{a} q$ and $\sigma(p, a, q) = 0$ otherwise, $\iota(p) = 1$ if $p \in I$ and $\iota(p) = 0$ otherwise, and $\tau(q) = 1$ if $q \in F$ and $\tau(q) = 0$ otherwise. Since $\mathcal{A}$ is deterministic, surely $\|\mathcal{A}'\| = \underline{L}$, while the automaton $\mathcal{A}'$ is clearly reversible with precisely one initial state i. As a result, it follows that $\underline{L} \in 1\mathrm{RevS}(S, \Sigma)$ and $L = \mathrm{supp}(\underline{L}) \in 1\mathrm{RevL}(S, \Sigma)$. $\square$

4   The Case of $\mathbb{F}_2$

We now explore in detail the *reversible weighted automata* over the *two-element field* $\mathbb{F}_2$. The field $\mathbb{F}_2$ is arguably the most important finite ring from the viewpoint of weighted automata theory, and weighted automata over $\mathbb{F}_2$ can also be seen as natural devices for describing languages. Indeed, any formal power series over $\mathbb{F}_2$ is a characteristic series of its support, and both can be identified as a result. The rational series over $\mathbb{F}_2$ can thus be seen as rational languages, and the realisation of a rational series by a weighted automaton over $\mathbb{F}_2$ corresponds to recognising a language in "parity mode", where a word w gets accepted when there is an odd number of successful runs on w in the automaton. In fact, the theory of series over $\mathbb{F}_2$ viewed as languages has largely been developed under the name *formal languages over* $\mathrm{GF}(2)$ [7,38], and weighted automata over $\mathbb{F}_2$ were also studied under the name *symmetric difference automata* [54,55].

Similar observations can be made in particular for reversible weighted automata over $\mathbb{F}_2$: the series from $\mathrm{RevS}(\mathbb{F}_2, \Sigma)$ might be, for any alphabet Σ, identified with languages from $\mathrm{RevL}(\mathbb{F}_2, \Sigma)$. It thus seems to be natural to ask about a characterisation of the language class $\mathrm{RevL}(\mathbb{F}_2)$ and about its relation to the usual class of reversible languages $\mathrm{RevL}(\mathbb{B})$. We prove in this section that $\mathrm{RevL}(\mathbb{F}_2)$ is in fact the *variety* generated by the positive variety $\mathrm{RevL}(\mathbb{B})$ – or equivalently, the *Boolean closure* of $\mathrm{RevL}(\mathbb{B})$.

Let us start by recording an easy observation that reversible weighted automata over $\mathbb{F}_2$ with *one initial state* realise precisely the characteristic series of languages recognised by reversible finite automata with one initial state.

Proposition 6. *Let Σ be an alphabet and $\mathcal{A}$ a reversible weighted automaton over $\mathbb{F}_2$ and Σ with one initial state. Then $\|\mathcal{A}\| = \underline{L}$ for some $L \in 1\mathrm{RevL}(\mathbb{B}, \Sigma)$ over $\mathbb{F}_2$. As a consequence, $1\mathrm{RevL}(\mathbb{F}_2, \Sigma) = 1\mathrm{RevL}(\mathbb{B}, \Sigma)$.*

Proof. Let $\mathcal{A} = (Q, \sigma, \iota, \tau)$ be a reversible weighted automaton with one initial state over $\mathbb{F}_2$ and Σ. As $\mathcal{A}$ is deterministic, the reversible finite automaton $\mathcal{A}' = (Q, \rightarrow, I, F)$ such that one has $p \overset{a}{\rightarrow} q$ for $p, q \in Q$ and $a \in \Sigma$ if and only if $\sigma(p, a, q) = 1$, while $I = \{p \in Q \mid \iota(p) = 1\}$ and $F = \{q \in Q \mid \tau(q) = 1\}$, clearly recognises $L = \mathrm{supp}(\|\mathcal{A}\|)$. Hence $L \in 1\mathrm{RevL}(\mathbb{B}, \Sigma)$ and $\|\mathcal{A}\| = \underline{L}$. The equality $1\mathrm{RevL}(\mathbb{F}_2, \Sigma) = 1\mathrm{RevL}(\mathbb{B}, \Sigma)$ then follows by Proposition 5. $\square$

Recall from [6] that the class of languages recognised by reversible finite automata with precisely one initial state is closed under intersection. Indeed, if $\mathcal{A}_1 = (Q_1, \rightarrow_1, \{i_1\}, F_1)$ and $\mathcal{A}_2 = (Q_2, \rightarrow_2, \{i_2\}, F_2)$ with $i_1 \in Q_1$ and $i_2 \in Q_2$ are reversible finite automata over an alphabet Σ, then the finite automaton $\mathcal{A} = (Q_1 \times Q_2, \rightarrow, \{(i_1, i_2)\}, F_1 \times F_2)$, with $\rightarrow$ given for all $p_1, q_1 \in Q_1, p_2, q_2 \in Q_2$, and $a \in \Sigma$ by $(p_1, p_2) \overset{a}{\rightarrow} (q_1, q_2)$ if and only if $p_1 \overset{a}{\rightarrow}_1 q_1$ and $p_2 \overset{a}{\rightarrow}_2 q_2$, is clearly reversible as well. Let us record this observation for later reference.

Proposition 7. (H. B. Axelsen, M. Holzer, and M. Kutrib [6]) *The set $1\mathrm{RevL}(\mathbb{B}, \Sigma)$ is closed under intersection for every alphabet Σ.*

We can now explore some basic closure properties of the class $\mathrm{RevL}(\mathbb{F}_2)$ of reversible languages over $\mathbb{F}_2$. First of all, Proposition 4 directly implies closure of $\mathrm{RevL}(\mathbb{F}_2)$ under symmetric difference. Somewhat more interesting is the closure of this class under all Boolean operations, which we now establish.

Proposition 8. *Let Σ be an alphabet. The set $\mathrm{RevL}(\mathbb{F}_2, \Sigma)$ is then closed under the Boolean operations.*

Proof. Let $L \in \mathrm{RevL}(\mathbb{F}_2, \Sigma)$. Then L is a support of some series in $\mathrm{RevS}(\mathbb{F}_2, \Sigma)$, which has to be the characteristic series $\underline{L}$ of L over $\mathbb{F}_2$. Thus $\underline{L} \in \mathrm{RevS}(\mathbb{F}_2, \Sigma)$. Moreover, it is easy to see that the characteristic series $\underline{\Sigma^*}$ of Σ^* over $\mathbb{F}_2$ is realised by a reversible weighted automaton over $\mathbb{F}_2$ and Σ with one state, so that $\underline{\Sigma^*} \in \mathrm{RevS}(\mathbb{F}_2, \Sigma)$. It thus follows by Proposition 4 that the series

$$\underline{L} + \underline{\Sigma^*} = \sum_{w \in \Sigma^*} \left((\underline{L}, w) + 1 \right) w$$

belongs to $\mathrm{RevS}(\mathbb{F}_2, \Sigma)$. As a result,

$$\mathrm{supp}(\underline{L} + \underline{\Sigma^*}) = \{ w \in \Sigma^* \mid (\underline{L}, w) + 1 \neq 0 \} = \{ w \in \Sigma^* \mid (\underline{L}, w) = 0 \} = \Sigma^* \setminus L$$

belongs to $\mathrm{RevL}(\mathbb{F}_2, \Sigma)$, and $\mathrm{RevL}(\mathbb{F}_2, \Sigma)$ is closed under complementation.

Next, let $L, K \in \mathrm{RevL}(\mathbb{F}_2, \Sigma)$. Again, this means that the characteristic series $\underline{L}, \underline{K}$ over $\mathbb{F}_2$ are in $\mathrm{RevS}(\mathbb{F}_2, \Sigma)$. By Proposition 3 and Proposition 6, there are $m, n \in \mathbb{N}$ and $L_1, \ldots, L_m, K_1, \ldots, K_n \in 1\mathrm{RevL}(\mathbb{B}, \Sigma)$ such that over $\mathbb{F}_2$, one has $\underline{L} = \underline{L_1} + \ldots + \underline{L_m}$ and $\underline{K} = \underline{K_1} + \ldots + \underline{K_n}$. Now, for every $w \in \Sigma^*$,

$$\begin{aligned}
(\underline{L \cap K}, w) &= (\underline{L_1} + \ldots + \underline{L_m}, w) \cdot (\underline{K_1} + \ldots + \underline{K_n}, w) \\
&= \left((\underline{L_1}, w) + \ldots + (\underline{L_m}, w) \right) \cdot \left((\underline{K_1}, w) + \ldots + (\underline{K_n}, w) \right) \\
&= \sum_{i=1}^{m} \sum_{j=1}^{n} (\underline{L_i}, w)(\underline{K_j}, w) = \sum_{i=1}^{m} \sum_{j=1}^{n} (\underline{L_i \cap K_j}, w)
\end{aligned}$$

over $\mathbb{F}_2$, so that

$$\underline{L \cap K} = \sum_{i=1}^{m} \sum_{j=1}^{n} \underline{L_i \cap K_j}.$$

As $L_i \cap K_j \in 1\mathrm{RevL}(\mathbb{B}, \Sigma)$ by Proposition 7 for $i = 1, \ldots, m$ and $j = 1, \ldots, n$, it follows by Proposition 5 that $\underline{L_i \cap K_j} \in 1\mathrm{RevS}(\mathbb{F}_2, \Sigma)$, and as a consequence, $\underline{L \cap K} \in \mathrm{RevS}(\mathbb{F}_2, \Sigma)$ by Proposition 3. As a result, $\mathrm{supp}(\underline{L \cap K}) = L \cap K$ has to be in $\mathrm{RevL}(\mathbb{F}_2, \Sigma)$, which is thus closed under intersection.

Finally, the closure of $\mathrm{RevL}(\mathbb{F}_2, \Sigma)$ under union follows by its closure under intersection and under complementation. $\qquad\square$

As a first step towards a characterisation of the class $\mathrm{RevL}(\mathbb{F}_2)$ of reversible languages over $\mathbb{F}_2$, let us prove that this class contains all reversible languages in the usual sense – that is, all languages from $\mathrm{RevL}(\mathbb{B})$.

Lemma 9. *Let Σ be an alphabet. Then* $\mathrm{RevL}(\mathbb{B}, \Sigma) \subseteq \mathrm{RevL}(\mathbb{F}_2, \Sigma)$.

Proof. Let $L \in \mathrm{RevL}(\mathbb{B}, \Sigma)$. Then, essentially by Proposition 3 applied in the case of the Boolean semiring, $L = L_1 \cup \ldots \cup L_n$ for some $n \in \mathbb{N}$ and languages $L_1, \ldots, L_n \in 1\mathrm{RevL}(\mathbb{B}, \Sigma)$. As $1\mathrm{RevL}(\mathbb{B}, \Sigma) = 1\mathrm{RevL}(\mathbb{F}_2, \Sigma)$ by Proposition 6 and as $1\mathrm{RevL}(\mathbb{F}_2, \Sigma) \subseteq \mathrm{RevL}(\mathbb{F}_2, \Sigma)$, we actually have $L_1, \ldots, L_n \in \mathrm{RevL}(\mathbb{F}_2, \Sigma)$. Thus $L = L_1 \cup \ldots \cup L_n$ has to be in $\mathrm{RevL}(\mathbb{F}_2, \Sigma)$ by Proposition 8. $\qquad\square$

We are now finally ready to prove the characterisation of the class $\mathrm{RevL}(\mathbb{F}_2)$ as the Boolean closure of the positive variety of reversible languages $\mathrm{RevL}(\mathbb{B})$.

Theorem 10. *The class* $\mathrm{RevL}(\mathbb{F}_2)$ *is the Boolean closure of* $\mathrm{RevL}(\mathbb{B})$.

Proof. The Boolean closure $\mathbf{B}(\mathrm{RevL}(\mathbb{B}))$ of $\mathrm{RevL}(\mathbb{B})$ is included in $\mathrm{RevL}(\mathbb{F}_2)$ by Lemma 9 and Proposition 8. For the converse, let $L \in \mathrm{RevL}(\mathbb{F}_2, \Sigma)$ for some alphabet Σ. Proposition 3 gives us $n \in \mathbb{N}$ and $r_1, \ldots, r_n \in 1\mathrm{RevS}(\mathbb{F}_2, \Sigma)$ such that $L = \mathrm{supp}(r_1 + \ldots + r_n)$. We prove that L is in $\mathbf{B}(\mathrm{RevL}(\mathbb{B}))$ by induction on n. If $n = 0$, then $L = \mathrm{supp}(0) = \emptyset$ is in $\mathbf{B}(\mathrm{RevL}(\mathbb{B}))$. Now, for any $k \in \mathbb{N}$, $n = k+1$, and $L_1 := \mathrm{supp}(r_1 + \ldots + r_k)$ in $\mathbf{B}(\mathrm{RevL}(\mathbb{B}))$, we see that $L_2 := \mathrm{supp}(r_{k+1})$ is in $\mathbf{B}(\mathrm{RevL}(\mathbb{B}))$, as it in fact belongs to $1\mathrm{RevL}(\mathbb{B}, \Sigma)$ by Proposition 6. Thus

$$L = \mathrm{supp}((r_1 + \ldots + r_k) + r_{k+1}) = (L_1 \cup L_2) \cap (\Sigma^* \setminus (L_1 \cap L_2))$$

belongs to $\mathbf{B}(\mathrm{RevL}(\mathbb{B}))$ as well. $\qquad\square$

The class $\mathrm{RevL}(\mathbb{F}_2)$ of all reversible languages over the field $\mathbb{F}_2$ thus forms a *variety of languages* given by the Boolean closure of the positive variety $\mathrm{RevL}(\mathbb{B})$ of the usual reversible languages in the sense of J.-É. Pin [43]. In other words, $\mathrm{RevL}(\mathbb{F}_2)$ is the *variety generated by* $\mathrm{RevL}(\mathbb{B})$. We thus see that introducing weights from $\mathbb{F}_2$ to reversible finite automata leads to describing a larger class of languages with better closure properties.

In fact, the Boolean closure of the positive variety of reversible languages $\mathrm{RevL}(\mathbb{B})$ is known to correspond, via Eilenberg's correspondence, to the pseudovariety $\mathbf{ECom} = \mathbf{J}_1 * \mathbf{G}$ of all finite monoids with commuting idempotents [4,18,19,37,42]. We thus obtain the following corollary, which can be seen both as a characterisation of the reversible languages over $\mathbb{F}_2$, as well as a new automata-theoretic characterisation of the pseudovariety $\mathbf{ECom}$.

Corollary 11. *Let Σ be an alphabet a $L \subseteq \Sigma^*$ a language. Then L belongs to* $\mathrm{RevL}(\mathbb{F}_2)$ *if and only if $ef = fe$ for any two idempotents e, f in the syntactic monoid M_L of L over Σ^*. Thus $L \in \mathrm{RevL}(\mathbb{F}_2)$ if and only if $M_L \in \mathbf{ECom}$.*

The pseudovariety $\mathbf{ECom}$ can also be described using the pseudoidentity $x^\omega y^\omega = y^\omega x^\omega$, which directly captures the property of commuting idempotents. In any case, the membership of a finite monoid in $\mathbf{ECom}$ is clearly decidable, which implies decidability of the membership of a rational language in $\mathrm{RevL}(\mathbb{F}_2)$.

5   Automata over Locally Finite Commutative Rings

In what follows, we show that the results just obtained for reversible weighted automata over $\mathbb{F}_2$ can actually be generalised to weighted automata over any nontrivial locally finite commutative ring.

We prove $\mathrm{RevL}(R) = \mathrm{RevL}(\mathbb{F}_2)$ for any such ring R, and in order to establish one of the inclusions between these two classes, we show that any characteristic series of a language from $\mathrm{RevL}(\mathbb{F}_2)$ over R can be realised by a reversible weighted automaton over R. In fact, local finiteness and commutativity of R are not necessary here.

Lemma 12. *Let R be a nontrivial ring and Σ an alphabet. Then for every $L \in \mathrm{RevL}(\mathbb{F}_2, \Sigma)$, the characteristic series $\underline{L}$ of L over R is in $\mathrm{RevS}(R, \Sigma)$, and L is in $\mathrm{RevL}(R, \Sigma)$.*

Proof. Let $L \in \mathrm{RevL}(\mathbb{F}_2, \Sigma)$; it suffices to show that $\underline{L} \in \mathrm{RevS}(R, \Sigma)$ over R. As $L \in \mathrm{RevL}(\mathbb{F}_2, \Sigma)$, it follows by Proposition 3 and by Proposition 6 that there exists some $n \in \mathbb{N}$ and languages $L_1, \dots, L_n \in 1\mathrm{RevL}(\mathbb{B}, \Sigma)$ such that $L = \mathrm{supp}(\underline{L_1} + \dots + \underline{L_n})$ over $\mathbb{F}_2$. This means that $w \in \Sigma^*$ is in L if and only if the set $X_w = \{i \in [n] \mid w \in L_i\}$ contains an odd number of elements. Thus the coefficients of the characteristic series $\underline{L}$ of L over R at $w \in \Sigma^*$ are given by

$$(\underline{L}, w) = \begin{cases} 1 & \text{if } |X_w| \text{ is odd,} \\ 0 & \text{if } |X_w| \text{ is even.} \end{cases} \tag{1}$$

With $-2 := -(1+1)$ in R, let us now consider the series $r \in R\langle\!\langle \Sigma^* \rangle\!\rangle$ defined by

$$r = \sum_{\emptyset \subsetneq X \subseteq [n]} (-2)^{|X|-1} \bigcap_{i \in X} \underline{L_i}$$

(in particular, note that we have $-2 = 0$ in case R is a ring of characteristic 2). Then $r \in \mathrm{RevS}(R, \Sigma)$ by virtue of Proposition 7, Proposition 5, Proposition 3, and Proposition 4. In order to prove that $\underline{L} \in \mathrm{RevS}(R, \Sigma)$, we now show that actually $r = \underline{L}$. Indeed, for $\emptyset \subsetneq X \subseteq [n]$ fixed and any $w \in \Sigma^*$, we have $\left(\bigcap_{i \in X} \underline{L_i}, w \right) = 1$ if $X \subseteq X_w$ and $\left(\bigcap_{i \in X} \underline{L_i}, w \right) = 0$ otherwise, while there are exactly $\binom{|X_w|}{k}$ sets $X \subseteq X_w$ of each size $k = 1, \dots, |X_w|$. As a result,

$$(r, w) = \sum_{k=1}^{|X_w|} \binom{|X_w|}{k} (-2)^{k-1}$$

for each $w \in \Sigma^*$ over R. However, over $\mathbb{Q}$ it holds that

$$\sum_{k=1}^{|X_w|} \binom{|X_w|}{k} (-2)^{k-1} = -\frac{1}{2} \left(\sum_{k=0}^{|X_w|} \binom{|X_w|}{k} (-2)^k - 1 \right) = \frac{1 - (-1)^{|X_w|}}{2}$$

by the Binomial theorem, which means that over $\mathbb{Z}$ we obtain

$$\sum_{k=1}^{|X_w|} \binom{|X_w|}{k}(-2)^{k-1} = \begin{cases} 1 & \text{if } |X_w| \text{ is odd,} \\ 0 & \text{if } |X_w| \text{ is even.} \end{cases}$$

Now, by applying the unique ring homomorphism from $\mathbb{Z}$ to R to both sides, we see that also over R we have

$$(r, w) = \sum_{k=1}^{|X_w|} \binom{|X_w|}{k}(-2)^{k-1} = \begin{cases} 1 & \text{if } |X_w| \text{ is odd,} \\ 0 & \text{if } |X_w| \text{ is even.} \end{cases}$$

Thus $r = \underline{L}$ by (1). $\qquad\square$

We now essentially establish the remaining inclusion in case the nontrivial ring R is *locally finite* and *commutative*.

Lemma 13. *Let R be a nontrivial locally finite commutative ring, Σ an alphabet, $L \in \mathrm{RevL}(R, \Sigma)$ a language, and M_L the syntactic monoid of L over Σ^*. Then the idempotents of M_L commute, i.e., $M_L \in \mathbf{ECom}$.*

Proof. As **ECom** is a pseudovariety of finite monoids and M_L divides the transition monoid of any deterministic finite automaton recognising L, it suffices to describe *some* deterministic finite automaton $\mathcal{D}$ over Σ recognising L such that the idempotents in the transition monoid of $\mathcal{D}$ commute.

As $L \in \mathrm{RevL}(R, \Sigma)$, there is a reversible weighted automaton $\mathcal{A}$ over R and Σ with state set $[n]$ for some $n \in \mathbb{N}$ such that $L = \mathrm{supp}(\|\mathcal{A}\|)$; let $\mathcal{P}_{\mathcal{A}} = (n, \mathbf{i}, \mu, \mathbf{f})$ be its linear representation. For $k = 1, \ldots, n$, let $\mathbf{e}_k = (a_1, \ldots, a_n) \in R^n$ be a vector with $a_k = 1$ and $a_j = 0$ for all $j \in [n] \setminus \{k\}$. As $\mathcal{A}$ always has to be over some finitely generated subring of R, we may assume that R is actually finite.

We may thus take $\mathcal{D} = (R^n, \cdot, \mathbf{i}, F)$, where $F = \{\mathbf{v} \in R^n \mid \mathbf{vf} \neq 0\}$, and $\mathbf{v} \cdot a = \mathbf{v}\mu(a)$ for all $\mathbf{v} \in R^n$ and $a \in \Sigma$. The automaton $\mathcal{D}$ then clearly recognises the language $L = \mathrm{supp}(\|\mathcal{A}\|)$ and its transition monoid can be represented as a monoid of all matrices $\mu(w)$ for $w \in \Sigma^*$ with matrix multiplication over R. By reversibility of $\mathcal{A}$, each of the matrices $\mu(a)$ for $a \in \Sigma$ contains at most one nonzero element in each row and column. The same property thus holds for all matrices $\mu(w)$ for $w \in \Sigma^*$. This means that if $\mu(w) = (a_{i,j})_{n \times n}$ is an idempotent, then $\mu(w)$ is a diagonal matrix, as $a_{i,j} \neq 0$ for some $i \neq j$ would imply $\mathbf{e}_i\mu(w) = a_{i,j}\mathbf{e}_j$ and $\mathbf{e}_i\mu(w)^2 = a_{i,j}\mathbf{e}_j\mu(w) \neq a_{i,j}\mathbf{e}_j$, as otherwise $a_{j,j}$ would have to be nonzero and the j-th column of $\mu(w)$ would contain two nonzero elements. This would mean $\mu(w)^2 \neq \mu(w)$ and $\mu(w)$ would not be idempotent. Any idempotent $\mu(w)$ is thus indeed a diagonal matrix, while diagonal matrices over commutative rings commute under matrix multiplication. $\qquad\square$

Theorem 14. *The class $\mathrm{RevL}(R)$ is the Boolean closure of $\mathrm{RevL}(\mathbb{B})$ for every nontrivial locally finite commutative ring R.*

Proof. Follows by Lemma 12, Lemma 13, and Theorem 10 coupled with the correspondence of the Boolean closure of $\mathrm{RevL}(\mathbb{B})$ to **ECom**. $\qquad\square$

Let us finally establish decidability of the existence of a reversible equivalent for weighted automata over effective locally finite commutative rings.

Proposition 15. *Let R be a nontrivial finite commutative ring, Σ an alphabet, and $r \in R\langle\!\langle \Sigma^* \rangle\!\rangle$ a series rational over R. Then $r \in \mathrm{RevS}(R, \Sigma)$ if and only if $\mathrm{supp}(r + x \cdot \underline{\Sigma^*}) \in \mathrm{RevL}(R, \Sigma)$ for all $x \in R$.*

Proof. If $r \in \mathrm{RevS}(R, \Sigma)$, then $r + x \cdot \underline{\Sigma^*} \in \mathrm{RevS}(R, \Sigma)$ for all $x \in R$ as well, hence $\mathrm{supp}(r + x \cdot \underline{\Sigma^*}) \in \mathrm{RevL}(R, \Sigma)$. If on the other hand $\mathrm{supp}(r + x \cdot \underline{\Sigma^*}) \in \mathrm{RevL}(R, \Sigma)$ for all $x \in R$, we see that $r = \sum_{x \in R} x \cdot (\underline{\Sigma^*} \setminus \mathrm{supp}(r - x \cdot \underline{\Sigma^*})) \in \mathrm{RevS}(R, \Sigma)$ by Theorem 14, Lemma 12, and Proposition 4 (observe that we have just expressed r as a specific recognisable step function [13, 14]). $\qquad\square$

Corollary 16. *The problem of checking reversibility of a rational series over an effective locally finite commutative ring R is decidable.*

Proof. If R is nontrivial and $\mathcal{A}$ is a weighted automaton over R, then $r = \|\mathcal{A}\|$ is a series over some finitely generated subring T of R, which is finite by local finiteness of R. Provided R is effective, one can compute all elements $x \in T$ and the syntactic monoids of the languages $\mathrm{supp}(r + x \cdot \underline{\Sigma^*})$. In case all these monoids turn out to be in **ECom**, we obtain $r \in \mathrm{RevS}(T, \Sigma)$ by Theorem 14 and Proposition 15; hence also $r \in \mathrm{RevS}(R, \Sigma)$. If at least one of these monoids is not in **ECom**, then Proposition 15 gives us $r \notin \mathrm{RevS}(T', \Sigma)$ for all finite subrings T' of R containing T. As R is locally finite, this implies $r \notin \mathrm{RevS}(R, \Sigma)$. $\qquad\square$

Acknowledgments. We would like to thank the anonymous reviewers for their careful reading of our manuscript and valuable comments.

References

1. Almeida, J.: Finite Semigroups and Universal Algebra. World Scientific (1994)
2. Ambainis, A., Freivalds, R.M.: 1-way quantum finite automata: strengths, weaknesses and generalizations. In: Foundations of Computer Science. FOCS 1998, pp. 332–341 (1998)
3. Angluin, D.: Inference of reversible languages. J. ACM **29**(3), 741–765 (1982)
4. Ash, C.J.: Finite semigroups with commuting idempotents. J. Aust. Math. Soc. (Ser. A) **43**(1), 81–90 (1987)
5. Axelsen, H.B., Holzer, M., Kutrib, M.: The degree of irreversibility in deterministic finite automata. In: Implementation and Application of Automata, CIAA, pp. 15–26 (2016)
6. Axelsen, H.B., Holzer, M., Kutrib, M.: The degree of irreversibility in deterministic finite automata. Int. J. Found. Comput. Sci. **28**(5), 503–522 (2017)
7. Bakinova, E., Basharin, A., Batmanov, I., Lyubort, K., Okhotin, A., Sazhneva, E.: Formal languages over GF(2). Inf. Comput. **283**, 104672 (2022)
8. Bala, S.: Which finitely ambiguous automata recognize finitely sequential functions? In: Mathematical Foundations of Computer Science, MFCS 2013, pp. 86–97 (2013)

9. Bell, J.P., Smertnig, D.: Computing the linear hull: deciding Deterministic? and Unambiguous? for weighted automata over fields. In: Logic in Computer Science, LICS 2023 (2023)

10. Benalioua, Y.I., Lhote, N., Reynier, P.A.: Minimizing cost register automata over a field. In: Mathematical Foundations of Computer Science, MFCS 2024, article 23 (2024)

11. Bennett, C.H.: Logical reversibility of computation. IBM J. Res. Dev. **17**(6), 525–532 (1973)

12. Berstel, J., Reutenauer, C.: Noncommutative Rational Series with Applications. Cambridge University Press, Cambridge (2011)

13. Droste, M., Gastin, P.: Weighted automata and weighted logics. Theoret. Comput. Sci. **380**(1–2), 69–86 (2007)

14. Droste, M., Gastin, P.: Weighted automata and weighted logics. In: Droste, M., Kuich, W., Vogler, H. (eds.) Handbook of Weighted Automata, chap. 5, pp. 175–211. Springer, Heidelberg (2009). https://doi.org/10.1007/978-3-642-01492-5_5

15. Droste, M., Kuich, W.: Handbook of Weighted Automata. Springer, Heidelberg (2009). https://doi.org/10.1007/978-3-642-01492-5

16. Droste, M., Kuske, D.: Weighted automata. In: Pin, J.É. (ed.) Handbook of Automata Theory, vol. 1, chap. 4, pp. 113–150. European Mathematical Society (2021)

17. García, P., de Parga, M.V., Cano, A., López, D.: On locally reversible languages. Theoret. Comput. Sci. **410**, 4961–4974 (2009)

18. Golovkins, M., Pin, J.É.: Varieties generated by certain models of reversible finite automata. In: Computing and Combinatorics, COCOON 2006, pp. 83–93 (2006)

19. Golovkins, M., Pin, J.É.: Varieties generated by certain models of reversible finite automata. Chicago J. Theoret. Comput. Sci. **2010**, article 2 (2010)

20. Guillon, B., Lavado, G.J., Pighizzini, G., Prigioniero, L.: Weakly and strongly irreversible regular languages. Int. J. Found. Comput. Sci. **33**(3–4), 263–284 (2022)

21. Holzer, M., Jakobi, S., Kutrib, M.: Minimal reversible deterministic finite automata. In: Developments in Language Theory, DLT 2015, pp. 276–287 (2015)

22. Holzer, M., Jakobi, S., Kutrib, M.: Minimal reversible deterministic finite automata. Int. J. Found. Comput. Sci. **29**(2), 251–270 (2018)

23. Holzer, M., Kutrib, M.: Reversible nondeterministic finite automata. In: Reversible Computation, RC 2017, pp. 35–51 (2017)

24. Jecker, I., Mazowiecki, F., Purser, D.: Determinisation and unambiguisation of polynomially-ambiguous rational weighted automata. In: Logic in Computer Science, LICS 2024 (2024)

25. Kiefer, S., Widdershoven, C.: Image-binary automata. In: Descriptional Complexity of Formal Systems, DCFS 2021, pp. 176–187 (2021)

26. Kiefer, S., Widdershoven, C.: Image-binary automata. Int. J. Found. Comput. Sci. (2024, to appear)

27. Klíma, O., Polák, L.: Forbidden patterns for ordered automata. In: Non-classical Models of Automata and Applications, NCMA 2018, pp. 99–115 (2018)

28. Klíma, O., Polák, L.: Forbidden patterns for ordered automata. J. Autom. Lang. Comb. **25**(2–3), 141–169 (2020)

29. Kostolányi, P.: Bideterministic weighted automata. In: Algebraic Informatics, CAI 2022, pp. 161–174 (2022)

30. Kostolányi, P.: Determinisability of unary weighted automata over the rational numbers. Theoret. Comput. Sci. **898**, 110–131 (2022)

31. Kostolányi, P.: Finite ambiguity and finite sequentiality in weighted automata over fields. In: Computer Science - Theory and Applications, CSR 2022, pp. 209–223 (2022)
32. Kostolányi, P.: Bideterministic weighted automata. Inf. Comput. **295B**, 105093 (2023)
33. Kostolányi, P.: Finitely ambiguous and finitely sequential weighted automata over fields. Theoret. Comput. Sci. **1012**, 114725 (2024)
34. Landauer, R.: Irreversibility and heat generation in the computing process. IBM J. Res. Dev. **5**(3), 183–191 (1961)
35. Lombardy, S.: On the construction of reversible automata for reversible languages. In: Automata, Languages and Programming, ICALP 2002, pp. 170–182 (2002)
36. Lombardy, S., Sakarovitch, J.: Sequential? Theoret. Comput. Sci. **356**, 224–244 (2006)
37. Margolis, S.W., Pin, J.É.: Inverse semigroups and varieties of finite semigroups. J. Algebra **110**(2), 306–323 (1987)
38. Okhotin, A., Radionova, M., Sazhneva, E.: GF(2)-operations on basic families of formal languages. Theoret. Comput. Sci. **995**, 114489 (2024)
39. Paul, E.: Finite sequentiality of unambiguous max-plus tree automata. In: Symposium on Theoretical Aspects of Computer Science, STACS 2019, pp. 55:1–55:17 (2019)
40. Paul, E.: Finite sequentiality of finitely ambiguous max-plus tree automata. In: Automata, Languages and Programming, ICALP 2020, pp. 137:1–137:15 (2020)
41. Paul, E.: Finite sequentiality of unambiguous max-plus tree automata. Theory Comput. Syst. **65**(4), 736–776 (2021)
42. Pin, J.É.: Hiérarchies de concaténation. RAIRO. Informatique théorique **18**(1), 23–46 (1984)
43. Pin, J.É.: On reversible automata. In: Latin American Symposium on Theoretical Informatics, LATIN 1992, pp. 401–416 (1992)
44. Pin, J.-E.: Syntactic Semigroups. In: Rozenberg, G., Salomaa, A. (eds.) Handbook of Formal Languages, pp. 679–746. Springer, Heidelberg (1997). https://doi.org/10.1007/978-3-642-59136-5_10
45. Radionova, M., Okhotin, A.: Decision problems for reversible and permutation automata. In: Implementation and Application of Automata, CIAA 2024, pp. 302–315 (2024)
46. Radionova, M., Okhotin, A.: A hierarchy of reversible finite automata. In: Implementation and Application of Automata, CIAA 2025, pp. 316–329 (2025)
47. Sakarovitch, J.: Elements of Automata Theory. Cambridge University Press, Cambridge (2009)
48. Salomaa, A., Soittola, M.: Automata-Theoretic Aspects of Formal Power Series. Springer, Heidelberg (1978). https://doi.org/10.1007/978-1-4612-6264-0
49. Schützenberger, M.P.: On the definition of a family of automata. Inf. Control **4**(2–3), 245–270 (1961)
50. Straubing, H., Weil, P.: Varieties. In: Pin, J.É. (ed.) Handbook of Automata Theory, vol. 1, chap. 16, pp. 569–614. European Mathematical Society (2021)
51. Tamm, H.: On transition minimality of bideterministic automata. Int. J. Found. Comput. Sci. **19**(3), 677–690 (2008)
52. Tamm, H., Ukkonen, E.: Bideterministic automata and minimal representations of regular languages. In: Implementation and Application of Automata, CIAA 2003, pp. 61–71 (2003)
53. Tamm, H., Ukkonen, E.: Bideterministic automata and minimal representations of regular languages. Theoret. Comput. Sci. **328**(1–2), 135–149 (2004)

54. van der Merwe, B., Tamm, H., van Zijl, L.: Minimal DFA for symmetric difference NFA. In: Descriptional Complexity of Formal Systems, DCFS 2012, pp. 307–318 (2012)
55. van Zijl, L.: On binary $\oplus$-NFAs and succinct descriptions of regular languages. Theoret. Comput. Sci. **328**(1–2), 161–170 (2004)

Generalized Wheeler Automata: Minimality

Nicola Cotumaccio$^{(\boxtimes)}$ (iD)

Department of Computer Science, University of Helsinki, Helsinki, Finland
`nicola.cotumaccio@helsinki.fi`

Abstract. *Wheeler automata* are a well-studied extension of the suffix array, the Burrows-Wheeler transform and the FM-index from strings to automata. The corresponding class of languages has proved to be one of the richest and most elegant subclasses of regular languages. In particular, every Wheeler language is recognized by a unique *minimal* Wheeler automaton, which in general is larger than the standard minimal automaton (Alanko et al., SODA 2020). In the seventies, Eilenberg introduced *generalized automata*, an extension of conventional automata in which edges can be labeled with strings. In general, a minimal generalized automaton is not unique (Giammarresi and Montalbano, STACS 1995), but the uniqueness can be retrieved through an appropriate parameterization (Cotumaccio, STACS 2024). Building on this parameterization, it is possible to combine Wheeler automata and generalized automata, thus obtaining *Wheeler generalized automata*. In this paper, we address a natural question: can we combine the previous results? Is there a unique minimal Wheeler generalized automaton? While we give an affirmative answer, we cannot naively combine the previous constructions. Unlike before, we define the unique minimal Wheeler generalized automaton using an algebraic construction somewhat similar to that used for *bisimulations*. More precisely, we can follow a bottom-up approach based on the *join* of equivalence relations, or we can follow a top-down approach based on an infinite *refinement* procedure. As a corollary, we show that, while being more expressive than conventional Wheeler automata, generalized Wheeler automata do not recognize every regular language.

Keywords: Wheeler automata · Minimal automaton · Myhill-Nerode equivalence

1   Introduction

Wheeler automata have received considerable attention in the literature in the last five years (see [19] for a recent survey). While originally introduced as a unifying framework in data compression [27], Wheeler automata have proved to be surprisingly elegant from a formal language theory perspective: the class of all languages recognized by some Wheeler automata (namely, the class of *Wheeler languages*) is one of the richest and most elegant subclasses of regular

M.-P. Béal and P. Caron (Eds.): DLT 2026, LNCS 16578, pp. 88–102, 2026.
https://doi.org/10.1007/978-3-032-28404-4_7

languages [2]. In particular, many properties of Wheeler languages mimic textbook properties of regular languages: non-determinism and determinism have the same expressive power, every Wheeler language is recognized by a unique minimal Wheeler deterministic automaton (which, in general, can be exponentially larger than the standard minimal deterministic automaton), and Wheeler languages admit a Myhill-Nerode theorem [3]. The state complexity of Wheeler languages has been investigated [23], as well as their relationship with locally testable languages [7]. In data compression, Wheeler automata provide an extension of the suffix array [33], the Burrows-Wheeler transform (BWT) [11] and the FM-index [26] from strings to automata [1,4,12,21,27]. Moreover, Wheeler automata subsume de Bruijn graphs, which are popular in bioinformatics to perform Eulerian sequence assembly [10], and generalize both the XBWT [25] and the eBWT [17,18,30,34].

Wheeler automata are a special case of conventional automata, but the same paradigm has been extended to other automata-based formalisms, such as *generalized automata* (which were first introduced by Eilenberger in 1974) [24]. In generalized automata, edges can be labeled with strings (and not only characters), see Fig. 1. Generalized automata have the same expressive power as conventional automata (that is, generalized automata recognize regular languages), but may represent a regular language more compactly (that is, using fewer states). However, in general a minimal generalized automaton is not unique. For example, every generalized deterministic automaton recognizing the regular language $\mathcal{L} = (a|b)a^2(aba|ba^2)^*a^2$ must have at least three states, and $\mathcal{L}$ is recognized by two non-isomorphic generalized deterministic automata having exactly three states [28,29]. Nonetheless, the uniqueness can be retrieved through an appropriate parameterization (see Fact 3 in Sect. 3). The parameter is a set of strings called $\mathcal{W}$ (in particular, the two non-isomorphic generalized deterministic automata with three states that recognize $\mathcal{L} = (b|a)a^2(ba^2|aba)^*a^2$ have distinct $\mathcal{W}$'s). The parameter $\mathcal{W}$ can also be used to extend Wheeler automata to the setting of generalized automata [15]. While generalized automata have the same expressive power as conventional automata, Wheeler generalized automata are more expressive than conventional Wheeler automata. For example, the regular language $\mathcal{L} = a(aa)^*$ is not recognized by any Wheeler automaton, but it is recognized by a Wheeler generalized automaton (see Sect. 2). This is useful because conventional Wheeler languages identify only a small subclass of regular languages: every Wheeler language must be star-free [3], and $\mathcal{L} = a(aa)^*$ is a classic example of a regular language that is not star-free [38].

1.1   Our Contribution

We have mentioned three settings where a minimal automaton is unique up to isomorphism: (i) conventional deterministic automata, (ii) Wheeler deterministic automata, and (iii) generalized deterministic automata (via the $\mathcal{W}$-parameterization), see Sect. 3 for details. In this paper, we address a natural question: can we combine the previous results? Is there a unique minimal

Wheeler generalized automaton? While we give an affirmative answer (Theorem 3), we cannot naively combine the previous constructions. Unlike before, we define the unique minimal Wheeler generalized automaton using an algebraic construction somewhat similar to that used for *bisimulations* (see, e.g., [5,32]). More precisely, we can follow a bottom-up approach based on the *join* of equivalence relations (Theorem 1), or we can follow a top-down approach based on an infinite *refinement* procedure (Theorem 2). Our main result (Theorem 3) not only shows that there exists a unique minimal Wheeler generalized automaton, but it also provides a full *Myhill-Nerode theorem* [35,36]. As a corollary, we show that, while being more expressive than conventional Wheeler automata, generalized Wheeler automata do not recognize every regular language (Example 1).

The paper is organized as follows. In Sect. 2, we discuss preliminaries. In Sect. 3, we outline our plan to prove Theorem 3. In Sect. 4, we present the bottom-up approach, and in Sect. 5, we present the top-down approach. In Sect. 6, we describe Theorem 3. In Sect. 7, we discuss our conclusions. Some proofs and some auxiliary results are deferred to the extended version of the paper.

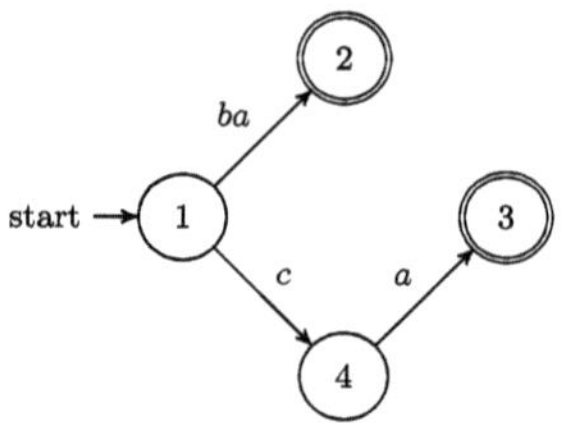

Fig. 1. A generalized automaton.

2   Preliminaries

Let us present the notation and the basic definitions used in the paper.

2.1   Relations

Let X be a set. A binary relation $\leq$ on X is a *partial order* if it is reflexive, antisymmetric and transitive. For every $x, y \in X$, we write $x < y$ if $(x \leq y) \wedge (x \neq y)$. The partial order $\leq$ is a *total order* if we have $(x \leq y) \vee (y \leq x)$ for every $x, y \in X$.

Let X be a set. A binary relation $\sim$ on X is an *equivalence relation* if it is reflexive, symmetric and transitive. For every $x \in X$, let $[x]_\sim = \{y \in X \mid x \sim y\}$. We say that $[x]_\sim$ is a $\sim$-*equivalence class*. The *index* of $\sim$ is the number of $\sim$-equivalence classes.

Let $\sim$ and $\sim'$ equivalence relation on X. We say that $\sim'$ *refines* $\sim$ if for every $x, y \in X$ such that $x \sim' y$ we have $x \sim y$. Equivalently, we say that $\sim'$ is *finer* than $\sim$ and $\sim$ is *coarser* than $\sim'$.

We denote by id_X the *finest* equivalence relation on X, that is, the equivalence relation on X whose equivalence classes are singletons.

2.2   Strings

Fix a finite alphabet Σ, and let Σ^* the set of all finite strings on Σ. We denote by ϵ the empty string and by Σ^+ the set $\Sigma^* \setminus \{\epsilon\}$ of all nonempty finite strings on Σ. If $\mathcal{L} \subseteq \Sigma^*$, let $\mathrm{Pref}(\mathcal{L})$ be the set of all prefixes of some string in $\mathcal{L}$. Note that if $\mathcal{L} \neq \emptyset$, then $\epsilon \in \mathrm{Pref}(\mathcal{L})$. We say that $\mathcal{L} \subseteq \Sigma^*$ is *prefix-free* if no string in $\mathcal{L}$ is a strict prefix of another string in $\mathcal{L}$. Note that if $\mathcal{L}$ is prefix-free and $\epsilon \in \mathcal{L}$, then $\mathcal{L} = \{\epsilon\}$. If $\mathcal{L} \subseteq \Sigma^*$, the *prefix-free kernel* of $\mathcal{L}$ is the set $\mathcal{K}(\mathcal{L})$ of all strings in $\mathcal{L}$ whose strict prefixes are all not in $\mathcal{L}$. Note that $\mathcal{K}(\mathcal{L})$ is always prefix-free, and $\mathcal{L}$ is prefix-free if and only if $\mathcal{L} = \mathcal{K}(\mathcal{L})$. If $\mathcal{L} \subseteq \Sigma^*$ and $\alpha \in \mathcal{L}$, let $\mathcal{L}_\alpha = \{\rho \in \Sigma^+ \mid \alpha\rho \in \mathcal{L}\}$. If $\mathcal{L} \subseteq \Sigma^*$ and $\mathcal{K}(\mathcal{L}_\alpha)$ is finite for every $\alpha \in \mathcal{L}$, then we say that $\mathcal{L}$ is *locally bounded*.

As customary in the literature on Wheeler automata, we fix a total order $\preceq$ on Σ, and we extend $\preceq$ to Σ^* *co-lexicographically* (that is, for every $\alpha, \beta \in \Sigma^*$, we have $\alpha \preceq \beta$ if and only if the reverse string α^R is lexicographically smaller than the reverse string β^R).

In our examples, Σ is the English alphabet, and $\preceq$ is the usual total order on Σ such that $a \prec b$, $b \prec c$, $c \prec d$, and so on. This implies, e.g., *caba* $\prec$ *abba* because *caba* is co-lexicographically smaller than *abba*.

2.3   Automata

A *generalized deterministic finite automaton (GDFA)* [28] is a 4-tuple $\mathcal{A} = (Q, E, s, F)$, where Q is a finite set of *states*, $E \subseteq Q \times Q \times \Sigma^*$ is a finite set of string-labeled *edges*, $s \in Q$ is the *initial state* and $F \subseteq Q$ is a set of *final states* (see Fig. 1). The 4-ple $\mathcal{A} = (Q, E, s, F)$ must satisfy the following properties:

- For every $u \in Q$, (i) no edge leaving u is labeled with ϵ, (ii) distinct edges leaving u have distinct labels, and (iii) the set of all strings labeling some edge leaving u is prefix-free (*determinism*).
- For every $u \in Q$, (i) u is reachable from the initial state and (ii) u is co-reachable, that is, u is either final or allows reaching a final state. (*reachability* and *co-reachability*).

Intuitively, we cannot define determinism by only imposing that distinct edges leaving the same state have distinct labels, because otherwise there may be two distinct ways of reading the same string starting from the initial state. Reachability and co-reachability are standard assumptions in automata theory, because if a state is not reachable or it is not co-reachable, then it can be removed without changing the recognized language (as long as the set of states does not become empty).

Note that every conventional deterministic automaton (that is, every DFA) is a GDFA. Conversely, given a GDFA, we can obtain a conventional automaton that recognizes the same language by decomposing every string-labeled edge into a path of character-labeled edges. Consequently, a language is regular if and only it is recognized by some GDFA.

Let $\mathcal{W}(\mathcal{A})$ be the set of all $\alpha \in \Sigma^*$ that can be read starting from s and following edges whose labels, when (fully) concatenated, yield α. In other words, for every $\alpha \in \Sigma^*$ we have $\alpha \in \mathcal{W}(\mathcal{A})$ if and only if there exist $t \geq 1$, $u_1, u_2, \ldots, u_t \in Q$ and $\alpha_1, \alpha_2, \ldots, \alpha_{t-1} \in \Sigma^*$ such that (i) $u_1 = s$, (ii) $(u_i, u_{i+1}, \alpha_i) \in E$ for every $1 \leq i \leq t-1$ and (iii) $\alpha = \alpha_1\alpha_2\alpha_3 \ldots \alpha_{t-1}$. If $\alpha \in \mathcal{W}(\mathcal{A})$, then the corresponding $t \geq 1$, $u_1, u_2, \ldots, u_t \in Q$ and $\alpha_1, \alpha_2, \ldots, \alpha_{t-1} \in \Sigma^*$ are all uniquely determined (intuitively, this is true because $\mathcal{A}$ is deterministic, see [15, Remark 9]). Consequently, we can denote by I_α the state u_t reached by α (when $\mathcal{A}$ is not clear from the context, we write $I_\alpha^\mathcal{A}$). Note that we always have $\epsilon \in \mathcal{W}(\mathcal{A})$ and $I_\epsilon = s$. As an example, in Fig. 1 we have $b \notin \mathcal{W}(\mathcal{A})$, $ca \in \mathcal{W}(\mathcal{A})$ and $I_{ca} = 3$.

Let $\mathcal{L}(\mathcal{A})$ be the language recognized by $\mathcal{A}$, that is, $\mathcal{L}(\mathcal{A}) = \{\alpha \in \Sigma^* \mid I_\alpha \in F\}$. For every $u \in Q$, let I_u be the set of all strings that can be read from the initial state to u by concatenating edge labels, that is, $I_u = \{\alpha \in \Sigma^* \mid I_\alpha = u\}$. Note that $\mathcal{W}(\mathcal{A}) = \bigcup_{u \in Q} I_u$.

We will use the following remark (see [15, Sect. 3]).

Remark 1. Let $\mathcal{A} = (Q, E, s, F)$ be a GDFA. For every $\alpha \in \mathcal{W}(\mathcal{A})$, we have $\mathcal{K}(\mathcal{W}(\mathcal{A})_\alpha) = \{\rho \in \Sigma^+ \mid \alpha\rho \in \mathcal{W}(\mathcal{A}), (I_\alpha, I_{\alpha\rho}, \rho) \in E\}$ (namely, the prefix-free kernel of $\mathcal{W}(\mathcal{A})_\alpha$ is equal to the set of all strings labeling some edge leaving the state I_α). In particular, $\mathcal{W}(\mathcal{A})$ is always locally bounded (because the set of all edges is finite). Moreover, we have $\mathcal{L}(\mathcal{A}) \cup \{\epsilon\} \subseteq \mathcal{W}(\mathcal{A}) \subseteq \mathrm{Pref}(\mathcal{L}(\mathcal{A}))$. Lastly, if $\mathcal{A}$ is a conventional deterministic automaton, then $\mathcal{W}(\mathcal{A}) = \mathrm{Pref}(\mathcal{L}(\mathcal{A}))$.

Given two GDFAs $\mathcal{A} = (Q, E, s, F)$ and $\mathcal{A}' = (Q', E', s', F)$, we say that $\mathcal{A}$ and $\mathcal{A}'$ are *isomorphic* if there exists a bijection $\phi : Q \mapsto Q'$ such that (i) for every $u, v \in Q$ and for every $\rho \in \Sigma^+$ we have $(u, v, \rho) \in E$ if and only if $(\phi(u), \phi(v), \rho) \in E'$, (ii) $\phi(s) = s'$ and (iii) for every $u \in Q$ we have $u \in F$ if and only if $\phi(u) \in F$.

Let us recall the definition of Wheeler GDFA [15]. To this end, given a GDFA $\mathcal{A}$, let us define the relation $\preceq_\mathcal{A}$.

Definition 1. *Let $\mathcal{A} = (Q, E, s, F)$ be a GDFA. Let $\preceq_\mathcal{A}$ be the reflexive relation on Q such that, for every $u, v \in Q$ with $u \neq v$, we have $u \prec_A v$ if and only if $(\forall \alpha \in I_u)(\forall \beta \in I_v)(\alpha \prec \beta)$.*

In other words, we have $u \prec_A v$ if and only if all strings reaching u (from the initial state) are co-lexicographically smaller than all strings reaching v (from the initial state).

The relation $\preceq_\mathcal{A}$ is always a partial order, but in general it is not a total order [15, Sect. 4]. We can then give the following definition:

Definition 2. *Let $\mathcal{A} = (Q, E, s, F)$ be a GDFA. We say that $\mathcal{A}$ is* Wheeler *if $\preceq_{\mathcal{A}}$ is a total order.*

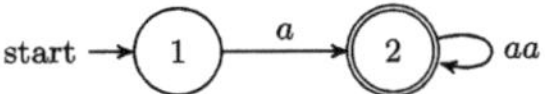

Fig. 2. A Wheeler GDFA $\mathcal{A}$ recognizing $\mathcal{L} = a(aa)^*$. The states are numbered following the total order $\preceq_{\mathcal{A}}$.

Note that the GDFA $\mathcal{A}$ in Fig. 1 is Wheeler (the states are numbered following the total order $\preceq_{\mathcal{A}}$). For example, we have $ba \in I_2$, $ca \in I_3$, $2 < 3$ and, consistently, $ba \prec ca$. Every conventional Wheeler DFA is a Wheeler GDFA [15], but Wheeler GDFAs are more expressive than conventional Wheeler automata: the Wheeler GDFA in Fig. 2 recognizes $\mathcal{L} = a(aa)^*$, but $\mathcal{L}$ is not recognized by any conventional Wheeler automaton (see [3, Example 3]).

Let $\mathcal{A}$ be a GDFA and let $\mathcal{W} \subseteq \Sigma^*$. We say that $\mathcal{A}$ is a *$\mathcal{W}$-GDFA* if we have $\mathcal{W}(\mathcal{A}) = \mathcal{W}$.

3 Our Plan

In this section, we present our plan to obtain the main result (Theorem 3).

Let us consider conventional DFAs, and fix $\mathcal{L} \subseteq \Sigma^*$.

- *(Fact 1)* If $\mathcal{L}$ is recognized by some conventional DFA (that is, if $\mathcal{L}$ is regular), then up to isomorphism there exists a unique minimal DFA recognizing $\mathcal{L}$ (see any textbook in automata theory, e.g., [31]).
- *(Fact 2)* If $\mathcal{L}$ is recognized by some conventional Wheeler DFA, then up to isomorphism there exists a unique minimal Wheeler DFA recognizing $\mathcal{L}$ [3].

Now let us consider GDFAs, and fix $\mathcal{L}, \mathcal{W} \subseteq \Sigma^*$.

- *(Fact 3)* If $\mathcal{L}$ is recognized by some $\mathcal{W}$-GDFA, then up to isomorphism there exists a unique minimal $\mathcal{W}$-GDFA recognizing $\mathcal{L}$ [15]. Note that Fact 3 is a generalization of Fact 1 because for conventional deterministic automata we must necessarily have $\mathcal{W} = \mathrm{Pref}(\mathcal{L})$ by Remark 1.

The main goal of the paper is to prove the natural extension of Fact 2: if $\mathcal{L}$ is recognized by some Wheeler $\mathcal{W}$-GDFA, then up to isomorphism there exists a unique minimal Wheeler $\mathcal{W}$-GDFA recognizing $\mathcal{L}$ (see Theorem 3).

The general idea to prove Fact 1, Fact 2 and Fact 3 is to define an appropriate *equivalence relation* for each of these three settings, and then use these equivalence relations to build the corresponding minimal automata. Consequently, to prove Theorem 3, we would like to define an appropriate equivalence relation analogously.

Let us consider conventional DFAs, and let us give a definition.

Definition 3. *Let $\mathcal{L} \subseteq \Sigma^*$. For every $\alpha, \beta \in \mathrm{Pref}(\mathcal{L})$, let $\alpha \sim_{\mathcal{L}} \beta$ if and only if the following is true: $\alpha \in \mathcal{L}$ if and only if $\beta \in \mathcal{L}$.*

We can use $\sim_{\mathcal{L}}$ to sketch how to prove Fact 1 and Fact 2.

- Let $\mathcal{L} \subseteq \Sigma^*$. Recall that an equivalence relation $\sim$ on $\mathrm{Pref}(\mathcal{L})$ is *right-invariant* if, for every $\alpha, \beta \in \mathrm{Pref}(\mathcal{L})$ such that $\alpha \sim \beta$ and for every $\phi \in \Sigma^*$, the following is true: (i) $\alpha\phi \in \mathrm{Pref}(\mathcal{L})$ if and only if $\beta\phi \in \mathrm{Pref}(\mathcal{L})$; (ii) if $\alpha\phi \in \mathrm{Pref}(\mathcal{L})$, then $\alpha\phi \sim \beta\phi$. Then, the equivalence relation used to prove Fact 1 is the coarsest *right-invariant* equivalence relation on $\mathrm{Pref}(\mathcal{L})$ that refines $\sim_{\mathcal{L}}$ (namely, the standard *Myhill-Nerode equivalence*). Intuitively, to obtain a well-defined DFA recognizing $\mathcal{L}$, we need to start from a right-invariant equivalence relation that refines $\sim_{\mathcal{L}}$, and the coarsest such equivalence relation induces the minimal DFA recognizing $\mathcal{L}$.
- Let $\mathcal{L} \subseteq \Sigma^*$. An equivalence relation $\sim$ on $\mathrm{Pref}(\mathcal{L})$ is *convex* if, for every $\alpha, \beta, \gamma \in \mathrm{Pref}(\mathcal{L})$ such that $\alpha \preceq \beta \preceq \gamma$ and $\alpha \sim \gamma$, we have $\alpha \sim \beta$ [3]. Then, the equivalence relation used to prove Fact 2 is the coarsest *right-invariant and convex* equivalence relation on $\mathrm{Pref}(\mathcal{L})$ that refines $\sim_{\mathcal{L}}$. Intuitively, in addition to the properties needed to prove Fact 1, we need convexity to ensure that the resulting DFA is Wheeler.

Now, let us consider GDFAs, and let us extend Definition 3 by considering an equivalence relation defined on an arbitrary subset $\mathcal{W}$ of Σ^* (and not necessarily on $\mathrm{Pref}(\mathcal{L})$).

Definition 4. *Let $\mathcal{L}, \mathcal{W} \subseteq \Sigma^*$. For every $\alpha, \beta \in \mathcal{W}$, we have $\alpha \sim_{\mathcal{L}, \mathcal{W}} \beta$ if and only if the following is true: $\alpha \in \mathcal{L}$ if and only if $\beta \in \mathcal{L}$.*

We can similarly extend right-invariance to equivalence relations defined on $\mathcal{W}$ [15].

Definition 5. *Let $\mathcal{W} \subseteq \Sigma^*$ and let $\sim$ be an equivalence relation on $\mathcal{W}$. We say that $\sim$ is* right-invariant *if, for every $\alpha, \beta \in \mathcal{W}$ such that $\alpha \sim \beta$ and for every $\phi \in \Sigma^*$, the following is true:*

- *$\alpha\phi \in \mathcal{W}$ if and only if $\beta\phi \in \mathcal{W}$;*
- *if $\alpha\phi \in \mathcal{W}$, then $\alpha\phi \sim \beta\phi$.*

Then, the equivalence relation used to prove Fact 3 is the coarsest *right-invariant* equivalence relation on $\mathcal{W}$ that refines $\sim_{\mathcal{L}, \mathcal{W}}$. Intuitively, when we work with GDFAs, we need to consider equivalence relations defined on $\mathcal{W}$ (and not necessarily on $\mathrm{Pref}(\mathcal{L})$) because, if $\mathcal{A}$ is a GDFA recognizing $\mathcal{L}$, we might have $\mathcal{W}(\mathcal{A}) \neq \mathrm{Pref}(\mathcal{L})$ (while, if $\mathcal{A}$ is a conventional DFA, we always have $\mathcal{W}(\mathcal{A}) = \mathrm{Pref}(\mathcal{L})$ by Remark 1).

We can now outline a plan to prove Theorem 3. Our first step is to extend convexity to equivalence relations defined on $\mathcal{W}$.

Definition 6. *Let $\mathcal{W} \subseteq \Sigma^*$ and let $\sim$ be an equivalence relation on $\mathcal{W}$. We say that $\sim$ is* convex *if, for every $\alpha, \beta, \gamma \in \mathcal{W}$ such that $\alpha \preceq \beta \preceq \gamma$ and $\alpha \sim \gamma$, we have $\alpha \sim \beta$.*

Then, it is natural to conjecture that the equivalence relation required to prove Theorem 3 is the coarsest *right-invariant and convex* equivalence relation on $\mathcal{W}$ that refines $\sim_{\mathcal{L},\mathcal{W}}$. As we will see, this conjecture is correct: by using the coarsest right-invariant and convex equivalence relation on $\mathcal{W}$ that refines $\sim_{\mathcal{L},\mathcal{W}}$, one can combine the proofs of Fact 2 and Fact 3 from the literature to prove Theorem 3. However, one subtle but important step is missing: we must still prove the existence of a (unique) coarsest right-invariant and convex equivalence relation on $\mathcal{W}$ that refines $\sim_{\mathcal{L},\mathcal{W}}$.

This is not as trivial as it may seem. As discussed in the extended version of the paper, the equivalence relations used to prove Fact 1, Fact 2 and Fact 3 can be defined explicitly, but we cannot simply proceed by analogy and adapt the previous definitions. To overcome this limitation, the key idea is that we do not need an explicit definition: we only need a suitable algebraic construction to prove the existence of a (unique) coarsest right-invariant and convex equivalence relation on $\mathcal{W}$ that refines $\sim_{\mathcal{L},\mathcal{W}}$. More generally, given an arbitrary equivalence relation $\sim$ on $\mathcal{W}$, we will prove that there exists a (unique) coarsest right-invariant and convex equivalence relation on $\mathcal{W}$ that refines $\sim$. We present two approaches: a bottom-up strategy and a top-down strategy.

4   A Bottom-up Strategy

Let $\mathcal{W} \subseteq \Sigma^*$, and let $\sim$ an equivalence relation of $\mathcal{W}$. Our goal is to prove that there exists a (unique) coarsest right-invariant and convex equivalence relation on $\mathcal{W}$ that refines $\sim$.

We present a bottom-up approach, that is, we intuitively start from $id_{\mathcal{W}}$ (the finest equivalence relation on $\mathcal{W}$) and, whenever possible, we merge equivalence classes. Our construction is somewhat similar to that used for bisimulations [5,32]. Indeed, the (maximal) bisimulation between two automata is defined algebraically by taking the union of all bisimulations between the two automata, and here we will consider all right-invariant and convex equivalence relations on $\mathcal{W}$ that refine $\sim$. However, one important difference is that, in general, the union of two equivalence relations is not an equivalence relation.

In detail, let X be a set, let $\sim_1$ and $\sim_2$ be two equivalence relations on X, and let $\sim$ be the union of $\sim_1$ and $\sim_2$ (that is, for every $x, y \in X$ we have $x \sim y$ if and only if $(x \sim_1 y) \vee (x \sim_2 y)$). In general, $\sim$ is not an equivalence relation because it need not be transitive. For example, if $X = \{1, 2, 3\}$, the $\sim_1$-equivalence classes are $\{1, 2\}$, $\{3\}$ and the $\sim_2$-equivalence classes are $\{1\}$, $\{2, 3\}$, then $\sim$ is not an equivalence relation because $1 \sim 2$ and $2 \sim 3$ but $1 \not\sim 3$. Nonetheless, it is still true that there exists a unique *finest* equivalence relation that is refined by both $\sim_1$ and $\sim_2$. Intuitively, such a relation is obtained by taking the transitive closure of $\sim_1$ and $\sim_2$. We now give the details of this construction for the general case where we start from a (possibly infinite) set of equivalence relations.

Definition 7. *Let X be a set. Let I be a nonempty index set and, for every $i \in I$, let $\sim_i$ be an equivalence relation on X. For every $x, y \in X$ let $x \left(\bigvee_{i \in I} \sim_i \right) y$ if and only if there exist $n \geq 2$ and $z_1, z_2, \ldots, z_n \in X$ such that (i) $z_1 = x$, (ii) $z_n = y$ and (iii) for every $2 \leq j \leq n$ there exists $i_j \in I$ such that $z_{j-1} \sim_{i_j} z_j$.*

In lattice theory, $\bigvee_{i \in I} \sim_i$ is called the *join* of the $\sim_i$'s.

Lemma 1. *[37, Chapter 1, Theorem 4.7] Let X be a set. Let I be a nonempty index set and, for every $i \in I$, let $\sim_i$ be an equivalence relation on X. Then, $\bigvee_{i \in I} \sim_i$ is the (unique) finest equivalence relation on X that is refined by every $\sim_i$.*

The following two lemmas show that $\bigvee_{i \in I} \sim_i$ inherits properties of the $\sim_i$'s.

Lemma 2. *Let X be a set, and let $\sim$ be an equivalence relation on X. Let I be a nonempty index set and, for every $i \in I$, let $\sim_i$ be an equivalence relation on X. If $\sim_i$ refines $\sim$ for every $i \in I$, then $\bigvee_{i \in I} \sim_i$ refines $\sim$.*

Lemma 3. *Let $\mathcal{W} \subseteq \Sigma^*$. Let I be a nonempty index set and, for every $i \in I$, let $\sim_i$ be an equivalence relation on $\mathcal{W}$.*

1. *If $\sim_i$ is right-invariant for every $i \in I$, then $\bigvee_{i \in I} \sim_i$ is right-invariant.*
2. *If $\sim_i$ is convex for every $i \in I$, then $\bigvee_{i \in I} \sim_i$ is convex.*

Proof. Let $\sim_* = \bigvee_{i \in I} \sim_i$.

1. Fix $\alpha, \beta \in \mathcal{W}$ such that $\alpha \sim_* \beta$, and fix $\phi \in \Sigma^*$. We must prove that (i) $\alpha\phi \in \mathcal{W}$ if and only if $\beta\phi \in \mathcal{W}$ and (ii) if $\alpha\phi \in \mathcal{W}$, then $\alpha\phi \sim_* \beta\phi$. Since $\alpha \sim_* \beta$, there exist $n \geq 2$ and $\xi_1, \xi_2, \ldots, \xi_n \in \mathcal{W}$ such that (i) $\xi_1 = \alpha$, (ii) $\xi_n = \beta$ and (iii) for every $2 \leq j \leq n$ there exists $i_j \in I$ such that $\xi_{j-1} \sim_{i_j} \xi_j$. We know that each $\sim_i$ is right-invariant, so for every $2 \leq j \leq n$ we have $\xi_{j-1}\phi \in \mathcal{W}$ if and only if $\xi_j\phi \in \mathcal{W}$. Since $\xi_1 = \alpha$ and $\xi_n = \beta$, we conclude that $\alpha\phi \in \mathcal{W}$ if and only if $\beta\phi \in \mathcal{W}$. Now assume that $\alpha\phi \in \mathcal{W}$. We must prove that $\alpha\phi \sim_* \beta\phi$. We know that each $\sim_i$ is right-invariant, so for every $2 \leq j \leq n$ we have $\xi_{j-1}\phi \in \mathcal{W}$ and $\xi_{j-1}\phi \sim_{i_j} \xi_j\phi$. Since $\xi_1 = \alpha$ and $\xi_n = \beta$, by the definition of $\sim_*$ we conclude that $\alpha\phi \sim_* \beta\phi$.
2. Fix $\alpha, \beta, \gamma \in \mathcal{W}$ such that $\alpha \preceq \beta \preceq \gamma$ and $\alpha \sim_* \gamma$. We must prove that $\alpha \sim_* \beta$. Since $\alpha \sim_* \gamma$, there exist $n \geq 2$ and $\xi_1, \xi_2, \ldots, \xi_n \in \mathcal{W}$ such that (i) $\xi_1 = \alpha$, (ii) $\xi_n = \gamma$ and (iii) for every $2 \leq j \leq n$ there exists $i_j \in I$ such that $\xi_{j-1} \sim_{i_j} \xi_j$. In particular, we have $\alpha \sim_* \xi_j$ for every $1 \leq j \leq n$.
   We only need to prove that, for every $1 \leq j \leq n$, if $\min\{\alpha, \xi_j\} \preceq \beta \preceq \max\{\alpha, \xi_j\}$, then $\alpha \sim_* \beta$ (the main claim will follow by choosing $j = n$ because $\xi_n = \gamma$ and $\alpha \preceq \beta \preceq \gamma$). We proceed by induction on j. For $j = 1$ we have $\xi_1 = a$, so $\beta = \alpha$ and the conclusion follows. Now assume that $2 \leq j \leq n$. In the rest of the proof, we suppose $\alpha \preceq \xi_{j-1}$ because the case $\xi_{j-1} \prec \alpha$ can be handled analogously. Note the following:
   - If $\alpha \preceq \beta \preceq \xi_{j-1}$, then $\alpha \sim_* \beta$. This follows from the inductive hypothesis.

- If $(\xi_{j-1} \preceq \beta \preceq \xi_j) \vee (\xi_j \preceq \beta \preceq \xi_{j-1})$, then $\alpha \sim_* \beta$. We prove this as follows. We know that $\xi_{j-1} \sim_{i_j} \xi_j$, so we obtain $\xi_{j-1} \sim_{i_j} \beta$ because $\sim_{i_j}$ is convex, and we infer $\xi_{j-1} \sim_* \beta$. From $\xi_{j-1} \sim_* \beta$ and $\alpha \sim_* \xi_{j-1}$ we conclude that $\alpha \sim_* \beta$.

To prove the inductive step, we need to show that, if $\min\{\alpha, \xi_j\} \preceq \beta \preceq \max\{\alpha, \xi_j\}$, then $\alpha \sim_* \beta$. We distinguish three cases, depending on the order between ξ_j and α and the order between ξ_j and ξ_{j-1}.

- Assume that $\xi_j \preceq \alpha \preceq \xi_{j-1}$. We must prove that, if $\xi_j \preceq \beta \preceq \alpha$, then $\alpha \sim_* \beta$. Assume that $\xi_j \preceq \beta \preceq \alpha$. Then, $\xi_j \preceq \beta \preceq \xi_{j-1}$, so $\alpha \sim_* \beta$.
- Assume that $\alpha \prec \xi_j \preceq \xi_{j-1}$. We must prove that, if $\alpha \preceq \beta \preceq \xi_j$, then $\alpha \sim_* \beta$. Assume that $\alpha \preceq \beta \preceq \xi_j$. Then, $\alpha \preceq \beta \preceq \xi_{j-1}$, so $\alpha \sim_* \beta$.
- Assume that $\alpha \preceq \xi_{j-1} \prec \xi_j$. We must prove that, if $\alpha \preceq \beta \preceq \xi_j$, then $\alpha \sim_* \beta$. Assume that $\alpha \preceq \beta \preceq \xi_j$. We distinguish two subcases:
  - Assume that $\beta \preceq \xi_{j-1}$. Then, we have $\alpha \preceq \beta \preceq \xi_{j-1}$, so $\alpha \sim_* \beta$.
  - Assume that $\xi_{j-1} \prec \beta$. Then, we have $\xi_{j-1} \preceq \beta \preceq \xi_j$, so $\alpha \sim_* \beta$.

Here is the main result of the section.

Theorem 1. *Let $\mathcal{W} \subseteq \Sigma^*$, and let $\sim$ be an equivalence relation on $\mathcal{W}$. Then, there exists a (unique) coarsest right-invariant and convex equivalence relation $\sim^{r,c}$ on $\mathcal{W}$ that refines $\sim$.*

Proof. Let $\mathcal{R}$ be the set of all right-invariant and convex equivalence relations on $\mathcal{W}$ that refine $\sim$. Note that $\mathcal{R} \neq \emptyset$ because $id_{\mathcal{W}} \in \mathcal{R}$. Let $\sim^{r,c} = \bigvee_{\sim' \in R} \sim'$. Let us prove that $\sim^{r,c}$ is the (unique) coarsest right-invariant and convex equivalence relation that refines $\sim$. By Lemma 2 and Lemma 3, $\sim^{r,c}$ is a right-invariant and convex equivalence relation that refines $\sim$ (so $\sim^{r,c} \in \mathcal{R}$) Now, let $\sim'$ be any right-invariant and convex equivalence relation that refines $\sim$. We are only left with showing that $\sim^{r,c}$ is coarser than $\sim'$. Let $\alpha, \beta \in \mathcal{W}$ such that $\alpha \sim' \beta$. We must prove that $\alpha \sim^{r,c} \beta$. Notice that $\sim' \in \mathcal{R}$, so the conclusion follows from the definition of $\sim^{r,c}$.

5   A Top-down Strategy

Let us show that the equivalence relation $\sim^{r,c}$ of Theorem 1 can also be obtained via a top-down strategy, that is, by appropriately refining $\sim$ (Theorem 2 below).

Let us recall a lemma showing that there exists a coarsest right-invariant equivalence relation refining $\sim$.

Lemma 4. *[15, Lemma 12] Let $\mathcal{W} \subseteq \Sigma^*$, and let $\sim$ be an equivalence relation on $\mathcal{W}$. For every $\alpha, \beta \in \mathcal{W}$, let $\alpha \sim^r \beta$ if and only if for every $\phi \in \Sigma^*$ the following is true:*

- *$\alpha\phi \in \mathcal{W}$ if and only if $\beta\phi \in \mathcal{W}$;*
- *if $\alpha\phi \in \mathcal{W}$, then $\alpha\phi \sim \beta\phi$.*

Then, $\sim_r$ is the (unique) coarsest right-invariant equivalence relation on $\mathcal{W}$ that refines $\sim$.

The next lemma shows that there exists a coarsest convex equivalence relation refining $\sim$.

Lemma 5. *Let $\mathcal{W} \subseteq \Sigma^*$, and let $\sim$ be an equivalence relation on $\mathcal{W}$. For every $\alpha, \beta \in \mathcal{W}$, let $\alpha \sim^c \beta$ if and only if for every $\gamma \in \Sigma^*$ such that $\min\{\alpha, \beta\} \preceq \gamma \preceq \max\{\alpha, \beta\}$ we have $\alpha \sim \gamma$. Then, $\sim^c$ is the (unique) coarsest convex equivalence relation on $\mathcal{W}$ that refines $\sim$.*

The main idea to prove Theorem 2 below is to define an *infinite* refinement procedure: starting from $\sim$, we consider $\sim^r, (\sim^r)^c, ((\sim^r)^c)^r, (((\sim^r)^c)^r)^c$, and so on. To this end, let us introduce the *intersection* of equivalence relations.

Definition 8. *Let X be a set. Let I be a nonempty index set and, for every $i \in I$, let $\sim_i$ be an equivalence relation on X. For every $x, y \in X$ let $x \left(\bigwedge_{i \in I} \sim_i \right) y$ if and only $x \sim_i y$ for every $i \in I$.*

While in general the union of equivalence relations is not an equivalence relation, the intersection of equivalence relations is always an equivalence relation, as shown in the next lemma.

Lemma 6. *[37, Chapter 1, Theorem 4.5] Let X be a set. Let I be a nonempty index set and, for every $i \in I$, let $\sim_i$ be an equivalence relation on X. Then, $\bigwedge_{i \in I} \sim_i$ is the (unique) coarsest equivalence relation on X that refines every $\sim_i$.*

In lattice theory, $\bigwedge_{i \in I} \sim_i$ is called the *meet* of the $\sim_i$'s.
We are now ready to define our infinite refinement procedure.

Theorem 2. *Let $\mathcal{W} \subseteq \Sigma^*$, and let $\sim$ be an equivalence relation on $\mathcal{W}$. We recursively define the equivalence relation $\sim_n$ for every $n \geq 0$ as follows. Let $\sim_0 = \sim$. For $n \geq 1$, if n is odd, let $\sim_n = \sim_{n-1}^r$, and if n is even, let $\sim_n = \sim_{n-1}^c$. Then, $\bigwedge_{n \geq 0} \sim_n = \sim^{r,c}$.*

6   The Main Result

We now have all the technical machinery required to prove our main result: after showing the existence of a (unique) coarsest right-invariant and convex equivalence relation on $\mathcal{W}$ that refines $\sim_{\mathcal{L},\mathcal{W}}$, the proof of Theorem 3 below is similar to the proofs of Fact 2 and Fact 3 (the details are in the extended version of the paper). Theorem 3 not only shows the uniqueness of a minimal $\mathcal{W}$-GDFA, but it also provides a full Myhill-Nerode theorem. In the statement of Theorem 3, $\sim_{\mathcal{L},\mathcal{W}}^{r,c}$ is the equivalence relation defined in Theorem 1 starting from $\sim_{\mathcal{L},\mathcal{W}}$.

Theorem 3. *Let $\mathcal{L} \subseteq \Sigma^*$ and let $\mathcal{W} \subseteq \Sigma^*$ be a locally bounded set such that $\mathcal{L} \cup \{\epsilon\} \subseteq \mathcal{W} \subseteq \mathrm{Pref}(\mathcal{L})$. The following are equivalent:*

1. $\mathcal{L}$ *is recognized by a Wheeler* $\mathcal{W}$*-GDFA.*
2. $\sim_{\mathcal{L},\mathcal{W}}^{r,c}$ *has finite index.*
3. *There exists a right-invariant and convex equivalence relation* $\sim$ *on* $\mathcal{W}$ *of finite index that refines* $\sim_{\mathcal{L},\mathcal{W}}$.

Moreover, if one of the above statements is true (and so all the above statements are true), then there exists a unique minimal Wheeler $\mathcal{W}$*-GDFA recognizing* $\mathcal{L}$ *(that is, two Wheeler* $\mathcal{W}$*-GDFAs recognizing* $\mathcal{L}$ *having the minimum number of states among all Wheeler* $\mathcal{W}$*-GDFAs recognizing* $\mathcal{L}$ *must be isomorphic).*

Let us use Theorem 3 to show that Wheeler GDFAs do not recognize every regular language.

Example 1. Let $\mathcal{L} = ad^* \mid cd^* \mid ad^*b$. Let us prove that there exists no Wheeler GDFA $\mathcal{A}$ such that $\mathcal{L}(\mathcal{A}) = \mathcal{L}$. Suppose for the sake of a contradiction that there exists a Wheeler GDFA $\mathcal{A}$ such that $\mathcal{L}(\mathcal{A}) = \mathcal{L}$. Let $\mathcal{W} = \mathcal{W}(\mathcal{A})$. By Remark 1, we have $\mathcal{L} \cup \{\epsilon\} \subseteq \mathcal{W} \subseteq \mathrm{Pref}(\mathcal{L})$, and $\mathcal{W}$ is locally bounded. By Theorem 3, $\sim_{\mathcal{L},\mathcal{W}}^{r,c}$ has finite index. Since $\mathcal{L} \subseteq \mathcal{W}$, we have $ad^n, cd^n, ad^n b \in \mathcal{W}$ for every $n \geq 0$.

We can obtain a contradiction by showing that $\sim_{\mathcal{L},\mathcal{W}}^{r,c}$ does not have finite index. To this end, we only need to show that, for every $n > m \geq 0$, we have $ad^m \not\sim_{\mathcal{L},\mathcal{W}}^{r,c} ad^n$. Fix $n > m \geq 0$. Suppose for the sake of a contradiction that $ad^m \sim_{\mathcal{L},\mathcal{W}}^{r,c} ad^n$. By Theorem 1 we know that $\sim_{\mathcal{L},\mathcal{W}}^{r,c}$ is convex. Since $ad^m, cd^m, ad^n \in \mathcal{W}$, $ad^m \prec cd^m \prec ad^n$ and $ad^m \sim_{\mathcal{L},\mathcal{W}}^{r,c} ad^n$, we obtain $ad^m \sim_{\mathcal{L},\mathcal{W}}^{r,c} cd^m$. By Theorem 1 we know that $\sim_{\mathcal{L},\mathcal{W}}^{r,c}$ is right invariant, so from $ad^m \sim_{\mathcal{L},\mathcal{W}}^{r,c} cd^m$ and $ad^m b \in \mathcal{W}$ we obtain $cd^m b \in \mathcal{W}$ and $ad^m b \sim_{\mathcal{L},\mathcal{W}}^{r,c} cd^m b$. By Theorem 1 we know that $\sim_{\mathcal{L},\mathcal{W}}^{r,c}$ refines $\sim_{\mathcal{L},\mathcal{W}}$, so from $ad^m b \sim_{\mathcal{L},\mathcal{W}}^{r,c} cd^m b$ we obtain $ad^m b \sim_{\mathcal{L},\mathcal{W}} cd^m b$. Since $ad^m b \in \mathcal{L}$, we conclude $cd^m b \in \mathcal{L}$. This is a contradiction because $cd^m b \notin \mathcal{L}$.

7   Conclusions and Future Work

We have shown that, in the setting of Wheeler generalized automata, there exists a unique minimal automaton up to isomorphism (Theorem 3). To obtain this result, we needed to develop new techniques (see Theorem 1 and Theorem 2).

Starting from our results, several research directions are possible.

- In the setting of conventional automata, Becker et al. showed how to efficiently determine whether a given language is Wheeler by examining the cycle structure of the minimal automata for the language [8]. We believe that, similarly, it should be possible to efficiently determine whether a given regular language can be recognized by some Wheeler $\mathcal{W}$-GDFA by examining the cycle structure of the minimal $\mathcal{W}$-GDFA for the language.
- In our paper, we have only focused on the deterministic setting. We have mentioned in the introduction that, in the setting of Wheeler automata, nondeterminism and determinism have the same expressive power [3]. Is the same

true for Wheeler generalized automata? We conjecture that the answer is negative, because non-deterministic (Wheeler) generalized automata do not appear to enjoy many properties, especially if one allows edges labeled with the empty string [15,16].
- Wheeler automata can be seen as the first level of a hierarchy containing all regular languages. The main idea to define this hierarchy is to use *partial orders* [20,22]. This approach has led to a fruitful research line (see, e.g., [6,9,13,14]). It is natural to wonder whether the same ideas can be extended to generalized automata.

Acknowledgments. Funded by the Helsinki Institute for Information Technology (HIIT).

References

1. Alanko, J., Cotumaccio, N., Prezza, N.: Linear-time minimization of Wheeler DFAs. In: 2022 Data Compression Conference (DCC), pp. 53–62. IEEE (2022)
2. Alanko, J., D'Agostino, G., Policriti, A., Prezza, N.: Regular languages meet prefix sorting. In: Proceedings of the Fourteenth Annual ACM-SIAM Symposium on Discrete Algorithms, pp. 911–930. SIAM (2020)
3. Alanko, J., D'Agostino, G., Policriti, A., Prezza, N.: Wheeler languages. Inf. Comput. **281**, 104820 (2021)
4. Alanko, J.N., Cenzato, D., Cotumaccio, N., Kim, S.H., Manzini, G., Prezza, N.: Computing the LCP array of a labeled graph. In: 35th Annual Symposium on Combinatorial Pattern Matching (2024)
5. Baier, C., Katoen, J.P.: Principles of Model Checking. MIT press (2008)
6. Becker, R., Cáceres, M., Cenzato, D., Kim, S.H., Kodric, B., Olivares, F., Prezza, N.: Sorting finite automata via partition refinement. In: 31st Annual European Symposium on Algorithms (ESA 2023), pp. 15–1. Schloss Dagstuhl–Leibniz-Zentrum für Informatik (2023)
7. Becker, R., Castiglione, G., D'Agostino, G., Policriti, A., Prezza, N., Restivo, A., Riccardi, B.: Universally Wheeler languages. In: International Conference on Developments in Language Theory, pp. 45–60. Springer (2025). https://doi.org/10.1007/978-3-032-01475-7_4
8. Becker, R., Cenzato, D., Kim, S.H., Kodric, B., Policriti, A., Prezza, N.: Optimal Wheeler language recognition. In: International Symposium on String Processing and Information Retrieval, pp. 62–74. Springer (2023). https://doi.org/10.1007/978-3-031-43980-3_6
9. Becker, R., Cotumaccio, N., Kim, S.H., Prezza, N., Tosoni, C.: Encoding co-lex orders of finite-state automata in linear space. In: 36th Annual Symposium on Combinatorial Pattern Matching (CPM 2025), pp. 15–1. Schloss Dagstuhl–Leibniz-Zentrum für Informatik (2025)
10. Bowe, A., Onodera, T., Sadakane, K., Shibuya, T.: Succinct de Bruijn graphs. In: International Workshop on Algorithms in Bioinformatics, pp. 225–235. Springer (2012). https://doi.org/10.1007/978-3-642-33122-0_18
11. Burrows, M., Wheeler, D.J.: A block-sorting lossless data compression algorithm (1994). Tech. rep

12. Conte, A., Cotumaccio, N., Gagie, T., Manzini, G., Prezza, N., Sciortino, M.: Computing matching statistics on Wheeler DFAs. In: 2023 Data Compression Conference (DCC), pp. 150–159. IEEE (2023)

13. Cotumaccio, N.: Graphs can be succinctly indexed for pattern matching in $O(|E|^2 + |V|^{5/2})$ time. In: 2022 Data Compression Conference (DCC), pp. 272–281 (2022)

14. Cotumaccio, N.: Prefix sorting DFAs: a recursive algorithm. In: 34th International Symposium on Algorithms and Computation (ISAAC 2023), pp. 22–1. Schloss Dagstuhl–Leibniz-Zentrum für Informatik (2023)

15. Cotumaccio, N.: A Myhill-Nerode theorem for generalized automata, with applications to pattern matching and compression. In: 41st International Symposium on Theoretical Aspects of Computer Science (2024)

16. Cotumaccio, N.: Fast pattern matching with epsilon transitions. In: International Workshop on Combinatorial Algorithms, pp. 242–255. Springer (2025). https://doi.org/10.1007/978-3-031-98740-3_18

17. Cotumaccio, N.: Improved circular dictionary matching. In: 36th Annual Symposium on Combinatorial Pattern Matching (CPM 2025), pp. 18–1. Schloss Dagstuhl–Leibniz-Zentrum für Informatik (2025)

18. Cotumaccio, N.: Sorting circular suffixes in linear time. In: International Conference on Developments in Language Theory, pp. 294–310. Springer (2025). https://doi.org/10.1007/978-3-032-01475-7_20

19. Cotumaccio, N., D'Agostino, G., Gibney, D., Policriti, A., Prezza, N., Thankachan, S.V.: Wheeler graphs and Wheeler languages. In: Expanding World of Compressed Data: A Festschrift for Giovanni Manzini's 60th Birthday, p. 12. Schloss Dagstuhl-Leibniz-Zentrum für Informatik (2025)

20. Cotumaccio, N., D'Agostino, G., Policriti, A., Prezza, N.: Co-lexicographically ordering automata and regular languages-part I. J. ACM **70**(4), 1–73 (2023)

21. Cotumaccio, N., Gagie, T., Köppl, D., Prezza, N.: Space-time trade-offs for the LCP array of Wheeler DFAs. In: International Symposium on String Processing and Information Retrieval, pp. 143–156. Springer (2023). https://doi.org/10.1007/978-3-031-43980-3_12

22. Cotumaccio, N., Prezza, N.: On indexing and compressing finite automata. In: Proceedings of the 2021 ACM-SIAM Symposium on Discrete Algorithms (SODA), pp. 2585–2599. SIAM (2021)

23. D'Agostino, G., Martincigh, D., Policriti, A.: Ordering regular languages and automata: complexity. Theoret. Comput. Sci. **949**, 113709 (2023)

24. Eilenberg, S.: Automata, Languages, and Machines. Academic press (1974)

25. Ferragina, P., Luccio, F., Manzini, G., Muthukrishnan, S.: Compressing and indexing labeled trees, with applications. J. ACM (JACM) **57**(1), 1–33 (2009)

26. Ferragina, P., Manzini, G.: Indexing compressed text. J. ACM (JACM) **52**(4), 552–581 (2005)

27. Gagie, T., Manzini, G., Sirén, J.: Wheeler graphs: a framework for BWT-based data structures. Theoret. Comput. Sci. **698**, 67–78 (2017)

28. Giammarresi, D., Montalbano, R.: Deterministic generalized automata. In: Annual Symposium on Theoretical Aspects of Computer Science, pp. 325–336. Springer (1995). https://doi.org/10.1007/3-540-59042-0_84

29. Giammarresi, D., Montalbano, R.: Deterministic generalized automata. Theoret. Comput. Sci. **215**(1–2), 191–208 (1999)

30. Hon, W.K., Lu, C.H., Shah, R., Thankachan, S.V.: Succinct Indexes for Circular Patterns. In: International Symposium on Algorithms and Computation, pp. 673–682. Springer (2011). https://doi.org/10.1007/978-3-642-25591-5_69

31. Hopcroft, J.E., Motwani, R., Ullman, J.D.: Introduction to Automata Theory, Languages, and Computation, 3rd edn. Addison-Wesley Longman Publishing Co., Inc., USA (2006)
32. Kozen, D.C.: Automata and Computability. Springer Science & Business Media (2007)
33. Manber, U., Myers, G.: Suffix arrays: a new method for on-line string searches. Siam J. Comput. **22**(5), 935–948 (1993)
34. Mantaci, S., Restivo, A., Rosone, G., Sciortino, M.: An extension of the Burrows-Wheeler transform. Theoret. Comput. Sci. **387**(3), 298–312 (2007)
35. Myhill, J.: Finite automata and the representation of events. WADD Tech. Rep. **57**, 112–137 (1957)
36. Nerode, A.: Linear automaton transformations. Proc. Am. Math. Soc. **9**(4), 541–544 (1958)
37. Sankappanavar, H.P., Burris, S.: A course in universal algebra. Graduate Texts Math **78**, 56 (1981)
38. Straubing, H.: Finite Automata, Formal Logic, and Circuit Complexity. Springer Science & Business Media (2012). https://doi.org/10.1007/978-1-4612-0289-9

Visibly Pushdown Languages in Groups

Laura Ciobanu[1,2(✉)] and Daniel Turaev[1]

[1] Institut für Mathematik, Technische Universität Berlin, 10623 Berlin, Germany
`ciobanu@math.tu-berlin.de`
[2] Department of Mathematics, Heriot-Watt University, EH14 4AS Edinburgh, UK

Abstract. In this paper we explore the connections between the class of Visibly Pushdown Languages (**VPL**) and the natural sets of words one can associate to a finitely generated group. We show that the word problem of a finitely generated group is **VPL** exactly when the group is finite.

We also show that free reduction does not preserve **VPL**, and that finding solutions to equations in a free group with **VPL** constraints (as reduced words) is undecidable. We explore the structure of sets whose full preimage is **VPL**, showing these are often recognisable sets. We conjecture that, in any group, this class is precisely the recognisable sets.

Keywords: Visibly Pushdown Languages · Groups · Equations · Language Constraints

Finitely generated groups are a rich and natural platform for formal languages, since all the group elements of a finitely generated group G can be represented as words over the finite alphabet that is the (inverse-closed) generating set of G. All generating sets in this paper will be assumed to be finite.

An important connection between groups and formal languages is given by the relation between the language complexity of the *word problem* of a group G and the algebraic nature of G, where the *word problem* of G over a generating set Σ (see Sect. 1) is the set of all the words over Σ that represent the trivial element in G. The celebrated results of Anisimov [4] and Muller-Schupp [20] establish that the word problem is regular exactly when a group is finite, and is context-free exactly when the group is virtually free (see Sect. 2).

In this paper we show that the word problem of a group is a Visibly Pushdown Language (**VPL**) exactly when the group is finite (Theorem 1), show that a Benois-type result about closure under free reduction does not hold for **VPL** (Proposition 6), and study **VPL**s as constraints on Word Equations (Proposition 10). We explore the structure of $\textbf{VPL}^\forall$ sets (Sect. 3.1), showing these frequently coincide with recognisable sets. We further conjecture that the class $\textbf{VPL}^\forall$ is precisely the class of recognisable sets over any group. Finally, we show that solving equations in free groups with certain **VPL** constrains is undecidable (Proposition 10), and suggest research directions on this topic.

© The Author(s), under exclusive license to Springer Nature Switzerland AG 2026
M.-P. Béal and P. Caron (Eds.): DLT 2026, LNCS 16578, pp. 103–116, 2026.
https://doi.org/10.1007/978-3-032-28404-4_8

Visibly Pushdown Languages are a class of languages situated between regular and context-free, with robust closure properties (see Proposition 1) akin to regular languages, but expressive power close to that of the context-free ones; they are the accept languages of Visibly Pushdown Automata, which were introduced in the 1980s under the name 'Input Driven Pushdown Automata' [19]. The first more in depth study was by Alur and Madhusudan [2,3] who came up with the name 'Visibly Pushdown Language'.

1   Preliminaries

Throughout this paper, X will refer to a finite generating set for a group, and Σ will refer to the inverse closure of X, that is, for each element $x \in X$, there are two formal symbols $x, x^{-1} \in \Sigma$, $x \neq x^{-1}$. We will write it as $\Sigma = X \cup X^{-1}$.

In general, Σ will also be the input alphabet of the automata we work with.

We will assume some familiarity with regular and context-free languages, as well as their associated finite-state automata and pushdown automata respectively. A comprehensive introduction can be found, for example, in [17]

1.1   VPLs and Their Closure Properties

Definition 1 (Visibly Pushdown Automaton). *A pushdown automaton on an alphabet Σ with stack alphabet Γ is called a* visibly pushdown automaton *if the alphabet Σ can be partitioned into three components, as $\Sigma = \Sigma_c \sqcup \Sigma_i \sqcup \Sigma_r$, such that the only allowed transitions upon reading $x \in \Sigma$ are as follows:*

- *if $x \in \Sigma_i$, then the state transition cannot depend on, or modify, the stack.*
- *if $x \in \Sigma_c$, then the state transition again cannot depend on the stack, but it modifies the stack by adding ('pushing') a single symbol.*
- *if $x \in \Sigma_r$, then the state transition must read the topmost symbol on the stack, and remove it ('pop' it) without pushing anything onto the stack.*

Furthermore, ε-transitions are not allowed. The three sets Σ_c, Σ_i, Σ_r are the sets of "call, internal, and response" letters, respectively.

By convention, we insist that a unique 'bottom of the stack' letter $\perp \in \Gamma$ exists, and the stack is empty if only this symbol is on the stack. The symbol $\perp$ cannot be pushed onto the stack by a call letter; $\perp$ may be popped by a response letter, but if this occurs it is not removed from the stack.

Remark 1. The above criterion for an automaton to be visibly pushdown may be defined just as well on deterministic or nondeterministic automata. A priori, these need not be equivalent, but it turns out that they are, as all visibly pushdown automata are determinisable (see Theorem 2 of [2]).

Further, by taking a partition $\Sigma_c = \Sigma_r = \emptyset$, $\Sigma_i = \Sigma$, we see that all finite-state automata are visibly pushdown automata.

Definition 2. (Visibly Pushdown Language). *A language is called* visibly pushdown *if it is the accept language[1] of a visibly pushdown automaton. We denote the class of Visibly Pushdown Languages by* **VPL**.

Example 1.

(i) For $\Sigma = \{a, b\}$, partitioned into $\Sigma_c = \{a\}, \Sigma_i = \emptyset, \Sigma_r = \{b\}$, the language $\{a^n b^n : n \in \mathbb{N}\}$ is **VPL**. Note that this language is not regular.

(ii) The language L over $\{a, b\}$ of words with equal number of a's as the number of b's is not **VPL** for any partition of the alphabet. Note that L is deterministic context-free.

Indeed, suppose V is a visibly pushdown automaton accepting L. Since L is not regular, exactly one of the two letters must be a call, and the other a response. Suppose a is a call, and b is a response. There are $n \neq m$ such that reading b^n and b^m will result in the same configuration of V. Since $b^n a^n$ is accepted by V, $b^m a^n$ must also be accepted by V – a contradiction.

(iii) Analogously, over the alphabet $\{a, a^{-1}\}$, the word problem of $\mathbb{Z}$ is not **VPL**, as this is the language of words with the same number of a's as a^{-1}'s.

Remark 1 and Example 1 show the following inclusion of language classes:

$$\mathbf{REG} \subsetneq \mathbf{VPL} \subsetneq \mathbf{DCF} \subsetneq \mathbf{CF},$$

where **REG** is the class of regular languages, **DCF** the class of deterministic context-free languages, and **CF** the class of context-free languages.

Proposition 1 (Theorem 1 of *[2]*). *The class of* **VPL***s over a fixed partition of Σ is closed under the operations of union, intersection, concatenation, Kleene Star, complement, and renaming. That is to say, if L_1, L_2 are* **VPL** *over a common partition $\Sigma = \Sigma_c \sqcup \Sigma_i \sqcup \Sigma_r$, and $f : \Sigma \to \Sigma'$ is a map sending calls to calls, responses to responses, and internals to internals, then*

$$L_1 \cup L_2, L_1 \cap L_2, L_1 \cdot L_2, L_1^\star, L_1^C, \ \text{and} \ f(L_1)$$

are all Visibly Pushdown Languages.

Note that the restriction to a common partition of the alphabet is necessary, and without it the same closure properties will not all hold.

Example 2. The languages $L_1 = \{a^n b^n c^*\}$ and $L_2 = \{a^* b^n c^n\}$ are **VPL**s, the first over the partition $\Sigma_c = \{a\}, \Sigma_r = \{b\}, \Sigma_i = \{c\}$, and the second over the partition $\Sigma_c = \{b\}, \Sigma_r = \{c\}, \Sigma_i = \{a\}$. The intersection $L_1 \cap L_2 = \{a^n b^n c^n\}$ is not context-free, let alone **VPL**. Therefore there is no partition of $\Sigma = \{a, b, c\}$ such that both L_1 and L_2 are both visibly pushdown over this partition.

---

[1] a VPA accepts a word if the final state when reading the word is an accepting state. Namely, there is no restrictions on the contents of the stack – it may be non-empty.

Similarly, the restriction that f respects the partition of the alphabet is essential, as the class **VPL** is not closed under general string morphisms, nor inverse morphisms. In fact, one can associate to any context-free language L, over an alphabet Σ, a Visibly Pushdown Language L' over a marked alphabet $\Sigma' = \Sigma_c \sqcup \Sigma_i \sqcup \Sigma_r$ with $\Sigma_c = \Sigma \times \{c\}$, $\Sigma_i = \Sigma \times \{i\}$, $\Sigma_r = \Sigma \times \{r\}$, such that under the string morphism f sending $(x, \alpha) \in \Sigma'$ to x, we get $f(L') = L$. Details of this construction can be found in [2].

1.2   Congruences on VPLs

The *syntactic congruence* of a language L is defined by

$$u_1 \approx_L u_2 \iff \forall v, w \in \Sigma^\star : (vu_1 w \in L \iff vu_2 w \in L).$$

Such a congruence may be defined for any language. We say a congruence has *finite index* if it has finitely many equivalence classes in $\Sigma^\star$. It is well known that the syntactic congruence of a language L is finite index if and only if L is regular. Similarly, the *Myhill-Nerode congruence* is defined as

$$u_1 \sim_L u_2 \iff \forall v \in \Sigma^\star : (u_1 v \in L \iff u_2 v \in L),$$

and also has finite index if and only if L is regular.

Congruences apply to **VPL**s as well [1], where the Myhill-Nerode and syntactic congruences are modified according to the partition of Σ.

Definition 3. *(i) Given a partition $\Sigma = \Sigma_c \sqcup \Sigma_i \sqcup \Sigma_r$ of an alphabet, the set of* matched-response *words, denoted $MR(\Sigma)$, is the set of words u with the property that in any prefix u' of u, the number of response symbols is at most the number of call symbols.*

(ii) Dually, the set of matched-call *words, denoted $MC(\Sigma)$, is the set of words u with the property that in any suffix u' of u, the number of call symbols is at most the number of response symbols.*

(iii) The set of well-matched *words is the set $WM(\Sigma) = MC(\Sigma) \cap MR(\Sigma)$.*

Where Σ is clear, we will write MR, MC, and WM respectively.

It is useful to consider languages consisting entirely of well-matched words. Adding this restriction to a **VPL** provides a strictly smaller class of languages. We will call a language L a "well-matched **VPL**" if $L \subseteq WM$ and $L \in$ **VPL**.

The following three congruences are defined in [1].[2] It is important to note that for the congruences $\sim_0$ and $\approx$, the congruences are not defined on all of $\Sigma^\star$.

$$u_1, u_2 \in \Sigma^\star, u_1 \equiv u_2 \iff \forall v \in MR : (u_1 v \in L \iff u_2 v \in L) \tag{1}$$

$$u_1, u_2 \in MC, u_1 \sim_0 u_2 \iff \forall v \in \Sigma^\star : (u_1 v \in L \iff u_2 v \in L) \tag{2}$$

$$u_1, u_2 \in WM, u_1 \approx u_2 \iff \forall v, w \in \Sigma^\star : (vu_1 w \in L \iff vu_2 w \in L) \tag{3}$$

---

[2] We have used here the same notation from the original paper.

Proposition 2 (Theorem 2 of [1]). *A language L is* **VPL** *if and only if the congruences $\equiv$, $\sim_0$, $\approx$ are all finite index.*

Note the similarity to the case of regular languages. Indeed, $\approx$ is precisely the syntactic congruence restricted to well-matched words. Likewise, $\sim_0$ and $\equiv$ are the Myhill-Nerode congruence, respectively restricted to MC words or restricted to MR suffixes.

1.3   Languages in Groups

There are many languages of interest in groups beyond the word problem. A key observation is that when representing elements by words in finitely generated monoids or groups, one often needs to decide which word(s) should represent a given element from potentially infinitely many choices. In free monoids and groups there are canonical choices, even unique ones in the monoid case. However, in non-free structures we need to pick either certain normal forms, or certain well-defined sets: we associate to subsets of a group G, generated by a set X, languages in the free monoid $\Sigma^\star = (X \cup X^{-1})^\star$ via the canonical projection $\pi : \Sigma^\star \to G$ sending each letter $a \in \Sigma$ to $a \in X \cup X^{-1} \subset G$ and extended in the natural way to a monoid homomorphism.

Given a language class $\mathcal{C}$, we associate two notions of $\mathcal{C}$-sets to a group G.

Definition 4. *A set $S \subseteq G$ is $\mathcal{C}^\forall$ if the full preimage $\pi^{-1}(S) \subseteq \Sigma^\star$ is in $\mathcal{C}$, and $\mathcal{C}^\exists$ if there is some $L \in \mathcal{C}$, such that $\pi(L) = S$.*

This notation is established in [7], and highlights the fact that these two language classes are defined by the following formula:

$$S \in \mathcal{C}^\square \iff \exists L \in \mathcal{C}, \forall s \in S, \square x \in \pi^{-1}(s) : (\pi(L) = S \wedge x \in L),$$

where $\square$ is $\forall$ and $\exists$, respectively.

When $\mathcal{C} = \mathbf{REG}$, the standard terminology in the literature is to call a $\mathbf{REG}^\forall$ set *recognisable*, and a $\mathbf{REG}^\exists$ set *rational*. Various (often contradictory) terms have also been used for $\mathbf{CF}^\forall$ and $\mathbf{CF}^\exists$ sets. The following algebraic characterisation of recognisable sets is due to Herbst and Thomas [16, Section 6].

Proposition 3 (Algebraic characterisation of recognisable sets). *In any group G, the recognisable sets are finite unions of cosets of a finite index normal subgroup of G. That is, if $\pi^{-1}(K)$ is regular for some $K \subseteq G$, then there exists a finite index normal subgroup N such that*

$$K = \bigcup_{i=1,\dots,n} N g_i : N \trianglelefteq_{f.i.} G, g_i \in G.$$

In the following we denote by $F(X)$ the free group generated by finite set X. As is standard, for any word w over $X \cup X^{-1}$, w is *reduced* if there are no occurrences of aa^{-1} in w, where $a \in X \cup X^{-1}$. In any free group, the set of reduced words is a regular language in $(X \cup X^{-1})^\star$.

Definition 5 (Reduction map) *Let $S \subseteq \Sigma^\star$ be a set, and let $\pi(S)$ be the image of S in $F(X)$. The reduction map r associates to S the set $r(S) \subset \Sigma^\star$ of reduced words with the same image as S. Or, in symbols:*

$$r(S) = \{w \in (X \cup X^{-1})^\star : w \text{ is reduced and } \pi(w) \in \pi(S)\}.$$

Proposition 4. *(1) A set $K \subset F(X)$ is rational (i.e. $\mathbf{REG}^\exists$) if and only if $r(K)$ is regular. [Benois, [5]]*
(2) A set $K \subset F(X)$ is $\mathbf{CF}^\vee$ if and only if $r(K)$ is context-free. [Herbst, [15]]

2   The Word Problem

For a group $G = \langle X \rangle^3$ with identity (or trivial) element 1, the *word problem* of G, written $WP(G, X)$, is the set of all $w \in \Sigma^*$ such that $\pi(w) = 1$ (recall that $\Sigma = X \cup X^{-1}$). In other words, $L = \pi^{-1}(1)$.

We recall that a group is *virtually free* if it has a free subgroup of finite index.

Proposition 5. *(1) A group $G = \langle X \rangle$ has regular word problem, i.e. $WP(G, X) \in \mathbf{REG}$, if and only if G is finite.[Anisimov, [4]]*
(2) A group $G = \langle X \rangle$ has context-free word problem, i.e. $WP(G, X) \in \mathbf{CF}$, if and only if G is virtually free.[Muller-Schupp, [20]]

We show in this section that considering word problems in the class $\mathbf{VPL}$ cannot strengthen the above results. In particular, we prove the following.

Theorem 1. *Let $G = \langle X \rangle$ be a non-trivial, finitely generated, group and let $\pi : \Sigma^* \to G$ be the standard projection map onto G. If $WP(G, X) \in \mathbf{VPL}$, then $WP(G, X)$ is regular and G is finite.*

We will need the following lemma to show this result.

Lemma 1. *Let G be virtually free, and assume $Y \subset G$ is a finite, inverse closed, generating set consisting of torsion[4] elements. Then either G is finite or there is some pair $x, y \in Y$ such that xy is of infinite order in G.*

The most elegant proof of the lemma relies on the fact that virtually free groups act non-trivially on simplicial trees with finite vertex and edge stabilisers, according to Bass-Serre theory. See [6] (among many references on Bass-Serre theory) for some background and the main lemma that we require.

Proof. Suppose that G is infinite and virtually free. Then G acts on some infinite tree T via automorphisms. If an element $g \in G$ fixes (at least) a point in the tree then we say g is elliptic and write $Fix(g)$ for the set of fixed points of g (see [6, Section 3]); otherwise g is hyperbolic and it acts as a translation, with no fixed points. (Automorphisms that invert an edge may exist, but we can easily modify the tree so that the action only has elliptic and hyperbolic elements.)

---

[3] We write $G = \langle X \rangle$ if X is a generating set for the group G.
[4] An element g of a group is called *torsion* if there is some $n \in \mathbb{N}_{n \geq 2}$ such that $g^n = 1$.

Since all elements in Y are torsion, they are elliptic.

An infinite, virtually free, group G does not have a global fixed point because all the stabilisers of the vertices (and edges) for the action of G on T are finite; to see this, note that having a global fixed point would imply the existence of a vertex stabiliser which is the whole group G, which is infinite, a contradiction. As G does not fix any vertex, the intersection $\cap_{z \in Y} Fix(z)$ of all the fixed point sets is empty. Since $Fix(z)$ is convex for any $z \in G$, there exist two elements $x, y \in Y$ such that $Fix(x) \cap Fix(y) = \emptyset$ (see [6, page 541]). Then by [6, Lemma 7] the element xy is hyperbolic, so of infinite order, which proves the lemma.

Proof. (Proof of Theorem 1) Since $WP(G, X)$ is in **VPL**, there is a partition of the alphabet Σ into call, response, and internal letters. Suppose some $a \in X$ has infinite order. Now consider the language $\{a, a^{-1}\}^\star \cap WP(G, X)$. This is precisely the word problem of $\mathbb{Z}$ over the generating set $\{a, a^{-1}\}$. Indeed, if it wasn't, then some non-trivial word $a^k, k \in \mathbb{Z} \setminus \{0\}$ would be in $WP(G, X)$, and hence a would have finite order. Now, since **VPL**s are closed under intersections, and $\{a, a^{-1}\}^\star$ is regular, if $WP(G, X)$ were **VPL** then so would $WP(\mathbb{Z}, \{a\})$ – a contradiction (see Example 1(iii)).

Therefore, for every $a \in X$ there is some $n_a \in \mathbb{N}$ such that $\pi(a^{n_a}) = 1$. So X is a finite subset of torsion elements of G. Since G has **VPL** word problem, and hence context-free word problem, we know by Proposition 5 that G is virtually free. Thus by Lemma 1, either G is finite or there are some $x, y \in X$ such that xy has infinite order. Note that this also means yx has infinite order, since xy and yx are conjugate.

Let m and n be the orders of x and y respectively, since x and y are torsion. Then for each $k \in \mathbb{N}$ consider the words

$$(xy)^k(y^{n-1}x^{m-1})^k \, , \, (yx)^k(x^{m-1}y^{n-1})^k \in \pi^{-1}(1) = WP(G, X).$$

Now, if one of x, y is a call, and the other a response, then one of $(xy)^k$ and $(yx)^k$ is an infinite family of well-matched (WM) words that are pairwise distinguished by suffixes – producing infinitely many equivalence classes with respect to the congruence $\approx$ and thus contradicting Proposition 2. If neither x nor y are calls, then $(xy)^k$ is an infinite family of matched-call (MC) words that are pairwise distinguished by suffixes – again a contradiction. If x and y are both calls, then $(y^{n-1}x^{m-1})^k$ is a family of matched-response (MR) suffixes, and $(xy)^k$ is an infinite family pairwise distinguished by MR suffixes – again a contradiction. Thus we cannot have the product of two generators having infinite order, so G must be finite.

We recall that the *co-word problem* of G is the complement of the word problem of G in $\Sigma^\star$, so it is the set of all $w \in \Sigma^\star$ such that $\pi(w) \neq 1$.

Corollary 1. *If G has **VPL** co-word problem, then G is finite.*

Proof. The class **VPL** is closed under complementation, and hence a group with **VPL** co-word problem also has **VPL** word problem.

Remark 2. As is the case for regular and context-free languages, the formal language complexity of the word problem is generating set independent for the **VPL** class. That is, if $WP(G, X)$ is **VPL** for some generating set X of G, it is **VPL** for any finite generating set of G, so we could write '$WP(G)$ is **VPL**'. For **REG** and **CF**, this independence comes from the closure of these classes under inverse morphisms. Since **VPL** is *not* closed under inverse morphisms, generating set independence is somewhat surprising.

3   Types of VPL Sets in Groups

One of the most useful properties of regular languages in the context of free groups is the fact that they are closed under performing free reductions. That is, one may assume by Proposition 4 that a rational set consists entirely of reduced words (over an inverse-closed generating set of a free group) without loss of generality. We consider here the interplay between **VPL**s and free reduction. We will denote by $\mathbf{VPL}^{red}$ the class of languages of reduced words accepted by some visibly pushdown automaton. A natural question then is whether this class coincides with $\mathbf{VPL}^{\exists}$ or $\mathbf{VPL}^{\forall}$, analogously to either Benois' or Herbst's Theorems (see Proposition 4). It turns out that neither case holds true.

Proposition 6. *The following inclusions of language classes are strict:*

$$\mathbf{VPL}^{\forall} \subsetneqq \mathbf{VPL}^{red} \subsetneqq \mathbf{VPL}^{\exists}.$$

Proof. We show this via two examples, one for each of the strict inclusions.

1. Consider the group $\mathbb{Z}$ with generating set $\{a, a^{-1}\}$. The set $\{0\} \subset \mathbb{Z}$ is a rational set (represented by ε), hence a $\mathbf{VPL}^{red}$ set by Benois' Theorem, but it is not $\mathbf{VPL}^{\forall}$, since the preimage of $\{0\}$ is $WP(\mathbb{Z}, \{a, a^{-1}\})$.
2. Now consider the free group F_2 of rank 2 over $\Sigma = \{a, b, a^{-1}, b^{-1}\}$ and let $L := \{(aaa^{-1})^n b^{2n} \mid n \in \mathbb{N}\} \subset \Sigma^\star$. With the partition $\Sigma_c = \{a\}$, $\Sigma_i = \{a^{-1}, b^{-1}\}$, $\Sigma_r = \{b\}$, the language L is the accept language of the VPA in Fig. 1. Thus $\pi(L) \subset F_2$ is a $\mathbf{VPL}^{\exists}$ set.

   However, the set of reduced words $r(L) = \{a^n b^{2n} \mid n \in \mathbb{N}\}$ is not **VPL** for any partition of the alphabet. Indeed, consider the infinite family $\{a^n b^n \mid n \in \mathbb{N}\}$. Each pair of words $a^i b^i$ and $a^j b^j$ for $i \neq j$ in this family can be pairwise distinguished from one another by the suffix b^j. If b is not a response, then $b^j \in MR$ for every j. Thus the congruence $\equiv$ is infinite index for $r(L)$. On the other hand, if b is a response, then $a^n b^n \in MC$ for each n, and hence the congruence $\sim_0$ is infinite index. Either way, by Proposition 2, $r(L)$ cannot be **VPL**.

3.1 $\mathbf{VPL}^{\forall}$ Sets in Groups

We now explore the structure of the class $\mathbf{VPL}^{\forall}$ over arbitrary groups. We conjecture, and make significant progress towards proving, the following.

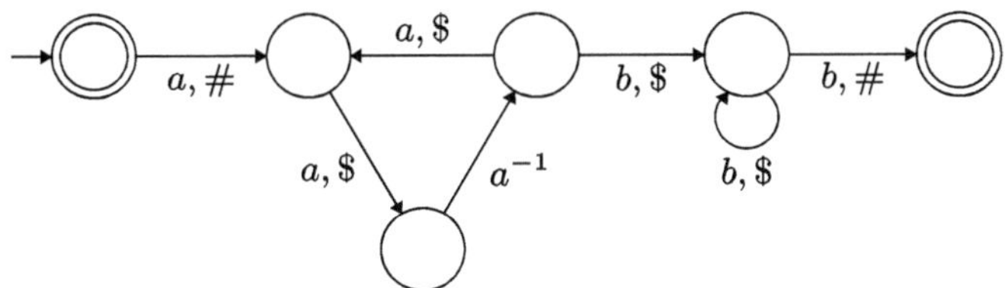

Fig. 1. A visibly pushdown automaton accepting $\{(aaa^{-1})^n b^{2n} \mid n \in \mathbb{N}\}$. Here, a is a call, and pushes the symbol after the comma onto the stack. Meanwhile b is a response, and pops the symbol after the comma from the stack.

Conjecture 1. In all groups, every $\mathbf{VPL}^\vee$ set is recognisable. In other words, the classes $\mathbf{VPL}^\vee$ and $\mathbf{REG}^\vee$ are equal.

We will call a partition of Σ into calls, responses, and internals *symmetric* if every call has infinite order, and for every call x, x^{-1} is a response, and for every response y, y^{-1} is a call.

Theorem 2. *Let G be a group generated by X. Suppose $\Sigma = \Sigma_c \sqcup \Sigma_i \sqcup \Sigma_r$ is not a symmetric partition of Σ. Then any $\mathbf{VPL}^\vee$ subset of G over this partition is a recognisable subset of G.*

Proof. The goal is to use the structure of the partition and the congruences outlined in Sect. 1.2 to show that the Myhill-Nerode congruence must have finite-index.

In order for our partition to not be symmetric, either there must exist some call x such that x^{-1} is not a response, or some response y such that y^{-1} is not a call, or some torsion element z which is either a call or a response.

Recalling the sets MR and MC from Definition 3, in the first of these cases, $\pi(MR) = G$. Indeed, for every $w \in \Sigma^\star$, consider the word $x^n x^{-n} w$, where $n = |w|$. If w has any unmatched responses, they will match to a call letter from the prefix $x^n x^{-n}$, which has no response letters in it. So $x^n x^{-n} w \in MR$ and $\pi(x^n x^{-n} w) = \pi(w)$.

Dually, in the second of the above cases, $\pi(MC) = G$. Indeed, for every $w \in \Sigma^\star$, consider the word $w y^n y^{-n}$, where $n = |w|$. If w has any unmatched calls, they will match to a response letter in the suffix $y^n y^{-n}$, which has no call letters in it. Hence $w y^n y^{-n} \in MC$ and $\pi(w y^n y^{-n}) = \pi(w)$.

Finally, in the third case of a call (resp. response) torsion element z, supposing $z^k = 1$, for each $w \in \Sigma^\star$ we can consider the word $z^{kn} w$ (resp. $w z^{kn}$), where $n = |w|$, to show that either $\pi(MR) = G$ or $\pi(MC) = G$, respectively.

Now, suppose $S \subseteq G$ is a $\mathbf{VPL}^\vee$ set, i.e. $L := \pi^{-1}(S) \in \mathbf{VPL}$. Consider the Myhill-Nerode congruence $\sim_L$ of the preimage language. Since L is a full preimage, $u_1 \sim_L u_2$ is equivalent to the statement $\forall v \in \Sigma^\star : \pi(u_1 v) \in S \iff \pi(u_2 v) \in S$. Notice here that, since π is a homomorphism, if for a given v_1, $\pi(u_1 v_1) \in S \iff \pi(u_2 v_1) \in S$ holds, and some v_2 is such that $\pi(v_2) = \pi(v_1)$, then $\pi(u_1 v_2) \in S \iff \pi(u_2 v_2) \in S$ also holds.

We therefore see that $u_1 \sim_L u_2$ is equivalent to the statement

$$\forall v \in W : \pi(u_1 v) \in S \iff \pi(u_2 v) \in S,$$

where $W \subseteq \Sigma^\star$ is a language such that $\pi(W) = G$.

We now treat the two cases arising from the partition not being symmetric.

- if $\pi(MR) = G$, then taking $W = MR$ above, we see that $u_1 \sim_L u_2$ if and only if $u_1 \equiv u_2$ (recall $\equiv$ from Sect. 1.2). Since $L \in \mathbf{VPL}$, $\equiv$ is finite-index by Proposition 2, and hence so is $\sim_L$. Thus L is regular and S is recognisable.
- if $\pi(MC) = G$, we first note that if u, v are such that $\pi(u) = \pi(v)$, then $u \sim_L v$. If $\pi(u) \neq \pi(v)$, then we consider $\tilde{u}, \tilde{v} \in MC$ such that $\pi(\tilde{u}) = \pi(u)$ and $\pi(\tilde{v}) = \pi(v)$. We see that $\tilde{u} \sim_L \tilde{v}$ is equivalent to $\tilde{u} \sim_0 \tilde{v}$. By transitivity, $u \sim_L v$ if and only if $\tilde{u} \sim_0 \tilde{v}$. Again by Proposition 2, $\sim_0$ is finite-index, and hence so is $\sim_L$. Thus S is recognisable.

Proposition 7. *Let G be a group, and suppose that $S \subseteq G$ is a $\mathbf{VPL}^\forall$ set. Suppose further that $\pi^{-1}(S)$ is a well-matched VPL, i.e. $\pi^{-1}(S) \subseteq WM$. Then S is recognisable.*

Proof. Suppose Σ_c is non-empty, and suppose a is a call. Then for $w \in \pi^{-1}(S)$, $wa^{-1}a \in \pi^{-1}(S)$, but this word is not well-matched since the final letter is a call. Likewise, suppose Σ_r is non-empty, and suppose b is a response. Then for $w \in \pi^{-1}(S)$, $bb^{-1}w \in \pi^{-1}(S)$, but this word is not well-matched since the first letter is a response.[5] Thus the only possible partition of Σ is $\Sigma_c = \Sigma_r = \emptyset$, $\Sigma_i = \Sigma$, and hence $\pi^{-1}(S)$ will be the accept language of a finite-state automaton.

Proposition 8. *Let $S \subseteq G$ be a finite set. Then S is $\mathbf{VPL}^\forall$ if and only if S is recognisable, if and only if G is finite.*

Proof. If G is finite there is nothing to prove. If G is infinite, S cannot be recognisable by Proposition 3. Thus the partition of Σ over which $\pi^{-1}(S)$ is $\mathbf{VPL}$ must have nonempty call and response alphabets, and by Theorem 2, the partition must be symmetric.

Now, take an element $g \in S$ with $w \in \Sigma^\star$ a reduced representative of g. Suppose $x \in \Sigma$ of infinite order is a call, with x^{-1} a response. By Proposition 2, the congruence $\sim_0$ must be finite index on any infinite family of MC words, including the family $\{x^{-k}, k \in \mathbb{N}\}$. Namely, for some infinite set $I \subset \mathbb{N}$ with $1 \in I$, for each $i, j \in I, x^{-i} \sim_0 x^{-j}$. Thus, since $x^{-1}xw \in \pi^{-1}(S)$, for each $i \in I$, $x^{-i}xw \in \pi^{-1}(S)$. Since S is finite and I is infinite, there must be some pair $i \neq j, i, j \in I$ such that $\pi(x^{-i}xw) = \pi(x^{-j}xw)$. Since π is a homomorphism, this implies that for some n, $\pi(x^{-n})$ is the identity in G, contradicting the fact that x has infinite order.

For free groups we prove an additional result about $\mathbf{VPL}^\forall$ subgroups.

Theorem 3. *Let $F(X)$ be a free group over the set X, and S a finitely generated subgroup of $F(X)$. Then S is a $\mathbf{VPL}^\forall$ set if and only if it has finite index in F, and hence is recognisable.*

---

[5] Here, a^{-1} and b^{-1} can refer to any choice of word corresponding to the inverse of a and b. Thus, the Proposition still holds for generating alphabets that are not inverse closed and hence may not have a letter corresponding to the inverses of a resp. b.

4   Word Equations with Language Constraints

The motivation for the present paper comes from studying equations in free groups with formal language constraints, in particular in relation to the recent results in [12]. We use the terminology 'word equations' when working in a free monoid and simply write 'equations' when working in a group.

Let $\mathcal{V}$ be a finite collection of variables and A a finite set of letters, and consider the word equation $U = V$ over $(A, \mathcal{V})$, where U, V are words in the free monoid $(A \cup \mathcal{V})^*$. A *solution* to $U = V$ is a morphism $\sigma \colon (\mathcal{V} \cup A)^* \to A^*$ that fixes A point-wise and such that $\sigma(U) = \sigma(V)$. We will refer to the question of whether a given word equation has solutions as (**WordEqn**). This problem is solvable by the groundbreaking work of Makanin [18] (and subsequent improvements).

A *word equation with rational constraints* is a word equation $U = V$ together with a regular language $R_Y \subseteq A^*$ for each variable $Y \in \mathcal{V}$. Then $\sigma \colon \mathcal{V} \to A^*$ is a *solution* if it is a solution to $U = V$ and also satisfies $\sigma(Y) \in R_Y$ for each $Y \in \mathcal{V}$. We will refer to the question of whether a given equation has solutions satisfying rational constraints as (**WordEqn**, **REG**$^{\exists}$), noting that in free monoids rational and regular sets coincide. This problem is decidable by [13]. One can analogously define word equations with context-free or **VPL** constraints in free monoids, or lift all definitions and questions to free groups, where we use the notation (**GpEqn**); background on these topics can be found in the surveys [8,11].

In the recent paper [12] it was shown that considering language classes only slightly more complex than the regular ones yields undecidability.

Proposition 9 (Theorem 4.4 of *[12]*). *In a free monoid, Word Equations with* **VPL** *constraints, and hence also with* **DCF** *or* **CF** *constraints, are undecidable.*

It is therefore interesting to consider **VPL** constraints in free groups as well. On the one hand, we see in Sect. 3.1 that **VPL**$^{\forall}$ sets behave very closely to recognisable sets and if Conjecture 1 holds, it would namely imply the decidability of (**GpEqn**) with **VPL**$^{\forall}$ constraints. On the other hand, in the case of **VPL**^{red} sets, that is, **VPL** sets consisting entirely of reduced words in the free group setting, a more straightforward analysis is possible.

Proposition 10. *In a free group, (GpEqn) with* **VPL**^{red} *constraints, and hence also with* **CF**$^{\forall}$ *constraints, are undecidable.*

Proof. In the free group $F(X)$ generated by $X = \{x_1, \ldots, x_n\}$ (X is the basic generating set and Σ is the inverse-closed one), the language of positive words $X^\star \subset F(X)$ is a regular language of reduced words. Namely $X^\star \in$ **VPL**^{red}. That means we can encode word equations over the free monoid $X^\star$ as equations over the free group $F(X)$ with the **VPL**^{red} constraint that solutions must be in $X^\star$. Since $X^\star$ is a set of reduced words by definition, every **VPL** in $X^\star$ is also a set of reduced words. Namely, every **VPL** constraint in $X^\star$ can be encoded as a **VPL**^{red} constraint in $F(X)$. Therefore decidability of (**GpEqn**) with **VPL**^{red} constraints in a free group would imply decidability of (**WordEqn**) with **VPL** constraints in the free monoid. This contradicts Proposition 9 above.

Since $\mathbf{VPL}^{red}$ is contained in $\mathbf{CF}^{\forall}$ by Proposition 4, $(\mathbf{GpEqn})$ with $\mathbf{CF}^{\forall}$ must also be undecidable.

Proposition 10 leads to two natural questions.

Question 1. If Conjecture 1 does not hold, and hence $\mathbf{REG}^{\forall} \subsetneq \mathbf{VPL}^{\forall}$, is $(\mathbf{GpEqn})$ with $\mathbf{VPL}^{\forall}$ constraints in free groups decidable?

Question 2. If $(\mathbf{GpEqn})$ with $\mathbf{VPL}^{\forall}$ constraints is decidable in free groups, is $(\mathbf{GpEqn})$ with $\mathbf{DCF}^{\forall}$ constraints also decidable?

5   Conclusions

Visibly Pushdown Languages have not been systematically studied in group theory before, so this paper explores the connections between this class of languages and the natural sets of words one can associate to a finitely generated group. The only work on this topic we are aware of is [14], where a non-standard version of the word problem is considered. In [14] the author shows that the 'nested word problem' for a group G over a generating set X is $\mathbf{VPL}$ if and only if G is virtually free. The *nested word problem* for G is a language L over an alphabet obtained by taking three copies X such that L projects onto the standard word problem $WP(G, X)$ (but is usually not in a bijective correspondence with $WP(G, X)$, nor is it the full word problem of G over the three copies of X).

So far, the study of regular, context-free, indexed, and n-counter languages within groups has been fruitful for algorithmic and structural questions. We hope that this paper can serve as a starting point for studying $\mathbf{VPL}$s in this setting as well. A key future direction would be to further explore Conjecture 1, and the questions that arose at the end of Sect. 4. Possibly the most important questions that this paper raises are about the boundary between decidability and undecidability of Word Equations when the constraints range between rational and context-free, as stated in Questions 1 and 2.

Many other questions can be asked for sets in finitely generated groups beyond the word problem. For example, are there groups with $\mathbf{VPL}$ (but not regular) normal forms, or $\mathbf{VPL}$ geodesics? Similarly, many types of conjugacy languages have been classified from the formal language point of view ([9,10]). Do any such conjugacy languages belong to the $\mathbf{VPL}$ class?

Acknowledgments. We thank Alex Levine and Joel Day for helpful discussions. This work was partially supported by the Heilbronn Institute for Mathematical Sciences (through a Small Grant) and the Deutsche Forschungsgemeinschaft (DFG, German Research Foundation) under Germany's Excellence Strategy – The Berlin Mathematics Research Center MATH+ (EXC-2046/1, EXC-2046/2, project ID: 390685689).

References

1. Alur, R., Kumar, V., Madhusudan, P., Viswanathan, M.: Congruences for visibly pushdown languages. In: Automata, languages and programming, LNCS, vol. 3580, pp. 1102–1114. Springer, Berlin (2005). https://doi.org/10.1007/11523468_89
2. Alur, R., Madhusudan, P.: Visibly pushdown languages. In: Proceedings of the 36th Annual ACM Symposium on Theory of Computing, pp. 202–211. ACM, New York (2004). https://doi.org/10.1145/1007352.1007390
3. Alur, R., Madhusudan, P.: Adding nesting structure to words. In: Developments in language theory, LNCS, vol. 4036, pp. 1–13. Springer, Berlin (2006). https://doi.org/10.1007/11779148_1
4. Anisimov, A.V.: The group languages. Kibernetika (Kiev) **4**, 18–24 (1971)
5. Benois, M.: Parties rationnelles du groupe libre. C. R. Acad. Sci. Paris Sér. A-B **269**, A1188–A1190 (1969)
6. Bucher, M., Talambutsa, A.: Exponential growth rates of free and amalgamated products. Israel J. Math. **212**(2), 521–546 (2016). https://doi.org/10.1007/s11856-016-1299-4
7. Carvalho, A., Nyberg-Brodda, C.F.: On linguistic subsets of groups and monoids (2025). https://arxiv.org/abs/2502.14329
8. Ciobanu, L.: Word equations, constraints, and formal languages. In: Developments in Language Theory, LNCS, vol. 14791, pp. 1–12. Springer, Cham (2024). https://doi.org/10.1007/978-3-031-66159-4_1
9. Ciobanu, L., Hermiller, S.: Conjugacy growth series and languages in groups. Trans. Amer. Math. Soc. **366**(5), 2803–2825 (2014). https://doi.org/10.1090/S0002-9947-2013-06052-7
10. Ciobanu, L., Hermiller, S., Holt, D., Rees, S.: Conjugacy languages in groups. Israel J. Math. **211**(1), 311–347 (2016). https://doi.org/10.1007/s11856-015-1274-5
11. Ciobanu, L., Levine, A.: Languages, groups, and equations. In: Languages and automata—GAGTA Book 3, pp. 63–93. De Gruyter, Berlin (2024)
12. Day, J.D., Ganesh, V., Grewal, N., Manea, F.: On the expressive power of string constraints. Proc. ACM Program. Lang. **7**(POPL) (2023). https://doi.org/10.1145/3571203
13. Diekert, V., Gutierrez, C., Hagenah, C.: The existential theory of equations with rational constraints in free groups is PSPACE-complete. Inform. Comput. **202**(2), 105–140 (2005). https://doi.org/10.1016/j.ic.2005.04.002
14. Henry, C.S.: The (nested) word problem. Inform. Process. Lett. **116**(11), 729–734 (2016). https://doi.org/10.1016/j.ipl.2016.06.013
15. Herbst, T.: On a subclass of context-free groups. RAIRO Inform. Théor. Appl. **25**(3), 255–272 (1991). https://doi.org/10.1051/ita/1991250302551
16. Herbst, T., Thomas, R.M.: Group presentations, formal languages and characterizations of one-counter groups. Theoret. Comput. Sci. **112**(2), 187–213 (1993). https://doi.org/10.1016/0304-3975(93)90018-O
17. Hopcroft, J.E., Ullman, J.D.: Introduction to Automata Theory, Languages, and Computation. Addison-Wesley Series in Computer Science, Addison-Wesley Publishing Co., Reading, MA (1979)
18. Makanin, G.S.: Equations in a free group. Izv. Akad. Nauk SSSR Ser. Mat. **46**(6), 1199–1273, 1344 (1982)

19. Mehlhorn, K.: Pebbling mountain ranges and its application to DCFL recognition. In: Automata, languages and programming (Proceedings of the 7th International Colloquium Noordwijkerhout, 1980), LNCS, vol. 85, pp. 422–435. Springer, Berlin-New York (1980). https://doi.org/10.1007/3-540-10003-2_89
20. Muller, D.E., Schupp, P.E.: Groups, the theory of ends, and context-free languages. J. Comput. System Sci. **26**(3), 295–310 (1983). https://doi.org/10.1016/0022-0000(83)90003-X

Centered Ascending Polyominoes

P. Massazza[1], S. Rinaldi[2], and Lama Tarsissi[3,4]($\boxtimes$)

[1] University of Insubria, Varese, Italy
`paolo.massazza@uninsubria.it`
[2] University of Siena, Siena, Italy
`rinaldi@unisi.it`
[3] SAFIR, Sorbonne University, Abu Dhabi, United Arab Emirates
`lama.tarsissi@sorbonne.ae`
[4] LIGM, Univ. Gustave Eiffel, CNRS, Marne-la-Vallée, France

Abstract. Polyomino enumeration remains a central and challenging problem in enumerative combinatorics. Among convex polyominoes, geometrically defined subclasses such as L-convex and Z-convex families exhibit markedly different enumerative behaviours, showing how geometric constraints influence the nature of the generating function. In this paper, we introduce a new class of convex polyominoes, called *centered ascending polyominoes*, which extends L-convex polyominoes and forms a proper subclass of the Z-convex ones. We provide their first exact enumeration with respect to the semiperimeter (size). Our approach is based on a geometric characterisation involving horizontal inclusion and northeast shift constraints between rows, leading to a decomposition into six natural subclasses. We construct a generating tree that uniquely produces objects of size $n + 1$ from those of size n. This yields a system of functional equations whose solution provides an explicit algebraic generating function, a closed counting formula, and the corresponding asymptotic estimate.

Keywords: polyomino enumeration · convex and k-convex polyominoes · generating functions

1 Introduction

A polyomino [15] is a geometrical figure consisting of a finite set of connected unit squares (called cells) in the plane $\mathbb{Z} \times \mathbb{Z}$, considered up to translations. The enumeration of polyominoes with respect to the semiperimeter or the area remains an open and very difficult problem. In order to model real situations, several subclasses of polyominoes based on convexity and connection have been introduced and studied [4]. In this paper, we are interested in enumerating polyominoes with respect to the semiperimeter (size).

The study of convex polyominoes (polyominoes where the intersection with an infinite horizontal or vertical stripe is a finite segment [4,11,12]) has produced a rich hierarchy of geometrically defined subclasses, ranging from L-convex [8–10]

© The Author(s), under exclusive license to Springer Nature Switzerland AG 2026
M.-P. Béal and P. Caron (Eds.): DLT 2026, LNCS 16578, pp. 117–129, 2026.
https://doi.org/10.1007/978-3-032-28404-4_9

and centered polyominoes to Z-convex [13] and 4-stack polyominoes [14]. These families exhibit remarkably different combinatorial behaviours, suggesting that geometric constraints have a strong impact on the enumerative complexity. We recall that L-convex and centered polyominoes have rational generating functions and growth of order 4^n [8,9,13], and the enumeration of these families can be obtained as a word-encoding in a regular language [1,9]. On the other hand, Z-convex [13,16], 4-stack [14], and convex polyominoes [4,11,12] display algebraic functions and asymptotics of order $n4^n$, $\sqrt{n}\,4^n$, and $n4^n$, respectively.

In this work we introduce a family of convex polyominoes called *centered ascending polyominoes*. They extend the class of L-convex polyominoes and form a proper subclass of the Z-convex ones. We provide the first exact enumeration of centered ascending polyominoes with respect to the size. Our analysis relies on a geometric characterisation of the class: in an ascending polyomino, any two rows must satisfy either a horizontal inclusion condition or a north–east shift constraint.

This property yields a decomposition into six natural subclasses, defined by the behaviour of the extremal rows and columns. We construct a generating tree describing how to uniquely generate centered ascending polyominoes of size $n + 1$ from those of size n, by listing all admissible local extensions for each subclass. The generating tree translates into a system of functional equations satisfied by the generating functions of these subclasses with respect to the size and additional parameters, following methods developed in the recent literature [2,3,12]. Solving the full system yields an explicit algebraic expression for the generating function of centered ascending polyominoes, a closed formula for their number, and the corresponding asymptotic estimate. The approach combines geometric decomposition with functional equation techniques.

2 Definitions and Basic Notions

A *path* in a polyomino P is a pair (q, β) where q is a cell of P and β a string over the alphabet $\{N, W, S, E\}$ of steps in the four directions, such that all the cells met along the path belong to P. The number of *changes of direction* in a path $\beta = \beta_1\beta_2\cdots\beta_r$ is defined as the number of indices i such that $\beta_i \neq \beta_{i+1}$, with $1 \leq i < r$. A path is *monotone* if $\beta \in \{N, W\}^+$ (NW-path) or $\beta \in \{N, E\}^+$ (NE-path) or $\beta \in \{S, E\}^+$ (SE-path) or $\beta \in \{S, W\}^+$ (SW-path). A polyomino is convex if and only if any two of its cells are joined by a monotone path ([7]). For two cells a, b, we write $a \searcharrow b$ (resp., $a \nearrow b$) if there exists a NW-path (resp., NE-path) joining a to b.

For any two cells a, b such that $a \nearrow b$, we define the *NE-distance* of a from b, denoted by $D_{\mathrm{NE}}(a, b)$, as the least number of changes of direction in any NE-path connecting a to b. The *NE-degree* of convexity of P, denoted by $D_{\mathrm{NE}}(P)$, is the least integer k such that for any two cells a, b such that $a \nearrow b$, we have $D_{\mathrm{NE}}(a, b) \leq k$. The NW-distance $D_{\mathrm{NW}}(a, b)$ and the NW-degree of convexity $D_{\mathrm{NW}}(P)$ are defined similarly.

Given $k \geq 0$, a convex polyomino P is said to be *k-convex* if any two of its cells can be joined by a monotone path in P with at most k changes of direction.

Obviously, a k-convex polyomino is also h-convex, for every $h \leq k$. We define the degree of convexity $D(P)$ of a convex polyomino P as the smallest integer k such that P is k-convex, that is, $D(P) = \max\{D_{\mathrm{NE}}(P), D_{\mathrm{NW}}(P)\}$. The authors of [7] proved the interesting following result:

Proposition 1. *Let P be a convex polyomino such that $D_{NW}(P) < D_{NE}(P)$. If $D_{NE}(P) > 2$, then $D_{NW}(P) = 1$.*

This property is important in the study and classification of convex polyominoes, as it describes how the degrees of convexity in the two directions may behave in a generic convex polyomino. In fact, a convex polyomino having degree of convexity greater than 2 must be 1-convex in one of the two directions. Hence, a convex polyomino may be either 2-convex in both directions or 1-convex in one direction and k-convex in the other one (k arbitrary). Figures 2 (a) and (b) show a polyomino where $D_{\mathrm{NW}}(P) = 1$ and $D_{\mathrm{NE}}(P) = 3$. We recall that 1-convex (resp., 2-convex) polyominoes are known as L-convex (resp., Z-convex) polyominoes [10] due to the shape of the monotone path connecting two cells a, b such that $D_{\mathrm{NW}}(a, b) = 1$ or $D_{\mathrm{NE}}(a, b) = 1$, see Fig. 1 (b) (resp., $D_{\mathrm{NW}}(a, b) = 2$ or $D_{\mathrm{NE}}(a, b) = 2$, see Fig. 1 (d)).

Lastly, within the class of L-convex polyominoes we have *stacks*. These are the convex polyominoes P which share exactly two adjacent corners with the minimal bounding rectangle including P, see Fig. 1 (a).

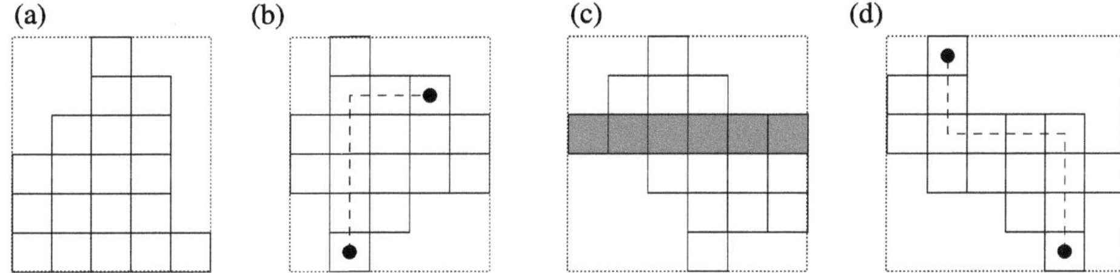

Fig. 1. (a) Stack polyomino; (b) L-convex polyomino; (c) Z-convex centered polyomino; (d) Z-convex (non centered) polyomino.

In this work, we are interested in enumerating polyominoes with respect to the semi-perimeter (size). So we recall that L-convex polyominoes have a rational generating function and grow like $\left(2 + \sqrt{2}\right)^n$. On the other hand, Z-convex polyominoes and convex polyominoes have the same rough asymptotic behavior $\frac{n}{8} 4^n$ [11] and algebraic generating functions. From these observations, the authors highlight the problem of studying classes intermediate between L-convex and Z-convex polyominoes, and of understanding how the geometric properties of the family determine the algebraicity or rationality of the generating function and its asymptotic behaviour.

For $x, y \geq 1$, let us denote by $C(x, y)$ the family of convex polyominoes for which the NE-degree of convexity is x and the NW-degree of convexity is y, see

Fig. 2 (a),(b). Notice that $C(0,0)$ only includes vertical and horizontal bars. The authors of [6] proved that Z-convex polyominoes (not L-convex) are given by the disjoint union of $C(1,2)$, $C(2,1)$, and $C(2,2)$, and that, for $k > 2$, k-convex polyominoes (not $(k-1)$-convex) are given by the disjoint union $C(1,k) \cup C(k,1)$.

A convex polyomino is said to be *h-centered*, (briefly, *centered*) if it contains at least one row touching both the left and the right side of its minimal bounding rectangle (see Fig. 1 (c)). We define:

$$A = \bigcup_{k \geq 1} C(k,1) \cup C(0,0) \quad (\text{resp.,} \quad D = \bigcup_{k \geq 1} C(1,k) \cup C(0,0))$$

as the family of *ascending* (resp. *descending*) convex polyominoes. It is immediate that for any given size, the number of ascending convex polyominoes is equal to the number of descending convex polyominoes. Their intersection is the family of L-convex polyominoes. Then, we have the following combinatorial identity:

Proposition 2. *For every $n \geq 2$, we have: $c_n = 2a_n + k_n - g_n$, where c_n (resp., a_n, k_n, g_n) denotes the number of convex (resp. ascending, $C(2,2)$, L-convex) polyominoes of size n.*

In this paper, we focus on the enumeration of ascending convex polyominoes which are also centered (briefly, centered ascending polyominoes). We believe that enumerating this family will contribute to a deeper understanding of both Z-convex polyominoes and convex polyominoes.

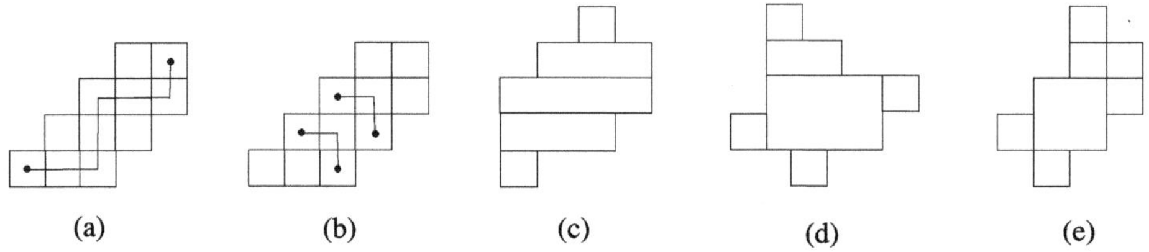

(a)   (b)   (c)   (d)   (e)

Fig. 2. (a),(b) Convex polyomino in $C(3,1)$. (c) Centered ascending polyomino; (d) Non-centered polyomino in $C(2,2)$. (e) Non-centered ascending polyomino in $C(2,1)$.

Centered ascending polyominoes can be characterized geometrically by examining the relative positioning of the rows. Let $R = [x_{\min}, x_{\max}] \times [y_{\min}, y_{\max}]$ be the minimal bounding rectangle of P. For any row r, let x_r and x'_r be the abscissas of the leftmost and rightmost vertical grid lines of r. Then the left-distance and right-distance of r are defined respectively by

$$d_\ell(r) = x_r - x_{\min}, \qquad d_r(r) = x_{\max} - x'_r.$$

We define the following relations on the rows of a convex polyomino:

i) the relation I, (*row inclusion*). Given two rows s, t, we say that $(s, t) \in I$ iff $x_s \leq x_t$ and $x'_t \leq x'_s$, or $x_s \geq x_t$ and $x'_t \geq x'_s$.

ii) the relation N, (*north-east shift*). Given two rows s, t, such that s is below t, we say that $(s, t) \in N$ if $x_s < x_t$ and $x'_s < x'_t$.

The following characterisation reveals that centered ascending polyominoes, denoted by $\mathcal{A}$, are intermediate between L-convex and centered polyominoes:

Proposition 3. *A centered polyomino is ascending iff for every pair of its rows* r_1, r_2, *with* r_1 *lying below* r_2, *one has* $(r_1, r_2) \in I$ *or* $(r_1, r_2) \in N$.

Given $P \in \mathcal{A}$, the set of rows with left and right distance equal to zero constitutes the supporting rectangle of maximal height, referred to as the *base* of P. We denote by $b(P)$ (resp., $\ell(P)$) the height (resp., the width) of the base of P. Let $\mathcal{L}$ (resp., $\mathcal{F}$) denote the family of centered ascending polyominoes P with one cell in the first column (resp., $b(P) > 1$). We indicate by $a_n = |\mathcal{A}(n)|$ (resp., $l_n = |\mathcal{L}(n)|$, $f_n = |\mathcal{F}(n)|$) the number of polyominoes of size n in $\mathcal{A}$ (resp., $\mathcal{L}$, $\mathcal{F}$).

3   Enumeration of Centered Ascending Polyominoes

We associate to a polyomino $P \in \mathcal{L}$ a label (w), where $w = w(P) \geq 1$ is an integer encoding the combinatorial information needed to describe its growth. If P is a *downstack* (that is, if P contains no cell above its base), then $w = \ell(P)$, see Fig. 5 (a). Otherwise, w is the left-distance of the row immediately above the base of P (Fig. 5 (b)). We distinguish downstacks by adding a bar to the label: thus, downstacks are labeled $(\overline{w})$, whereas non-downstacks are labeled (w).

Proposition 4. *For* $n \geq 2$ *we have:*

1. $f_n = a_{n-1}$;
2. $a_n = l_{n+1} - l_n$.

Proof.   1. There is a simple bijection between $\mathcal{A}(n-1)$ and $\mathcal{F}(n)$: take $P \in \mathcal{A}(n-1)$ and add a row of length $\ell(P)$ above its base. You obtain $P' \in \mathcal{F}(n)$. Conversely, if $Q \in \mathcal{F}(n)$ (i.e., $b(Q) > 1$), you can delete the upper row of its base and obtain $Q' \in \mathcal{A}(n-1)$.

2. We prove that for $n > 2$ we have $l_n = l_{n-1} + a_{n-1}$ by showing a decomposition of the set $\mathcal{L}(n)$ into two disjoint families, counted respectively by l_{n-1} and a_{n-1}. Let $P \in \mathcal{L}(n)$ and distinguish two cases:

   (a) $(w(P) \geq 2)$: Let P' be the polyomino obtained by deleting the leftmost cell of the base. It is immediate that $P' \in \mathcal{A}(n-1)$. In other words, P can be obtained in exactly one way by applying the operation *"add a cell to the left of the top-left corner of the base"* to a suitable $P' \in \mathcal{A}(n-1)$. The number of such polyominoes P is therefore a_{n-1};

   (b) $(w(P) = 1)$: Let P' be the polyomino obtained by deleting the upper row B_u of the rectangle of width $\ell(P) - 1$ immediately above the base of P (see Fig. 3); observe that B_u is not null because of the label $w = 1$. It is immediate that P' belongs to $\mathcal{L}(n-1)$. In other words, P can be obtained in exactly one way by applying the operation *"add a row of length $\ell(P) - 1$ and left-distance 1 on top of the base"* to a suitable $P' \in \mathcal{A}(n-1)$. The number of such polyominoes P is therefore l_{n-1}.

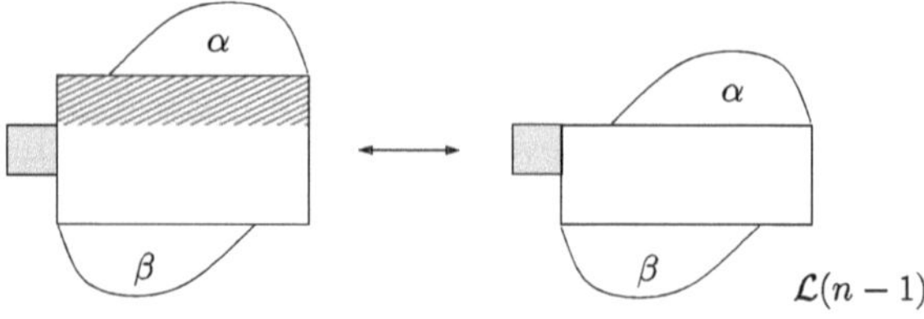

Fig. 3. A polyomino in $\mathcal{L}(n)$ obtained as in case (b).

Generating Tree for $\mathcal{L}$-polyominoes. We provide exact enumeration of $\mathcal{L}$ using the method of generating trees: we introduce operations that generate $\mathcal{L}$-polyominoes of size $n + 1$ from those of size n, then, we define subclasses of $\mathcal{L}$-polyominoes and, for each subclass, specify which operations can be applied, so that the generation of $\mathcal{L}$-polyominoes of size $n + 1$ is unique.

- **Left Cell.** This operation can be applied to any $\mathcal{L}$-polyomino. It attaches a new cell to the left of the base of P. This yields a polyomino P' satisfying $w(P') = w(P) + 1$ (see Fig. 4 (lc)). Observe that P' is a downstack if and only if P is a downstack. Let

$$\mathcal{L}^{\oplus} = \{P \in \mathcal{L} : \exists\, P' \in \mathcal{L} \text{ st } P \text{ is obtained from } P' \text{ textbythe Left Cell operation}\}.$$

Equivalently, $\mathcal{L}^{\oplus}$ consists of those $\mathcal{L}$-polyominoes that extend at least two cells to the left, that is, $w(P) > 1$ and $d_\ell(r_d) > 1$, where r_d is the row immediately below the base (the condition $w(P) > 1$ alone is not sufficient for a polyomino to belong to $\mathcal{L}^{\oplus}$).

- **Right Cell.** This operation can be applied to all $\mathcal{L}$-polyominoes except those in $\mathcal{L}^{\oplus}$, in order to avoid ambiguity. It attaches a new cell to the right side of the base of P. If P is a downstack, then the resulting polyomino P' is still a downstack and satisfies $w(P') = w(P) + 1$. Otherwise, P is not a downstack and one has $w(P') = w(P)$ (see Fig. 4 (rc)).

- **Up-Row.** This operation adds a row of left–distance 1 and length $\ell(P) - 1$ above the base of P. It applies only to polyominoes in $\mathcal{L}^{\oplus}$ and produces an $\mathcal{L}$-polyomino P' which is neither a downstack nor an element of $\mathcal{L}^{\oplus}$ since $w(P') = 1$ (see Fig. 4 (ur)).

- **Down-Row.** This operation adds a row of left–distance 1 and length $\ell(P) - 1$ below the base of P. It applies only to polyominoes $P \in \mathcal{L}$ such that the row immediately above the base (if any) has a right-distance $u > 0$. It produces an $\mathcal{L}$-polyomino P', with $w(P') = w(P)$, which is not an element of $\mathcal{L}^{\oplus}$ and has at least two cells in the rightmost column (see Fig. 4 (dr)).

- **Shift.** This operation can be applied to all $\mathcal{L}$-polyominoes not belonging to $\mathcal{L}^{\oplus}$. For simplicity, we distinguish two cases:

(1) if P is a downstack (that is, $w(P) = \ell(P)$ and the left–distance of the row below the base is 1 since $P \notin \mathcal{L}^{\oplus}$) the *Shift* operation produces $\ell(P) - 1$ polyominoes obtained by adding a row with right–distance 0 and length ranging from 1 to $\ell(P) - 1$ on top of the uppermost row of the base of P (see Fig. 5(a)).

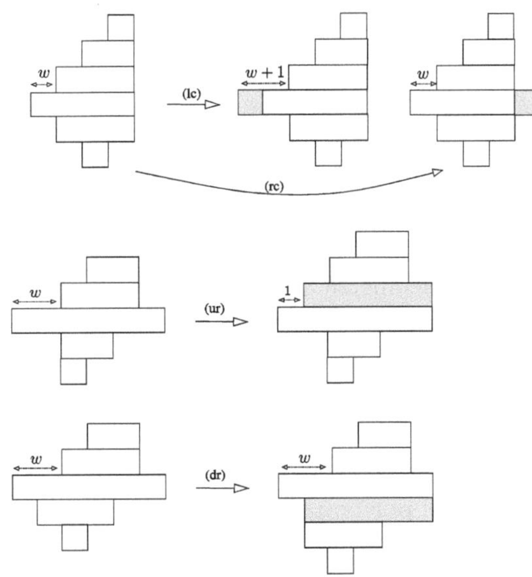

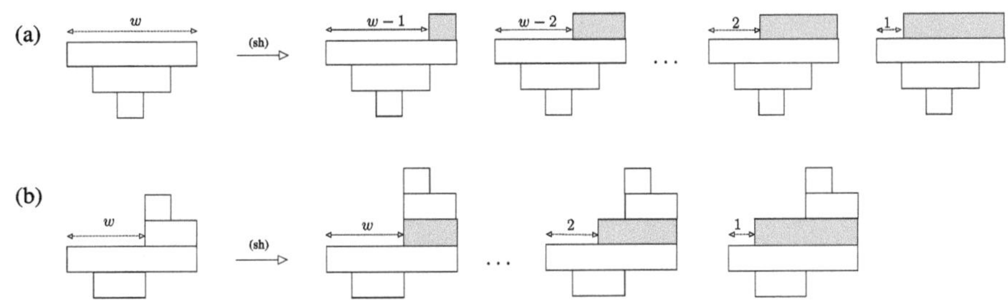

Fig. 4. The application of the operations *Left Cell* (lc), *Right Cell* (rc), *Up-Row* (ur) and *Down-Row* (dr) to non-downstack $\mathcal{L}$-polyominoes.

Fig. 5. The application of the operation *Shift*: (a) to a downstack; (b) to an $\mathcal{L}$-polyomino which is not a downstack.

The polyominoes P' obtained are neither downstacks nor elements of $\mathcal{L}^{\oplus}$, and satisfy the relation $1 \leq w(P') < w(P)$.

(2) Otherwise, the *Shift* operation adds a row with right–distance 0 and length ranging from $\ell(P) - w(P)$ to $\ell(P) - 1$ on top of the uppermost row of the base of P (see Fig. 5(b)). It therefore produces $\ell(P) - w(P)$ polyominoes P', which are neither downstacks nor elements of $\mathcal{L}^{\oplus}$, and with $w(P) \leq w(P') < \ell(P)$.

Observe that the condition $P \notin \mathcal{L}^{\oplus}$ is necessary to avoid ambiguity in generation. Indeed, if the *Shift* operation were applied to a polyomino $P \in \mathcal{L}^{\oplus}$, two situations would arise. If the added row has length smaller than $\ell(P) - 1$, the resulting polyomino would still belong to $\mathcal{L}^{\oplus}$ and would already be obtained by the operation *Left Cell*. If the added row has length $\ell(P) - 1$, the resulting polyomino would not belong to $\mathcal{L}^{\oplus}$, but it would coincide with the polyomino obtained by applying the operation *Up-Row*.

Theorem 1. *Let P be a $\mathcal{L}$-polyomino of size $n + 1$. Then, there is only one $\mathcal{L}$-polyomino P' of size n such that P can be obtained from P' by applying one among the operations* Left Cell, Right Cell, Up-Row, Down-Row *and* Shift.

The statement follows by defining, for each $\mathcal{L}$-polyomino P of size $n + 1$, a unique predecessor $\phi(P)$ of size n obtained by removing a suitable cell or row. A case analysis based on the geometric position of P within its minimal bounding rectangle shows that exactly one of the operations *Left Cell, Right Cell, Up-Row, Down-Row*, or *Shift* applies, and that these cases are mutually exclusive and exhaustive.

The full proof is omitted for brevity.   □

Succession Rule. We divide $\mathcal{L}$ into classes according to the geometric properties of their elements, so that all polyominoes within a given class are subject to the same set of operations. This classification is based on three properties:

- whether P is a downstack or not;
- whether P belongs to $\mathcal{L}^{\oplus}$ or not; if $P \in \mathcal{L}^{\oplus}$, its label carries the subscript l;
- if P is not a downstack, consider the right-distance u of the row immediately above the base. If $u > 0$, then the label of P carries a prime, namely $(w)'$.

Downstacks: For downstack polyominoes we use the label $(\overline{w})_l$ or $(\overline{w})$.

- **Downstacks in $\mathcal{L}^{\oplus}$:** Let us denote by S_l family of downstacks in $\mathcal{L}^{\oplus}$, with label $(\overline{w})_l$. To each such polyomino, we may apply the operations *Left Cell, Down-Row*, and *Up-Row*, leading to the productions (see Fig. 6):

$$(\overline{w})_l \;\rightarrow\; (\overline{w+1})_l \; (\overline{w}) \; (1).$$

- **Downstacks not in $\mathcal{L}^{\oplus}$:** Let us denote by S family of downstacks not in $\mathcal{L}^{\oplus}$, with label $(\overline{w})$. We may apply the operations *Left Cell, Right Cell, Down-Row*, and *Shift*, leading to the following productions (see Fig. 6):

$$(\overline{w}) \;\rightarrow\; (\overline{w+1})_l \; (\overline{w+1}) \; (\overline{w}) \; (1) \; (2) \; \dots \; (w-1).$$

Non-downstack polyominoes: We further distinguish between polyominoes in $\mathcal{L}^{\oplus}$ and not, and right-distance $u > 0$ or $u = 0$.

- $P \in \mathcal{L}^{\oplus}$, $u = 0$: Let us denote by N_l this family of polyominoes, with label (w). To a polyomino in N_l we may apply the operations *Left Cell* and *Up-Row* leading to the production $(w)_l \;\rightarrow\; (w+1)_l \; (1)$, (see Fig. 7).

- $P \in \mathcal{L}^{\oplus}$, $u > 0$: Let us denote by N_l' this family of polyominoes, with label $(w)_l'$. To a polyomino in N_l' we may apply the operations *Left Cell*, *Down-Row*, and *Up-Row*. Then, we have $(w)_l' \rightarrow (w+1)_l' (w)' (1)$.

- $P \notin \mathcal{L}^{\oplus}$, $u = 0$: Let us denote by N this family of polyominoes, with label (w). To a polyomino in N we may apply the operations *Left Cell*, *Right Cell* and *Shift*, leading to the following productions (see Fig. 8):

$$ (w) \rightarrow (w+1)_l (w)' (1) \ldots (w). $$

- $P \notin \mathcal{L}^{\oplus}$, $u > 0$: Let us denote by N' this family of polyominoes, with label $(w)'$. To a polyomino in N' we may apply the operations *Left Cell*, *Right-Cell*, *Down-Row* and *Shift*, leading to the following productions , (see Fig. 8):

$$ (w)' \rightarrow (w+1)_l' (w)' (w)' (1) \ldots (w). $$

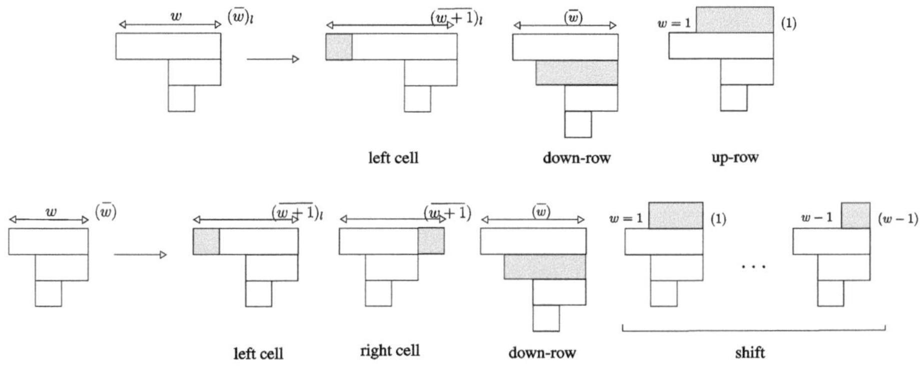

Fig. 6. The productions of downstack polyominoes with labels $(\overline{w})_l$ and $(\overline{w})$.

Solving the Functional Equation. Using standard techniques ([2, 4, 12]), the generating tree can be translated into a system of functional equations satisfied by the generating functions of the classes of $\mathcal{L}$-polyominoes described above (Fig. 9).

For any given class K of centered ascending polyominoes, we denote by $K(x; t) = \sum_{P \in K} x^{w(P)} t^{s(P)}$, the generating function of K according to the parameter w and the size s. Concerning downstacks we have:

$$ S_l(x) = x^2 t^3 + xt \, S_l(x) + xt \, S(x) \tag{1} $$

$$ S(x) = t \, S_l(x) + t(1 + x) \, S(x) \tag{2} $$

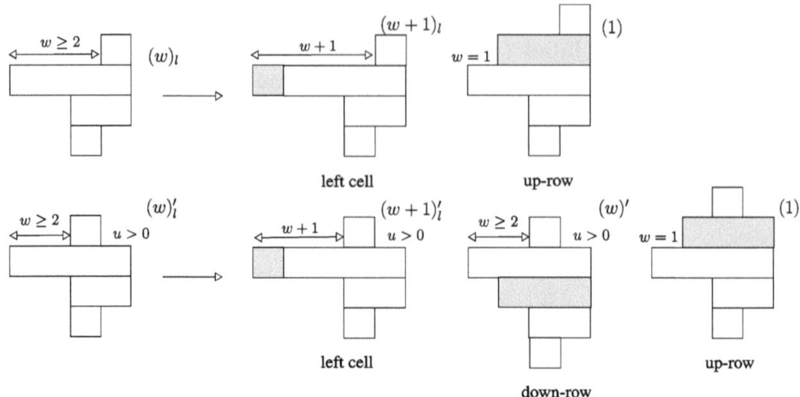

Fig. 7. The productions of a non-downstack polyomino in $\mathcal{L}^{\oplus}$.

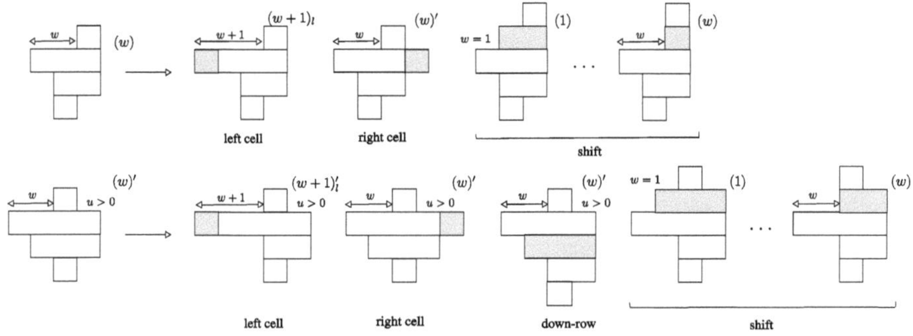

Fig. 8. The productions of polyominoes in N (up) and in N' (down).

The solution of Eqs. (1), (2) by standard substitutions gives the (rational) known generating functions of families of downstack (stack) polyominoes:

$$S_l(x) = \frac{x^2 t^3 \left(1 - xt - t\right)}{t^2 x^2 - 2tx - t + 1} \qquad S(x) = \frac{x^2 t^4}{t^2 x^2 - 2tx - t + 1}. \tag{3}$$

The equations for the other families are:

$$N_l(x) = xt\, N_l(x) + xt N(x) \tag{4}$$

$$N_l'(x) = xt\, N_l'(x) + xt N'(x) \tag{5}$$

$$N(x) = xt\, S_l(1) + xt\, N_l(1) + xt\, N_l'(1) + \frac{tx}{1-x}\left(S(1) - \frac{1}{x} S(x)\right)$$

$$+ \frac{tx}{1-x}\left(N(1) - N(x)\right) + \frac{tx}{1-x}\left(N'(1) - N'(x)\right) \tag{6}$$

$$N'(x) = t\, N_l'(x) + t\, N(x) + 2t\, N'(x) \tag{7}$$

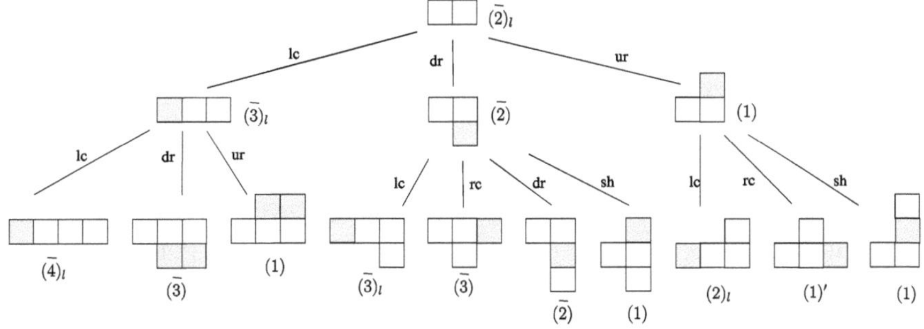

Fig. 9. The first levels of the generating tree of $\mathcal{L}$-polyominoes.

Now, using (4), (5), (7), and (3), we can express (6) in terms of the unknowns $N(x)$ and $N(1)$, precisely:

$$N(x) \cdot ker(x) = \alpha(x)\, N(1) + \beta(x),\tag{8}$$

where $ker(x) = (2t - 1)\left(t\,x^2 - x + 1\right)\left(t^2 x^2 - 2tx - t + 1\right)\left(t^2 - 3t + 1\right)(t - 1)$.
We can solve Eq. (8) by a neat application of the *kernel method* [3,4]. Indeed,
the *kernel* $ker(x)$ vanishes for:

$$x_1 = \frac{1 + \sqrt{1 - 4t}}{2t}, \quad x_2 = \frac{1 - \sqrt{1 - 4t}}{2t} \quad x_3 = \frac{1 - \sqrt{t}}{t}, \quad x_4 = \frac{1 + \sqrt{t}}{t}.$$

Of the four solutions, only $x_2(t)$ is a proper formal power series (and it is
precisely the generating function of *Catalan numbers*). Now, substituting $v = x_2(t)$ in (8), we have the equation $0 = \alpha(\,x_2(t)\,N(1)) + \beta(x_2(t))$, with the single
unknown $N(1)$, the generating function of the class N according to the size:

$$N(1) = \frac{\left(\sqrt{1 - 4t}\,t - 2t^2 - \sqrt{1 - 4t} - t + 1\right)\left(\sqrt{1 - 4t} + 4t - 1\right)t^2\,(1 - t)}{4\,(1 - 4t)\,(1 - 2t)}.\tag{9}$$

By substituting $N(1)$ in the system above, we obtain the generating functions
of the classes N_l, N_l', N. Finally, summing the generating functions of all the
classes, we obtain the generating function of $\mathcal{L}$-polyominoes:

$$\mathcal{L}(1) = \frac{t^2\left(\sqrt{1 - 4t} + 4t - 1\right)}{2(1 - 4t)} = t^3 + 3t^4 + 10t^5 + 35t^6 + 126t^7 + 462t^8 + 1716t^9 + \ldots$$

Proposition 5. *The number l_n of centered ascending convex polyominoes in $\mathcal{L}$
is:*

$$l_{n+2} = \binom{2n - 1}{n} = \frac{1}{2}\binom{2n}{n}.$$

Using Propositions 4 and 5, we have:

Theorem 2. *The number a_n of centered ascending polyominoes is:*

$$a_{n+2} = \frac{3n+1}{n+1}\binom{2n-1}{n} = \frac{3n+1}{2}\, C_n \quad n \geq 0, \tag{10}$$

where $C_n = \frac{1}{n+1}\binom{2n}{n}$ is the nth Catalan number.

The sequence a_n is in OEIS with reference A097613 [17]. Using Stirling's approximation, $\binom{2n}{n} \sim \frac{4^n}{\sqrt{\pi n}}$, we have that $a_n \sim \frac{7}{4} \cdot \frac{4^n}{\sqrt{\pi n}}$. We point out that the generating functions of ascending centered polyominoes are algebraic (as that of Z-convex, convex ones), while the generating functions for L-convex and centered polyominoes are rational.

Further Work. This work provides a structural characterization, and thus a generating tree, for centered ascending polyominoes. Although the counting formula for $\mathcal{L}$-polyominoes is well known and widely used (and the expression for a_n is relatively simple), their enumeration is nontrivial, and no bijective proof of these formulas is known.

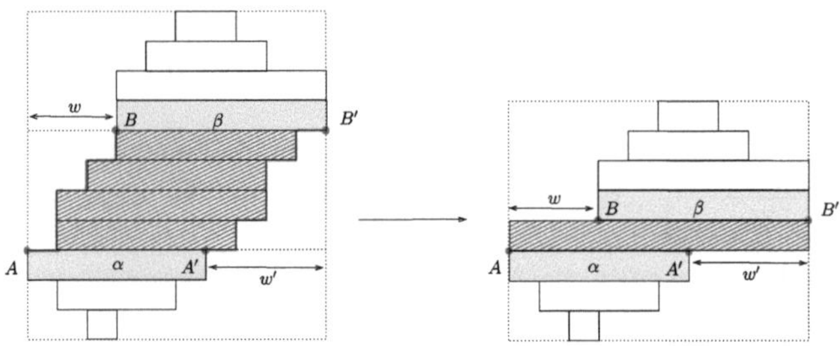

Fig. 10. An ascending non-centered polyomino and its unique decomposition into an ascending centered polyomino and a pair of non-intersecting lattice paths.

Our next goal is to extend this approach to the enumeration of general ascending polyominoes. The key idea is that every non-centered polyomino admits a canonical decomposition showing that the centered subclass is, in a precise sense, the combinatorially most complex one (see Fig. 10). Let P be an ascending polyomino that is not horizontally centered. Then some rows touch only the left boundary and others only the right boundary of its bounding rectangle. Let α be the highest row touching the left side and β the lowest row touching the right side. Removing the rows above α and below β, and inserting a full row between them, produces a centered ascending polyomino Γ. The deleted part is not arbitrary: its boundaries form a unique pair of nonintersecting lattice paths connecting the extremal vertices of α and β. Hence every polyomino decomposes into a centered ascending polyomino and a pair of nonintersecting lattice paths. This suggests that the enumeration of general ascending polyominoes factors into two independent components: the centered class and the admissible path pairs.

References

1. Battaglino, D., Fedou, J.M., Frosini, A., Rinaldi, S.: Encoding centered polyominoes by means of a regular language. In: Mauri, G., Leporati, A. (eds.) Developments in Language Theory. DLT 2011. LNCS, vol 6795. Springer, Berlin, Heidelberg (2011). https://doi.org/10.1007/978-3-642-22321-1_40
2. Banderier, C., Bousquet-Mélou, M., Denise, A., Flajolet, P., Gardy, D., Gouyou-Beauchamps, D.: Generating functions for generating trees. Discr. Math. **246**(1–3), 29–55 (2002)
3. Barcucci, E., Del Lungo, A., Pergola, E., Pinzani, R.: ECO: a methodology for the enumeration of combinatorial objects. J. Diff. Eq. App. **5**, 435–490 (1999)
4. Bousquet-Mélou, M.: A method for the enumeration of various classes of column convex polygons. Disc. Math. **154**, 1–25 (1996)
5. Boussicault, A., Rinaldi, S., Socci, S.: The number of directed k-convex polyominoes. Discr. Math. **343**(3), 111731 (2020)
6. Brocchi, S., Castiglione, G., Massazza, P.: On computing the degree of convexity of polyominoes. Electron. J. Comb. **22**(1), 7 (2015)
7. Brocchi, S., Castiglione, G., Massazza, P.: On the exhaustive generation of k-convex polyominoes. Theoret. Comput. Sci. **664**, 54–66 (2017)
8. Castiglione, G., Frosini, A., Munarini, E., Restivo, A., Rinaldi, S.: Combinatorial aspects of L-convex polyominoes. European J. Combin. **28**, 1724–1741 (2007)
9. Castiglione, G., Frosini, A., Restivo, A., Rinaldi, S.: Enumeration of L-convex polyominoes by rows and columns. Theor. Comp. Sci. **347**, 336–352 (2005)
10. Castiglione, G., Restivo, A.: Reconstruction of L-convex polyominoes. Electron. Notes Discrete Math. **12** (2003)
11. Delest, M., Viennot, X.: Algebraic languages and polyominoes enumeration. Theor. Comp. Sci. **34**, 169–206 (1984)
12. Del Lungo, A., Duchi, E., Frosini, A., Rinaldi, S.: On the generation and enumeration of some classes of convex polyominoes. Electron. J. Combinatorics **11**, #R60 (2004)
13. Duchi, E., Rinaldi, S., Schaeffer, G.: The number of Z-convex polyominoes. Adv. Appl. Math. **4**, 54–72 (2008)
14. Fedou, J.M., Frosini, A., Rinaldi, S.: Enumeration of 4-stack polyominoes. Theor. Comp. Sci. **502**, 88–97 (2013)
15. Golomb, S.W., S.W.: Checker boards and polyominoes. Amer. Math. Monthly **61**, 675–682 (1954)
16. Guttmann, T., Massazza, P.: Asymptotics of Z-convex polyominoes. RAIRO-Theor. Inf. Appl. **58**, 12 (2024)
17. OEIS Foundation Inc.: The on-line encyclopedia of integer sequences (2011). https://oeis.org/

The Joker Game Only Characterizes History-Determinism in Parity Automata With up to Two Priorities

Dorian Guyot[(✉)] and Ulrich Ultes-Nitsche

University of Fribourg, Fribourg, Switzerland
{dorian.guyot,uun}@unifr.ch

Abstract. The recently proved 2-token theorem [10] shows that history-determinism, a "mild" form of non-determinism, is characterized by the 2-token game on parity automata. This result yields a practical tool for testing the history-determinism of parity automata, but the authors also show that for basic parity automata (Büchi and co-Büchi), solving a simpler game (the joker game) is enough. Unfortunately, this does not extend very far up the parity hierarchy: a counterexample with parity index $[1,3]$ is provided, answering the question for this parity index and all above it. However, the question was left open for parity automata with index $[0,2]$. We answer it negatively by providing a counterexample.

Keywords: Parity automata · History-determinism · Games on Graphs

1 Introduction

First introduced by Henzinger and Piterman [7], history-deterministic automata allow a mild form of nondeterminism, where the nondeterministic choices can be resolved on the fly by looking at the prefix read so far. These automata turn out to retain some of the properties that make deterministic automata interesting for synthesis or verification tasks [3, 6, 9].

The idea of history-determinism is captured by the history-determinism (HD) game. In this game, two players (Adam and Eve) take turns picking letters and transitions, respectively. Adam builds a word, and Eve builds a run on that word. Eve is declared the winner if her run is accepting, or if Adam's word is not in the language. If Eve wins the game and therefore has a strategy to win, this strategy can be used to resolve the nondeterministic choices in the automaton, making it possible to build an accepting run in a unique way for any word in the language and thus recover some of the practicality of deterministic automata.

The complexity of testing for history-determinism in parity automata has also been studied. Solving the HD game directly leads to an EXPTIME [7] algorithm, as it involves the determinisation of the automaton. However, there was a significant complexity gap, as for a long time, the best known lower bound

M.-P. Béal and P. Caron (Eds.): DLT 2026, LNCS 16578, pp. 130–138, 2026.
https://doi.org/10.1007/978-3-032-28404-4_10

was that of solving parity games, which is in $\text{NP} \cap \text{co-NP}$ [8] and can be solved in quasi-polynomial time [5]. More recently, Prakash [11] provided a better lower bound and showed that this problem is in fact NP-hard. PTIME algorithms were found for co-Büchi automata [8], using the so-called joker game to build another automaton used to test the history-determinism of the original one, and for Büchi automata [2], using the 2-token game, which is a game similar to the HD game, but where Adam also has tokens (two to be exact) and constructs two runs. Later, it was proven that the 2-token game also characterizes HD for co-Büchi automata [4]. This led to the 2-token conjecture, which states that the 2-token game characterizes history-determinism on all parity games. The 2-token conjecture has been proven recently by Lehtinen and Prakash [10]. Their proof additionally provides a way to decide HD for parity automata in PTIME for a fixed parity index and in PSPACE otherwise.

In the recent work of Acharya, Jurdzinski, and Prakash [1], they show that the joker game is a sufficient condition for HD for Büchi and later [12] also for co-Büchi automata (which are the parity automata with indices $[0, 1]$ and $[1, 2]$ respectively). This makes it practically relevant, as the size of the arena of joker games is $\mathcal{O}(n^2)$ while that of the 2-token games is $\mathcal{O}(n^3)$ for n states. However, this does not generalize very well to other parity automata: they provide an automaton of index $[1, 3]$ where the joker game no longer suffices (Eve wins the joker game, but the automaton is not history-deterministic). This result also automatically holds for all parity indices above $[1, 3]$, but the case of $[0, 2]$-automata was left unanswered. We provide a counterexample with parity index $[0, 2]$. This completes the picture and shows that among all parity automata, the joker game characterizes history-determinism only on Büchi and co-Büchi automata.

2   Preliminaries

Let $\mathbb{N} = \{0, 1, 2, ...\}$ be the set of natural numbers. For $i, j \in \mathbb{N}$ s.t. $i < j$, we write $[i, j]$ for the set $\{i, i+1, ..., j\} \subset \mathbb{N}$ and $[j]$ for $\{0, 1, ..., j\} \subset \mathbb{N}$. An *alphabet* Σ is a finite set of *letters*. The set of words of finite length and the set of words of countably infinite length over Σ are denoted by Σ^* and Σ^ω, respectively. A *language* $\mathcal{L} \subseteq \Sigma^\omega$ is a set of words.

2.1   Games

An *arena* is a directed graph $G = (V, E)$ where V is partitioned into vertices owned by Adam: V_A; and vertices owned by Eve: V_E. Some vertex $v_0 \in V$ is designated as the initial vertex of the game. Vertices in arenas are also called *positions*.

A *play* on this arena is an infinite path built in the following way: it starts with a token in v_0 and is built by adding edges in infinitely many rounds: the player owning the vertex on which the token is currently located chooses one of the outgoing edges, adds it to the path and moves the token along this edge.

A *game* $\mathcal{G} = (G, \mathcal{L})$ consists of an arena $G = (V, E)$ and a winning condition specified through a language $\mathcal{L} \subseteq E^\omega$. A play ρ is won by Eve if $\rho \in \mathcal{L}$ and by Adam otherwise. A *strategy* for Eve is a function from the set of play prefixes where the last transition ends on a vertex $v_e \in V_E$ that maps any such play prefix to a transition leaving v_e. A strategy is winning if all the plays induced by it are winning. If there exists a winning strategy for Eve on a game $\mathcal{G}$, we say that Eve wins $\mathcal{G}$.

2.2   Automata

Parity Automata. A parity automaton $\mathcal{A} = (Q, \Sigma, q_0, \Delta)$ with index $[i, j]$ is a directed graph with a finite number of vertices (*states*) Q. The set of edges Δ contains the *transitions*. Transitions are labeled by *letters* in Σ and *priorities* in $[i, j]$ for $i, j \in \mathbb{N}$, so $\Delta \subseteq Q \times \Sigma \times [i, j] \times Q$. A transition from state p to q on letter a with priority c is denoted $p \xrightarrow{a:c} q$. Parity automata have an *initial state* $q_0 \in Q$.

A *run* on $w \in \Sigma^\omega$ is an infinite path in the automaton that starts in q_0 and follows transitions labeled with the letters of w sequentially. Parity automata are equipped with a winning condition that defines which runs are *accepting* and which are *rejecting*. The winning condition for parity automata is the parity condition: if the lowest priority occurring infinitely often in a run of the automaton on $w \in \Sigma^\omega$ is even, the run is accepting. Otherwise, the run is rejecting. A word $w \in \Sigma^\omega$ is accepted by a parity automaton $\mathcal{A}$ if it admits an accepting run on w. The set of all words accepted by automaton $\mathcal{A}$ is the *language* of $\mathcal{A}$, denoted $\mathcal{L}(\mathcal{A})$. An automaton is called *deterministic* if, from any state $q \in Q$ and on any given letter $a \in \Sigma$, there is at most a single transition.

Parity Index. The parity index $[i, j]$ of a parity automaton $\mathcal{A}$ denotes the range of priorities that $\mathcal{A}$ uses. It is always possible to relabel $\mathcal{A}$'s parity index such that $i \in \{0, 1\}$ by shifting all priorities down by steps of 2 without altering the behavior of $\mathcal{A}$. A Büchi automaton is a parity automaton with parity index $[0, 1]$ and a co-Büchi automaton is a parity automaton with index $[1, 2]$.

Parity indices form a hierarchy by means of inclusion (see Fig. 1).

2.3   Games on Automata

Games are a tool that arises naturally when considering problems like controller synthesis, where a controller (the first player) has to find a winning strategy against an environment (the second player). The controller wins if it is able to find such a strategy. Transposed to automata, this idea makes it possible to characterize properties of automata by the existence of such a strategy when playing certain games on them. One way to play games on automata is to use one or more tokens, which are moved from state to state by the players along the transitions of the automaton while adding a winning condition to the runs that these tokens create. Here, we list the games relevant to this work:

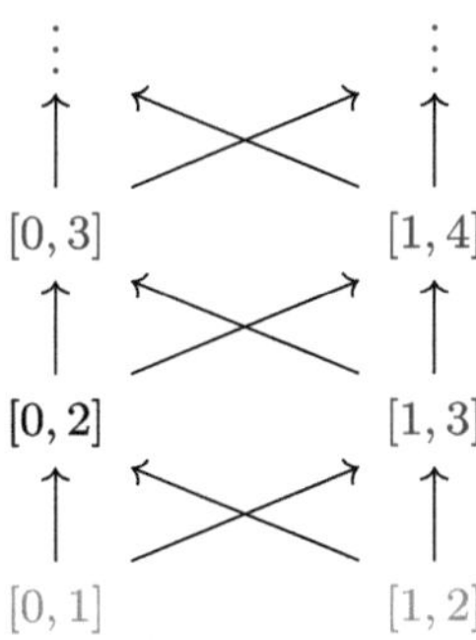

Fig. 1. The parity index hierarchy. Arrows indicate inclusion. The color of an index indicates if the joker game characterizes history determinism on automata with this index: green for yes ($[0,1]$, $[1,2]$), blue for no ($[1,3]$ and up) and black for our result ($[0,2]$). (Color figure online)

Definition 1 (History-determinism game). *The history-determinism game on $\mathcal{A} = (Q, \Sigma, q_0, \Delta)$ is played by the players Adam and Eve in infinitely many rounds with positions in Q. Initially, the token is at state q_0, and at round $i \in \mathbb{N}$, the token is at state $q_i \in Q$. Each round proceeds as follows:*

1. *Adam picks a letter $a_i \in \Sigma$;*
2. *Eve picks a transition $q_i \xrightarrow{a_i : c_i} q_{i+1} \in \Delta$.*

Eve wins the game if the following condition is met: if the word built by Adam is in $\mathcal{L}(\mathcal{A})$, then Eve's run is accepting.

We say that $\mathcal{A}$ is *history-deterministic* (HD) whenever Eve wins the history-determinism game on $\mathcal{A}$.

Definition 2 (one-token game). *The one-token game on $\mathcal{A} = (Q, \Sigma, q_0, \Delta)$ is played by the players Adam and Eve in infinitely many rounds with positions in $Q \times Q$. The initial position is (q_0, q_0). At round $i \in \mathbb{N}$, the token is at position $(q_i, p_i) \in Q \times Q$, and the round proceeds as follows:*

1. *Adam picks a letter $a_i \in \Sigma$;*
2. *Eve picks a transition $q_i \xrightarrow{a_i : c_i} q_{i+1} \in \Delta$;*
3. *Adam picks a transition $p_i \xrightarrow{a_i : c_i'} p_{i+1} \in \Delta$.*

Eve wins the game if the following condition is met: if Adam's run is accepting, then Eve's run is accepting.

The one-token game was expanded by Kuperberg and Skrzypczak to the joker game [8] by giving Adam the possibility to play a special joker move instead of his normal move each round, but only finitely often:

Definition 3 (Joker game). *The joker works as the one-token game, but the third step becomes*

1. *Adam either picks a transition $p_i \xrightarrow{a_i : c_i'} p_{i+1} \in \Delta$ or plays a joker and picks a transition from Eve's position: $q_i \xrightarrow{a_i : c_i''} p_{i+1} \in \Delta$.*

Eve wins the joker game if the following condition is met: if Adam's "run" is accepting and he played finitely many jokers, then Eve's run is accepting. (Adam's run is not technically a run, as playing a joker allows him to jump to q_i before picking a transition)

Remark 1. These two games are not only similar; they are also related: if Eve wins the joker game on an automaton $\mathcal{A}$, then she also wins the one-token game on a simulation-equivalent sub-automaton $\mathcal{B}$ of $\mathcal{A}$ *from everywhere*, i.e. from all reachable positions in the one-token game's arena [12, Theorem 3.34].

3   When are Joker Games Enough?

Eve winning the joker game on a (co-)Büchi automaton is equivalent to this automaton being HD [10]. However, there exists an automaton with parity index $[1, 3]$ where this is not the case (i.e., Eve wins the joker game, but the automaton is not HD) [1]. This automaton is automatically also a counterexample for all automata higher up in the parity index hierarchy. Remember that Büchi and co-Büchi automata are parity automata with parity index $[0, 1]$ and $[1, 2]$ respectively. The question of joker games characterizing history-determinism is thus settled for all parity indices, except $[0, 2]$, which is left open.

The counterexample provided in [1] is the automaton $\mathcal{A}_{[1,3]}$ shown in Fig. 2. This automaton is not history-deterministic because Adam wins the HD game by picking letter a whenever Eve's token is in q_0 and letter b whenever her token is in q_1. Note that any word that Adam builds admits an accepting run because $\mathcal{L}(\mathcal{A}_{[1,3]}) = \Sigma^\omega$. However, Eve still wins the joker game on $\mathcal{A}_{[1,3]}$ with the following strategy: if Adam's token is not on the same state as hers, she moves to Adam's state. Adam can play only finitely many jokers and take only finitely many 1-transitions; hence, this strategy ensures that Eve's and Adam's runs eventually become identical, securing a victory for Eve.

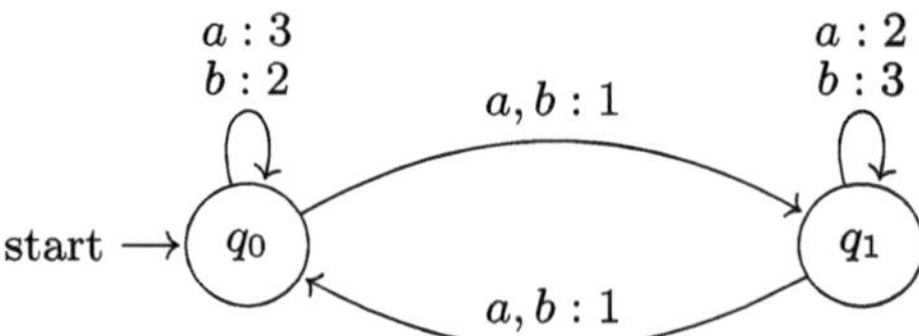

Fig. 2. The automaton $\mathcal{A}_{[1,3]}$ provided by Acharya, Jurdzinski and Prakash [1]. This automaton is not HD, but Eve still wins the joker game on it. It is worth noting that this automaton is also an example of a residual automaton that is not HD [2].

Lemma 1 ([1]) *There is a* [1,3] *automaton that is not history-deterministic, and on which Eve wins the Joker game.*

As a complementary result, we show that the joker game does not characterize history-determinism on parity automata with index $[0,2]$ either. Our counterexample $\mathcal{A}_{[0,2]}$ is shown in Fig. 3. As before, this automaton is not HD, but Eve still wins the joker game. Intuitively, it works in a similar way as $\mathcal{A}_{[1,3]}$:

- It is not HD because it is universal, but there is a way to force Eve to alternate between q_0 and q_1 and therefore taking 1-transitions, while denying her access to q_2 and its 0-transitions.
- Eve wins the joker game because, by seeing Adam's token, she can "chase" it to try to merge her run with Adam's when appropriate or progress towards q_2, which would give her a 0-transition.

Lemma 2. *There is a* $[0,2]$ *automaton that is not history-deterministic and on which Eve wins the Joker game.*

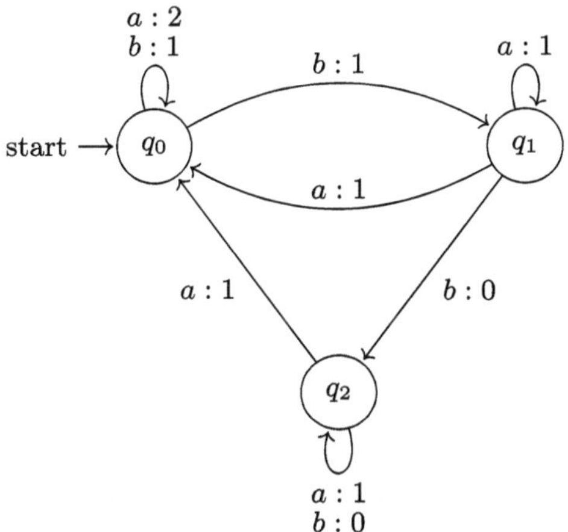

Fig. 3. The automaton $\mathcal{A}_{[0,2]}$.

Proof. We show that the automaton $\mathcal{A}_{[0,2]} = (Q, \Sigma, q_0, \Delta)$ depicted in Fig. 3 is not HD, but Eve still wins the joker game on it.

First, we show that $\mathcal{A}_{[0,2]}$ is universal, i.e. $\mathcal{L}(\mathcal{A}_{[0,2]}) = \Sigma^\omega$. For this, we partition Σ^ω into words with finitely many b's (possibly none) and words with infinitely many b's. Words with finitely many b's admit the accepting run staying in q_0. Words with infinitely many b's are accepted by a run that moves towards q_1 on the first b, to q_2 with the second b and then stays in q_2. There are self-loops on $a = \Sigma \setminus \{b\}$ at each state, so the progress towards q_2 is never lost.

We then show that $\mathcal{A}_{[0,2]}$ is not HD. Adam has the following strategy: when Eve is in q_0, choose b and when Eve is in q_1, choose a. This works because if Eve stays in q_0 she only takes the self loop with priority 1, so she eventually has to move to q_1. When she does, she again only takes the self loop with priority 1 and eventually has to move back to q_0. The run she builds this way uses only transitions with priority 1 and is thus rejecting. But since $\mathcal{A}_{[0,2]}$ is universal, the word admits an accepting run.

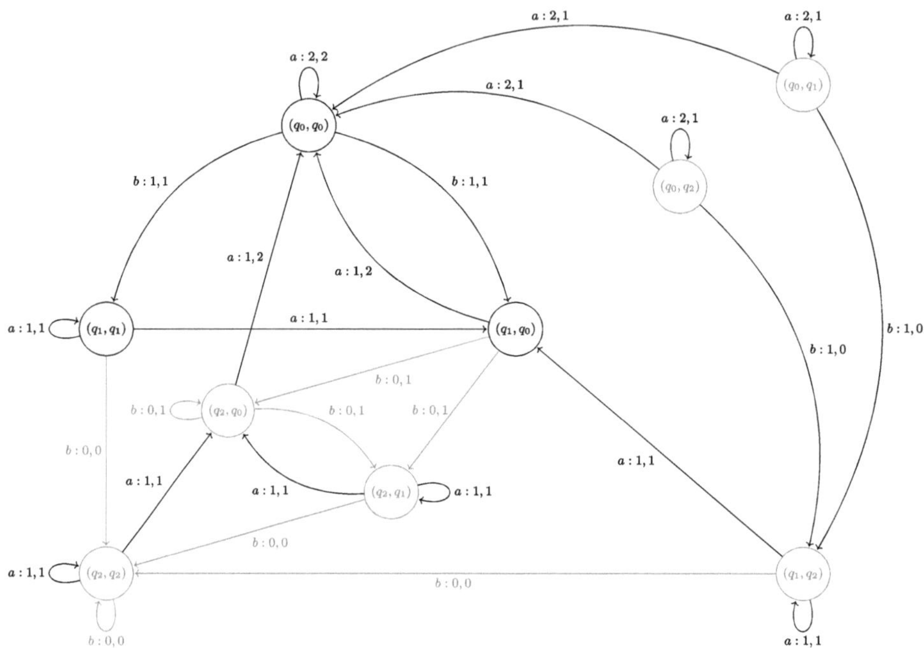

Fig. 4. A simplified view of the arena of the one-token game on $\mathcal{A}_{[0,2]}$. The vertices represent positions as (q_E, q_A), where q_E is the state Eve's token is on and q_A is the state Adam's token is on. Each transition is labeled with two priorities: one for Eve and one for Adam. Eve's moves are restricted to her strategy (and therefore all remaining choices are resolved by Adam). The grayed transitions are those where Eve has a 0 priority. The grayed states are those that are not part of a cycle of black edges other than a self-loop benefiting Eve.

We then show that Eve wins the one-token game on $\mathcal{A}$ from all positions in $Q \times Q$ with the following positional strategy:

- In q_0:
  - If Adam picks a: stay in q_0.
  - If Adam picks b: move to q_1.
- In q_1:
  - If Adam picks a: If Adam is in q_0, move to q_0. Otherwise, stay in q_1.

- If Adam picks b: move to q_2.
- In q_2:
  - If Adam picks a: If Adam is in q_0, move to q_0. Otherwise, stay in q_2.
  - If Adam picks b: stay in q_2.

This is enough to prove that Eve wins the joker game on $\mathcal{A}_{[0,2]}$, since when Adam has played all his (finitely many) jokers, the joker game plays exactly as the one-token game, but possibly starting at some other position than (q_0, q_0). Figure 4 is the arena of the one-token game on $\mathcal{A}$ from all positions (as a 2-dimensional parity game: there are two priorities on each transition and the winning condition can be expressed as a boolean combination of parity conditions), where Eve's moves are restricted to her strategy. Adam can only use the grayed transitions finitely often, as they provide Eve with a 0-transition. This means that eventually, Adam has to find a winning cycle in the black transitions. All grayed positions are not in Adam's winning region, as they are either not in a black cycle or the only black cycle they are part of is a self-loop that would make Eve win. We can also see that there is no black loop for Adam's token from a black position with an even maximum priority, except the self-loop in (q_0, q_0), which also benefits Eve. Eve therefore wins the one-token game on $\mathcal{A}_{[0,2]}$ from any position in $Q \times Q$ and therefore also the joker game on $\mathcal{A}_{[0,2]}$. $\qquad\square$

This allows us to conclude with the following result:

Theorem 1. *For parity automata, the joker game characterizes history-determinism only at the Büchi and co-Büchi levels.*

4   Conclusion

We show that the joker game only characterizes history-determinism in parity automata at the Büchi and co-Büchi level by providing a counterexample for the parity index $[0, 2]$, and therefore complete the picture painted in [10] by Lehtinen and Prakash.

References

1. Acharya, R., Jurdziński, M., Prakash, A.: Lookahead games and efficient determinisation of history-deterministic Büchi automata. In: Bringmann, K., Grohe, M., Puppis, G., Svensson, O. (eds.) 51st International Colloquium on Automata, Languages, and Programming (ICALP 2024). Leibniz International Proceedings in Informatics (LIPIcs), vol. 297, pp. 124:1–124:18. Schloss Dagstuhl – Leibniz-Zentrum für Informatik, Dagstuhl (2024). https://doi.org/10.4230/LIPIcs.ICALP.2024.124
2. Bagnol, M., Kuperberg, D.: Büchi good-for-games automata are efficiently recognizable. In: Ganguly, S., Pandya, P. (eds.) 38th IARCS Annual Conference on Foundations of Software Technology and Theoretical Computer Science (FSTTCS 2018). Leibniz International Proceedings in Informatics (LIPIcs), vol. 122, pp. 16:1–16:14. Schloss Dagstuhl – Leibniz-Zentrum für Informatik, Dagstuhl (2018). https://doi.org/10.4230/LIPIcs.FSTTCS.2018.16

3. Boker, U., Kuperberg, D., Kupferman, O., Skrzypczak, M.: Nondeterminism in the presence of a diverse or unknown future. In: Fomin, F.V., Freivalds, R., Kwiatkowska, M., Peleg, D. (eds.) ICALP 2013. LNCS, vol. 7966, pp. 89–100. Springer, Heidelberg (2013). https://doi.org/10.1007/978-3-642-39212-2_11
4. Boker, U., Kuperberg, D., Lehtinen, K., Skrzypczak, M.: On succinctness and recognisability of alternating good-for-games automata. CoRR arxiv:2002.07278 (2020)
5. Calude, C.S., Jain, S., Khoussainov, B., Li, W., Stephan, F.: Deciding parity games in quasi-polynomial time. SIAM J. Comput. **51**(2), STOC17-152–STOC17-188 (2022). https://doi.org/10.1137/17M1145288
6. Colcombet, T.: The theory of stabilisation monoids and regular cost functions. In: Albers, S., Marchetti-Spaccamela, A., Matias, Y., Nikoletseas, S., Thomas, W. (eds.) ICALP 2009. LNCS, vol. 5556, pp. 139–150. Springer, Heidelberg (2009). https://doi.org/10.1007/978-3-642-02930-1_12
7. Henzinger, T.A., Piterman, N.: Solving games without determinization. In: Ésik, Z. (ed.) CSL 2006. LNCS, vol. 4207, pp. 395–410. Springer, Heidelberg (2006). https://doi.org/10.1007/11874683_26
8. Kuperberg, D., Skrzypczak, M.: On determinisation of good-for-games automata. In: Halldórsson, M.M., Iwama, K., Kobayashi, N., Speckmann, B. (eds.) ICALP 2015. LNCS, vol. 9135, pp. 299–310. Springer, Heidelberg (2015). https://doi.org/10.1007/978-3-662-47666-6_24
9. Kupferman, O., Safra, S., Vardi, M.Y.: Relating word and tree automata. Ann. Pure Appl. Logic **138**(1), 126–146 (2006). https://doi.org/10.1016/j.apal.2005.06.009
10. Lehtinen, K., Prakash, A.: The 2-token theorem: recognising history-deterministic parity automata efficiently. In: Proceedings of the 57th Annual ACM Symposium on Theory of Computing, STOC '25, pp. 1839–1850. Association for Computing Machinery, New York (2025).https://doi.org/10.1145/3717823.3718310
11. Prakash, A.: Checking history-determinism is np-hard for parity automata. In: Kobayashi, N., Worrell, J. (eds.) Foundations of Software Science and Computation Structures, pp. 212–233. Springer, Cham (2024). https://doi.org/10.1007/978-3-031-57228-9_11
12. Prakash, A.: History-deterministic parity automata: games, complexity, and the 2-token theorem (2025). https://arxiv.org/abs/2501.12302

Context-Free, Conjunctive and Boolean Grammars and SCYK Automata

Étienne Grandjean$^{(\boxtimes)}$, Théo Grente, and Véronique Terrier

Normandie Univ, UNICAEN, ENSICAEN, CNRS, GREYC, 14000 Caen, France
{etienne.grandjean,theo.grente,veronique.terrier}@unicaen.fr

Abstract. In this paper, we study the generalized grammars that Okhotin designed to enrich context-free grammars with logical connectives, conjunction and negation, in addition to the implicit disjunction: Boolean grammars and their positive restriction called conjunctive grammars. Building on the inductive principle underlying the Sakai-Cocke-Younger-Kasami algorithm, we introduce a new parallel inductive process, called the SCYK automaton, whose local version, called the linear SCYK automaton, is essentially the trellis automaton. We prove that a language is generated by a Boolean grammar if and only if it is recognized by a SCYK automaton. Furthermore, we refine this equivalence for subclasses of Boolean grammars: conjunctive and context-free grammars and their linear restrictions.

Keywords: Boolean grammar · Conjunctive grammar · Context-free grammar · Sakai-Cocke-Younger-Kasami algorithm · Trellis automaton

1 Introduction

One criterion for the importance of a class of formal languages is that it possesses equivalent and natural definitions in various domains: grammars, automata, logic, algebra, computational complexity. For example, it is well known that the class of regular languages can be characterized in each of the five frameworks mentioned above [4,6,8,16]. On the other hand, a language is context-sensitive if and only if it belongs [4,7] to the complexity class $\mathsf{NSPACE}(n)$, and it is context-free if and only if it is recognized by a non-deterministic pushdown automaton [4,6,16].

In this paper, we study the generalized grammars that Okhotin designed to enrich context-free grammars with logical connectives, conjunction and negation, in addition to the implicit disjunction: Boolean grammars and their positive restriction called conjunctive grammars [1,5,9–11,14]. We are also interested in linear versions of grammars: linear context-free grammars, and linear Boolean grammars, which are equivalent to linear conjunctive grammars.

As Okhotin showed [11], the classical algorithm independently discovered by Sakai, Cocke, Younger and Kasami [4,6,15,18], which we call the SCYK algorithm, which determines whether an input word is generated by a fixed context-free grammar in Chomsky normal form, can be adapted to the more general

M.-P. Béal and P. Caron (Eds.): DLT 2026, LNCS 16578, pp. 139–152, 2026.
https://doi.org/10.1007/978-3-032-28404-4_11

Boolean grammars with the same complexity. The SCYK algorithm essentially consists of the inductive construction, for each factor of the input word, of the set of non-terminals of the grammar that generate that factor.

Okhotin [12] also proved that a language L is generated by a linear conjunctive grammar (or, equivalently, a linear Boolean grammar) if and only if L is recognized by a trellis automaton, or equivalently, by a one-way cellular automaton [2,3,17]. The proof that a language generated by a linear conjunctive grammar is recognized by a trellis automaton uses the inductive principle underlying the SCYK algorithm.

Building on the same inductive principle, this paper introduces a new parallel inductive process, called the SCYK automaton, whose local version, called the linear SCYK automaton, is essentially the trellis automaton. Our main result is that a language is generated by a Boolean grammar if and only if it is recognized by a SCYK automaton (see Sects. 3.1 and 3.2). Furthermore, we refine this equivalence for subclasses of Boolean grammars in two ways:

- *General grammars:* A language is generated by a conjunctive (resp. context-free) grammar *iff* it is recognized by an increasing (resp. additive) SCYK automaton, see Sects. 3.3 and 3.4.
- *Linear grammars:* A language is generated by a linear conjunctive (resp. linear context-free) grammar *iff* it is recognized by a linear increasing (resp. linear additive) SCYK automaton, see Sect. 4.

2   Preliminaries

We assume that the reader is familiar with the following classic concepts from formal language theory: context-free and linear context-free grammars and their Chomsky and Greibach normal forms [4,6,16]. A context-free grammar $G = (\Sigma, N, R, S)$ is said to be *proper* if it has neither ε-rules, i.e. rules of the form $A \to \varepsilon$, nor unit rules $A \to B$, for $A, B \in N$. Two crucial properties of a proper grammar are that for each $A \in N$, the set of words $L(A)$ it generates does not contain the empty word, $L(A) \subseteq \Sigma^+$, and, as Kozen wrote in his book [6], "every step in the derivation makes demonstrable progress towards the terminal string in the sense that either the sentential form gets strictly stronger or a new terminal symbol appears." It is well known that any context-free grammar G can be transformed into a proper context-free grammar G' such that $L(G') = L(G)\backslash\{\varepsilon\}$, see e.g. Lemma 21.3 in [6]

The main grammar we are studying in this paper is Boolean grammar invented by Okhotin [11]. In his papers [11,13] and his survey [14], Okhotin defined this notion in the most general sense, with rules of the form $A \to \alpha_1 \& \ldots \& \alpha_k \& \neg\beta_1 \& \ldots \& \neg\beta_\ell$ for any strings $\alpha_i, \beta_i \in (\Sigma \cup N)^*$, which can be the empty string or some $B \in N$; next, by translating each Boolean grammar into a system of language equations, he defined several semantics of that grammar which make it possible to determine if it is correct, that is, generates a (unique) language.

In this paper, we focus on *proper Boolean grammars*, in the same way we consider *proper* context-free grammars, that is: Boolean grammars that do not possess rules of the forms $A \rightarrow \cdots \& \varepsilon \& \cdots$, $A \rightarrow \cdots \& \neg\varepsilon \& \cdots$, $A \rightarrow \cdots \& B \& \cdots$ and $A \rightarrow \cdots \& \neg B \& \cdots$, in other words, there is no ε-conjunct $\varepsilon, \neg\varepsilon$ and no unit conjunct $B, \neg B$, with $B \in N$.

We now give our definition of proper Boolean grammars, which we simply call *Boolean grammars*, with its natural inductive semantics.

Definition 1. *A* Boolean grammar *is of the form* $G := (\Sigma, N, R, S)$ *where* Σ *is the* alphabet, N *is the set of* non-terminals, $S \in N$ *is the* start symbol, *and* R *is a set of* rules *of the form*

$$A \rightarrow \alpha_1 \& \cdots \& \alpha_k \& \neg\beta_1 \& \cdots \& \neg\beta_\ell \tag{1}$$

where $A \in N$ *and* $k + \ell \geq 1$, *and* $\alpha_i, \beta_i \in (\Sigma \cup N)^+ \setminus N$, *that is,* $\alpha_i, \beta_i \in \Sigma \cup \bigcup_{m \geq 2}(\Sigma \cup N)^m$.

A conjunctive grammar *is a Boolean grammar without negation, that is, with* $\ell = 0$ *in (1).*

Examples of Rules: $A \rightarrow b$, $A \rightarrow CD \& EFG \& \neg JK \& \neg aH$, $A \rightarrow \neg a \& \neg bC$. (Concatenation takes precedence over Boolean operations.)

Semantics. To each string $\alpha \in (\Sigma \cup N)^+ \setminus N$, we associate the *generated* set $L(\alpha) \subseteq \Sigma^+$ of (non-empty) words, defined inductively from the rules (1): for $\alpha = s_1 \ldots s_m \in (\Sigma \cup N)^m$, $m \geq 2$, we get $L(\alpha) := L(s_1) \ldots L(s_m)$ with, for $a \in \Sigma$, $L(a) := \{a\}$, and, for $A \in N$,

$$L(A) := \bigcup_{(A \rightarrow \alpha_1 \& \ldots \& \alpha_k \& \neg\beta_1 \& \ldots \& \neg\beta_\ell) \in R} \left(\bigcap_{i=1}^{k} L(\alpha_i) \cap \bigcap_{i=1}^{\ell} (\Sigma^+ \setminus L(\beta_i)) \right) \tag{2}$$

This definition is well founded. Indeed, each α_i or β_i either is a letter $a \in \Sigma$, or is a string $\alpha = s_1 \ldots s_m \in (\Sigma \cup N)^m$ with $m \geq 2$. For $n \geq 1$, let $L_n(\alpha)$ be the set of words in $L(\alpha)$ of length n. For each $\alpha = s_1 \ldots s_m \in (\Sigma \cup N)^m$ with $m \geq 2$, and for each $n \geq 2$, we get

$$L_n(\alpha) = \bigcup_{i_1, \ldots, i_m \geq 1 \text{ such that } i_1 + \cdots + i_m = n} L_{i_1}(s_1) \cdots L_{i_m}(s_m) \tag{3}$$

with therefore each $i_j < n$, which justifies the following definition of $L_n(\alpha)$ for $\alpha \in \Sigma \cup \bigcup_{m \geq 2}(\Sigma \cup N)^m$ by recurrence on $n \geq 1$.

For $n \geq 1$, the sets $L_n(A)$ for $A \in N$ are defined by the following equation (which is Eq. (2) where $L(A), L(\alpha_i)$, $L(\beta_i)$ and Σ^+ are replaced by $L_n(A), L_n(\alpha_i), L_n(\beta_i)$ and Σ^n, respectively):

$$L_n(A) := \bigcup_{(A \rightarrow \alpha_1 \& \ldots \& \alpha_k \& \neg\beta_1 \& \ldots \& \neg\beta_\ell) \in R} \left(\bigcap_{i=1}^{k} L_n(\alpha_i) \cap \bigcap_{i=1}^{\ell} (\Sigma^n \setminus L_n(\beta_i)) \right) \tag{4}$$

Finally, for all $n \geq 1$ and $\alpha \in (\Sigma \cup N)^m$ with $m \geq 2$, the set $L_n(\alpha)$ can be computed from the sets $L_h(\beta)$ for $h < n$ and $\beta \in (\Sigma \cup N)^+ \setminus N$ thanks to Eq. (3), completed by the equation $L_1(a) := \{a\}$ for $a \in \Sigma$, and Eq. (4) for $A \in N$.

The *language generated* by a Boolean (resp. conjunctive) grammar G, called a *Boolean language* (resp. *conjunctive language*), is defined as $L(G) := L(S)$.

Theorem 1 (Chomsky normal form). *For each Boolean grammar* $G = (\Sigma, N, R, S)$, *we can construct another Boolean grammar* $G' = (\Sigma, N', R', S)$, *said in Chomsky (or binary) normal form or normal Boolean grammar, which is equivalent to it (that is, it generates the same language as G) and whose rules are of one of the following two forms:*

- Short rule: $A \to a$, *with* $a \in \Sigma$;
- Long rule: $A \to B_1 C_1 \ \& \cdots \& \ B_k C_k \ \& \ \neg D_1 E_1 \ \& \cdots \& \ \neg D_\ell E_\ell$, *also written* $A \to \bigwedge_{i \in [1,k]} B_i C_i \wedge \bigwedge_{i \in [1,\ell]} \neg D_i E_i$, *with* $k \geq 1$, $\ell \geq 0$, *and* $B_i, C_i, D_i, E_i \in N'$.

Note that the words generated by a short (resp. long) rule have a length of 1 (resp. at least 2).

Example of Normalization: The rule $A \to CD \ \& \ \neg EFG \ \& \ \neg JK \ \& \ \neg aH$ is replaced by the long rule $A \to CD \ \& \ \neg UG \ \& \ \neg JK \ \& \ \neg VH$ supplemented by the rules $U \to EF$ (long rule) and $V \to a$ (short rule), which define the new non-terminals U and V, meaning that $L(U) = L(EF)$ and $L(V) = \{a\}$.

Proof (Theorem 1). This is the same procedure used to convert a context-free grammar into Chomsky normal form. [4,6,16]. $\qquad\square$

Convention 1. *The fact that the list of rules of a grammar whose left-hand side is A, is $A \to \alpha_1, \dots, A \to \alpha_q$, with $\alpha_i \in (\Sigma \cup N)^+$, $i \in [1,q]$, is expressed by the language equation $L(A) = \bigcup_{i \in [1,q]} L(\alpha_i)$.*
We will freely translate this equation into the propositional equivalence $A \iff \psi_A$ where $\psi_A := \bigvee_{i \in [1,q]} \alpha_i$.

3   Binary Normal Forms and SCYK Automata

3.1   From a Boolean Grammar to a SCYK Automaton

Let $G = (\Sigma, N, R, S)$ be a normal Boolean grammar. From the "binary" character of the grammar, we will deduce, for each word $w \in \Sigma^+$, an inductive computation of the set of non-terminals that generate it, that is $\{A \in N \mid w \in L(A)\}$. This set will be called the *state* of w.

For a given input word $w = w_1 \cdots w_n \in \Sigma^n$, $n \geq 1$, and indices $1 \leq x \leq y \leq n$, we denote by $\langle x, y \rangle$ the state of its factor $w_{x,y} := w_x \cdots w_y$, in other words, $\langle x, y \rangle := \{A \in N \mid w_{x,y} \in L(A)\}$. This notation evokes a trellis automaton [2, 3,9,12,14,17] acting on the word $w = w_1 \cdots w_n$: for $1 \leq x \leq y \leq n$, the state of the cell (x, y) of the trellis automaton is $\langle x, y \rangle$. In particular, we have the equivalence $w \in L(G) \iff S \in \langle 1, n \rangle$.

The structure of the normal Boolean grammar gives us the following induction:

- *Initial states:* For $1 \leq x \leq n$, we have
  $\langle x, x \rangle = \{A \in N \mid (A \to w_x) \in R\}$;
- *Computed states:* For $1 \leq x < y \leq n$, the state $\langle x, y \rangle$ is the set
  of non-terminals $A \in N$ such that there is a rule
  $A \to B_1 C_1 \& \cdots \& B_k C_k \& \neg D_1 E_1 \& \cdots \& \neg D_\ell E_\ell$ of R which satisfies the
  following conjunction:

$$\bigwedge_{i \in [1,k]} \exists z \in [x, y[\, : B_i \in \langle x, z \rangle \wedge C_i \in \langle z + 1, y \rangle \wedge$$

$$\bigwedge_{i \in [1,\ell]} \neg(\exists z \in [x, y[\, : D_i \in \langle x, z \rangle \wedge E_i \in \langle z + 1, y \rangle).$$

The following notations will be convenient. Let $m := y - x + 1$ denote the cardinality of the interval $[x, y]$, for $1 \leq x \leq y \leq n$. We call the *prefix memory* $\mathcal{P}_{x,y}$ of (x, y) the $(m - 1)$-vector $(\langle x, x \rangle, \langle x, x + 1 \rangle, \ldots, \langle x, y - 1 \rangle)$. Similarly, we call the *suffix memory* $\mathcal{S}_{x,y}$ of (x, y) the $(m - 1)$-vector $(\langle y, y \rangle, \langle y - 1, y \rangle, \ldots, \langle x + 1, y \rangle)$. To associate each proper prefix $w_{x,z}$ of $w_{x,y}$ with the complementary suffix $w_{z+1,y}$, we introduce the product

$$\mathcal{P}_{x,y} \otimes \mathcal{S}_{x,y} := \bigcup_{z \in [x, y[} \langle x, z \rangle \times \langle z + 1, y \rangle$$

where $\times$ denotes the Cartesian product. With this product, we can write that for all $1 \leq x < y \leq n$, the state $\langle x, y \rangle$ is the set of $A \in N$ such that there exists a rule $A \to B_1 C_1 \& \cdots \& B_k C_k \& \neg D_1 E_1 \& \cdots \& \neg D_\ell E_\ell$ of R which satisfies the following conjunction:

$$\bigwedge_{i \in [1,k]} (B_i, C_i) \in \mathcal{P}_{x,y} \otimes \mathcal{S}_{x,y} \wedge \bigwedge_{i \in [1,\ell]} (D_i, E_i) \notin \mathcal{P}_{x,y} \otimes \mathcal{S}_{x,y} \tag{5}$$

This justifies the following definition.

Definition 2. *A SCYK automaton is a tuple $\mathcal{A} := (\Sigma, Q, I, \delta, \mathcal{F})$ where Σ is the alphabet, Q is the set of* state atoms, *$\mathscr{P}(Q)$ is the set of* states, *$I : \Sigma \to \mathscr{P}(Q)$ is the* input function, *$\delta : \mathscr{P}(Q^2) \to \mathscr{P}(Q)$ is the* transition function, *$\mathcal{F} \subseteq \mathscr{P}(Q)$ is the set of* accepting states.

When acting on an input word $w = w_1 \ldots w_n \in \Sigma^n$, $n \geq 1$, the SCYK automaton $\mathcal{A}$ computes for each cell (x, y), $1 \leq x \leq y \leq n$, its memory $\mathcal{M}(x, y) := (\langle x, y \rangle, \mathcal{P}_{x,y}, \mathcal{S}_{x,y})$ composed of

- *the state $\langle x, y \rangle \in \mathscr{P}(Q)$ of (x, y),*
- *the prefix memory $\mathcal{P}_{x,y} := (\langle x, x \rangle, \langle x, x + 1 \rangle, \ldots, \langle x, y - 1 \rangle)$,*
- *and the suffix memory $\mathcal{S}_{x,y} := (\langle y, y \rangle, \langle y - 1, y \rangle, \ldots, \langle x + 1, y \rangle)$.*

Each cell memory $\mathcal{M}(x, y)$ is computed by the following induction:

- $\mathcal{M}(x,x) := (I(w_x), \lambda, \lambda)$, *where* $\lambda = \mathcal{P}_{x,x} = \mathcal{S}_{x,x}$ *is the empty list;*
- $\mathcal{M}(x,y) := (\delta(\mathcal{P}_{x,y} \otimes \mathcal{S}_{x,y}), \mathcal{P}_{x,y}, \mathcal{S}_{x,y})$ *if* $x < y$
  with $\mathcal{P}_{x,y} = (\mathcal{P}_{x,y-1}, \langle x, y-1 \rangle)$ *and* $\mathcal{S}_{x,y} = (\mathcal{S}_{x+1,y}, \langle x+1, y \rangle)$.

The input word $w = w_1 \cdots w_n$ *is* accepted *by the SCYK automaton* $\mathcal{A}$ *iff* $\langle 1, n \rangle \in \mathcal{F}$. *The* language $L(\mathcal{A})$ *recognized by* $\mathcal{A}$ *is the set of words that it* accepts.

Thus, for $1 \leq x < y \leq n$, the memory $\mathcal{M}(x,y)$ of a cell (x,y) is of *linear size* and depends only on the memories $\mathcal{M}(x, y-1)$ and $\mathcal{M}(x+1, y)$ of its children.

3.2 The Case of Boolean Grammars

Boolean languages can be recognized by SCYK automata.

Proposition 1. *Let* $L \subseteq \Sigma^+$ *be a Boolean language. Then there is a SCYK automaton* $\mathcal{A}$ *such that* $L = L(\mathcal{A})$.

Proof. Let $G = (\Sigma, N, R, S)$ be a normal Boolean grammar such that $L = L(G)$. To the grammar G we associate the SCYK automaton $\mathcal{A} := (\Sigma, Q, I, \delta, \mathcal{F})$ where Q is N, the input function $I : \Sigma \to \mathscr{P}(Q)$ is defined as $I(a) := \{A \in Q \mid (A \to a) \in R\}$, the set $\mathcal{F}$ is $\{U \subseteq Q \mid S \in U\}$, and the transition function $\delta : \mathscr{P}(Q^2) \to \mathscr{P}(Q)$ is given, for each $\mathcal{E} \subseteq Q^2$, by

$$\delta(\mathcal{E}) := \{A \in Q \mid \text{there exists a rule}$$
$$(A \to B_1 C_1 \ \& \cdots \& \ B_k C_k \ \& \ \neg D_1 E_1 \ \& \cdots \& \ \neg D_\ell E_\ell) \text{ in } R \tag{6}$$
$$\text{such that } \bigwedge_{i \in [1,k]} (B_i, C_i) \in \mathcal{E} \ \wedge \ \bigwedge_{i \in [1,\ell]} (D_i, E_i) \notin \mathcal{E}\}.$$

In particular, for $1 \leq x < y \leq n$ and $A \in Q$, we have $A \in \delta(\mathcal{P}_{x,y} \otimes \mathcal{S}_{x,y})$ iff there is in R a rule $A \to B_1 C_1 \ \& \cdots \& \ B_k C_k \ \& \ \neg D_1 E_1 \ \& \cdots \& \ \neg D_\ell E_\ell$ such that

$$\bigwedge_{i \in [1,k]} (B_i, C_i) \in \mathcal{P}_{x,y} \otimes \mathcal{S}_{x,y} \ \wedge \ \bigwedge_{i \in [1,\ell]} (D_i, E_i) \notin \mathcal{P}_{x,y} \otimes \mathcal{S}_{x,y}, \text{ which means}$$

$A \in \langle x, y \rangle$, according to condition (5). This proves $\langle x, y \rangle = \delta(\mathcal{P}_{x,y} \otimes \mathcal{S}_{x,y})$, as expected.

We also have, for $1 \leq x \leq n$, $I(w_x) = \{A \in Q \mid (A \to w_x) \in R\} = \langle x, x \rangle$.

Finally, for each $U \subseteq Q$, we have $S \in U \iff U \in \mathcal{F}$, which gives in particular $S \in \langle 1, n \rangle \iff \langle 1, n \rangle \in \mathcal{F} \iff w \in L(\mathcal{A})$.

Thus, we have proven $L = L(G) = L(\mathcal{A})$. $\qquad\qquad\square$

The converse of Proposition 1 is true.

Proposition 2. *Each language* $L \subseteq \Sigma^+$ *recognized by a SCYK automaton is a Boolean language.*

Thus, SCYK automata *exactly* characterize Boolean languages.

Theorem 2. *A language* $L \subseteq \Sigma^+$ *is a Boolean language iff it is recognized by a SCYK automaton.*

Proof (Proposition 2). Suppose $L = L(\mathcal{A})$ for a SCYK automaton $\mathcal{A} := (\Sigma, Q, I, \delta, \mathcal{F})$. Let us construct a Boolean grammar $G = (\Sigma, N, R, S)$. First, for each $A \in I(a)$, create the rule $A \to a$. For each node (x, y) such that $1 \le x < y \le n$, we have $\langle x, y \rangle = \delta(\mathcal{P}_{x,y} \otimes \mathcal{S}_{x,y})$ with $\mathcal{P}_{x,y} \otimes \mathcal{S}_{x,y} := \bigcup_{z \in [x,y[} \langle x, z \rangle \times \langle z+1, y \rangle$. If the assertion $\exists z \in [x, y[\ \ B \in \langle x, z \rangle \wedge C \in \langle z+1, y \rangle$ is abbreviated to $BC \in \langle x, y \rangle$ we obtain the equality

$$\langle x, y \rangle = \delta(\{(B, C) \in Q^2 \mid BC \in \langle x, y \rangle\}) \tag{7}$$

(Intuitively, for each $A \in Q$ and all $1 \le x < y \le n$, the truth value of the assertion $A \in \langle x, y \rangle$ is determined by the set of pairs $(B, C) \in Q^2$ such that $BC \in \langle x, y \rangle$.) Equality (7) becomes the equivalence

$$A \in \langle x, y \rangle \iff \delta_A(\{(B, C) \in Q^2 \mid BC \in \langle x, y \rangle\}) = 1 \tag{8}$$

for each $A \in Q$, where the Boolean function $\delta_A : \mathcal{P}(Q^2) \to \{0, 1\}$ is defined as the "projection" of δ on A, which means $\delta_A(\mathcal{E}) = 1 \iff A \in \delta(\mathcal{E})$, for each $\mathcal{E} \subseteq Q^2$. Therefore, because of equivalence (8), there is for each $A \in Q$, a propositional formula ψ_A (dependent only on A) on the family of variables $(p_{BC})_{(B,C) \in Q^2}$ such that the assertion $A \in \langle x, y \rangle$ is equivalent to $\psi_A((p_{BC})_{(B,C) \in Q^2})$ where each p_{BC} represents the assertion $BC \in \langle x, y \rangle$. Let $\bigvee_{i \in [1,r]} \psi_i$ be a disjunctive normal form of ψ_A. Each ψ_i is a conjunction of literals $\psi_i = \bigwedge_{j \in [1,k_i]} p_{B_j^i C_j^i} \wedge \bigwedge_{j \in [1,\ell_i]} \neg p_{D_j^i E_j^i}$ where $k_i + \ell_i \ge 1$ and $B_j^i, C_j^i, D_j^i, E_j^i \in Q$. We obtain the equivalence, for all $1 \le x < y \le n$,

$$A \in \langle x, y \rangle \iff \bigvee_{i \in [1,r]} \left(\bigwedge_{j \in [1,k_i]} B_j^i C_j^i \in \langle x, y \rangle \wedge \bigwedge_{j \in [1,\ell_i]} D_j^i E_j^i \notin \langle x, y \rangle \right)$$

This justifies the long rules $A \to BB \ \wedge \ \bigwedge_{j \in [1,k_i]} B_j^i C_j^i \ \wedge \ \bigwedge_{j \in [1,\ell_i]} \neg D_j^i E_j^i$, for $i \in [1, r]$ and a new symbol B, defined by the rules $B \to aB$ and $B \to a$, for all $a \in \Sigma$, to constrain the lengths of the words produced to be at least equal to 2 in the case $k_i = 0$ because $L(BB) = \Sigma\Sigma^+$.

We set $N := Q \cup \{B, S\}$ and add, for each set $U \in \mathcal{F}$, the "rule"

$$R_U : S \to \bigwedge_{A \in U} A \ \wedge \ \bigwedge_{A \in Q \setminus U} \neg A.$$

We therefore obtain, by definition of $L(\mathcal{A})$,

$$L(\mathcal{A}) = L(S) = \bigcup_{U \in \mathcal{F}} \left(\bigcap_{A \in U} L(A) \cap \bigcap_{A \in Q \setminus U} (\Sigma^+ \setminus L(A)) \right) \tag{9}$$

However, the "rule" R_U is not of the required form (1). To address this problem by making the notation and understanding intuitive, we write equality (9) in the following logical form:

$$S \iff \bigvee_{U \in \mathcal{F}} \left(\bigwedge_{A \in U} A \ \wedge \ \bigwedge_{A \in Q \setminus U} \neg A \right) \tag{10}$$

We will do the same for other non-terminals. For example, if the set of rules whose left-hand side is A is $\{A \to CD \ \& \ \neg EF, \ A \to a\}$, then this set of rules is logically expressed by the equivalence $A \iff \psi_A$ where $\psi_A := (CD \wedge \neg EF) \vee a$. More generally, for each $A \in Q$, ψ_A is a propositional formula in disjunctive normal form on the "variables" $a \in \Sigma$ and CD, for $C, D \in Q$.

Replacing the "rules" R_U by equivalent rules of the form (1): If we substitute each A by ψ_A in the right-hand side of equivalence (10), then this equivalence becomes $S \iff \bigvee_{U \in \mathcal{F}} \varphi_U$ where $\varphi_U := \bigwedge_{A \in U} \psi_A \ \wedge \ \bigwedge_{A \in Q \setminus U} \neg \psi_A$. Let $\delta_U = \bigvee_{i \in I} \gamma_i$ be a disjunctive normal form of φ_U. Each γ_i is a conjunction of the form $\bigwedge_j (\neg) B_j C_j \wedge \bigwedge_j (\neg) a_j$ with $B_j, C_j \in Q$ and $a_j \in \Sigma$. For each $U \in \mathcal{F}$, the rule R_U is therefore equivalent to the set of rules $S \to \gamma_i$, for $i \in I$, which can be easily put into the required form (1). $\qquad\square$

3.3 The Case of Conjunctive Grammars

If we consider Theorem 2, a natural question arises: Which class of automata characterizes conjunctive languages, that is to say "positive" Boolean languages?

Definition 3. *A SCYK automaton $\mathcal{A} := (\Sigma, Q, I, \delta, \mathcal{F})$ is* increasing *if the following two conditions are met:*

1) *The function δ is increasing: for all $\mathcal{E} \subseteq \mathcal{E}' \subseteq Q^2$, we have $\delta(\mathcal{E}) \subseteq \delta(\mathcal{E}')$;*
2) *The set $\mathcal{F}$ is closed by superset: if $U \subseteq V \subseteq Q$, then $U \in \mathcal{F}$ implies $V \in \mathcal{F}$.*

Theorem 3. *A language $L \subseteq \Sigma^+$ is conjunctive iff it is recognized by an increasing SCYK automaton.*

Proof. $\Rightarrow$: First, we observe in the proof of Proposition 1 that if a grammar $G = (\Sigma, N, R, S)$ is conjunctive in normal form, i.e. has long rules of the form $A \to B_1 C_1 \ \& \ldots \& \ B_k C_k$ and $Q := N$, we have for each $\mathcal{E} \subseteq Q^2$, $\delta(\mathcal{E}) := \{A \in Q \mid$ there is a rule $A \to B_1 C_1 \ \& \ldots \& \ B_k C_k$ such that $\bigwedge_{i \in [1,k]} (B_i, C_i) \in \mathcal{E}\}$. This implies that for all $\mathcal{E} \subseteq \mathcal{E}' \subseteq Q^2$, we have $\delta(\mathcal{E}) \subseteq \delta(\mathcal{E}')$. Secondly, note that the set $\mathcal{F} := \{U \subseteq Q \mid S \in U\}$ is closed by superset.

$\Leftarrow$: We are simply presenting the modifications compared to the proof of Proposition 2. Suppose that L is recognized by an increasing SCYK automaton $\mathcal{A} = (\Sigma, Q, I, \delta, \mathcal{F})$. We have for $1 \leq x < y \leq n$ the equality $\langle x, y \rangle = \delta(\{(B, C) \in Q^2 \mid BC \in \langle x, y \rangle\})$. Since δ is an increasing function, the propositional formula ψ_A on the variables $(p_{BC})_{(B,C) \in Q^2}$

equivalent to the assertion $A \in \langle x, y \rangle$ is positive, that is, it contains only connectives $\wedge$ and $\vee$. It then possesses a positive disjunctive normal form $\bigvee_{i \in [1,r]} \bigwedge_{j \in [1,k_i]} p_{B_j^i C_j^i}$. Thus, we get the equivalence, for $1 \leq x < y \leq n$, $A \in \langle x, y \rangle \iff \bigvee_{i \in [1,r]} \bigwedge_{j \in [1,k_i]} B_j^i C_j^i \in \langle x, y \rangle$, which gives the rules

$$A \to \bigwedge_{j \in [1,k_i]} B_j^i C_j^i, \text{ for } i \in [1,r]$$

which are conjunctive. Let $U_1, \ldots, U_k$ be the list of minimal sets (minimal for inclusion) that belong to $\mathcal{F}$. Since $\mathcal{F}$ is closed by superset, we have the equivalence

$$\langle 1, n \rangle \in \mathcal{F} \iff \exists i \in [1, k] \; U_i \subseteq \langle 1, n \rangle \iff \bigvee_{i \in [1,k]} \bigwedge_{A \in U_i} A \in \langle 1, n \rangle.$$

This justifies the conjunctive rules $S \to \bigwedge_{A \in U_i} A$, for all $i \in [1, k]$. $\qquad\square$

3.4   The Case of Ordinary (context-Free) Grammars

What about the context-free languages? They have a similar characterization!

Definition 4. *A SCYK automaton* $\mathcal{A} = (\Sigma, Q, I, \delta, \mathcal{F})$ *is* additive *if the following two conditions are met:*

1) the transition function δ is additive: *for all subsets $\mathcal{E}, \mathcal{E}' \subseteq Q^2$, $\delta(\mathcal{E} \cup \mathcal{E}') = \delta(\mathcal{E}) \cup \delta(\mathcal{E}')$;*
2) the set $\mathcal{F}$ satisfies the equivalence, for all subsets $U, V \subseteq Q$, $U \cup V \in \mathcal{F} \iff U \in \mathcal{F} \vee V \in \mathcal{F}$.

Remark 1. It is easy to see that condition 2 is equivalent to the following:
   2') for all $U \subseteq Q$, U belongs to $\mathcal{F}$ iff U contains (at least) one of the singleton sets $\{A_1\}, \ldots, \{A_m\}$ that belong to $\mathcal{F}$.

Theorem 4. *A language $L \subseteq \Sigma^+$ is context-free iff it is recognized by an additive SCYK automaton.*

Proof. Here again, we only give the modifications with respect to the proofs of Propositions 1 and 2.
   $\Rightarrow$: Let $G = (\Sigma, N, R, S)$ be a context-free grammar in normal form, with rules of G of the form $A \to a$ or $A \to BC$, such that $L = L(G)$. Thus, the transition function δ of the SCYK automaton $\mathcal{A} = (\Sigma, Q, I, \delta, \mathcal{F})$ that recognizes L satisfies, for each $\mathcal{E} \subseteq Q^2$, $\delta(\mathcal{E}) = \{A \in Q \mid \text{there is a rule } A \to BC \text{ in } R \text{ such that } (B, C) \in \mathcal{E}\}$. This implies $\delta(\mathcal{E} \cup \mathcal{E}') = \delta(\mathcal{E}) \cup \delta(\mathcal{E}')$ for all $\mathcal{E}, \mathcal{E}' \subseteq Q^2$. In addition, the set $\mathcal{F} := \{U \subseteq Q \mid S \in Q\}$ satisfies condition 2.

   $\Leftarrow$: Let $\mathcal{A} = (\Sigma, Q, I, \delta, \mathcal{F})$ be an additive SCYK automaton such that $L = L(\mathcal{A})$. Since δ is additive, we obtain for each input $w = w_1 \ldots w_n$ and all $1 \leq x < y \leq n$, $\langle x, y \rangle = \bigcup_{BC \in \langle x, y \rangle} \delta(\{(B, C)\})$. Thus, we have for each $A \in Q$ the equivalence

$$A \in \langle x, y \rangle \iff \exists (B, C) : A \in \delta(\{(B, C)\}) \wedge BC \in \langle x, y \rangle. \tag{11}$$

Now, consider any $A \in Q$. Let $(B_1, C_1), \ldots, (B_k, C_k)$ be the list of pairs (B, C) such that $A \in \delta(\{(B, C)\})$. Equivalence (11) becomes $A \in \langle x, y \rangle \iff \bigvee_{i=1}^{k} B_i C_i \in \langle x, y \rangle$. This corresponds to the context-free rules $A \to B_i C_i$ for $1 \leq i \leq k$. In addition, by condition 2' of Remark 1, we get $\mathcal{F} = \{U \subseteq Q \mid \bigvee_{i=1}^{m} A_i \in U\}$ for a list $A_1, \ldots, A_m$ of elements of Q. Therefore, we take $N := Q \cup \{S\}$ and add the context-free rules $S \to A_i$, for $1 \leq i \leq m$. $\square$

4   Linear Grammars and Linear SCYK Automata

Like a linear context-free grammar, a Boolean grammar $G = (\Sigma, N, R, S)$ is said to be *linear* if each string over the alphabet $N \cup \Sigma$ that appears in the right-hand side of a rule contains at most one non-terminal. In other words, each string α_i, β_i occurring in any rule (1) of the grammar given in Definition 1 above includes at most one letter in N.

The following theorem, implicit in paper [14], Sect. 4, expresses that negations appearing in a linear Boolean grammar can be eliminated.

Theorem 5. *Each linear Boolean grammar is equivalent to a linear conjunctive grammar.*

Intuitively, the linearity of a grammar allows us to eliminate negations by replacing each of them by a finite disjunction of cases.

Definition 5. *A linear SCYK automaton is a SCYK automaton $\mathcal{A} := (\Sigma, Q, I, \delta, \mathcal{F})$ where the prefix memory $\mathcal{P}_{x,y}$ (resp. suffix memory $\mathcal{S}_{x,y}$) only contains the two states of prefixes (resp. suffixes) of sizes 1 and $y - x$: $\mathcal{P}_{x,y} := (\langle x, x \rangle, \langle x, y - 1 \rangle)$ and $\mathcal{S}_{x,y} := (\langle y, y \rangle, \langle x + 1, y \rangle)$.*

Apart from this difference, a linear SCYK automaton behaves exactly like a SCYK automaton. In particular, we have, for all $1 \leq x < y \leq n$, $\langle x, y \rangle := \delta(\mathcal{P}_{x,y} \otimes \mathcal{S}_{x,y}) = \delta(\langle x, x \rangle \times \langle x + 1, y \rangle \cup \langle x, y - 1 \rangle \times \langle y, y \rangle).$

As in the general case, linear restrictions of Boolean grammars (or, equivalently, conjunctive grammars) and context-free grammars can be characterized by linear SCYK automata.

Proposition 3. *Let $L \subseteq \Sigma^+$ be a language.*

1. *If L is linear context-free then there is an additive linear SCYK automaton $\mathcal{A}$ such that $L = L(\mathcal{A})$.*
2. *If L is linear conjunctive then there is an increasing linear SCYK automaton $\mathcal{A}$ such that $L = L(\mathcal{A})$.*
3. *If L is linear Boolean then there is a linear SCYK automaton $\mathcal{A}$ such that $L = L(\mathcal{A})$.*

Proof. Let $G = (\Sigma, N, R, S)$ be a linear grammar in Greibach normal form such that $L = L(G)$. We associate to the grammar G a linear SCYK automaton $\mathcal{A} := (\Sigma, Q, I, \delta, \mathcal{F})$ such that $Q := N \cup \Sigma$. A state atom $a \in \Sigma$ will appear in a state $\langle x, x \rangle$ when $w_x = a$. The input function $I : \Sigma \to Q$ is defined as $I(a) := \{A \in Q \mid (A \to a) \in R\} \cup \{a\}$, $\mathcal{F} := \{U \subseteq Q \mid S \in U\}$, and the transition function $\delta : \mathscr{P}(Q^2) \to \mathscr{P}(Q)$ is given, for each $\mathcal{E} \subseteq Q^2$, by:

1. for a linear context-free grammar:

$$\delta(\mathcal{E}) := \{A \in Q \mid \text{there exists a rule } (A \to bC) \text{ such that } (b, C) \in \mathcal{E}$$
$$\text{or a rule } (A \to Bc) \text{ such that } (B, c) \in \mathcal{E}\}$$

2. for a linear conjunctive grammar:

$$\delta(\mathcal{E}) := \{A \in Q \mid \text{there exists a rule } (A \to bC_1 \wedge \cdots \wedge bC_k \wedge$$
$$B_1 c \wedge \cdots \wedge B_\ell c) \text{ such that } \bigwedge_{i \in [1,k]} (b, C_i) \in \mathcal{E} \wedge \bigwedge_{i \in [1,\ell]} (B_i, c) \in \mathcal{E}\}.$$

3. for a linear Boolean grammar:

$$\delta(\mathcal{E}) := \{A \in Q \mid \text{there exists a rule } (A \to bC_1 \wedge \cdots \wedge bC_k \wedge B_1 c \wedge \cdots \wedge B_\ell c$$
$$\wedge \neg a_1 D_1 \wedge \cdots \wedge \neg a_m D_m \wedge \neg E_1 b_1 \wedge \cdots \wedge \neg E_p b_p) \text{ such that}$$
$$\bigwedge_{i \in [1,k]} (b, C_i) \in \mathcal{E} \wedge \bigwedge_{i \in [1,\ell]} (B_i, c) \in \mathcal{E} \wedge \bigwedge_{i \in [1,m]} (a_i, D_i) \notin \mathcal{E} \wedge \bigwedge_{i \in [1,p]} (E_i, b_i) \notin \mathcal{E}\}.$$

The proof that a linear SCYK automaton recognizing a linear context-free (resp. conjunctive) language is additive (resp. increasing) is the same as the proof of Theorem 4 (resp. Theorem 3). $\qquad\square$

Proposition 4. *Each language $L \subseteq \Sigma^+$ recognized by a linear SCYK automaton $\mathcal{A}$ is:*

1. *linear context-free if $\mathcal{A}$ is additive;*
2. *linear conjunctive if $\mathcal{A}$ is increasing;*
3. *linear Boolean.*

Proof. We only mention the differences with the proof of the general case (proofs of Proposition 2 and Theorems 3 and 4). Suppose $L = L(\mathcal{A})$ for a linear SCYK automaton $\mathcal{A} := (\Sigma, Q, I, \delta, \mathcal{F})$. We want to define the rules of a linear grammar $G = (\Sigma, N, R, S)$ such that $L = L(G)$. We recall we take $\mathcal{P}_{x,y} \otimes \mathcal{S}_{x,y} := \langle x, x \rangle \times \langle x + 1, y \rangle \cup \langle x, y - 1 \rangle \times \langle y, y \rangle$. Let us define for each $A \in Q$ the set $A_\Sigma := \{a \in \Sigma \mid A \in I(a)\}$. Then we have the equivalence: $(B, C) \in \mathcal{P}_{x,y} \otimes \mathcal{S}_{x,y} \iff (w_x \in B_\Sigma \wedge w_{x+1,y} \in L(C)) \vee (w_{x,y-1} \in L(B) \wedge w_y \in C_\Sigma)$. Thus, the property $(B, C) \in \mathcal{P}_{x,y} \otimes \mathcal{S}_{x,y}$ is expressed by the disjunction $\left(\bigvee_{b \in B_\Sigma} bC \vee \bigvee_{c \in C_\Sigma} Bc \right)$. This justifies the following equivalences:

1. If δ is additive: For each $A \in Q$, let $\mathcal{E}_A$ be the set of pairs $(B, C) \in Q^2$ such that $A \in \delta(\{(B, C)\})$. For all $A \in Q$ we have:

$$A \iff \bigvee_{(B,C) \in \mathcal{E}_A} \left(\bigvee_{b \in B_\Sigma} bC \vee \bigvee_{c \in C_\Sigma} Bc \right)$$

150   É. Grandjean et al.

2. If δ is increasing: For each $A \in Q$, let Min_A be the set of minimal sets (for inclusion) $\mathcal{E} \subseteq Q^2$ such that $A \in \delta(\mathcal{E})$. We have:

$$A \iff \bigvee_{\mathcal{E} \in Min_A} \bigwedge_{(B,C) \in \mathcal{E}} \left(\bigvee_{b \in B_\Sigma} bC \vee \bigvee_{c \in C_\Sigma} Bc \right)$$

3. For general δ: The assertion $(D, E) \notin \mathcal{P}_{x,y} \otimes \mathcal{S}_{x,y}$ is expressed by the conjunction $\bigwedge_{d \in D_\Sigma} \neg dE \wedge \bigwedge_{e \in E_\Sigma} \neg De$. Thus, for all $A \in Q$ we have:

$$A \iff \bigvee_{\substack{\mathcal{E} \subseteq Q^2 \\ A \in \delta(\mathcal{E})}} \left(\bigwedge_{(B,C) \in \mathcal{E}} \left(\bigvee_{b \in B_\Sigma} bC \vee \bigvee_{c \in C_\Sigma} Bc \right) \wedge \bigwedge_{(D,E) \notin \mathcal{E}} \left(\bigwedge_{d \in D_\Sigma} \neg dE \wedge \bigwedge_{e \in E_\Sigma} \neg De \right) \right)$$

Like in the general case, it is easy to transform the equivalences obtained in case 1 (resp. case 2, case 3) into rules of a context-free (resp. conjunctive, Boolean) grammar.   □

From Proposition 3 and Proposition 4, we deduce the following theorem.

Theorem 6. *A language $L \subseteq \Sigma^+$ is:*

1. *linear context-free iff it is recognized by an additive linear SCYK automaton;*
2. *linear conjunctive iff it is recognized by an increasing linear SCYK automaton;*
3. *linear Boolean iff it is recognized by a linear SCYK automaton.*

From Theorem 5 and items 2 and 3 of Theorem 6, we deduce the following corollary.

Corollary 1. *For any linear SCYK automaton $\mathcal{A}$, there is an increasing linear SCYK automaton $\mathcal{A}'$ such that $L(\mathcal{A}') = L(\mathcal{A})$.*

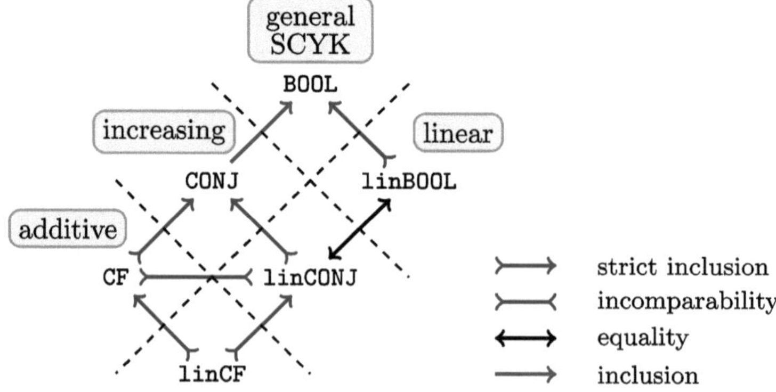

Fig. 1. The SCYK automaton and its restrictions

5   Conclusion

In this paper, we were interested in Boolean grammars and their restrictions. We proposed a device with memory and a dot product, the SCYK automaton, that recognizes the class of Boolean languages. We showed that the different variants of SCYK automata (where the transition function is increasing or additive; where the transition function uses a finite amount of memory) characterize the different restrictions of Boolean grammars (conjunctive or context-free; linear rules). See Fig. 1.

We implicitly revisited the known inclusions as any additive SCYK automaton is increasing and any increasing SCYK automaton is general, in one part, and any linear SCYK has reduced memory.

Regarding strict inclusions, one question is whether the SCYK automaton and its variants can help to determine the limits between the classes?

- It works well for the linear increasing SCYK automata: they are equivalent to trellis automata whose limits are quite well known.
- For the additive SCYK automata that recognize context-free languages, it should be not difficult to show a direct equivalence with non-deterministic pushdown automata. Now, how can the pumping lemma or the cardinality properties known for context-free languages be understood in terms of additive SCYK automata?
- The main question relating Boolean grammars is whether conjunctive grammars and Boolean grammars are equally expressive. In terms of SCYK automata, this corresponds to asking whether the increasing SCYK automata and the (general) SCYK automata have the same recognition power. Now, the SCYK automata can serve as a tool for comparing Boolean languages with other complexity classes. We wonder also which properties of SCYK automata capture Boolean grammars with unambiguous syntax.

References

1. Aizikowitz, T., Kaminski, M.: Conjunctive grammars and alternating pushdown automata. Acta Informatica **50**(3), 175–197 (2013)
2. Čulík II, K., Gruska, J., Salomaa, A.: Systolic trellis automata. I. Int. J. Comput. Math. **15**(3–4), 195–212 (1984)
3. Dyer, C.R.: One-way bounded cellular automata. Inf. Control **44**(3), 261–281 (1980)
4. Hopcroft, J.E., Motwani, R., Ullman, J.D.: Introduction to Automata Theory, Languages, and Computation, 3rd edn, Pearson International Edition. Addison-Wesley (2007)
5. Kountouriotis, V., Nomikos, C., Rondogiannis, P.: Well-founded semantics for Boolean grammars. Inf. Comput. **207**(9), 945–967 (2009)
6. Kozen, D.: Automata and Computability. Undergraduate Texts in Computer Science. Springer, New York (1997)
7. Kuroda, S.-Y.: Classes of languages and linear-bounded automata. Inf. Control **7**(2), 207–223 (1964)

8. Libkin, L.: Elements of Finite Model Theory. Texts in Theoretical Computer Science. An EATCS Series. Springer, New York (2004)
9. Okhotin, A.: Conjunctive grammars. J. Autom. Lang. Comb. **6**(4), 519–535 (2001)
10. Okhotin, A.: Boolean grammars. In: Ésik, Z., Fülöp, Z. (eds.) DLT 2003. LNCS, vol. 2710, pp. 398–410. Springer, Heidelberg (2003). https://doi.org/10.1007/3-540-45007-6_32
11. Okhotin, A.: Boolean grammars. Inf. Comput. **194**(1), 19–48 (2004)
12. Okhotin, A.: On the equivalence of linear conjunctive grammars and trellis automata. Theor. Inform. Appl. **38**(1), 69–88 (2004)
13. Okhotin, A.: Fast parsing for Boolean grammars: a generalization of Valiant's algorithm. In: Gao, Y., Lu, H., Seki, S., Yu, S. (eds.) DLT 2010. LNCS, vol. 6224, pp. 340–351. Springer, Heidelberg (2010). https://doi.org/10.1007/978-3-642-14455-4_31
14. Okhotin, A.: Conjunctive and Boolean grammars: the true general case of the context-free grammars. Comput. Sci. Rev. **9**, 27–59 (2013)
15. Sakai, I.: Syntax in universal translation. In: Proceedings of the International Conference on Machine Translation and Applied Language Analysis, vol. II, pp. 593–608 (1961)
16. Sipser, M.: Introduction to the Theory of Computation. PWS Publishing Company (1997)
17. Terrier, V.: Language recognition by cellular automata. In: Rozenberg, G., Bäck, T., Kok, J.N. (eds.) Handbook of Natural Computing, pp. 123–158. Springer, Heidelberg (2012). https://doi.org/10.1007/978-3-540-92910-9_4
18. Younger, D.H.: Context-free language processing in time n^3. In: 7th Annual Symposium on Switching and Automata Theory, pp. 7–20. IEEE Computer Society (1966)

Minimizing Streaming String Transducers: An Algebraic Approach

Yahia Idriss Benalioua[1]([✉]) [iD], Nathan Lhote[2] [iD], and Pierre-Alain Reynier[2] [iD]

[1] University of Warsaw, Warsaw, Poland
y.benalioua@uw.edu.pl
[2] Aix Marseille Univ, CNRS, LIS, Marseille, France
{nathan.lhote,pierre-alain.reynier}@lis-lab.fr

Abstract. In this work, we study minimization of rational functions given as appending streaming string transducers (aSST for short). We rely on an algebraic presentation of these functions, known as bimachines, to address the minimization of both states and registers of aSST.

First, we show a bijection between a subclass of aSST and bimachines, which maps the numbers of states and registers of the aSST to two natural parameters of the bimachine. Using known results on the minimization of bimachines, this yields a PTime (resp. NP) procedure to minimize this subclass of aSST with respect to registers (resp. both states and registers). In a second step, we introduce a new model of bimachines, named asynchronous bimachines, which allows to lift the bijection to the whole class of aSST. Based on this, we prove that register minimization with a fixed underlying automaton is NP-complete.

Keywords: Rational functions · Streaming String Transducers · Bimachines · Minimization

1 Introduction

Important connections between computation, mathematical logic, and algebra have been established for regular languages of finite words. A language is regular iff it is recognized by a finite automaton, iff it is definable in monadic second-order logic (MSO) with one successor [8,11,21], and iff its syntactic monoid, a canonical monoid attached to every language, is finite (see for instance [20]). While automata are well-suited to study the algorithmic properties of regular languages, the algebraic view has provided effective characterizations of regular languages and its subclasses. Most notably, the problem of deciding whether a regular language is first-order definable amounts to checking whether its syntactic monoid, which is computable from any finite automaton recognizing the language, is aperiodic, which is decidable.

At the computational level, word functions are defined by (one-way) transducers, which extend automata with outputs on their transitions. The non-determinism of automata may result in mapping an input to different output

© The Author(s), under exclusive license to Springer Nature Switzerland AG 2026
M.-P. Béal and P. Caron (Eds.): DLT 2026, LNCS 16578, pp. 153–167, 2026.
https://doi.org/10.1007/978-3-032-28404-4_12

words. This yields the class of rational relations, long studied in the literature, but for which numerous problems are undecidable, starting with equivalence. When the relation is functional, we obtain the class of *rational functions* [7], which behaves much better regarding decidability.

While non-determinism is important for expressiveness, it is not realistic regarding actual implementations. An alternative model, known as *streaming string transducers* [1,2], has been introduced to remedy this problem. The underlying automaton is deterministic, but the model is no longer finite-state: it is equipped with finitely many registers that are used to store partial output words, and that can be updated during the execution.

For automata based models, a very important problem is to simplify the models. For instance, deterministic machines are essential in order to derive efficient evaluation algorithms. Similarly, reducing the size of the models allows to reduce the computation time of most algorithms. For the class of regular languages, classical algebraic tools such as Myhill-Nerode congruence not only allow to minimize automata, but also yield canonical models, which can be used to derive learning algorithms. When considering automata models extended with registers, the minimization may be with respect to the number of registers, the number of states, or both (see e.g. [4] for cost register automata over a field).

In this work, we consider the class of rational functions, and are interested in minimization problems for streaming string transducers. In [9], the problem of minimizing the number of registers has been solved for this class, and shown to be PSpace-complete. More precisely, the approach is based on the presence of some pattern in a transducer realizing the function, and, equivalently, on some machine independent property of the function itself. This result allows to identify the minimal number of registers needed to realize the function, but comes with a blow-up of the number of states.

Our objective is to follow an algebraic approach to address minimization of streaming string transducers, w.r.t. both the number of states and registers. Indeed, Reutenauer and Schutzenberger have introduced in [18] congruences that allow to derive an algebraic characterization of rational functions, and a representation of these functions by an alternative model known as bimachine (the name comes from the fact that it is represented as a pair of automata with outputs, known as a left and a right automaton). Unlike other models, this algebraic characterization does not yield a unique, but a finite family of canonical bimachines. Our contributions can be summarized as follows: (see also the table)

1. We prove a strong equivalence between bimachines and a subclass of aSST (denoted aSST$_{\mathsf{iffo}}$), which maps the number of registers (resp. states) of the aSST to the size of the right (resp. left) automaton of the bimachine. Using known results on the minimization of bimachines, we can minimize the corresponding subclass of aSST with respect to registers in PTime, and with respect to states and registers in NP.

2. We introduce a new model of bimachines, named asynchronous bimachines, which allows to lift the previous equivalence to the whole class of aSST. Based

on this, we consider register minimization with fixed underlying automaton, and show that (a slight generalization of) this problem is NP-COMPLETE.

Due to lack of space, omitted proofs can be found in [5].

	Fixed Underlying Automaton Register Minimization	Register Minimization	State-Register Minimization
aSST	**NP-complete** (for aSST with partial updates)	PSPACE-complete [9]	open
aSST$_{\mathsf{iffo}}$	**PTIME**	**PTIME**	**NP**

2 Models of Transducers

Words, Languages. An alphabet Σ is a finite set. An element $\sigma \in \Sigma$ is called a *letter* and a finite sequence of letters is called a *word*. The length of a word w will be denoted by $|w|$. The word of length 0 is called the empty word and is denoted by ϵ. The number of occurrences of a letter σ in a word w will be denoted by $|w|_\sigma$.

Given two words $w = a_1 \ldots a_n$ and $w' = a'_1 \ldots a'_m$, we will denote by $w \cdot w'$ or simply ww' their *concatenation* $a_1 \ldots a_n a'_1 \ldots a'_m$ and the concatenation of w with itself n times will be denoted by w^n ($w^0 = \epsilon$ for all w).

The set of all words over Σ is denoted by Σ^*. Together with concatenation, Σ^* is a monoid called the free monoid over Σ. Its neutral element is the empty word ϵ. Any subset $L \subseteq \Sigma^*$ is called a *language*. The class of rational (or regular) languages over Σ is denoted $\mathrm{Rat}(\Sigma^*)$.

Let u and v be two words of Σ^*. We will write $u \leq v$ if u is a prefix of v, *i.e.* if there exists $w \in \Sigma^*$ such that $v = uw$. We will denote w by $u^{-1}v$. The longest common prefix of u and v is denoted by $u \wedge v$. We generalize this definition to languages in the natural way and define $\bigwedge L \in \Sigma^*$ as the longest common prefix of all the words of L for all nonempty language $L \subset \Sigma^*$ and $\bigwedge \varnothing = \epsilon$.

Automata. A finite automaton $\mathcal{A}$ on an alphabet Σ is a tuple (Q, I, Δ, F) where Q is a finite set of *states*, $I \subseteq Q$ and $F \subseteq Q$ are the sets of *initial* and *final* states respectively and $\Delta \subseteq Q \times \Sigma \times Q$ is the set of *transitions* of the automaton.

A *run* of $\mathcal{A}$ on a word $w = a_1 \ldots a_n \in \Sigma^*$ is a finite sequence of states $(q_i)_{i \in [\![0,n]\!]}$ such that $(q_{i-1}, a_i, q_i) \in \Delta$ for all $i \in [\![1,n]\!]$. It will be denoted by $q_0 \xrightarrow{w}_\mathcal{A} q_n$ or $q_0 \xrightarrow{w} q_n$ when $\mathcal{A}$ is clear from the context. We say that the run is *accepting* if $q_0 \in I$ and $q_n \in F$ and define the language *recognized* by $\mathcal{A}$, denoted by $[\![\mathcal{A}]\!]$, as the set of words for which there exists an accepting run of $\mathcal{A}$.

$\mathcal{A}$ is called *deterministic* (resp. *complete*) if, for all $p \in Q$ and $\sigma \in \Sigma$, there exists at most (resp. at least) one state $q \in Q$ such that $(p, \sigma, q) \in \Delta$. A deterministic automaton is also required to have a single initial state. We say that $\mathcal{A}$ is codeterministic if the mirror automaton of $\mathcal{A}$, obtained by reversing transitions and swapping initial and final states, is deterministic. Last, $\mathcal{A}$ is *unambiguous* if, for all $w \in \Sigma^*$, there exists at most 1 accepting run of $\mathcal{A}$ on w.

In the following, we will use the abbreviations NFA and DFA for *nondeterministic finite automaton* and *deterministic finite automaton* respectively.

Given a complete DFA $\mathcal{A}$, its transitions induce an action from words on states, that we denote by $\cdot_{\mathcal{A}}$, which, given a state q and a word w, returns $q \cdot_{\mathcal{A}} w$ the unique state reached from q by reading the word w. It also induces a right congruence on Σ^*, denoted by $\sim_{\mathcal{A}}$, defined by $u \sim_{\mathcal{A}} v$ if and only if $q_0 \cdot_{\mathcal{A}} u = q_0 \cdot_{\mathcal{A}} v$, where q_0 is the initial state of $\mathcal{A}$. Conversely, given a right congruence of finite index $\sim$ on Σ^* and choosing a set F of equivalence classes, we can define a DFA $\mathcal{A}_\sim$ with $\Sigma^*/\sim$ as its set of states, $[\epsilon]_\sim$ as its initial state, F as its set of final states, and, for all $u \in \Sigma^*$ and $\sigma \in \Sigma$, $[u]_\sim \cdot_{\mathcal{A}_\sim} \sigma = [u\sigma]_\sim$.

Given two binary relations R and R' on Σ^*, we say that R *finer* than R' (or R' is coarser than R) whenever, for all $u, v \in \Sigma^*$, if uRv then $uR'v$. Moreover, a DFA $\mathcal{A}$ is called *finer* than a relation R whenever its associated congruence $\sim_{\mathcal{A}}$ is finer than R. In particular, $\mathcal{A}$ is called *finer* than another DFA $\mathcal{A}'$ whenever $\sim_{\mathcal{A}}$ is finer than $\sim_{\mathcal{A}'}$. Intuitively, in this case, there exist a morphism between the states of $\mathcal{A}$ and those of $\mathcal{A}'$ "merging" states together.

All the definitions above are adapted as expected when $\mathcal{A}$ is codeterministic and co-complete.

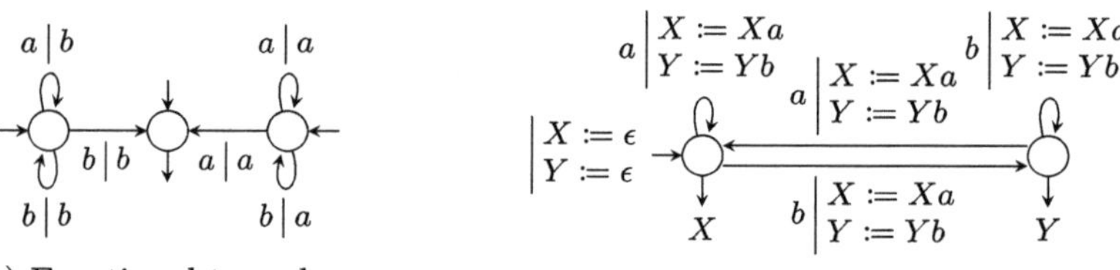

(a) Functional transducer

(b) Streaming String Transducer

Fig. 1. A functional transducer and an equivalent aSST

Transducers. are defined over an input (Σ) and an output (Γ) alphabet.

Definition 1. *A finite-state transducer from Σ^* to Γ^* is a tuple $\mathcal{T} = (Q, i, o, f)$ where Q is a finite set of* states *and $i\colon Q \to \Gamma^*$, $o\colon Q \times \Sigma \times Q \to \Gamma^*$ and $f\colon Q \to \Gamma^*$ are partial functions called respectively the* initial output, output *and* final output *functions of $\mathcal{T}$.*

The *underlying automaton* of $\mathcal{T}$ is defined as the NFA $\mathcal{A} = (Q, I, \Delta, F)$ where $I = \mathrm{dom}(i)$, $\Delta = \mathrm{dom}(o)$ and $F = \mathrm{dom}(f)$. $\mathcal{T}$ is said *sequential* (resp. *cosequential*) if $\mathcal{A}$ is a deterministic (resp. codeterministic).

A run of $\mathcal{T}$ on a word $w \in \Sigma^*$ is a run $\rho = (q_i)_{i \in [\![0,n]\!]}$ of $\mathcal{A}$ on w with transitions $(q_{i-1}, a_i, q_i) \in Q \times \Sigma \times Q$ for all $i \in [\![1,n]\!]$, such that $w = a_1 \cdots a_n$, together with an *output* word of Γ^* defined by the concatenation

$$o(\rho) = i(q_0)\, o(q_0, a_1, q_1) \cdots o(q_{n-1}, a_n, q_n)\, f(q_n)$$

$\mathcal{T}$ *realizes* a relation $[\![\mathcal{T}]\!] \subseteq \Sigma^* \times \Gamma^*$ defined by

$$[\![\mathcal{T}]\!] = \{(u,v) \in \Sigma^* \times \Gamma^* \mid \exists \text{ a run } \rho \text{ of } \mathcal{T} \text{ on } u \text{ with output } o(\rho) = v\}$$

In particular, when $[\![\mathcal{T}]\!]$ is a partial function, $\mathcal{T}$ is called a *functional* transducer. This defines the class of *rational functions*, as those realizable by functional transducers. It is worth noting that a function is rational if and only if it is realizable by an unambiguous one (see *e.g.* [6, Theorem IV.4.2]).

Example 1. Let $\Sigma = \Gamma = \{a,b\}$. The functional transducer depicted on Fig. 1a realizes a function mapping each word $w \in \Sigma^*$ to $\sigma^{|w|}$ where σ is the last letter of w (ϵ is mapped to itself). It is, in particular, a cosequential transducer.

Streaming String Transducers. have originally been introduced with general updates allowing to capture the class of regular functions [1]. However, the subclass of *appending* streaming string transducers (aSST) were shown to have the same expressiveness as rational functions (see *e.g.* [3, Theorem 4]). Hence, in the following, we will focus on aSST. We will not elaborate on SST with more general register updates, but the interested reader can find characterizations of their expressiveness in [1] and [15] (see also [14, Section 5]).

Definition 2 (Appending expressions). *Given a finite set of variables $\mathcal{X}$ and an output alphabet Γ, consider the set of expressions* $\mathrm{Exp}(\mathcal{X})$ *of the form*[1] $X \cdot w$, *where $X \in \mathcal{X}$ and $w \in \Gamma^*$.*

A substitution over $\mathcal{X}$ is a map $s\colon \mathcal{X} \to \mathrm{Exp}(\mathcal{X})$. It can be extended into a map from $\mathrm{Exp}(\mathcal{X})$ to $\mathrm{Exp}(\mathcal{X})$ by substituting each variable X in the expression given as an input by $s(X)$. We can compose substitutions by identifying them with their extension. Valuations are maps of the form $v\colon \mathcal{X} \to \Gamma^*$, which can also be extended to $\mathrm{Exp}(\mathcal{X})$. The set of substitutions will be denoted by $\mathrm{Sub}(\mathcal{X})$, the set of valuations by $\mathrm{Val}(\mathcal{X})$ and, given a substitution s and a variable X, we will write $X := e$ whenever $s(X) = e$ for some expression e.

Definition 3 (Appending Streaming String Transducer). *An appending streaming string transducer from Σ^* to Γ^* is a tuple $\mathcal{S} = (Q, \mathcal{X}, q_0, v_0, \delta^s, \delta^r, \gamma)$ where Q is a finite set of* states, *$\mathcal{X}$ is a finite set of* registers, *$q_0 \in Q$ is the initial* state, *$v_0 \in \mathrm{Val}(\mathcal{X})$ is the registers' initial valuation, $\delta^s\colon Q \times \Sigma \to Q$ is the* transition function, *$\delta^r\colon Q \times \Sigma \to \mathrm{Sub}(\mathcal{X})$ is the register update function, and $\gamma\colon Q \to \mathrm{Exp}(\mathcal{X})$ is a partial output function.*

We define the *underlying automaton* of $\mathcal{S}$ as the DFA $\mathcal{A}$ with the same set of states, initial state and transition function as $\mathcal{S}$. Its set of final states is $\mathrm{dom}(\gamma)$.

The runs of $\mathcal{S}$ will be runs of $\mathcal{A}$ together with valuations mapping each register to its current value. *i.e.* we define the configurations of $\mathcal{S}$ as pairs $(q,v) \in$

---

[1] We will usually omit the $\cdot$ from the expressions and omit the w when $w = \epsilon$.

$Q \times \mathrm{Val}(\mathcal{X})$. Given a word $w = a_1 \ldots a_n \in \Sigma^*$, the run of $\mathcal{S}$ on w is the sequence of configurations $(q_i, v_i)_{i \in [\![0,n]\!]}$ where q_0 is the initial state, v_0 is the initial valuation and, for all $i \in [\![1,n]\!]$, $q_i = \delta^s(q_{i-1}, a_i)$ and $v_i = v_{i-1} \circ \delta^r(q_{i-1}, a_i)$. The function $[\![\mathcal{S}]\!] \colon \Sigma^* \to \Gamma^*$ *realized* by $\mathcal{S}$ is defined by $[\![\mathcal{S}]\!](w) = v_n(\gamma(q_n))$.

Lemma 1 ([3]). *Let $f \colon \Sigma^* \to \Gamma^*$ be a partial function. f is a rational function iff it can be realized by an aSST.*

Intuitively, from a transducer to an aSST, one uses one register per state and performs a classical subset construction. Conversely, for each register, one builds a run that simulates its value. This leads to an unambiguous transducer.

We will consider an extension of this model, in which we allow the register update function to output partial substitutions, *i.e.* partial maps from $\mathcal{X}$ to $\mathrm{Exp}(\mathcal{X})$. We call these machines aSST *with partial updates*. The semantics of aSST can be lifted to this model by considering partial valuations.

Example 2. The aSST depicted on Fig. 1b realizes the same function as the transducer of Example 1 depicted on Fig. 1a.

Problems Considered. The first problem we consider, which has been studied before in [9], is the register minimization problem. It can be stated as follows:
Problem: Register Minimization (REG MIN for short)
Input: a rational function f and $k \in \mathbb{N}$
Question: Does there exist an aSST $\mathcal{S}$ that realizes f and has k registers?

A more constrained version, which has been studied in [4] for Cost-register automata over a field, can be defined as follows:
Problem: State-Register Minimization (STT-REG MIN for short)
Input: a rational function f and $k, n \in \mathbb{N}$
Question: Does there exist an aSST $\mathcal{S}$ with k registers and n states s.t. $[\![\mathcal{S}]\!] = f$?

Let f be a rational function. The *register complexity* (resp. *state complexity*) of f is the minimal number of registers (resp. states) needed for an aSST to realize it. The *state-register complexity* of f is the set of pairs (n, k) such that there exists an aSST with n states and k registers realizing f, and such that any aSST realizing f with n' states and k' registers with $(n, k) \neq (n', k')$ has either strictly more states ($n' > n$) or strictly more registers ($k' > k$).

3 Algebraic Toolbox

Let us now consider functional transducers from Σ^* to Γ^* for two finite alphabets Σ and Γ, and introduce the tools that allow to obtain canonical and minimal representations of the functions they realize.

Bimachines. Introduced by Schützenberger in [19] and named by Eilenberg in [10], *bimachines* read their input simultaneously from left to right using their left automaton and in the opposite direction using their right automaton and produces the output based on the states they reach during the run. They realize the same functions as functional transducers and are defined more formally as follows:

Definition 4 (Bimachine). *A bimachine on two finite alphabets Σ and Γ is a tuple $\mathcal{B} = (\mathcal{L}, \mathcal{R}, \lambda, \omega, \rho)$ where $\mathcal{L} = (L, l_{\mathrm{i}}, \delta_{\mathcal{L}}, F_{\mathcal{L}})$ is the* left automaton *of $\mathcal{B}$. It is a deterministic finite automaton. Dually, $\mathcal{R} = (R, I_{\mathcal{R}}, \delta_{\mathcal{R}}, r_{\mathrm{f}})$ is its* right automaton. *It is a codeterministic finite automaton. $\lambda\colon I_{\mathcal{R}} \to \Gamma^*$, $\omega\colon L \times \Sigma \times R \to \Gamma^*$, and $\rho\colon F_{\mathcal{L}} \to \Gamma^*$ are respectively its* left output, output *and* right output *functions.*

Defining $\omega(l, \epsilon, r) = \epsilon$ and using the identity $\omega(l, uv, r) = \omega(l, u, v \cdot_{\mathcal{R}} r) \cdot \omega(l \cdot_{\mathcal{L}} u, v, r)$, for all $u, v \in \Sigma^*$, $l \in L$ and $r \in R$, ω is extended to words by induction. The function $[\![\mathcal{B}]\!]\colon \Sigma^* \to \Gamma^*$ *realized* by $\mathcal{B}$ is then defined, for all $w \in \Sigma^*$, by $[\![\mathcal{B}]\!](w) = \lambda(w \cdot_{\mathcal{R}} r_{\mathrm{f}}) \, \omega(l_{\mathrm{i}}, w, r_{\mathrm{f}}) \, \rho(l_{\mathrm{i}} \cdot_{\mathcal{L}} w)$.

Any functional transducer can be converted into a bimachine and vice versa [10, Theorem XI.7.1] (see also [6, Theorem IV.5.1]).

Theorem 1. *A partial function $f\colon \Sigma^* \to \Gamma^*$ is rational if and only if there exists a bimachine $\mathcal{B}$ such that $[\![\mathcal{B}]\!] = f$.*

In contrast with DFA, rational functions does not always have a unique minimal transducer realizing them. However, Reutenauer and Schützenberger showed in [18] that they can still be represented in a canonical way using bimachines.

Syntactic Congruences. Given a language $L \subseteq \Sigma^*$, and a word $u \in \Sigma^*$, we define the left residual of L w.r.t. u as the language $u^{-1}L = \{v \in \Sigma^* \mid uv \in L\}$. Right residuals are defined symmetrically. We define the equivalence relation $\sim_L$ as $u \sim_L v$ iff $u^{-1}L = v^{-1}L$. It is well known that $\sim_L$ is a right congruence, called the *Myhill-Nerode congruence* of L, and that L is a regular language iff $\sim_L$ has finite index k. In addition, when this holds, the minimal DFA recognizing L has precisely k states. We consider the distance on words defined, for all $u, v \in \Sigma^*$, by $\|u, v\| = |u| + |v| - 2|u \wedge v|$. Two words are close for this distance if they share a long common prefix. Based on this distance, congruences characterizing rational transductions were introduced in [18].

Definition 5. *Let $f\colon \Sigma^* \to \Gamma^*$ be a partial function.*

The left syntactic congruence $\leftsquigarrow_f$ *of f is defined, for all $u, v \in \Sigma^*$, by $u \leftsquigarrow_f v$ if and only if the two following conditions are verified:* $\mathrm{dom}(f)u^{-1} = \mathrm{dom}(f)v^{-1}$ *and* $\sup_{w \in \mathrm{dom}(f)u^{-1}} \|f(wu), f(wv)\| < \infty$.

The right syntactic congruence $\rightsquigarrow_f$ *of f is defined symmetrically, using a distance based on the longest common suffix.*

Theorem 2 ([18]). *Let $f\colon \Sigma^* \to \Gamma^*$ be a partial function. T.f.a.e.:*

(1) f is a rational transduction

(2) $\leftsquigarrow_f$ has a finite index and $f^{-1}(L) \in \mathrm{Rat}(\Sigma^*)$ for all $L \in \mathrm{Rat}(\Gamma^*)$.

(3) $\rightsquigarrow_f$ has a finite index and $f^{-1}(L) \in \mathrm{Rat}(\Sigma^*)$ for all $L \in \mathrm{Rat}(\Gamma^*)$.

Definition 6. *We define the* canonical right automaton $\mathcal{R}_f$ *of a rational function f as the codeterministic automaton $\mathcal{A}_{\leftsquigarrow_f}$ obtained from the left syntactic congruence $\leftsquigarrow_f$ with $\{[w]_{\leftsquigarrow_f} \mid w \in \mathrm{dom}(f)\}$ as its set of initial states. Its canonical left automaton $\mathcal{L}_f$ is defined symmetrically using $\rightsquigarrow_f$.*

Definition 7. *A bimachine $\mathcal{B}$, with left and right automata $\mathcal{L}$ and $\mathcal{R}$ respectively, realizing a function f is called* minimal *if no other bimachine $\mathcal{B}'$, with left and right automata $\mathcal{L}'$ and $\mathcal{R}'$ respectively, realizing f is such that $\mathcal{L}$ is finer than $\mathcal{L}'$ and $\mathcal{R}$ is finer than $\mathcal{R}'$.*

For a total function f, [18] shows the existence of a canonical minimal bimachine, with $\mathcal{R}_f$ as its right automaton. Moreover, $\mathcal{R}_f$ is minimal among all the right automata of the bimachines realizing f. The same is true for $\mathcal{L}_f$, which allows to define a left canonical bimachine with a minimal left automaton. This result was later extended to all rational functions in [12] where the canonical bimachines were shown to be computable in polynomial time. The paper also provides PTIME algorithms to obtain the minimal right automaton associated with a given left automaton and vice-versa, allowing the efficient minimization of any given bimachine. We sum up the results relevant to this paper as follows:

Theorem 3 ([12,18]). *Bimachines can be minimized in PTIME. Moreover, given a rational function f, there exist a computable finite family of minimal bimachines realizing f.*

Remark 1. Unlike Definition 4 (which is the same as in [18]) [12] requires from both automata of a bimachine to recognize the domain of the function. This is important for minimization problems of the next sections.

4   Relating Streaming String Transducers and Bimachines

We start by introducing a subclass of aSST that will allow us to obtain an equivalence with bimachines.

Let Σ and Γ be two finite alphabets and let $\mathcal{S} = (Q, \mathcal{X}, q_0, v_0, \delta^s, \delta^r, \gamma)$ be an aSST from Σ^* to Γ^*. By swapping the contents of the registers if necessary, we may assume without loss of generality that $\mathcal{S}$ has a fixed output register. Meaning that there exists a register $X_{\mathrm{out}} \in \mathcal{X}$ such that, for all $q \in \mathrm{dom}(\gamma)$, $\gamma(q) = X_{\mathrm{out}}w$ for some $w \in \Gamma^*$. Using the terminology of [13],

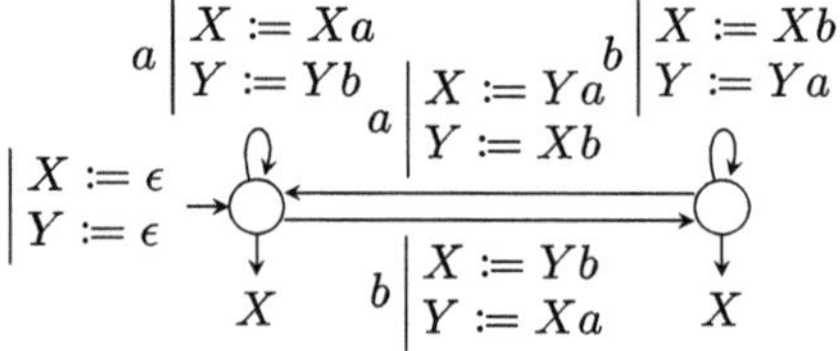

Fig. 2. An aSST with a fixed output register and dependent flows.

given a register update $X := Yw$ of $\mathcal{S}$, the register Y is said to *flow* to X. We will say that $\mathcal{S}$ has *independent flows* if all the flows of the register updates of its transitions only depend on the letter read rather than both the letter and the state of its underlying automaton, or, more formally:

Definition 8. *An aSST $\mathcal{S} = (Q, \mathcal{X}, q_0, v_0, \delta^s, \delta^r, \gamma)$ is said to have independent flows if for all $X \in \mathcal{X}$ and $\sigma \in \Sigma$, there exists $Y \in \mathcal{X}$ such that for all $q \in Q$, $\delta^r(q, \sigma)(X) = Yw$ for some word[2] $w \in \Gamma^*$.*

Example 3. The aSST depicted on Fig. 1b has independent flows while the equivalent aSST of Fig. 2 has a fixed output register but flows depending on its states.

Barring some considerations concerning the domains, this restriction, in conjunction with the fixed output register assumption, allows us to show a one-to-one correspondence between the aSST verifying these conditions and the bimachines realizing the same transduction.

In the following, we consider the class of aSST that have independent flows, and a fixed output register. We denote it by $\mathrm{aSST_{iffo}}$.

Proposition 1. *Let $f \colon \Sigma^* \to \Gamma^*$ be a total function, and k, n be two integers.*

There exists an $\mathrm{aSST_{iffo}}$ that realizes f with n states and k registers iff there exists a bimachine that realizes f whose left automaton has n states, and whose right automaton has k states.

The idea is that the left automaton, corresponds to the underlying automaton of the aSST, while the right automaton specifies the flows of the registers. The condition of independence of the flows is required since the transitions of the right automaton of a bimachine do not depend on the states reached in its left one. As a consequence, we obtain:

Corollary 1. *Let f be a total rational function. The register complexity of f, relative to the class $\mathrm{aSST_{iffo}}$, is the number of states of the right automaton of a bimachine with a minimal right automaton realizing f. The state-register complexity of f, relative to the class $\mathrm{aSST_{iffo}}$, is the following set:*

$$\{(|\mathcal{L}|, |\mathcal{R}|) \mid \mathcal{B} = (\mathcal{L}, \mathcal{R}, \lambda, \omega, \rho) \text{ is a minimal bimachine realizing } f\}$$

Since bimachine minimization can be performed in PTime, as we have seen in Sect. 3, we can then solve the register minimization problem for this class of aSST by computing a bimachine with a minimal right automaton and converting it into an aSST using the construction above. Similarly, in order to solve STT-REG MIN, one can non-deterministically guess a bimachine $\mathcal{B}$ of the expected size, check that it is minimal, and that it recognizes f. By definition, the size of $\mathcal{B}$ is polynomial in n and k, which we assume are given in unary (as was done in [9]) and minimality and equivalence can be checked in PTime. We obtain:

Theorem 4. REG MIN *(resp.* STT-REG MIN*) is decidable in* PTime *(resp. NP) for* $\mathrm{aSST_{iffo}}$ *realizing total functions.*

Minimization results of [12,18] require from the right automaton of the bimachine to recognize the domain of the function, which is a condition that is not required from an aSST since its states and register updates are not necessarily

---

[2] Which can depend on the state q.

correlated. Thus, the characterization of the register complexity of Corollary 1 cannot easily be extended to partial functions. Moreover, even for total functions, the register complexity for the class $\mathrm{aSST_{iffo}}$ can be higher than the one for general aSST. For instance, considering the function of Example 1, one can verify that the register complexity of this function relative to the class $\mathrm{aSST_{iffo}}$ is three, while it can be a realized by an aSST with two registers (Fig. 1b).

5   A New Model of Bimachines

To drop the independent flow restriction, we will modify bimachines as follows: instead of using both automata to read the input simultaneously in both directions, what we will call an *asynchronous bimachine* will first read the input using its left automaton, and then, using its right automaton, it will read the run of the left automaton and produce the output. This is also reminiscent of the proof of [6, Theorem IV.5.2] of Elgot and Mezei's decomposition.

Definition of the Model. An asynchronous bimachine on two finite alphabets Σ and Γ is a tuple $\mathcal{B} = (\mathcal{L}, \mathcal{R}, \lambda, \omega, \rho)$ where its *left automaton* is $\mathcal{L} = (L, l_\mathrm{i}, \delta_\mathcal{L}, F_\mathcal{L})$, a deterministic finite automaton on the alphabet Σ, its *right automaton* is $\mathcal{R} = (R, I_\mathcal{R}, \delta_\mathcal{R}, r_\mathrm{f})$, a codeterministic finite automaton on the alphabet $L \times \Sigma$, and $\lambda \colon I_\mathcal{R} \to \Gamma^*$, $\omega \colon R \times (L \times \Sigma) \times R \to \Gamma^*$, and $\rho \colon F_\mathcal{L} \to \Gamma^*$ are respectively its *left output*, *output* and *right output* functions.

A run of $\mathcal{B}$ on a word $w = a_1 \ldots a_n \in \Sigma^*$ is a finite sequence $(l_j, r_j)_{j \in [\![0,n]\!]}$ of elements of $L \times R$ such that $l_0 = l_\mathrm{i}$, $(l_j)_{j \in [\![0,n]\!]}$ is the run of $\mathcal{L}$ on w, $r_n = r_\mathrm{f}$ and $(r_j)_{j \in [\![0,n]\!]}$ is the run of $\mathcal{R}$ on the word $w_\mathcal{L} = (l_0, a_1) \ldots (l_{n-1}, a_n) \in (L \times \Sigma)^*$. The transduction $[\![\mathcal{B}]\!] \colon \Sigma^* \to \Gamma^*$ *realized* by $\mathcal{B}$ is defined by

$$[\![\mathcal{B}]\!](w) = \lambda(r_0) \prod_{j=1}^{n} \omega\big(r_{j-1}, (l_{j-1}, a_j), r_j\big) \, \rho(l_n)$$

As described above, we can view $[\![\mathcal{B}]\!]$ as the composition of a function realized by a sequential transducer $\mathcal{T}_\mathcal{L}$ annotating the input by the states of $\mathcal{L}$ and a cosequential transducer $\mathcal{T}_\mathcal{R}$ reading these annotated words.

Correspondence with Streaming String Transducers. Using asynchronous bimachines we manage to characterize the full class of aSST:

Theorem 5. *A rational function can be realized by an aSST with n states and k registers if and only if it can be realized by an asynchronous bimachine having a left automaton with n states and a right automaton with k states.*

The conversions are the same as those of Proposition 1. The equality of the domains and the removal of the restriction on the flows of the aSST are direct consequences of the definition of asynchronous bimachines.

Corollary 2. *The register complexity of a rational function f, relative to the class aSST, is the minimal number of states needed by the right automaton of an asynchronous bimachine to realize f.*

Minimization of Asynchronous Bimachines. By Corollary 2, to solve REG MIN, we have to find an asynchronous bimachine with a right automaton having the minimal number of states among all the bimachines realizing the same function. However, adapting the known algorithms of [12,18] for the minimization of bimachines to asynchronous bimachines seems rather challenging.

A first step would be to fix the left automaton and search for the smallest possible right automaton that can be associated with it. This corresponds to minimizing the number of registers of a given aSST without changing its underlying automaton. We formalize this problem as follows:

Problem: Fixed Underlying Automaton Register Minimization (FA-REG MIN)

Input: an aSST S realizing a function f, and $k \in \mathbb{N}$

Question: Does there exist an aSST S' that realizes f, has k registers and s.t. S and S' have the same underlying automaton?

Viewing an asynchronous bimachine $\mathcal{B}$ as the composition of a sequential transducer $\mathcal{T}_\mathcal{L}$ and a cosequential one $\mathcal{T}_\mathcal{R}$ as above, we want to find a cosequential transducer $\mathcal{T}$ with the minimal number of states such that $[\![\mathcal{T}]\!](w) = [\![\mathcal{T}_\mathcal{R}]\!](w)$ for all $w \in \mathcal{T}_\mathcal{L}(\Sigma^*)$. Without loss of generality, we can consider $[\![\mathcal{T}]\!]$ to be a total function. Furthermore, up to mirroring the input word, we can consider $\mathcal{T}_\mathcal{R}$ sequential and search for a sequential transducer.

We have just reduced the problem of minimizing the right automaton of an asynchronous bimachine, with a fixed left automaton, to the problem of extension of sequential transducers, defined as:

Problem: Extension of sequential transducers

Input: a sequential transducer $\mathcal{T}$ and $k \in \mathbb{N}$

Question: Does there exist a sequential transducer $\mathcal{T}'$ with k states such that $[\![\mathcal{T}']\!](w) = [\![\mathcal{T}]\!](w)$ for all $w \in \mathrm{dom}([\![\mathcal{T}]\!])$?

This problem has been studied in previous works [16,17], and shown to be NP-complete. Regarding the problem FA-REG MIN, one can propose an NP procedure which follows a classical guess-and-check scheme. Adapting the hardness proofs developed for the extension of sequential transducers, we prove:

Proposition 2. *Solving problem* FA-REG MIN *for an* aSST *with partial updates is* NP-COMPLETE.

This shows that FA-REG MIN is hard, and we are thus interested in identifying an algorithm to solve it using an automata-based approach. In the particular case of letter-to-letter transducers, or, rational functions with a prefix-closed domain, we show that the problem of extension of a sequential transducer is equivalent to a refinement problem for deterministic automata.

Given a DFA $\mathcal{A} = (Q, q_0, \delta, F)$, we will call a precongruence on Q any reflexive and symmetric relation $\approx$ on Q such that for all $p, q \in Q$ and $\sigma \in \Sigma$, if $p \approx q$ then $p \cdot_\mathcal{A} \sigma \approx q \cdot_\mathcal{A} \sigma$. Any precongruence $\approx$ gives rise to a relation on Σ^*, that we will also denote by $\approx$, defined for all $u, v \in \Sigma^*$ by $u \approx v$ if and only if $q_0 \cdot_\mathcal{A} u \approx q_0 \cdot_\mathcal{A} v$. We can view $\approx$ as a compatibility relation on the states of $\mathcal{A}$, which is not necessarily transitive, but ensures that any two runs of $\mathcal{A}$ on the same word starting from two compatible states ends in two states that are also compatible.

We want to obtain a DFA with a smaller number of states by merging together the states of $\mathcal{A}$ that are compatible. More formally, we define:

Problem: Minimal refinement Problem

Input: a precongruence $\approx$ on the states of a DFA and $k \in \mathbb{N}$

Question: Does there exist a DFA finer than $\approx$ with at most k states?

Proposition 3. *The problem of extension of sequential transducers, for rational functions realizable by a letter-to-letter transducer and those with a prefix-closed domain, is equivalent to the minimal refinement problem.*

We will now describe a construction, we call the *subset expansion*, allowing the enumeration of all the minimal DFA refining a given precongruence. It is similar to the classical subset construction for the determinization of NFA, but the subsets are formed with states that are pairwise compatible, representing all the merges of states allowed by the precongruence. We define it as follows.

Definition 9. *Let $\mathcal{A} = (Q_A, q_0, \delta_A, F_A)$ be a DFA and let $\approx$ be a precongruence on Q. We define the NFA $\mathcal{E} = (Q_\mathcal{E}, I_\mathcal{E}, \Delta_\mathcal{E}, F_\mathcal{E})$ of the subset expansion of $\mathcal{A}$ relative to $\approx$ as follows: $Q_\mathcal{E} = \{P \subseteq Q_A \,|\, \forall p, q \in P, \, p \approx q\}$, $I_\mathcal{E} = \{P \in Q_\mathcal{E} \,|\, q_0 \in P\}$, $\Delta_\mathcal{E} = \{(P, \sigma, Q) \in Q_\mathcal{E} \times \Sigma \times Q_\mathcal{E} \,|\, \forall p \in P, \, p \cdot_A \sigma \in Q\}$ and $F_\mathcal{E} = Q_\mathcal{E}$.*

Given two automata $\mathcal{A} = (Q, I, \Delta, F)$ and $\mathcal{A}' = (Q', I', \Delta', F')$, we call $\mathcal{A}'$ a *subautomaton* of $\mathcal{A}$ if $Q' \subseteq Q$, $I' \subseteq I$, $\Delta' \subseteq \Delta$ and $F' \subseteq F$. Note that all the deterministic subautomata of $\mathcal{E}$ are finer than $\approx$ by definition. We prove:

Theorem 6. *Let $\mathcal{A}$ be a DFA, let $\approx$ be a precongruence on its states and let $\mathcal{E}$ be the automaton of the subset expansion of $\mathcal{A}$ relative to $\approx$.*

If $\mathcal{B}$ is a DFA finer than $\approx$ then there exists a deterministic subautomaton $\mathcal{B}'$ of $\mathcal{E}$ such that $\sim_\mathcal{B}$ is finer than $\sim_{\mathcal{B}'}$. Moreover, if $\mathcal{B}$ is a minimal automaton among all the DFA finer than $\approx$ then $\mathcal{B}'$ is equal to $\mathcal{B}$ up to state renaming.

We can then find all the solutions to the minimal refinement problem by enumerating the deterministic subautomata of the automaton of the subset expansion having a number of states bounded the given integer k. Thanks to the equivalence of Proposition 3 and the correspondence between aSST and asynchronous bimachines of Theorem 5, this allows us to find the minimal number of registers needed by an aSST with a fixed underlying automaton having register updates of the form $X := Y\sigma$ with $\sigma \in \Sigma$ or such that the domain of the function it realizes is suffix-closed[3].

In the general case of the extension problem for arbitrary sequential transducers, the possible delays between the outputs produced from each pairs of states prevents the definition of a suitable compatibility relation and, to the best of our knowledge, the problem remains open.

---

[3] Note the inversion of the reading direction since the right automaton we seek to minimize is actually codeterministic.

6   Discussion and Perspectives

On the Tradeoff Between States and Registers. Let f be a rational function. The *state complexity* of f is at least the finite index of the Myhill Nerode congruence associated with $\mathrm{dom}(f)$. Moreover, thanks to our correspondence with bimachines, it is at most the index of the right syntactic congruence $\leadsto_f$ of f.

Regarding the *register complexity*, we recall the result of [9], with a precision on the number of states of the equivalent SST:

Theorem 7 ([9]). *Let f be a rational function realized by a functional transducer with N states, and $k \in \mathbb{N}$ be an integer. One can decide whether there exists an aSST with k registers that realizes f. In addition, when this is the case, then one can build such an aSST, and its number of states is in $2^{O(N^k)}$.*

This theorem shows that reducing the number of registers may induce an *exponential blowup* of the number of states. We show, using the next example, that this bound is tight, by providing an example for which an exponential blowup of the number of states occurs.

Example 4. Let $\Sigma = \{a, b\}$, $N \in \mathbb{N}$, and consider the following function: $f :$ $u\sigma \in \Sigma^{N+1} \mapsto \sigma u$, where $\sigma \in \Sigma$ is a letter, and $u \in \Sigma^N$ is a word on Σ of length N. This function can be realized by an aSST with two registers X_a, X_b and $O(N)$ states. It simply copies the input word u in both registers, but on the first transition, it adds an a in X_a and a b in X_b. On the last transition, it records the letter to decide which register to output. It uses $O(N)$ states to check the length of the input word.

This function can also be realized by an aSST with a single register, but one can show that this requires to store u in the states, so as to produce the whole output on the last letter of the input, yielding an exponential number of states.

Perspectives. Using the equivalence with bimachines, we have obtained strong and efficient minimization results for aSST$_{\mathsf{iffo}}$ with total domain. On the other side, the model of asynchronous bimachines allows to capture the full class of aSST, but is more difficult to minimize. This has led us to the study of the register minimization for a fixed underlying automaton. Next, the converse should be investigated. *i.e.* minimizing the left automaton with respect to a fixed right automaton and checking the canonicity of the asynchronous bimachine obtained by composing the two constructions. As future work, another challenging research direction is the state-register minimization problem for the full class of aSST.

References

1. Alur, R., Cerný, P.: Expressiveness of streaming string transducers. In: Lodaya, K., Mahajan, M. (eds.) IARCS Annual Conference on Foundations of Software Technology and Theoretical Computer Science, FSTTCS 2010, December 15-18, 2010, Chennai, India. LIPIcs, vol. 8, pp. 1–12. Schloss Dagstuhl - Leibniz-Zentrum für Informatik (2010). https://doi.org/10.4230/LIPICS.FSTTCS.2010.1

2. Alur, R., Cerný, P.: Streaming transducers for algorithmic verification of single-pass list-processing programs. In: Ball, T., Sagiv, M. (eds.) Proceedings of the 38th ACM SIGPLAN-SIGACT Symposium on Principles of Programming Languages, POPL 2011, Austin, TX, USA, January 26-28, 2011, pp. 599–610. ACM (2011). https://doi.org/10.1145/1926385.1926454

3. Alur, R., D'Antoni, L., Deshmukh, J.V., Raghothaman, M., Yuan, Y.: Regular functions and cost register automata. In: 28th Annual ACM/IEEE Symposium on Logic in Computer Science, LICS 2013, New Orleans, LA, USA, June 25-28, 2013, pp. 13–22. IEEE Computer Society (2013). https://doi.org/10.1109/LICS.2013.65

4. Benalioua, Y.I., Lhote, N., Reynier, P.: Minimizing cost register automata over a field. In: Královic, R., Kucera, A. (eds.) 49th International Symposium on Mathematical Foundations of Computer Science, MFCS 2024, August 26-30, 2024, Bratislava, Slovakia. LIPIcs, vol. 306, pp. 23:1–23:15. Schloss Dagstuhl - Leibniz-Zentrum für Informatik (2024). https://doi.org/10.4230/LIPICS.MFCS.2024.23

5. Benalioua, Y.I., Lhote, N., Reynier, P.A.: Minimizing streaming string transducers: an algebraic approach (2026). https://arxiv.org/abs/2604.11567

6. Berstel, J.: Transductions and Context-Free Languages, Teubner Studienbücher : Informatik, vol. 38. Teubner (1979). https://www.worldcat.org/oclc/06364613

7. Berstel, J., Boasson, L.: Transductions and context-free languages. Ed. Teubner, pp. 1–278 (1979)

8. Büchi, J.R.: Weak second-order arithmetic and finite automata. Zeitschrift für Mathematische Logik und Grundlagen der Mathematik **6**(1–6), 66–92 (1960)

9. Daviaud, L., Reynier, P.A., Talbot, J.M.: A generalised twinning property for minimisation of cost register automata. In: Grohe, M., Koskinen, E., Shankar, N. (eds.) Proceedings of the 31st Annual ACM/IEEE Symposium on Logic in Computer Science, LICS '16, New York, NY, USA, July 5-8, 2016, pp. 857–866. ACM (2016). https://doi.org/10.1145/2933575.2934549

10. Eilenberg, S.: Automata, Languages, and Machines A. Pure and Applied Mathematics, vol. A. Academic Press (1974). https://www.worldcat.org/oclc/310535248

11. Elgot, C.C.: Decision problems of finite automata design and related arithmetics. Trans. Am. Math. Soc. **98**(1), 21–51 (1961)

12. Filiot, E., Gauwin, O., Lhote, N.: Logical and algebraic characterizations of rational transductions. Logical Methods Comput. Sci. **15**(4) (2019). https://doi.org/10.23638/LMCS-15(4:16)2019

13. Filiot, E., Krishna, S.N., Trivedi, A.: First-order definable string transformations. In: Raman, V., Suresh, S.P. (eds.) 34th International Conference on Foundation of Software Technology and Theoretical Computer Science, FSTTCS 2014, December 15-17, 2014, New Delhi, India. LIPIcs, vol. 29, pp. 147–159. Schloss Dagstuhl - Leibniz-Zentrum für Informatik (2014). https://doi.org/10.4230/LIPICS.FSTTCS.2014.147

14. Filiot, E., Reynier, P.A.: Transducers, logic and algebra for functions of finite words. ACM SIGLOG News **3**(3), 4–19 (2016). https://doi.org/10.1145/2984450.2984453

15. Filiot, E., Reynier, P.A.: Copyful streaming string transducers. Fundam. Inform. **178**(1–2), 59–76 (2021). https://doi.org/10.3233/FI-2021-1998

16. Pfleeger, C.P.: State reduction in incompletely specified finite-state machines. IEEE Trans. Comput. **22**(12), 1099–1102 (1973). https://doi.org/10.1109/T-C.1973.223655

17. Reusch, B., Merzenich, W.: Minimal coverings for incompletely specified sequential machines. Acta Informatica **22**(6), 663–678 (1986). https://doi.org/10.1007/BF00263650

18. Reutenauer, C., Schützenberger, M.P.: Minimization of rational word functions. SIAM J. Comput. **20**(4), 669–685 (1991). https://doi.org/10.1137/0220042
19. Schützenberger, M.P.: A remark on finite transducers. Inf. Control **4**(2–3), 185–196 (1961). https://doi.org/10.1016/S0019-9958(61)80006-5
20. Straubing, H.: Finite Automata, Formal Logic, and Circuit Complexity. Birkhäuser, Boston, Basel and Berlin (1994)
21. Trakhtenbrot, B.A.: Finite automata and logic of monadic predicates (in Russian). Dokl. Akad. Nauk SSSR **140**, 326–329 (1961)

Closure Operations on Picture Languages and Their Relation to Floor Plans

Stefano Crespi Reghizzi[1], Antonio Restivo[2], and Pierluigi San Pietro[1(✉)]

[1] Politecnico di Milano - DEIB and CNR - IEIIT, Milano, Italy
{stefano.crespireghizzi,pierluigi.sanpietro}@polimi.it
[2] Dipartimento di Matematica e Informatica, Università di Palermo, Palermo, Italy
antonio.restivo@unipa.it

Abstract. We investigate two classical operations—tiling (T) closure and horizontal/vertical concatenation (C) closure—which extend the 1D Kleene star to two-dimensional picture languages. We show that several properties which are trivial for word languages become undecidable in the context of recognizable picture languages (REC): specifically, whether a local language is T-closed or C-closed, and whether a recognizable language is indecomposable. We prove that the minimal basis for both T- and C-closed languages is unique; however, we demonstrate a recognizable language whose basis is non-recognizable, a significant departure from the 1D case. We study picture-based encodings of floor plans (rectangular partitions) and prove that the language of all floor plan pictures is local, while the subfamily of guillotine floor plans is not. The recognizability of guillotine floor plans is a sufficient condition for REC to be closed under C-closure. While this remains an open problem, we show that the complement of the guillotine language is recognizable and that strong equivalence classes of structurally similar floor plans are in REC.

1 Introduction

The study of picture languages seeks natural two-dimensional analogues of classical one-dimensional results. When concepts such as recognizability, closure, or minimal generating sets are extended from words to pictures, their behavior often diverges in surprising ways. This paper investigates some fundamental properties of two classical Kleene-like operations: the tiling closure (T-closure) [12], and the closure under horizontal and vertical concatenation [6], (C-closure). For recognizable picture languages, Simplot [12] proved the strict inclusion of the C-closure within the T-closure, independently using the windmill floor plan as a witness. He also proved that the T-closure of a recognizable language is recognizable, and stated as an open question the same problem for the C-closure. Some variants of the C-closure operations have been introduced in [10]. The T-closure is the essential operation of tile-rewriting picture grammars [5,11].

First, in Sect. 3, we address problems concerning closure operations that are classic for word languages but have not yet been considered for 2D languages. We show that several decidable problems for recognizable word languages become

© The Author(s), under exclusive license to Springer Nature Switzerland AG 2026
M.-P. Béal and P. Caron (Eds.): DLT 2026, LNCS 16578, pp. 168–180, 2026.
https://doi.org/10.1007/978-3-032-28404-4_13

undecidable already at the level of local and recognizable (REC) picture languages [6]. Specifically, it is undecidable whether a given local language is a fixed point of the T-closure or C-closure, and it is undecidable whether a recognizable language is T-indecomposable or C-indecomposable. We then focus on the basis (minimal generating set) of closed picture languages—which for word languages was studied already in [4] and more recently in [3]—and prove it to be unique for both closures. In another divergence from word languages, we prove the basis of a recognizable picture language is not necessarily recognizable.

Second, in Sect. 4, with the goal of studying the C-closure of recognizable languages, we examine a well-known class of 2D discrete-geometry objects: the floor plans or rectangular partitions. These are used to represent planar graphs and applied in CAD systems for laying out VLSI circuits (see [2] for a historical introduction). Floor plans are conceptually similar to pictures defined by closure operations, but to our knowledge, they have not been studied as picture languages until now. To represent them, we reify the rectangular partition into blocks by using contours that encode block edges with reserved symbols. Floor plans can then be studied as picture languages, focusing on fundamental sub-classes[1]. We analyze (i) triangulated floor plans, (ii) guillotine decomposable floor plans, and (iii) the floor plans containing the windmill pattern.

We show that (i) is a local picture language while (ii) is not, and that (ii) coincides with the C-closure of the trivially partitioned floor plans. The question of whether the class REC is closed under C-closure remains open, but we prove this problem can be solved positively if the language of guillotine floor plans is recognizable. As another positive result, we show that the complement of guillotine floor plans is recognizable, as is the set of strongly equivalent floor plans (where blocks preserve horizontal and vertical contacts).

2   Notations and Preliminaries

Throughout this paper, all alphabets are assumed to be finite. A *picture* is a rectangular array of letters over an alphabet. If a picture p has size (m, n), where m is the number of rows and n is the number of columns, we write $|p|_{row}$ for m and $|p|_{col}$ for n. The *domain* of a picture p of size (m, n) is the set $\{(1, 1), \ldots, (m, n)\}$. Any connected rectangular subset of the domain of p is a *subpicture*.

For an alphabet Σ, the set of pictures of size (m, n) is denoted by $\Sigma^{m,n}$, and the set of nonempty finite pictures by Σ^{++}. A *picture language* (omitting "picture" for brevity) is a subset of Σ^{++}. Let p, q be two pictures. Their *horizontal* or *H-concatenation* $p \oplus q$ is defined if $|p|_{row} = |q|_{row}$, resulting in a picture of size $(|p|_{row}, |p|_{col} + |q|_{col})$. The *vertical* or *V-concatenation* $p \ominus q$ is defined similarly. Each concatenation is associative and naturally extends to languages.

Tilings and Closures. A *rectangularly partitioned picture* (rpp) is a pair $\langle p, r \rangle$ where $p \in \Sigma^{++}$ is a picture and r is a *rectangular partition* of the domain of p,

---

[1] The terminology varies, e.g., triangulated floor plans have also been called rectangular duals of planar triangulated graphs, rectangulations [9], or floor plans [2,8].

i.e., $r = \{B_1, B_2, \ldots, B_k\}$, $k \geq 1$, where the *blocks* B_i are (disjoint) subpictures that pave picture p. If all B_i belong to a language L, then r is called a *tiling* of p with L. The *tiling closure* [12] (T-closure) L^{++} of a language L is the set $\{p \in \Sigma^{++} \mid \exists$ a tiling of p with $L\}$. Intuitively, L^{++} is the set of all pictures that can be exactly tiled, i.e., paved without gaps or overlaps, with pictures in L. The *concatenation closure* (*C-closure*) $L^{\ominus\oplus+}$, of a language L is the smallest language that includes L and is closed under both H- and V-concatenations. If $p \in L^{\ominus\oplus+}$, then there exists a rpp $\langle p, r \rangle$ such that all blocks of r are in L and p can be obtained by H- and V-concatenations of those blocks. Then, r is called a *guillotine decomposition* of p into L. A guillotine decomposition is a special case of tiling. Tilings are often depicted as rpp's, e.g., this tiling is called a windmill

and this one is a guillotine decomposition:

The inclusions $L \subseteq L^{\ominus\oplus+} \subseteq L^{++}$ hold for any language L and are proper in some cases, e.g., if the only tiling with L of a picture p is the above windmill, then $p \in L^{++} \smallsetminus L^{\ominus\oplus+}$.

Local and Recognizable Languages. The set of all 2×2 subpictures of a picture p is denoted $B_{2,2}(p)$; its elements are called *tiles*. A picture p surrounded by a border of the reserved symbol $\#$ is called *bordered* and is denoted $\widehat{p}$; it has size $(|p|_{row} + 2, |p|_{col} + 2)$. Bordering naturally extends to a language.

Given a tile set $\mathsf{T} \subseteq (\Sigma \cup \{\#\})^{2,2}$, let $L(\mathsf{T})$ denote the language $\{p \in \Sigma^{++} \mid B_{2,2}(\widehat{p}) \subseteq \mathsf{T}\}$, i.e., every 2×2 subpicture of p is covered by a tile of T. This causes the tiles to overlap, rather than just cover p as in the definition of T-closure. A language $L \subseteq \Sigma^{++}$ is *local* if there is a tile set $\mathsf{T} \subseteq (\Sigma \cup \{\#\})^{2,2}$ such that $L = L(\mathsf{T})$. *Domino* languages are a special case of local languages, defined by overlapping tiles that are either horizontal (1×2) or vertical (2×1) instead of square (2×2).

A *tiling system* (TS) [7] is a quadruple $(\Sigma, \Gamma, \mathsf{T}, \pi)$ where Σ and Γ are respectively the *terminal* and *local* alphabets, $\mathsf{T} \subseteq (\Gamma \cup \{\#\})^{2,2}$ is a finite set of tiles, and $\pi : \Gamma \to \Sigma$ is a projection (extended homomorphically to pictures). A language $L \subseteq \Sigma^{++}$ is *recognized* by such a TS if $L = \pi(L(\mathsf{T}))$, i.e., L is the projection of the local language defined by the tile set. The family of all *tiling recognizable* languages is denoted by REC. It includes all local languages and it enjoys many closure properties, notably under union, intersection, H-concatenation, V-concatenation [6]. The class REC does not change if domino local languages are considered instead of local languages.

3   Some Properties of the Closures of Tiling Recognizable Languages

First, we address a classical theme in formal language theory: the existence and uniqueness of minimal generating sets. We prove that every language closed under tiling closure or under concatenation closure admits a unique minimal generating basis. We also show that, unlike the one-dimensional case, basis extraction from a language need not preserve recognizability: there exist REC picture languages whose minimal generating set is not recognizable.

For word languages, if L is local (regular), then L^* is regular and it is decidable (indeed, in polynomial time) whether $L^* = L$. In 2D, the situation is fundamentally different for both decidability and recognizability properties and for both tiling and concatenation closures. The results clarify the algebraic structure induced by T- and C-closures and highlight a significant departure from the well-behaved world of word languages.

In what follows we treat in parallel the tiling closure L^{++} and the concatenation closure $L^{\ominus\oplus+}$ whenever possible, indicating explicitly where the two notions diverge. A language L is *T-closed* if $L = L^{++}$, and *C-closed* if $L = L^{\ominus\oplus+}$.

Proposition 1. *It is undecidable whether a local language is T- or C-closed.*

Proof. The proof is by reduction from the undecidability [6] of the emptiness problem for local languages. Consider an arbitrary instance of such a problem, i.e., a set T of tiles over an alphabet Σ. Extend Σ with a new letter \$ and define $L = L(\mathsf{T}) \cup \$^{++}$. This language is still local since the alphabets are disjoint (choose the tile set so that no 2×2 tile mixes \$ with symbols of Σ; hence, any accepted picture is either entirely over \$ or entirely over Σ and satisfying T). If $L(\mathsf{T}) = \varnothing$, then $L^{++} = \$^{++} = L$. Conversely, if $L(\mathsf{T}) \neq \varnothing$, then given $q \in L(\mathsf{T})$, we have $q \oplus \$^{|q|_{row},1} \in L^{++} \smallsetminus L$. The same concrete witness works for C-closure. Then: $L^{++} = L$ iff $L(\mathsf{T}) = \varnothing$, and checking $L^{++} = L$ is undecidable. The above argument applies also to $L^{\ominus\oplus+}$, since $\$^{++} = \$^{\ominus\oplus+}$. $\square$

If L is a word language, then the T- and C-closure collapse into the Kleene star L^*; L is star-closed if $L = L^*$, i.e., if L is a submonoid of Σ^*. In this case, there exists a *basis*, i.e., a minimal language B such that $L = B^*$. B is unique and equals the set of indecomposable words [4]. If L is recognizable (regular), then its basis is also recognizable. We now extend the notion of basis to pictures, in the two cases of the T- and C-closures, and then check whether the definition verifies the fundamental properties of indecomposability and uniqueness.

Definition 1 (Basis, indecomposability).   *We say that $p \in L$ is T-indecomposable if the only tiling of p with L consists of p itself; p is C-indecomposable if p can be written neither as $p_1 \oplus p_2$ nor $p_1 \ominus p_2$ for some $p_1, p_2 \in L$.*
A set B is a T-basis of a T-closed language L if $L = B^{++}$ (generation) and $b \notin (B \smallsetminus \{b\})^{++}$ for all pictures $b \in B$ (minimality). Similarly, B is a C-basis of a C-closed L if $L = B^{\ominus\oplus+}$ and $b \notin (B \smallsetminus \{b\})^{\ominus\oplus+}$ for all pictures $b \in B$.

In other words, if B is a T-basis (resp. a C-basis) of L, then B generates the entire language L via T-closure (resp. C-closure), and no proper subset of B can do so. Notice that picture indecomposability is closely related with the theory of pictures codes, e.g., [1].

Theorem 1. *It is undecidable whether a recognizable language L satisfies $L = \mathrm{IND}_T(L)$ or $L = \mathrm{IND}_C(L)$.*

Proof. We reduce to the undecidable emptiness problem for REC. Let p be a picture over a fresh alphabet.

T-Indecomposability. Define $L' = L^{++} \cup \{p\}$. By Simplot's theorem [12] on the closure of REC under tiling, $L^{++} \in$ REC effectively, hence $L' \in$ REC effectively. If $L = \varnothing$ then $L^{++} = \varnothing$ and $L' = \{p\}$, which is clearly T-indecomposable. If $L \neq \varnothing$, pick $q \in L$; then $q \oplus q \in L^{++} \subseteq L'$, so L' contains a T-decomposable picture and thus is not T-indecomposable. Hence L' is T-indecomposable iff $L = \varnothing$, so the problem is undecidable.

C-Indecomposability. Let $L' = L \cup (L \oplus L) \cup \{p\}$, which is in REC. If $L = \varnothing$ then $L' = p$, which is C-indecomposable; otherwise, there exists $q \in L$ such that both q and $q \oplus q$ are in L', therefore $q \oplus q$ is not C-indecomposable. Hence L' is C-indecomposable iff $L = \varnothing$. $\qquad\square$

Lemma 1 (Basis). *If a language L is T-closed, then the set $\mathrm{IND}_T(L)$ is a T-basis for L. If L is C-closed, then the set $\mathrm{IND}_C(L)$ is a C-basis for L.*

Proof. **T-case.** Let $B = \mathrm{IND}_T(L) \subseteq L$. We verify the two defining properties of a T-basis.

(Generation). We show $B^{++} = L$ by proving both inclusions.

$(\subseteq)$ Since $B \subseteq L$, monotonicity of tiling closure yields $B^{++} \subseteq L^{++}$. As L is T-closed, $L^{++} = L$, hence $B^{++} \subseteq L$.

$(\supseteq)$ We prove that all $p \in L$ belong to B^{++} by induction on $\mathrm{area}(p) = |p|_{\mathrm{row}} \cdot |p|_{\mathrm{col}}$. Let $p \in L$. If $p \in B$ (i.e., p is T-indecomposable), then trivially $p \in B^{++}$. Otherwise, p is T-decomposable, hence there exists a tiling of p with $k \geq 2$ pictures $p_1, \ldots, p_k \in L$. Since $k \geq 2$ and the tiles are nonempty, each tile has smaller area than p, i.e., $\mathrm{area}(p_i) < \mathrm{area}(p)$ for all i. By induction hypothesis, each $p_i \in B^{++}$, so each p_i can be tiled by pictures in B. Substituting these tilings into the tiling of p yields a tiling of p entirely with elements of B. Hence $p \in B^{++}$. This proves $B^{++} \supseteq L$, and hence $B^{++} = L$.

(Minimality). Take $b \in \mathrm{IND}_T(L)$ and set $B' = B \smallsetminus \{b\}$. We show $b \notin (B')^{++}$, which is exactly the minimality condition for a T-basis. If, by contradiction, $b \in (B')^{++}$, then there is a tiling of b with B'. Since $b \notin B'$, such a tiling cannot consist of a single tile, but of at least two tiles from $B' \subseteq L$. Hence b is tiled by two or more pictures from L, contradicting that $b \in B = \mathrm{IND}_T(L)$. Thus $b \notin (B')^{++}$.

Since both minimality and generation are satisfied, the set $\mathrm{IND}_T(L)$ is a T-basis for L.

C-case. The C-case is analogous. Minimality is immediate from the definition. For generation, one argues by induction on the area of p: if $p \in L$ is C-decomposable, then $p = p_1 \oplus p_2$ or $p = p_1 \ominus p_2$ with $p_1, p_2 \in L$ of smaller area, and the induction hypothesis applies. $\qquad\square$

Theorem 2. *Both the T-basis of a T-closed language and the C-basis of a C-closed language are unique.*

Proof. Let L be T-closed. By Lemma 1, $\mathrm{IND}_T(L)$ is a T-basis of L. Let B be a different T-basis of L. We prove that indeed $B = \mathrm{IND}_T(L)$, a contradiction. Therefore, the T-basis is unique. *(1)* $B \subseteq \mathrm{IND}_T(L)$. Assume for contradiction that some $b \in B$ is not T-indecomposable, therefore it admits a tiling with other pictures from L. So, after removing b from L, the picture b remains in the T-closure of the result: $b \in (L \smallsetminus \{b\})^{++}$. In a proper tiling of b, every tile has strictly smaller area than b, so when those tiles are expanded using B, none of them can use b. Then $b \in (B \smallsetminus \{b\})^{++}$. Let $B' = B \smallsetminus \{b\}$. From the above, $b \in B'^{++}$, and therefore $L = B^{++} = (B' \cup \{b\})^{++} \subseteq (B'^{++})^{++} = B'^{++}$. The reverse inclusion $B'^{++} \subseteq B^{++}$ also holds, hence $B'^{++} = L$, contradicting that B is a T-basis. Hence, every $b \in B$ must in fact be T-indecomposable. Thus $B \subseteq \mathrm{IND}_T(L)$.

(2) $\mathrm{IND}_T(L) \subseteq B$. Let $p \in \mathrm{IND}_T(L)$. Since $L = B^{++}$, picture p admits a tiling with elements of B. But p is indecomposable, so this tiling cannot be proper: it must consist of a single tile, namely p itself. Hence $p \in B$.

The case of the C-basis is analogous to the T-closed one. The removal of any element of $\mathrm{IND}_C(L)$ prevents generating L, and any C-basis must contain all C-indecomposable pictures. The uniqueness argument is unchanged once "tiling" is replaced by "guillotine decomposition": $\mathrm{IND}_C(L)$ is the unique C-basis of L. $\square$

We denote by $\mathrm{Base}_T(L)$ $(\mathrm{Base}_C(L))$ the T-basis of a T-closed language (the C-basis of a C-closed language L). Notice $\mathrm{Base}_T(L) \subseteq \mathrm{Base}_C(L)$ and $\mathrm{Base}_T(L) = \mathrm{IND}_T(L)$, $\mathrm{Base}_C(L) = \mathrm{IND}_C(L)$.

The recognizability of the basis does not extend to picture languages.

Theorem 3 (Basis of T-closure non-recognizable). *There exists a T-closed language $L \in \mathrm{REC}$ such that its T-basis $\mathrm{Base}_T(L)$ is not REC.*

Proof. Let $\Sigma = \{0,1\}$. Introduce four new edge symbols N, E, S, W. For a picture p we denote by $\bar{p}$ the $(|p|_{row}+2) \times (|p|_{col}+2)$ picture obtained by surrounding p with a border of width one so that the top edge is a (nonempty) row of N's, the bottom edge a row of S's, the left edge a column of W's, and the right edge a column of E's. This can be compactly written as $\bar{p} = N^{\oplus+} \ominus (W^{\ominus+} \oplus p \oplus E^{\ominus+}) \ominus S^{\oplus+}$ where $N^{\oplus+}$ is the closure of N under horizontal concatenation. We extend the border operation to languages by $\overline{L} = \{\bar{p} \mid p \in L\}$.

A *double square* is a picture of the form $w \ominus w'$, where $w, w' \in \Sigma^{n,n}$ for some $n \geq 2$. Let $C \subseteq \Sigma^{++}$ be the set of double squares with unequal halves, i.e., such that $w \neq w'$. It is known that $C \in \mathrm{REC}$ (from Theorem 7.5 in [6]). Clearly, $\overline{C}$ is also in REC: the 2×2 tile set can enforce the border pattern, so occurrences of bordered pictures are locally verifiable. Note that occurrences of $\overline{C}$ in any picture are necessarily cell-disjoint: since $\{N, E, S, W\} \cap \Sigma = \varnothing$, the directional border of one occurrence cannot occur within the interior of another, nor can opposing edge symbols (e.g., N and S) overlap. This ensures that the markers act as a rigid "skeleton" for the $\overline{C}$ subpictures, preventing overlapping or nested occurrences that could complicate counting. It may help to look at Fig. 1, left.

Define two languages over $\Sigma \cup \{N, E, S, W\}$:

$$L = \{\, p \mid p \text{ contains } at\ least \text{ one subpicture in } \overline{C}\,\}, \quad \text{and}$$
$$B = \{\, p \mid p \text{ contains } exactly \text{ one subpicture in } \overline{C}\,\} \tag{1}$$

L is T-closed by construction, since any tiling of pictures that each contain at least one occurrence of $\overline{C}$ will result in a picture that preserves this property.

We first show that $L = B^{++}$. Let $p \in L$ and consider the set of *complete* occurrences of $\overline{C}$ in p (which are non-overlapping). Because occurrences are disjoint, there exists a rectangular partition of p into blocks $\{b_1, b_2, \ldots, b_k\}$ such that each block contains exactly one complete occurrence of $\overline{C}$ (this is always possible for any set of disjoint sub-rectangles). Thus each block belongs to B, and therefore $p \in B^{++}$. Hence $L \subseteq B^{++}$, and the reverse inclusion is immediate from $B \subseteq L$, so $L = B^{++}$.

To see that B is minimal, fix $b \in B$. If b can be tiled with pictures from $B \setminus \{b\}$, then such tiling must contain at least two parts. Therefore, b contains two subpictures in $\overline{C}$, a contradiction with the definition of B.

Since $\overline{C} \in \mathrm{REC}$, a TS can non-deterministically guess a putative position for an occurrence of $\overline{C}$ and locally verify it. The local 2×2 tiles can also enforce the directional-border pattern around the guessed block (including edges); therefore, the guessed occurrence can be locally checked while the rest of the picture is unconstrained. Thus, the same nondeterministic tiling construction shows $L \in \mathrm{REC}$.

Suppose, towards a contradiction, $B \in \mathrm{REC}$. Consider the language:

$$L_2 = \{p_1 \oplus p_2 \mid p_1 \in \overline{C},\ p_2 \text{ is any bordered double square}\}.$$

$L_2 \in \mathrm{REC}$ since a TS can enforce the border constraints and verify that the left part is in $\overline{C}$ and the right part is a bordered double square (which is recognizable). The local tile set checks the directional-border pattern on both parts and verifies that the left part belongs to $\overline{C}$ and the right part is a bordered double square.

Let $D = \{w \ominus w \mid w \text{ is a square picture}\}$. Assuming toward contradiction that $B \in \mathrm{REC}$, closure under intersection gives $B \cap L_2 \in \mathrm{REC}$. The pictures in this intersection are precisely those formed by a picture from $\overline{C}$ concatenated with a bordered double square that does not contain any picture from $\overline{C}$—meaning it must be an element of $\overline{D}$. Therefore, $B \cap L_2 = \{c \oplus d \mid c \in \overline{C}, d \in \overline{D}\}$.

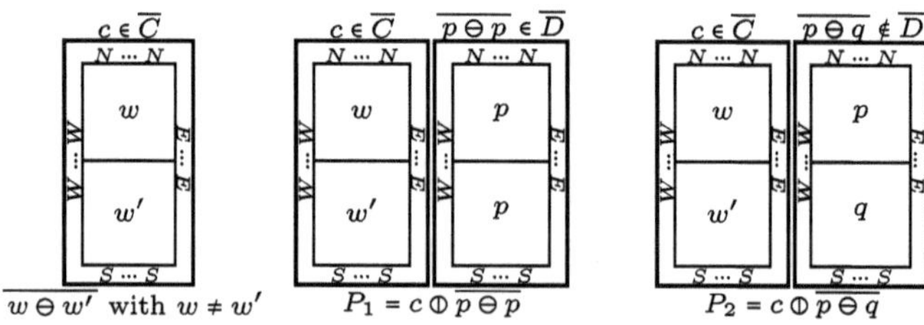

Fig. 1. Left: A bordered double square $\overline{w \ominus w'}$ with $w, w' \in \Sigma^{n,n}, n \geq 2, w \neq w'$: the outer band is the directional frame. Middle and Right: The two pictures used in the syntactic-equivalence argument. In both cases the left block is a bordered double square $c \in \overline{C}$ with unequal halves; the right block is a bordered double square with equal halves in P_1 and with unequal halves in P_2.

However, from the proof of Theorem 7.5 of [6], we know that $D \notin \text{REC}$. We rely on the fact that for any $L \in \text{REC}$, there is a finite number of tiling configurations on the contour of any subpicture (Syntactic Equivalence Lemma 9.2 of [6]). Assume, for contradiction, that $B \cap L_2$ is in REC and is recognized by a tiling system, TS. Therefore, there exist at least two distinct square pictures, p and q of the same size (with $p \neq q$), that are syntactically equivalent with respect to TS. This means TS cannot distinguish between p and q in any context. Now, fix $c \in \overline{C}$ and consider $P_1 = c \oplus \overline{p \ominus p}$. By definition, P_1 is in $B \cap L_2$ and must be accepted by TS.

Next, construct a new picture $P_2 = c \oplus \overline{p \ominus q}$ by replacing the second square p in P_1 with the square q. Since p and q are syntactically equivalent for TS, the tiling system cannot detect this swap and must also accept P_2.

However, by definition, P_2 is not in $B \cap L_2$ because its bottom part, $\overline{p \ominus q}$, is not an element of $\overline{D}$ (as $p \neq q$). This is a contradiction: the TS accepts a picture that is not in the language it supposedly recognizes. Therefore, our assumption that $B \cap L_2$ is in REC must be false. Since $L_2 \in \text{REC}$ and REC is closed under intersection, the only way $B \cap L_2$ can fail to be in REC is if B is not in REC. $\square$

Lemma 2. *Let languages B and L be defined as in formula (1) in the proof of Theorem 3. Then $L = B^{\ominus\oplus+}$.*

Proof. Let $k(p) \geq 1$ be the number of *complete* occurrences of elements of $\overline{C}$ in p. (Recall that by construction occurrences of $\overline{C}$ are pairwise disjoint.) We prove by induction on $k(p)$ that $p \in B^{\ominus\oplus+}$. The base case $k(p) = 1$ is trivial: p contains exactly one complete occurrence and hence $p \in B$. For the inductive step, let $k = k(p) \geq 2$ and assume the claim holds for all pictures with fewer than k complete occurrences. Consider two distinct complete occurrences o_1, o_2 of $\overline{C}$ in p. Since they are pairwise disjoint, there is always a horizontal or vertical cut line separating o_1 from o_2. Cut p along such a line to obtain two rectangular subpictures p_1, p_2 with o_i occurring in p_i for $i = 1, 2$. Thus either $p = p_1 \oplus p_2$ or $p = p_1 \ominus p_2$.

A complete occurrence lying strictly on one side of the cut remains complete on that side, while any occurrence intersecting the cut becomes *broken* and therefore does not count as a complete occurrence in either side. Hence each of p_1, p_2 contains at least one complete occurrence (namely o_1 or o_2) and at most $k - 1$ complete occurrences; in particular $k(p_1) \leq k - 1$ and $k(p_2) \leq k - 1$. Apply the induction hypothesis to each p_i to express it as a concatenation of elements of B. Concatenating these decompositions along the cut yields a decomposition of p into B. Hence $p \in B^{\ominus\oplus+}$. $\square$

Theorem 4 (Basis non-recognizable under T- and C-closure). *There exists $L \in \text{REC}$ that is both C-closed and T-closed and whose bases $\text{Base}_C(L)$ and $\text{Base}_T(L)$ are not in REC.*

Proof (Sketch). Let L and B be as in Theorem 3. Since $L = B^{++}$ and $L = B^{\ominus\oplus+}$ by Lemma 2, L is both T-closed and C-closed with the same candidate basis B. Minimality of B with respect to $L = B^{++}$ carries over to $L = B^{\ominus\oplus+}$, so B is the basis of L under C-closure as well. The proof of Theorem 3 showed $B \notin \text{REC}$. $\square$

4   Floor Plans as Picture Languages

A floor plan in the digital plane can be formalized as a rectangular partitioned picture $\langle p, r \rangle$ such that the alphabet of p is unary (e.g., it contains only the blank symbol b). Thus, a floor plan is a rectangle subdivided into smaller rectangular blocks by horizontal and vertical line segments. If we add the extra condition that no four blocks meet at the same point, we obtain the subfamily of *triangulated floor plans* (*tfp*). Even when restricted to the digital, rather than continuous, plane, a floor plan is not a picture in the sense used in this paper, because the segments are mathematical abstractions (infinitely thin lines) rather than symbols occupying cells of a rectangular array. This motivates the introduction of a picture-based encoding: by reifying block contours as symbols, floor plans can be studied as (local/recognizable) picture languages and connected to the closure operations investigated above. To illustrate, the middle picture in Fig. 2 is a floor plan partitioned into five blocks.

When necessary to prevent confusion, we call *floor plan picture* (fpp) the representation of a floor plan by means of a picture as in Fig. 2. From the survey contained in [2] we select the definitions of the floor plan families we study.

Definition 2. *The* floor plan picture families *are the following.*

L_{basic} *the basic* floor plan family, *consists of the set of fpp's where the partition is trivial.*

L_{fp} *is the whole* floor plan family, *consists of all the fpps.*

L_{tfp} *the triangulated fpp family, consists of the fpps such that no four blocks share a vertex.*

L_{guil} *the guillotine* floor plan family, *consists of the triangulated floor plans whose partition is a guillotine decomposition.*

A windmill *in a floor plan is a quadruple of segments forming one of the following patterns:* ⊢, ⊣ *(e.g., in the second fpp of Fig. 2).*

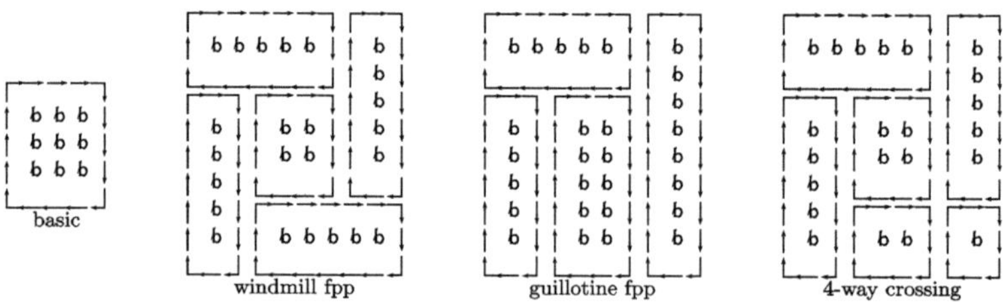

Fig. 2. Floor plans represented as pictures. From left to right, an fpp in L_{basic}, a windmill pattern, a guillotine fpp in L_{guil}, and a fpp in $L_{fp} \setminus L_{tfp}$.

Figure 2, from left to right, shows a basic floor plan, a windmill fpp, a guillotine one, and a floor plan with a four-way junction. Each block is delimited by a clockwise-oriented rectangular contour.

A fpp is called guillotine if its partition into basic blocks is guillotine. The following characterization of guillotine floor plans (e.g., in [2]) is well-known.

Theorem 5. *A tfp is guillotine if and only if it avoids the windmills.*

Different choices are possible for representing a floor plan as a picture. Our fpp's materialize the four vertices of a block as corner symbols, and the four sides by means of contour lines that use different symbols on each side. The block interiors are filled with the same symbol, typically the blank "b". The fpp alphabet is $\Xi = \{b, \rightarrow, \rightarrow, \downarrow, \uparrow, \ulcorner, \urcorner, \llcorner, \lrcorner\}$.

Any alternative representation of floor plans as pictures, e.g., by assigning distinct colors to any two neighboring blocks, would not change our constructions and proofs, provided the set of resulting pictures is locally definable.

Having transformed floor plans into fpp's, we proceed to study the language families they belong to, and how they are related with the families produced by means of the closure operations.

Theorem 6 (Locality of floor plans). L_{basic}, L_{fp}, *and* L_{tfp} *are local.*

Proof. (Hint.) For simplicity, we describe the tile set for pictures of size at least 3×3; the finitely many smaller cases can be defined by adding finitely many additional local constraints. It suffices to prove the statement for L_{fp}, because it immediately follows for L_{basic} and L_{tfp}.
The 2×2 tiles defining the local languages use the alphabet Ξ introduced above. All the tiles needed occur in the fpp's of Fig. 2. We reproduce some typical tiles in Table 1. Every 2×2 subpicture of any picture in L_{fp} is in the specified tile set.

Table 1. The tile names are self-explanatory. The tiles that define L_{basic} are in the first two rows, while for L_{fp} the same tiles are used, plus many others needed for the edges and corners of internal blocks; the third row reproduces two of them. To define L_{fp}–but not L_{tfp}– we need the additional 4-corner tile.

Interior	Top edge	Bottom edge	Left edge	Right edge
$b\,b$ $b\,b$	$\rightarrow\,\rightarrow$ $b\,b$	$b\,b$ $\leftarrow$	$\uparrow\,b$ $\uparrow\,b$	$b\,\downarrow$ $b\,\downarrow$

Top-left corner	Top-right corner	Bottom-left corner	Bottom-right corner	
$\ulcorner\!\rightarrow$ $\uparrow\,b$	$\rightarrow\!\urcorner$ $b\,\downarrow$	$\uparrow\,b$ $\llcorner\!\leftarrow$	$b\,\downarrow$ $\leftarrow\!\lrcorner$	

Straight border	Border over corners	more corner and border tiles	4-corner tile
$\leftarrow\,\leftarrow$ $\rightarrow\,\rightarrow$	$\leftarrow\,\leftarrow$ $\urcorner\,\ulcorner$	$\cdots$	$\lrcorner\llcorner$ $\urcorner\ulcorner$

Conversely, we prove that any picture defined by such tiles is a floor plan. This follows from several easy properties that ensure the picture is rectangularly partitioned. We list the essential ones: (i) the arrow and the corner symbols of a picture occur in the rectangular boundaries and nowhere else, (ii) two rectangles cannot overlap, and (iii) a rectangle cannot be nested inside another one.   □

The 2×2 tiles occurring in pictures of the local language L_{fp} are the same as the 2×2 tiles occurring in pictures of $L_{\mathrm{basic}}^{\ominus\oplus+}$; if two languages have the same set of 2×2 subpictures, then the local language generated by that tile set is the same. Thus, since $L_{\mathrm{fp}} \neq L_{\mathrm{basic}}^{\ominus\oplus+}$, the latter cannot be local:

Corollary 1. *The language $L_{\mathrm{basic}}^{\ominus\oplus+}$ is not local.*

Properties of Floor Plans as Picture Languages. The following properties of the T-closure and of the C-closure of fpp languages are obvious.

- The whole set of floor plans, L_{fp}, coincides with the T-closure of the basic fpp language, i.e., $L_{\mathrm{fp}} = L_{\mathrm{basic}}^{++}$.
- The set of triangulated floor plans is strictly included in the T-closure of the basic fpp language, i.e., $L_{\mathrm{tfp}}\ s \nsubseteq L_{\mathrm{basic}}^{++}$, since the T-closure introduces four-corner joints.
- L_{fp}–but not L_{tfp}–is closed under T-closure, and its (minimal) basis is L_{basic}.
- The C-closure $L_{\mathrm{basic}}^{\ominus\oplus+}$ is the set of guillotine floor plans.

It is not known if the fpp language $L_{\mathrm{basic}}^{\ominus\oplus+}$ is in REC. The importance of this open problem is highlighted by the following theorem: if guillotine floor plans are recognizable, then REC is closed under C-closure.

Theorem 7 (Implication of C-closure). *If $L_{\mathrm{basic}}^{\ominus\oplus+}$ is in REC, then the C-closure of any language $R \in \mathrm{REC}$ over an arbitrary alphabet Σ is in REC.*

Proof (sketch). Let $R \subseteq \Sigma^{++}$ be recognized by a TS $(\Sigma, \Gamma_R, \mathsf{T}_R, \pi_R)$, i.e., $R = \pi_R(L(\mathsf{T}_R))$. Assume that $L_C = L_{\mathrm{basic}}^{\ominus\oplus+} \subseteq \Xi^{++}$ is recognized by a TS $(\Xi, \Gamma_C, \mathsf{T}_C, \pi_C)$, i.e., $L_C = \pi_C(L(\mathsf{T}_C))$. Consider the product alphabet $\Gamma = \Gamma_R \times \Gamma_C$ and use it to define a local language. Intuitively, in a picture over Γ the second component (in Γ_C) is a local witness for a guillotine rectangular partition (the 'shape'), while the first component (in Γ_R) is a local witness for the content. A *local* language $L(\mathsf{T}) \subseteq \Gamma^{++}$ is defined by a set of 2×2 tiles $\mathsf{T} \subseteq (\Gamma \cup \{\#\})^{2,2}$ such that: 1) the projection to Γ_C of a picture in $L(\mathsf{T})$ belongs to $L(\mathsf{T}_C)$ (hence it is a valid local witness for some picture in L_C after applying π_C), and 2) whenever a rectangular subpicture corresponds (in the projection to Ξ) to a block of the guillotine partition, the restriction of the projection to Γ_R to its interior satisfies T_R, with the rectangular contour treated as the external border next to an (imaginary) border of $\#$ symbols.

The second condition is enforceable by 2×2 tiles because the contour alphabet locally identifies the interior/border of each block in a guillotine partition.

Define the terminal projection $\pi : \Gamma \to \Sigma$ by $\pi(g, h) = \pi_R(g)$ for all $(g, h) \in \Gamma_R \times \Gamma_C$, and let $G_R = \pi(L(\mathsf{T}))$. Since $L(\mathsf{T})$ is local, $(\Sigma, \Gamma, \mathsf{T}, \pi)$ is a TS and therefore $G_R \in \mathrm{REC}$. By construction, G_R consists exactly of those pictures obtainable by a finite sequence of horizontal and vertical guillotine cuts whose interiors are labeled by pictures of R, i.e., $G_R = R^{\ominus\oplus+}$. Hence $R^{\ominus\oplus+} \in \mathrm{REC}$. $\square$

Remark. The proof of Theorem 7 is essentially parametric in the "shape" language: it only relies on the fact that $L_{\mathrm{basic}}^{\ominus\oplus+}$ is a recognizable language of floor plans, and decorates each block with a picture of R via a TS that uses a product alphabet. Informally, any closure operator that depends only on the *shape* of the underlying floor plan and not on the internal content of the blocks, admits this construction and therefore belongs to REC. We do not attempt to formalize this general notion here, but we just observe that if we replace $L_{\mathrm{basic}}^{\ominus\oplus+}$ in the proof of Theorem 7 with the tiling closure language L_{basic}^{++}, the very same construction yields a TS recognizing R^{++}. In this way, one obtains an alternative, more direct proof of the classical result [12] that $R^{++} \in$ REC whenever $R \in$ REC.

It is not known if indeed $L_{\mathrm{basic}}^{\ominus\oplus+}$ is in REC, but we can show a class of recognizable guillotine languages, based on the studies that formalize a collection of floor plans by means of equivalence relations. Here we consider the *strong equivalence relation* [2], omitting for brevity the coarser relation known as weak equivalence. Denote by $E_p \subseteq L_{\mathrm{fp}}$ the equivalence class

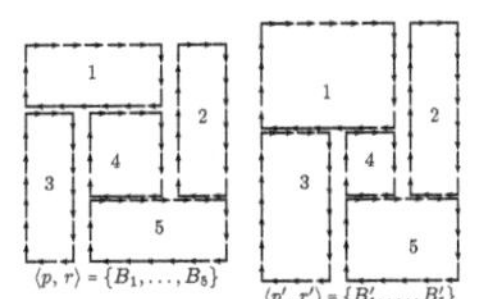

of p. There are infinitely many equivalence classes in L_{fp}. The theorem below follows by enumerating the partition blocks as $1, \ldots, n$, and defining a local language over the Cartesian product of the contour alphabet and $\{1, \ldots, n\}$. A local test over the first component enforces that a picture is in L_{fp}, and a local test over the second component enforces the contact requirements. A projection deletes the enumeration component.

Theorem 8 (Equivalence class is REC). *Let $E_p \subset L_{fp}$ be the strong equivalence class of a picture p. Then the floor plan language E_p is REC.*

If p in E_p is guillotine then all pictures in E_p are guillotine, because destroying guillotine decomposability would require changing at least one block-block contact. Therefore, each equivalence class of a guillotine picture is a guillotine language in REC.

Finally, we show that the complement of the guillotine language is in REC. The idea of the proof is that it is possible to define a TS recognizing a language W over alphabet $\varXi$, with one windmill. The intersection $L_{\mathrm{tfp}} \cap W$ is the set of triangulated fpp's without guillotine. The proof is completed by suitable union and intersection constructions. This result does not suffice to solve the open question if the $L_{\mathrm{basic}}^{\ominus\oplus+}$ belongs to REC, since REC is not closed under complement.

Theorem 9. *The complement $\varXi^{++} \setminus L_{guil}$ of the language L_{guil} is in REC.*

5   Conclusion

We investigated tiling (T) and concatenation (C) closures in 2D, proving that while they admit unique minimal bases, recognizability is not necessarily preserved and key fixed-point properties are undecidable in REC. By reifying floor plans via a contour-based encoding, we linked these abstract closures to geometric rectangular partitions. This framework provides a path for future research into shape-based closure operators, decision problems, and structural equivalence within the picture language framework.

Acknowledgement. We thank the anonymous referees for their suggestions.

References

1. Anselmo, M., Giammarresi, D., Madonia, M.: Picture codes and deciphering delay. Inf. Comput. 253, 358–370 (2017). https://api.semanticscholar.org/CorpusID: 3302927
2. Asinowski, A., Cardinal, J., Felsner, S., Fusy, É.: Combinatorics of rectangulations: old and new bijections. Comb. Theory 5(1) (2025). https://doi.org/10.5070/C65165025
3. Berstel, J., Perrin, D., Reutenauer, C.: Codes and Automata. Cambridge University Press, Encyclopedia of Mathematics and its Applications (2009)
4. Brzozowski, J.A.: Roots of star events. J. ACM 14(3), 466–477 (1967). https://doi.org/10.1145/321406.321409
5. Cherubini, A., Crespi-Reghizzi, S., Pradella, M., San Pietro, P.: Picture languages: tiling systems versus tile rewriting grammars. Theor. Comput. Sci. 356(1–2), 90–103 (2006). https://doi.org/10.1016/j.tcs.2006.01.038
6. Giammarresi, D., Restivo, A.: Two-dimensional languages. In: Rozenberg, G., Salomaa, A. (eds.) Handbook of Formal Languages, vol. 3, pp. 215–267. Springer, Berlin (1997). https://doi.org/10.1007/978-3-642-59126-6_4
7. Giammarresi, D., Restivo, A.: Recognizable picture languages. Int. J. Pattern Recognit Artif Intell. 6(2&3), 241–256 (1992). https://doi.org/10.1142/S021800149200014X
8. He, B.D.: A simple optimal binary representation of mosaic floorplans and baxter permutations. Theor. Comput. Sci. 532, 40–50 (2014). https://doi.org/10.1016/J.TCS.2013.05.007
9. Lai, Y., Leinwand, S.M.: A theory of rectangular dual graphs. Algorithmica 5(4), 467–483 (1990). https://doi.org/10.1007/BF01840399
10. Matz, O.: Regular expressions and context-free grammars for picture languages. In: Reischuk, R., Morvan, M. (eds.) STACS 97, 14th Annual Symposium on Theoretical Aspects of Computer Science, Lübeck, Germany, 1997, Lecture Notes in Computer Science, vol. 1200, pp. 283–294. Springer, Berlin, Heidelberg (1997). https://doi.org/10.1007/BFb0023466
11. Pradella, M., Cherubini, A., Crespi-Reghizzi, S.: A unifying approach to picture grammars. Inf. Comput. 209(9), 1246–1267 (2011). https://doi.org/10.1016/J.IC.2011.07.001
12. Simplot, D.: A characterization of recognizable picture languages by tilings by finite sets. Theor. Comput. Sci. 218(2), 297–323 (1999)

Complexity of Linear Subsequences of Fibonacci-Automatic Sequences

Delaram Moradi[1], Narad Rampersad[2], and Jeffrey Shallit[1(✉)]

[1] School of Computer Science, University of Waterloo, 200 University Ave. W., Waterloo, ON N2L 3G1, Canada
{delaram.moradi,shallit}@uwaterloo.ca

[2] Department of Math/Stats, University of Winnipeg, 515 Portage Ave., Winnipeg, MB R3B 2E9, Canada
n.rampersad@uwinnipeg.ca

Abstract. We construct automata with input(s) in Fibonacci representation (also known as Zeckendorf representation) recognizing some basic arithmetic relations and study their number of states. We also consider some basic operations on Fibonacci-automatic sequences and discuss their state complexity. As a consequence of our results, we improve a bound in a recent paper of Bosma and Don. We also discuss the state complexity and runtime complexity of using a reasonable interpretation of Büchi arithmetic to actually construct some of the studied automata recognizing relations.

1   Introduction

In this paper, we discuss the state complexity of specific arithmetic relations (such as addition and multiplication) when the inputs are given in Fibonacci representation (also known as Zeckendorf representation). We also study the state complexity of shifts and linear subsequences of Fibonacci-automatic sequences, and the computational complexity of forming the automata using an interpretation of Büchi arithmetic. In a previous paper [12] we studied the same questions for k-automatic sequences, but the case of Fibonacci representation presents many new and different challenges.

In Sect. 2, we provide the necessary background for the rest of the paper. In Sect. 3, we construct automata of $O(\log c)$ states recognizing $Y = X + c$ and $Y = X - c$ where X, Y are inputs, and $c \geq 0$ is a constant. We construct an automaton of $O(n^2)$ states recognizing the relation $Y = nX + c$ where $n \geq 1$ is a natural number, $0 \leq c < n$, and the numbers X and Y are inputs. In Sect. 4, we show that if $(h(i))_{i \geq 0}$ is a fixed Fibonacci-automatic sequence generated by an m-state automaton, then the shifted subsequence $(h(i+c))_{i \geq 0}$ can be generated by a DFAO of $O(m^2 c^2)$ states, and the linear subsequence $(h(ni + c))_{i \geq 0}$ can be generated by a DFAO of $O(m^2 n^4)$ states (again for $n \geq 1$ and $0 \leq c < n$).

D. Moradi and J. Shallit—Research supported by NSERC grant RGPIN-2024-03725.

N. Rampersad—Research supported by NSERC grants RGPIN-2019-04111 and RGPIN-2025-04076.

As a consequence, we improve a bound in a recent paper by Bosma and Don [3] concerning the size of morphisms for linear subsequences of the Fibonacci word.

An example of a system using an interpretation of Büchi arithmetic is `Walnut`, a free software that can carry out computations with automatic sequences and statements about them phrased in first-order logic. For more information about `Walnut`, see, for example, [13,17]. In Sect. 5, we show how the automaton for recognizing $Y = nX$ is created in polynomial time using an interpretation of Büchi arithmetic.

In this paper, we occasionally refer to $O(\log i)$ where i can be 0 or 1. In these cases, we adopt the usual convention that $O(\log i) = O(1)$ for $i \in \{0, 1\}$.

2   Background

2.1   Finite Automata

In this paper we use the familiar model of deterministic finite automaton, as discussed in [10], and some variations on it. A deterministic finite automaton (DFA) consists of a finite set of states Q, an input alphabet Σ, an initial state q_0, a set of accepting states A, and a transition function $\delta : Q \times \Sigma \to Q$ that specifies the next state of the automaton, based on the current state and the current input letter. We extend the domain of δ to $Q \times \Sigma^*$ in the usual manner. Acceptance is defined by whether completely processing an input causes the DFA to enter an accepting state. Usually the transition function δ is taken to be a total function from $Q \times \Sigma$ to Q, but in this paper, we allow it to be a partial function in order to gracefully handle dead states. A state q of a DFA is called *dead* if it is not possible to reach any accepting state from q by a (possibly empty) path. A minimal DFA for a language L clearly has at most one dead state. By convention, we do not count or display dead states in this paper. A state is called *accessible* if it is reachable from the start state, and *co-accessible* if it is not dead.

Another model we need is the nondeterministic finite automaton (NFA), which is similar to the DFA, except that now the transition function also allows the automaton to transition to multiple (or no) states on the same input letter, and defines acceptance based on the existence of at least one path that enters an accepting state after completely processing the input.

We also use the notion of deterministic finite automaton with output (DFAO), which is like a DFA, except that outputs are associated with states. The output corresponding to an input is the output associated with the last state reached. A DFAO is called a *Fibonacci DFAO* if its input(s) are in Fibonacci representation (see Sect. 2.2). A Fibonacci DFAO $M = (Q, \{0, 1\}, \Delta, \delta, q_0, \tau)$ generates a *Fibonacci-automatic sequence* $(h(i))_{i \geq 0}$ in the following way: the input is an msd-first (most-significant-digit first) representation x of i in the Fibonacci numeration system, and the output $h(i)$ is $\tau(\delta(q_0, x))$. We assume that there are no transitions in the DFAO corresponding to two consecutive inputs of 1; in other words, if there is a state q reachable on 1 from some other state, then $\delta(q, 1)$ is undefined. For more information about such sequences, see [14].

Another consideration involves leading zeros in the input. What if, for example, the automaton gives a different result on input 00101 as it does on input 101? To avoid this issue, the designed automata satisfy the following criterion: they have a self-loop on the initial state q_0 on input 0 (or, if there are multiple inputs, on input $[0, 0, \ldots, 0]$). This convention is crucial when we are working with multiple integer inputs of different lengths.

For a given automatic sequence $(h(i))_{i \geq 0}$ there is a unique associated *interior sequence* $(h'(i))_{i \geq 0}$ taking its values in Q, defined by taking the minimal DFAO $(Q, \Sigma, \Delta, \delta, q_0, \tau)$ and replacing τ by the identity map on Q. Here "unique" means up to renaming of the letters. It follows that $(h(i))_{i \geq 0}$ is the image of $(h'(i))_{i \geq 0}$ under the coding $q \to \tau(q)$.

Example 1. The most famous Fibonacci-automatic sequence is undoubtedly the infinite Fibonacci word $\mathbf{f} = 01001010 \cdots$ [2], defined as the fixed point of the morphism $0 \to 01$, $1 \to 0$. It can be generated by a DFAO of 2 states, as illustrated in Fig. 1.

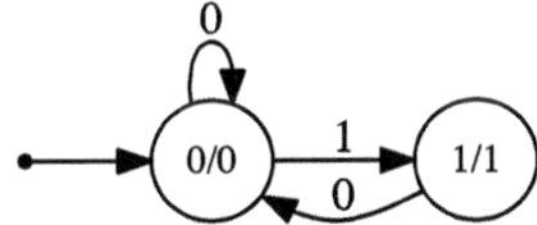

Fig. 1. Automaton for the Fibonacci word $\mathbf{f}$.

The *state complexity* of a formal language is the number of states in the minimal DFA recognizing it and we say the state complexity of an automatic sequence is the number of states in the minimal DFAO generating it.

The DFA and NFA models are described in detail in standard reference works, such as Hopcroft and Ullman [10, Chaps. 2 and 3]. For the DFAO, see, for example, [1, Chapter 5].

Many details have been omitted in this conference paper due to paper length limitations; the details can be found in the arXiv paper [12]. We thank three referees for their careful reading and for suggesting many improvements.

2.2   Fibonacci Representation

We let $\mathbb{N} = \{0, 1, 2, \ldots\}$. Define the Fibonacci numbers, as usual, by $F_0 = 0$, $F_1 = 1$, and $F_n = F_{n-1} + F_{n-2}$ for $n \geq 0$.

Fibonacci representation was introduced by Lekkerkerker [11] and Zeckendorf [20], although it can be found in an earlier, much more general form in a paper of Ostrowski [15]. The Fibonacci representation of $n \in \mathbb{N}$ writes n as a sum of distinct Fibonacci numbers F_i for $i \geq 2$; that is, $n = \sum_{2 \leq i \leq t} e_i F_i$ for $e_i \in \{0, 1\}$. To ensure uniqueness of the representation, no two consecutive Fibonacci numbers can be used; that is, in a valid representation we have $e_i e_{i+1} = 0$ for $2 \leq i < t$. A representation can be associated with the binary word of its coefficients $e_t e_{t-1} \cdots e_2$; note that representations are written with the most significant bit at the left. We need a way to convert an arbitrary binary word $x = e_t \cdots e_2$ to the integer it represents. For this we write $[x]_F = \sum_{2 \leq i \leq t} e_i F_i$. For example, we have $[01001]_F = 6$.

We say a word z is a *valid right extension* of a word x if x is a prefix of z and z contains no occurrence of 11.

Sometimes we will need representations of pairs of natural numbers. To do so, we use the alphabet $\Sigma_2 \times \Sigma_2$, where $\Sigma_2 = \{0, 1\}$. This may require padding the shorter representation with leading zeros. Thus, for example, the word $[0, 1][0, 0][1, 1][0, 0][1, 0]$ represents the pair $[4, 11]$. If w and x are words of the same length over the alphabet Σ_2, then by $w \times x$ we mean the element $y \in (\Sigma_2 \times \Sigma_2)^*$ such that the first component is w and the second is x.

We now state three facts about Fibonacci representation. The first is easy and the second can be found in [16]. The third follows from the fact that the Fibonacci representation results from the greedy algorithm. Let $\alpha = (1 + \sqrt{5})/2$ and $\beta = (1 - \sqrt{5})/2$, the zeros of the polynomial $X^2 - X - 1$.

Lemma 1.

(a) Let $x \in \Sigma_2^*$. Then $[x00]_F = [x]_F + [x0]_F$.
(b) Let x be a valid Fibonacci representation. Then $-\beta^2 < [x0]_F - \alpha[x]_F < -\beta$.
(c) If x and y are valid Fibonacci representations, then $x < y$ in radix order if and only if $[x]_F < [y]_F$.

We will also need a lemma about the subword complexity (also called "factor complexity") of a Fibonacci-automatic sequence $\mathbf{x}$. The subword complexity of $\mathbf{x}$ is a function $\rho_{\mathbf{x}}(n)$ counting the number of distinct subwords of length n appearing in $\mathbf{x}$.

Lemma 2. *Let $\mathbf{x}$ be a Fibonacci-automatic sequence generated by an m-state DFAO with msd-first input. Then $\rho_{\mathbf{x}}(n) = O(nm^2)$.*

Proof. See [1, Theorem 10.3.1]. The proof is stated there for base k, but the same proof works for Fibonacci-automatic sequences. $\qquad\square$

3   Recognizing Relations

3.1   Addition

Theorem 1. *For all integers $c \geq 0$, there exists an msd-first Fibonacci automaton with at most $O(\log c)$ states accepting $x \times y$ where $[x]_F + c = [y]_F$.*

Proof. We assume $c \geq 1$, since the case $c = 0$ is trivial. We design a Fibonacci automaton M_c that reads input $x \times y$ in parallel and accepts if and only if $[y]_F = [x]_F + c$. We modify the same approach we used previously for the case of base-k representation [12]. Namely, defining the difference $D(x, y) = [y]_F - [x]_F$, we first find a range of values I containing c such that if $D(x, y)$ ever falls outside I, then $D(x', y')$ stays outside I for all right extensions x' of x and y' of y.

Suppose $D(x,y) < 0$. Then $[y]_F < [x]_F$, and by Lemma 1 (c) we see that $y < x$ under the radix order. Hence, if y' is a right-extension of y and x' is a right-extension of x, we have $y' < x'$ under the radix order, and so $[y'] < [x']$ and hence $D(x',y') < 0$.

Now suppose $D(x,y) > c$. We have

$$D(xa,yb) = [yb] - [xa]_F = [y0]_F - [x0]_F + b - a > (\alpha[y]_F - \beta^2) - (\alpha[x]_F - \beta) - 1$$
$$= \alpha D(x,y) - \beta^2 + \beta - 1 \geq \alpha(c+1) - \beta^2 + \beta - 1 = \alpha c - \beta^2 > c.$$

Therefore, we can take the range I to be $[0,c]$.

Our automaton $M_c = (Q, \Sigma_2, q_0, \delta, A)$ is constructed as follows.

$$Q \subseteq \{[a,b,d',d] \ : \ a,b \in \{0,1\} \text{ and } d,d' \in I\},$$
$$q_0 = [0,0,0,0]_F,$$
$$\delta([a',b',d'',d'],[a,b]) = [a,b,d',d] \text{ where } a,b,a',b \in \{0,1\}, aa', bb' \neq 11, \text{ and}$$
$$d = d' + d'' + b' - a' + b - a,$$
$$A = \{[a,b,d',d] \ : \ a,b \in \{0,1\} \text{ and } d = c\}.$$

To understand the transition rule, suppose processing the last two symbols of an input $xa'a \times yb'b'$ leads to the following transitions:

$$[a'',b'',d''',d''] \ \xrightarrow{[a',b']} \ [a',b',d'',d'] \ \xrightarrow{[a,b]} \ [a,b,d',d].$$

Then $D(x,y) = [y]_F - [x]_F$, $D(xa',yb') = [y0]_F - [x0]_F + b' - a'$, and $D(xa'a,yb'b) = [y00]_F - [x00]_F + 2b' - 2a' + b - a$. The correctness of the transition rule now follows from Lemma 1 and an easy induction on the length of the input.

Our next task is to define Q more precisely. Define $d' = D(x,y)$, $d = D(xa,yb)$, and $\varepsilon_x = [x0]_F - \alpha[x]_F$. Thus

$$d = D(xa,yb) = [yb]_F - [xa]_F = [y0]_F - [x0]_F + b - a$$
$$= \alpha[y]_F + \varepsilon_y - \alpha[x]_F - \varepsilon_x + b - a = \alpha d' + \varepsilon_y - \varepsilon_x + b - a.$$

Rewriting, we get

$$d' - (d/\alpha) = \frac{\varepsilon_x - \varepsilon_y + a - b}{\alpha}. \tag{1}$$

There are now three cases to consider. By Lemma 1 (b) we know $-\beta^2 < \varepsilon_x, \varepsilon_y < -\beta$. So there are the following cases:

Case 1: $a = b$. Then $d/\alpha - 1 < d' < d/\alpha + 1$.

Case 2: $a = 0$, $b = 1$. Then $d/\alpha - 2 < d' < d/\alpha$.

Case 3: $a = 1$, $b = 0$. Then $d/\alpha < d' < d/\alpha + 2$.

This gives eight possible forms of states corresponding to a particular d:

$$\{[0,0,\lfloor d/\alpha \rfloor, d], [0,0,\lceil d/\alpha \rceil, d], [1,1,\lfloor d/\alpha \rfloor, d], [1,1,\lceil d/\alpha \rceil, d],$$
$$[0,1,\lfloor d/\alpha \rfloor, d], [0,1,\lfloor d/\alpha \rfloor - 1, d], [1,0,\lceil d/\alpha \rceil, d], [1,0,\lceil d/\alpha \rceil + 1, d]\}.$$

Starting from an accepting state, of the form $[a, b, d', c]$, we determine the possible predecessors of this state, and the possible predecessors of those, etc. We group the predecessors together in levels. The states of the form $[a, b, d', c]$ are at level 0, and the prececessors of these are at level 1, and so forth. We now have to bound the number of states in each level, and the number of levels.

States in level i are of the form $[a, b, d', d]$ and $d \in [D_i, D_i + r_i]$ for certain D_i, r_i. Note that $D_0 = c$ and $r_0 = 0$. Since the smallest possible d' in a level i state is $\lfloor D_i/\alpha \rfloor - 1$ and the largest possible is $\lceil (D_i + r_i)/\alpha \rceil + 1$, it follows that $r_{i+1} < r_i/\alpha + 4$. Starting with $r_0 = 0$, using this inequality four times shows that $r_i \leq 8$ for all i. Thus there are at most $8 + 1 = 9$ possible values of d among all states of any given level.

To estimate the number of possible levels, note that Eq. (1) gives $d' < (d + 2)/\alpha$ and thus $D_{i+1} + r_{i+1} < (D_i + 10)/\alpha < D_i/1.1$ provided $D_i \geq 22$. Thus there are at most $\log_{1.1} c = O(\log c)$ levels until $D_i \leq 22$. When the largest d in a level is smaller than 22, we cease our computation of levels and simply include all states $[a, b, d', d]$ with $d \leq 22$ (a constant number).

This shows that there are only $O(\log c)$ co-accessible states, and we are done. $\square$

By exchanging the role of x and y we immediately get the following.

Theorem 2. *For all integers $c \geq 0$, there exists an msd-first Fibonacci automaton with at most $O(\log c)$ states accepting $x \times y$ where $[x]_F - c = [y]_F$.*

3.2   Multiplication

Let $n \geq 1$ be a natural number and $0 \leq c < n$. We provide an upper bound on the number of states required for an automaton to recognize the relation $Y = nX + c$.

We let $L_{n,c}$ be the language of all words over $\Sigma_2 \times \Sigma_2$ where the first component represents an integer $[x]_F$ and the second component represents $n[x]_F + c$. More precisely,

$$L_{n,c} := \{x \times y \in (\Sigma_2 \times \Sigma_2)^* \ : \ [y]_F = n[x]_F + c\}.$$

For example, Fig. 2 depicts a minimal DFA recognizing $L_{2,0}$ computed by `Walnut`. Here the state numbered 0 is the initial state (denoted by the headless arrow) and the accepting states 0, 3, 6, and 8 are indicated by double circles.

Theorem 3. *There is a DFA of $O(n^2)$ states that accepts $L_{n,c}$.*

A result related to Theorem 3 was proved by Charlier et al. [6]. They proved that $2n^2$ states are necessary and sufficient for a DFA to accept multiples of the constant n with input in the Fibonacci numeration system. However, despite the close relationship of their result and ours, we see no direct way to get our result from theirs.

Our proof for the case of Fibonacci representation is modeled on the proof of the case of base k in [12], but there are some subtleties that make it more

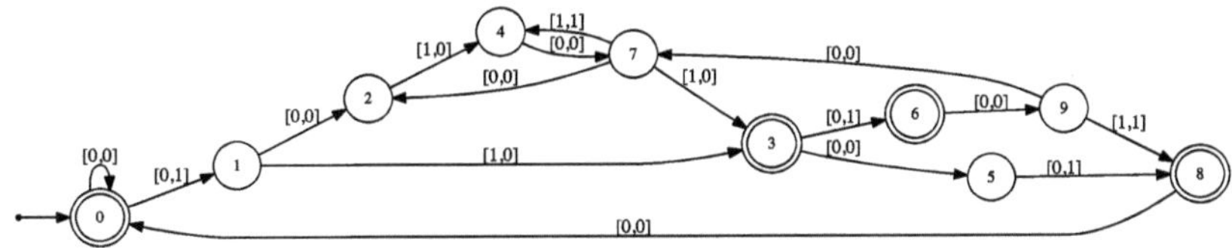

Fig. 2. Multiplication by 2 in Fibonacci representation.

complicated. We use an automaton $M_{n,c}$ to keep track of the difference function $D(x,y) := [y]_F - n[x]_F$. (We use the same letter D as in the previous section, but now it means something different.) The complications arise from four things:

- In base k, reading a single letter a from the input changes the integer read from n to $kn + a$. But the formula is more involved for Fibonacci representation.
- It is no longer sufficient to maintain just a single d, the current difference. We also need to maintain another related quantity, d', the previous difference.
- We have to ensure that the inputs are valid Fibonacci representations. This means we have to store, in the state, the last letter read for both input words.
- It is no longer so simple to determine the interval I_n for d and d', outside of which no right extension would be accepted.

Proof details can be found in the arXiv paper [12]. A referee points out that another proof of Theorems 1–3 can be obtained using the ideas in [7,8]. The advantage to our methods is their explicit nature that could lead to tight estimates.

4   Linear Subsequences of Fibonacci-Automatic Sequences

4.1   Shifts of Fibonacci-Automatic Sequences

Theorem 4. *Let $(h(i))_{i\geq 0}$ be a Fibonacci-automatic sequence generated by a DFAO of m states. For an integer constant $c \geq 0$, there is a DFAO of $O(m^2 c^2)$ states generating $(h(i+c))_{i\geq 0}$.*

Proof. Let $(h'(i))_{i\geq 0}$ be the interior sequence of $(h(i))_{i\geq 0}$ generated by $M = (Q, \Sigma_2, q_0, \delta, \Delta, \tau)$. We use a method similar to the one used for base-k in [12] and create an automaton M' that on input x reaches a state of the form $[[h'([x]_F), a_0'], \ldots, [h'([x]_F + c), a_c']]$ where a_i' is the last letter in $([x]_F + i)$, and after reading the letter a transitions to the next state $[[h'([xa]_F), a_0], \ldots, [h'([xa]_F + c), a_c]]$.

The automaton $M' = (Q', \Sigma_2, q_0', \delta', \Delta, \tau')$ is constructed as follows. The states in Q' are of the form $[[p_0, a_0], \ldots, [p_c, a_c]]$ where $p_i \in Q$ and $a_i \in \Sigma_2$. The initial state q_0' is $[[h'(0), 0], \ldots, [h'(c), a_c]]$ where a_c is the last letter of (c). Furthermore, we have $\tau'([[p_0, a_0], \ldots, [p_c, a_c]]) = \tau(p_c)$.

The transition function δ' is more complicated and requires explanation. If the first argument of δ' is $[[p_0, a_0], \ldots, [p_c, a_c]]$ and a, it computes the result as follows. First, it creates a list. Then it goes through each $[p_i, a_i]$ in order; if $a_i = 0$, then $[\delta(p_i, 0), 0]$ and $[\delta(p_i, 1), 1]$ are added to the list; otherwise only $[\delta(p_i, 0), 0]$ is added to the list. Finally, if $a = 0$, the function returns the 0-th to c-th element of this list; otherwise it outputs the first to $(c+1)$-th element.

Now let us find the number of states in M'. Since $p_0, \ldots, p_c$ form a length-$(c+1)$ contiguous subsequence of $(h'(i))_{i \geq 0}$, by Lemma 2 there are $O(m^2 c)$ possibilities for the word formed by the first components of the states. Furthermore, the $a_0, \ldots, a_c$ form a length-$(c+1)$ factor of the infinite Fibonacci word $\mathbf{f}$ (since $\mathbf{f}[i]$ is just the last bit of the Fibonacci representation of i), and it is well-known that there are exactly $c+2$ factors of $\mathbf{f}$ of length $c+1$. So the automaton M' for $(h(i+c))_{i \geq 0}$ has $O(m^2 c^2)$ states. $\qquad\square$

4.2   Linear Subsequences

Theorem 5. *Let $n \geq 1$ and $0 \leq c < n$, and let $(h(i))_{i \geq 0}$ be a Fibonacci-automatic sequence generated by a DFAO of m states. There is a DFAO of $O(m^2 n^4)$ states recognizing the linear subsequence $(h(ni + c))_{i \geq 0}$.*

Example 2. Let us find the minimal DFAO for $\mathbf{f}[2i]$. We can use the free software `Walnut` for this, using the commands

```
def f2 "?msd_fib F[2*i]=@1":
combine F2 f2=1:
```

This gives us the DFAO in Fig. 3.

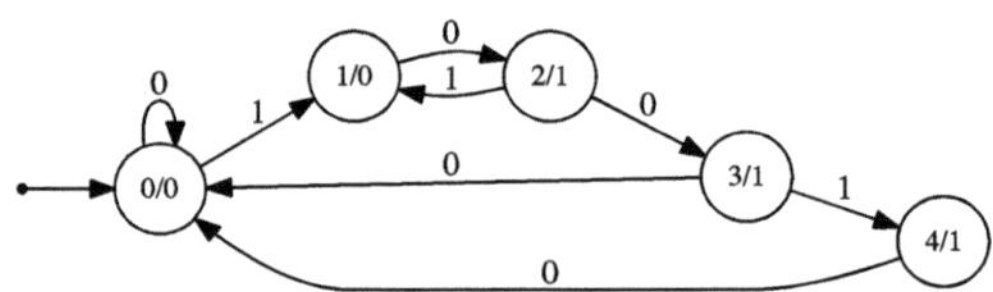

Fig. 3. A Fibonacci DFAO for $\mathbf{f}[2i]$.

Note that `Walnut`'s output is a minimized automaton. So Fig. 3 is the minimized version of the automaton built by the construction provided in this paper.

Remark 1. We do not know examples where the n^4 bound is actually attained. It seems possible that a tight upper bound is $O(n^2)$, but we do not know how to prove this.

For the special case where $c = 0$ and $h(i) = \mathbf{f}[i]$, the infinite Fibonacci word, the smallest DFAO for the linear subsequence $\mathbf{f}[ni]$ has a number of states given by OEIS sequence A385021 (calculated with `Walnut`). The first 10 terms, corresponding to $1 \leq n \leq 10$, are $2, 5, 10, 17, 27, 36, 52, 65, 78, 103$.

4.3   Application to a Problem of Bosma and Don

As is well-known, a k-automatic sequence can also be generated as the image (under a coding) of the fixed point $\gamma^\omega(0)$ of a prolongable morphism γ that maps each letter of some alphabet Σ to a word of Σ^* of length k [1]. Here "prolongable" means that the image of the letter 0 begins with 0. In fact, the appropriate morphism can be deduced immediately from the automaton, as follows: the letters of the alphabet are the states q of the automaton; the value $\gamma(q)$ is given by the states reached on inputs $0, 1, \ldots, k - 1$. The coding is defined by the output associated with each state. Roughly the same thing is true for Fibonacci-automatic sequences, except now the values of $\gamma(q)$ are given by the outputs on 0 and 1 (if the latter is possible by the rules of the numeration system).

Bosma and Don [3] studied the linear subsequences of the Fibonacci word **f**. They were mostly concerned with the size of the smallest morphism generating the sequence $(\mathbf{f}[ni + c])_{i \geq 0}$, and they were able to prove an exponential upper bound on the size of this morphism. Here the size of the morphism is the total number of symbols required to write it down, that is, $\sum_{a \in \Sigma} |\gamma(a)|$.

Example 3. From the automaton in Fig. 3 we can read off the following representation for th $(\mathbf{f}[2i])_{i \geq 0}$ in terms of a morphism γ and coding τ:

$$\gamma : 0 \to 01, \quad 1 \to 2, \quad 2 \to 31, \quad 3 \to 04, \quad 4 \to 0$$
$$\tau : 0, 1 \to 0, \quad 2, 3, 4 \to 1.$$

When we iterate the morphism γ we get the sequence $01231040 \cdots$ and after applying the coding τ we get $00110000 \cdots$. This example is mentioned in Bosma and Don [3].

As a consequence of our results, we can reduce the size of the needed morphism from exponential to polynomial.

Corollary 1. *Let $n \geq 1$ and $0 \leq c < n$, and let $(h(i))_{i \geq 0}$ be a Fibonacci-automatic sequence generated by a DFAO of m states. There is a morphism of size $O(m^2 n^4)$ generating the subsequence $(h(ni + c))_{i \geq 0}$.*

Proof. As we saw in Theorem 5, the resulting DFAO has $O(m^2 n^4)$ states. Hence the corresponding morphism that one can read off directly from the DFAO has size $O(m^2 n^4)$. $\qquad\square$

5   Construction by Büchi Arithmetic

Similar to what had been previously discussed for the case of base-k [12], given a sequence that is Fibonacci-automatic, we can compute the automaton for a linear subsequence using the free software `Walnut` [13,17]. `Walnut` uses an interpretation of Büchi arithmetic (the logical theory of the natural numbers together with addition and the function $V_k(n)$, the largest power of k dividing n, for some fixed $k \geq 2$) and implements a decision procedure due to Büchi [4,5] and takes a first-order logical statement and translates it into the corresponding automaton. The users of such software are interested in knowing how long it takes to algorithmically create the corresponding automata; in this section we study this time complexity. We use the total number of transitions of an automaton M as a proxy for the computational complexity of constructing M.

In particular, in the current version of `Walnut`, the algorithms for creating the automata discussed in previous sections are not used. Instead, more general techniques, suitable for any addable numeration system, are used. This means that the size of intermediate automata may differ from the minimal automata we constructed, and the running time can similarly be larger. Sometimes the complexity bounds in this section include the addition of $O(\log i)$ or $O(i^j)$ components for some constant i; these addition components are not significant compared to multiplication in $O(\log i)$ or $O(i^j)$ components for some constant i.

Furthermore, the minimization steps in the implementation for an n-state automaton can use an algorithm running in $O(n \log n)$ time by Hopcroft [9] or Valmari [18]. The algorithm by Hopcroft can be slightly modified to be applied to DFAOs as described by van Spaendonck [19]. Note that in an n-state DFA, the number of transitions is $O(n)$. The minimization is applied to each DFA or DFAO created. In this paper, we sometimes only apply partial minimizations and ignore the potential effects of fully minimizing automata in order to be able to analyze the runtime complexity and obtain an upper bound; the upper bound on state complexity and runtime complexity will remain correct regardless and the runtime complexity obtained reflects the computational cost of a complete minimization.

Consider the automaton M_{add} from Mousavi et al. [14] that accepts input $x \times y \times z$ if $[x]_F + [y]_F = [z]_F$. On input $x \times y \times z$ this automaton leads to the state denoted by the sequence $([x0^n] + [y0^n] - [z0^n])_{n \geq 0}$. We modify this automaton to keep track of some extra information. Consider the automaton M'_{add} created from M_{add} that on input $xa \times yb \times ze$ leads to the state denoted by $[s, s', e]$ where $s = ([xa0^n] + [yb0^n] - [ze0^n])_{n \geq 0}$ and $s' = ([x0^n] + [y0^n] - [z0^n])_{n \geq 0}$.

Suppose we have $M_{\text{add}} = (Q, \Sigma_2^3, \delta, q_0, A)$. We create $M'_{\text{add}} = (Q', \Sigma_2^3, \delta', q_0', A')$ as follows.

$$Q' = Q \times Q \times \Sigma_2,$$
$$\delta'([s, s', e], [a, b, e]) = [\delta(s, [a, b, e]), s, e],$$
$$q_0' = [q_0, q_0, 0],$$
$$A' = \{[s, s', e] \ : \ s \in A\}.$$

We use $M'_{\text{add}}(x, y, z)[i]$ to show the i-th component of the state M'_{add} reaches on input $x \times y \times z$ for $i \in \{0, 1, 2\}$; for example, $M'_{\text{add}}(x, y, z)[0]$ is $([x0^n] + [y0^n] - [z0^n])_{n \geq 0}$. The automaton M'_{add} has $O(1)$ states.

We start with the automaton recognizing the relation $[z]_F = n[x]_F$ where n is a fixed constant. There are a few possible recursions for implementing this automaton.

- $\exists y \ M_{n-1}(x, y) \wedge M_{\text{add}}(x, y, z).$
- $\exists y_1, y_2 \ M_{\lfloor n/2 \rfloor}(x, y_1) \wedge M_{\lceil n/2 \rceil}(x, y_2) \wedge M_{\text{add}}(y_1, y_2, z).$

However, a third recursive implementation that follows is more efficient than the two above.

If n is even, we recursively compute a DFA $M_{n/2}(x, y)$ recognizing $[y]_F = (n/2)[x]_F$ and then use the first-order expression

$$\exists y \; M_{n/2}(x, y) \land M'_{\text{add}}(y, y, z),$$

which is translated into an automaton by a direct product construction for the $\land$, and projection of the 2-nd coordinate to remove transitions on y.

If n is odd, we recursively compute a DFA $M_{n-1}(x, y)$ recognizing $[y]_F = (n-1)[x]_F$ and use the expression

$$\exists y \; M_{n-1}(x, y) \land M'_{\text{add}}(x, y, z).$$

In both cases, the translation could conceivably generate a nondeterministic automaton (by the projection) and determinizing it could take exponential time. However, we show that this is not the case. For now, let us ignore the cost of the minimization steps.

Theorem 6. *The DFA M_n for recognizing $[y]_F = n[x]_F$ constructed by the binary method has $O(n^2)$ states corresponding to all possible $[a, b, d, d']$ where $a, b \in \Sigma_2$ and $d, d' \in I_n = [-n, 2n - 1]$.*

Proof. We prove by induction on n that we can create the automaton M_n with states of the form $[a, b, d, d']$ such that on input $xa \times yb$ the automaton reaches a state where $d = [yb]_F - n[xa]_F$ and $d' = [y]_F - [x]_F$. The base case is $n = 1$. In this case, a simple 1-state automaton recognizes $[y]_F = [x]_F$. The single state corresponds to $[a, b, d, d'] = [0, 0, 0, 0]$.

For the induction step, assume the result is true for all $n' < n$; we prove it for n. There are two cases, corresponding to $n \equiv 0, 1 \pmod 2$. We only prove the first here.

By the induction hypothesis, we have constructed a DFA $M_{n/2}$ with $O(n^2)$ states corresponding to all possible $[a, b, d, d']$ where $a, b \in \Sigma_2$ and $d, d' \in I_{n/2} = [-(n/2), 2(n/2) - 1]$. In this case, we use the formula

$$\exists y \; M_{n/2}(x, y) \land M'_{\text{add}}(y, y, z).$$

First we create an automaton M'_n that is the product construction of $M_{n/2}$ and M'_{add}. So this automaton has $O(n^2)$ states. Consider an input $xa \times yb \times ze$ for M'_n leading to some state $[q, p]$ where q is from $M_{n/2}$ and p is from M'_{add}. By the construction of M'_n, we know $q = [a, b, d, d']$ where $d = [yb]_F - (n/2)[xa]_F$, $d' = [y]_F - (n/2)[x]_F$ and $p = [s, s', e]$ where $s = (2[yb0^n] - [ze0^n])_{n \geq 0}$, $s' = (2[y0^n] - [z0^n])_{n \geq 0}$. So we have $2d - s(0) = [ze] - n[xa]_F$ and $2d' - s'(0) = [z]_F - n[x]_F$. Therefore, for each state reachable by $xa \times yb \times ze$ in M'_n a tuple $[a, e, \bar{d}, \bar{d}']$ can be attributed where $\bar{d} = [ze] - n[xa]_F$ and $\bar{d}' = [z]_F - n[x]_F$. Furthermore, two input with the same xa, ze cannot lead to states with different $[a, e, d, d']$ attributed to them.

Next, we project away the y component. The resulting automaton is an NFA, and we use subset construction to create a DFA. Consider a state from the DFA reachable on $xa \times ze$. This state consists of some pairs $[q, p]$ from the NFA such that $q = [a, b, d, d']$, $p = [s, s', e]$ and we can create $[a, e, 2d - s(0), 2d' - s'(0)]$ from $[q, p]$ where $[ze] - n[xa]_F = 2d - s(0)$ and $[z]_F - n[x]_F = 2d' - s'(0)$.

In each state of the DFA, using an argument similar to the one used in the proof of Theorem 5, there are $O(n^4)$ possibilities for the $q = [a, b, d, d']$ parts. All $[q, p]$ in this state have the same $2d - s(0)$ and $2d' - s'(0)$ ($O(n^2)$ possibilities). If we look at the transition table from Mousavi et al. [14] and consider all $\delta_{\text{add}}(s', [b, b, e]) = s$ where δ_{add} is the transition function of M_{add}, then for each b the $(s'(0), s(0))$ pairs obtained from transitions are unique. So considering we have already determined the b, d, d' of each $[q, p]$ and we have fixed the $2d - s(0)$ and $2d' - s'(0)$, the s, s' are now determined as well ($O(1)$ possibilities). So there are $O(n^6)$ states in the DFA.

From the proof of Theorem 3, we know the automaton recognizing $n[x]_F = [z]_F$ on input $x \times z$ only needs to keep track of a difference $[z]_F - n[x]_F$ in range $I_n = [-n, 2n - 1]$. So after some minimization, the DFA $M_n = (Q, \Sigma_k^2, \delta, q_0, A)$ can be formally defined as follows.

$$Q = \{[a, e, d, d'] \; : \; a, e \in \Sigma_2, \; d, d \in I_n]\},$$
$$q_0 = [0, 0, 0, 0],$$
$$\delta([a', e', d, d'], [a, e]) = [a, e, g, d] \text{ where } g = d + d' + e' - na' + e - na,$$
$$A = \{[a, e, d, d'] \; : \; d = 0\}.$$

Theorem 7. *Let $n \geq 1$, $0 \leq c < n$ be fixed constants. Using translation of a first-order statement using Büchi arithmetic, an automaton for recognizing $[y]_F = n[x]_F + c$ on input $x \times y$ is constructed by* `Walnut` *in $O(n^6 \log^2 n)$ time.*

6   Open Problems

Problem 1. Find a tight lower bound for the problem in Theorem 5.

In a previous paper [12], we studied topics similar to this paper, but with base-k input instead of input in Fibonacci representation. We can attempt to study these topics in other numeration systems and solve similar problems. For example, we state the following as an open problem.

Problem 2. What is the state complexity of $(r(i+c))_{i \geq 0}$ where r is the Tribonacci word, as a function of c?

References

1. Allouche, J.-P., Shallit, J.: Automatic Sequences: Theory, Applications. Cambridge University Press, Generalizations (2003)

2. Berstel, J.: Fibonacci words – a survey. In: Rozenberg, G., Salomaa, A. (eds.) The Book of L, pp. 13–27. Springer, Berlin, Heidelberg (1986). https://doi.org/10.1007/978-3-642-95486-3_2

3. Bosma, W., Don, H.: Constructing morphisms for arithmetic subsequences of Fibonacci. In: Capretta, V., Krebbers, R., Wiedijk, F. (eds.) Logics and Type Systems in Theory and Practice. Lecture Notes in Computer Science, vol. 14560, pp. 100–110. Springer, Cham (2024). https://doi.org/10.1007/978-3-031-61716-4_6

4. Bruyère, V., Hansel, G., Christian, M., Villemaire, R.: Logic and p-recognizable sets of integers. Bull. Belgian Math. Soc. Simon Stevin $\mathbf{1}$(2), 191–238 (1994)

5. Büchi, J.R.: Weak second-order arithmetic and finite automata. Zeitschrift für Mathematische Logik und Grundlagen der Mathematik, 6:66–92, 1960. Reprinted. In: Mac Lane, S., Siefkes D. (eds.) The Collected Works of J. Richard Büchi, pp. 398–424, Springer, New York, NY (1990). https://doi.org/10.1007/978-1-4613-8928-6_22

6. Charlier, É., Rampersad, N., Rigo, M., Waxweiler, L.: The minimal automaton recognizing $m\mathbb{N}$ in a linear numeration system. INTEGERS Electron. J. Comb. Number Theory $\mathbf{11B}$(A4) (2011)

7. Frougny, C.: Numeration systems. In: Lothaire, M. (ed.) Algebraic Combinatorics on Words, pp. 230–268. Cambridge University Press (2002)

8. Grabner, P.J., Steiner, W.: Redundancy of minimal weight expansions in Pisot bases. Theoret. Comput. Sci. $\mathbf{412}$(45), 6303–6315 (2011)

9. Hopcroft, J.: An $n \log n$ algorithm for minimizing states in a finite automaton. In: Kohavi, Z., Paz, A. (eds.) Theory of Machines and Computations, pp. 189–196. Academic Press (1971)

10. Hopcroft, J.E., Ullman, J.D.: Introduction to Automata Theory, Languages, and Computation. Addison-Wesley Series in Computer Science. Addison-Wesley Publishing Company (1979)

11. Lekkerkerker, C.G.: Voorstelling van natuurlijke getallen door een som van getallen van Fibonacci. Simon Stevin $\mathbf{29}$, 190–195 (1952)

12. Moradi, D., Rampersad, N., Shallit, J.: Complexity of linear subsequences of k-automatic sequences. arXiv:2512.10017 (2026)

13. Mousavi, H.: Automatic theorem proving in Walnut. arXiv:1603.06017 (2021)

14. Mousavi, H., Schaeffer, L., Shallit, J.: Decision algorithms for Fibonacci-automatic words, I: basic results. RAIRO Theor. Inf. Appl. $\mathbf{50}$(1), 39–66 (2016)

15. Ostrowski, A.: Bemerkungen zur Theorie der diophantischen Approximationen. Abhandlungen aus dem Mathematischen Seminar der Universität Hamburg $\mathbf{1}$(1), 77–98 (1922). Reprinted. In: Collected Mathematical Papers, vol. 3 pp. 57–80. Birkhäuser Verlag (1984)

16. Reble, D.: Zeckendorf vs. Wythoff representations: comments on A007895 (2008). https://oeis.org/A007895

17. Shallit, J.: The Logical Approach to Automatic Sequences: Exploring Combinatorics on Words with Walnut, Volume 482 of London Mathematical Society Lecture Note Series. Cambridge University Press (2023)

18. Valmari, A.: Fast brief practical DFA minimization. Inf. Process. Lett. $\mathbf{112}$(6), 213–217 (2012)

19. van Spaendonck, F.: Automatic sequences: the effect of local changes on complexity. Master's thesis, Radboud University, Netherlands (2020)

20. Zeckendorf, E.: Représentation des nombres naturels par une somme de nombres de Fibonacci ou de nombres de Lucas. Bull. Soc. Roy. Sci. Liège $\mathbf{41}$(3–4), 179–182 (1972)

Visibly Recursive Automata

Kévin Dubrulle[1,2]($\boxtimes$) , Véronique Bruyère[1] , Guillermo A. Pérez[2] ,
and Gaëtan Staquet[3]

[1] Université de Mons, Mons, Belgium
{kevin.dubrulle,veronique.bruyere}@umons.ac.be
[2] Universiteit Antwerpen, Antwerp, Belgium
guillermo.perez@uantwerpen.be
[3] Nantes Université, École Centrale Nantes, CNRS, LS2N, Nantes, France
gaetan.staquet@ec-nantes.fr

Abstract. As an alternative to visibly pushdown automata, we introduce visibly recursive automata (VRAs), composed of a set of classical automata that can call each other. VRAs are a strict extension of so-called systems of procedural automata, a model proposed in 2021 by Frohme and Steffen. We study the complexity of standard language-theoretic operations and classical decision problems for VRAs. Since the class of deterministic VRAs forms a strict subclass in terms of expressiveness, we propose a (weaker) notion that does not restrict expressive power and that we call codeterminism. Codeterminism comes with many desirable algorithmic properties that we demonstrate by using it, e.g., as a stepping stone towards implementing complementation of VRAs.

Keywords: Visibly pushdown languages · Modular model of computation · Automata theoretic properties

1 Introduction

Almost all computer systems are programmed by defining multiple functions that call each other, sometimes recursively. When modeling such systems—for verification and model checking purposes, for instance—it is thus important to take into account the *calls* to functions (i.e., jump into a different part of the code) and their *returns* (i.e., jump back to the position immediately following that before the corresponding call). These behaviors can be represented by context-free grammars or, equivalently, pushdown automata (PDAs) [20].

While *context-free languages* (CFLs), i.e., the family of languages accepted by PDAs, have model checking tools [2,8,26,28], many properties of interest for CFLs are undecidable, for example: checking the equivalence of two PDAs and universality of a PDA. Thus, along the years, some restrictions have been considered to obtain positive decidability results, such as in [10–12,16]. Among these

---

K. Dubrulle is a *FRIA grantee* of the Belgian *Fonds de la Recherche Scientifique*–FNRS;
G. A. Pérez is supported by the FWO "SynthEx" project (G0AH524N).

M.-P. Béal and P. Caron (Eds.): DLT 2026, LNCS 16578, pp. 194–208, 2026.
https://doi.org/10.1007/978-3-032-28404-4_15

restrictions, here, we focus on *visibly pushdown languages* (VPLs), recognized by *visibly pushdown automata* (VPAs) [4] and *nested word automata* [1,5]. We consider only VPAs, as they are more commonly used in the literature.

VPAs split the alphabet into three disjoint subsets: one set of symbols is only used for *calls* whose reading triggers a push on the stack, a second set only for *returns* whose reading pops the top of the stack, and the last set contains the *internal* symbols with no influence on the stack of calls. Thus, it is the type of symbol that dictates the operation to be applied on the stack. Thanks to this restriction, VPLs are closed under several operations, including Boolean operations [4]. Moreover, some undecidable problems for CFLs such as universality, language equivalence, and inclusion are decidable for VPLs [4].

VPAs have been used in practice to verify XML and JSON documents against their schemas in a streaming context [9,23,25]. In particular, in [9], we learned (in Angluin's active-learning setup [6,21]) a VPA modeling a given JSON schema from a sample of good and bad documents. We implemented our algorithm in a prototype tool, showing that an automaton-based approach is feasible. However, VPAs suffer from a significant drawback: they tend to be large and, thus, complex to construct and learn. While a variation of VPA specialized for JSON schemas is studied in [25], we conjecture that it would be more efficient to build an automaton defined as a *collection of smaller automata*, rather than a single large automaton. In [15], we studied the validation of JSON documents using such a model (introduced in the sequel) and we obtained [13] much smaller automata and faster validation times. Whilst other work treat models that strictly include VPAs [2,19,29], we here focus on "modularizing" VPAs. Previous work considered two ways for modularization: *k-module single-entry automata* (*k*-SEVPAs) [3] and *systems of procedural automata* (SPAs) [18].

The set of call symbols of a k-SEVPA is partitioned into k classes, the automaton has a main module and k distinct interconnected submodules, one for each class. The transitions labeled by call and return symbols manage these interconnections, by hard-coding the stack manipulations. While VPAs and k-SEVPAs are equivalent (for any value of k) and there exists a unique minimal k-SEVPA for a given language, this minimal automaton may have exponentially more states than a VPA accepting the same language. Nonetheless, this family has some active learning algorithms [21,22]. Our VPA-based approach from [9] actually relies on 1-SEVPAs. Due to their sizes, our algorithm for JSON documents is slower than state-of-the-art JSON validators, as highlighted in [25].

The submodules of an SPA are classical finite automata (FAs) that are not interconnected, and the transitions do not manipulate the stack directly. Instead, each call symbol is associated with a specific automaton. Whenever the call symbol corresponding to an FA $\mathcal{A}$ is read by the SPA, a call to this FA is performed, with a jump to its initial state. Later on, when a sub-word is accepted by $\mathcal{A}$ and followed by a return symbol, the SPA goes back to the state from which it reads the call symbol. In [18], an active learning algorithm for SPAs is also presented. Interestingly, SPAs are strictly less expressive than VPAs.

Contributions. In this paper[1], we introduce a new kind of modular automata we call *visibly recursive automata* (VRAs). Similarly to SPAs [18], a VRA is composed of multiple FAs. However, in contrast to that work, we lift the restriction that each call symbol corresponds to a unique automaton. Instead, we allow multiple FAs to share a common call symbol. We argue that VRAs form a strict superset of SPAs, and we show that they are equivalent to VPAs with polynomial size translations (see Theorem 1).

We claim that, like for [18], since we construct smaller FAs, each serving a specific purpose, VRAs are easier to construct and understand: one can focus on each part of the system individually. This is much closer to how programs are engineered and implemented: each function serves a specific role and can call other functions to achieve its goal. It also mirrors the way in which JSON schemas are structured,[2] i.e., in a modular way. Furthermore, we conjecture that this decomposition will allow for more efficient learning than what is possible for VPAs (each part is an FA that can be learned more easily than VPAs [6,21, 22]) and enable a modular and compositional learning algorithm more generally applicable than that for SPAs (an open challenge of active automata learning [17]).

Table 1. Summary of our complexity results, and comparison with known bounds for VPAs, where $|\mathcal{A}|$ is the number of states and transitions of either a VRA, or a VPA $\mathcal{A}$. We highlight where VRAs have better complexity.

		VRA	**VPA** [4]
Operations (Result size)	Concatenation ($\widetilde{L}(\mathcal{A}_1) \cdot \widetilde{L}(\mathcal{A}_2)$)	$\mathcal{O}(|\mathcal{A}_1| + |\mathcal{A}_2|)$	$\mathcal{O}(|\mathcal{A}_1| + |\mathcal{A}_2|)$
	Kleene-$*$ ($\widetilde{L}(\mathcal{A}_1)^*$)	$\mathcal{O}(|\mathcal{A}_1|)$	$\mathcal{O}(|\mathcal{A}_1|)$
	Union ($\widetilde{L}(\mathcal{A}_1) \cup \widetilde{L}(\mathcal{A}_2)$)	$\mathcal{O}(|\mathcal{A}_1| + |\mathcal{A}_2|)$	$\mathcal{O}(|\mathcal{A}_1| + |\mathcal{A}_2|)$
	Intersection ($\widetilde{L}(\mathcal{A}_1) \cap \widetilde{L}(\mathcal{A}_2)$)	$\mathcal{O}(|\mathcal{A}_1| \cdot |\mathcal{A}_2|)$	$\mathcal{O}(|\mathcal{A}_1| \cdot |\mathcal{A}_2|)$
	Complementation ($\overline{\widetilde{L}(\mathcal{A}_1)}$)	$2^{\mathcal{O}(|\mathcal{A}_1|)}$	$2^{\mathcal{O}(|\mathcal{A}_1|^2)}$
Decision problems (Runtime)	Emptiness ($\widetilde{L}(\mathcal{A}_1) \overset{?}{=} \varnothing$)	$\mathcal{O}(|\mathcal{A}_1|)$	$\mathcal{O}(|\mathcal{A}_1|^3)$
	Universality ($\widetilde{L}(\mathcal{A}_1) \overset{?}{=} WM(\widetilde{\Sigma})$)	$2^{\mathcal{O}(|\mathcal{A}_1|)}$	$2^{\mathcal{O}(|\mathcal{A}_1|^2)}$
	Inclusion ($\widetilde{L}(\mathcal{A}_1) \overset{?}{\subseteq} \widetilde{L}(\mathcal{A}_2)$)	$\mathcal{O}(|\mathcal{A}_1|) \cdot 2^{\mathcal{O}(|\mathcal{A}_2|)}$	$\mathcal{O}(|\mathcal{A}_1|^3) \cdot 2^{\mathcal{O}(|\mathcal{A}_2|^2)}$
	Equivalence ($\widetilde{L}(\mathcal{A}_1) \overset{?}{=} \widetilde{L}(\mathcal{A}_2)$)	$2^{\mathcal{O}(|\mathcal{A}_1|+|\mathcal{A}_2|)}$	$2^{\mathcal{O}(|\mathcal{A}_1|^2+|\mathcal{A}_2|^2)}$

Our long-term objective is to obtain efficient active learning algorithms for VRAs. In this work, we focus on a first step in that direction. Namely, we study the complexity of the usual language-theoretic operations for languages accepted by VRAs, as well as the classical decision problems for VRAs. Some of our algorithms leverage the interreduction between VRAs and VPAs, but mostly we provide direct algorithms with better complexity. Concerning the language-theoretic operations, our main result is the complementation closure that requires translating any VRA into a *codeterministic*[3] *and complete* one (determinism does

---

[1] Detailed proofs of the results are deferred in the long version of this paper [14].

[2] See https://json-schema.org/understanding-json-schema/structuring.

[3] We borrow terminology and draw inspiration from [7] for this notion.

not help as deterministic VRAs form a strict subclass). Table 1 summarizes our results. For the decision problems (see Theorem 4), we highlight that the complexity for VRAs is consistently lower than for VPAs. For the operations over the languages (see Theorem 3), we obtain the same complexities as for VPAs with the exception of complementation, where we again get a lower complexity.

2   Visibly Recursive Automaton Model

In this section, we present the *visibly recursive automaton* model and provide a comparison with some other models. Visibly recursive automata are composed of several classical *finite automata* and accept *well-matched words*.

2.1   Preliminaries

Definition 1 (Finite automaton). *A* finite automaton *(FA) is a tuple* $\mathcal{A} = \langle \Sigma, Q, I, F, \delta \rangle$ *where* Σ *is the* input alphabet; Q, *a finite set of* states; $I \subseteq Q$, *a set of* initial *states;* $F \subseteq Q$, *a set of* final *states; and* $\delta \subseteq Q \times \Sigma \times Q$, *a set of* transitions. *The* size *of an FA* $\mathcal{A}$, *denoted by* $|\mathcal{A}|$, *is* $|Q| + |\delta|$.

We denote by $L(\mathcal{A})$ the *language* of $\mathcal{A}$ composed of all accepted words. An FA $\mathcal{A}$ is *deterministic* (DFA) if $|I| = 1$, and, for all $q \in Q$, $a \in \Sigma$, there is at most one transition $(q, a, p) \in \delta$. It is *complete* if, for all $q \in Q$, $a \in \Sigma$, there exists a transition $(q, a, p) \in \delta$. Any FA $\mathcal{A}$ can be transformed into an equivalent complete DFA $\mathcal{B}$ with $|\mathcal{B}| = 2^{\mathcal{O}(|\mathcal{A}|)}$ [20]. The empty word is denoted ε.

Definition 2 (Well-matched word). *A* pushdown alphabet $\widetilde{\Sigma} = \Sigma_{int} \cup \Sigma_{call} \cup \Sigma_{ret}$ *is the union of three pairwise disjoint finite alphabets, which are, respectively, the set of* internal, call, *and* return *symbols. The set* $WM(\widetilde{\Sigma})$ *of* well-matched words *over* $\widetilde{\Sigma}$ *is the smallest set satisfying:*

- *Let* $w \in \Sigma_{int}^*$, *then* $w \in WM(\widetilde{\Sigma})$.
- *Let* $w \in WM(\widetilde{\Sigma})$, $c \in \Sigma_{call}$ *and* $r \in \Sigma_{ret}$, *then* $c \cdot w \cdot r \in WM(\widetilde{\Sigma})$.
- *Let* $w_1, w_2 \in WM(\widetilde{\Sigma})$, *then* $w_1 \cdot w_2 \in WM(\widetilde{\Sigma})$.

The *depth* of $w \in WM(\widetilde{\Sigma})$, denoted by $depth(w)$, is the deepest level of unmatched call symbols at any point in the word. Any $w \in WM(\widetilde{\Sigma})$ can be decomposed as $w = u_0 c_1 w_1 r_1 u_1 \ldots c_n w_n r_n u_n$ for some $n \in \mathbb{N}$, with $u_i \in \Sigma_{int}^*$, $c_i \in \Sigma_{call}$, $r_i \in \Sigma_{ret}$ and $w_i \in WM(\widetilde{\Sigma})$ such that $depth(w_i) < depth(w)$, for all i. Note that if $w \in \Sigma_{int}^*$, then $n = 0$ and $depth(w) = 0$.

Given $S \subseteq D$, we denote by $\overline{S}$ the complement of S in D, i.e., $\overline{S} = D \setminus S$. By convention, the intersection over an empty family of subsets of D is equal to D: if $(S_i)_{i \in I}$ is a family of subsets of D and $I = \varnothing$, then $\bigcap_{i \in I} S_i = D$.

2.2   Visibly Recursive Automata

A *visibly recursive automaton* (VRA), inspired by the formalisms from [18,19], is composed of several FAs that can call each other by reading specific call symbols of a pushdown alphabet. Each of these FAs is identified by a unique *procedural symbol*. By convention, we use capital letters to denote procedural symbols. See Fig. 1a for a first example.

Definition 3 (Procedural alphabet). *A procedural alphabet* Σ_{proc} *w.r.t.* $\widetilde{\Sigma} = \Sigma_{int} \cup \Sigma_{call} \cup \Sigma_{ret}$ *is a set of* procedural symbols. *With* Σ_{proc} *we associate a linking function* $f : \Sigma_{proc} \to \Sigma_{call} \times \Sigma_{ret}$. *Let* f_{call} *and* f_{ret} *be the functions such that* $f(J) = \langle f_{call}(J), f_{ret}(J) \rangle$, *for all* $J \in \Sigma_{proc}$.

Definition 4 (Visibly recursive automaton). *A* visibly recursive automaton (VRA) *is a tuple* $\mathcal{A} = \langle \widetilde{\Sigma}, \Sigma_{proc}, \Lambda, \mathcal{A}^S \rangle$, *where:*

- $\widetilde{\Sigma} = \Sigma_{int} \cup \Sigma_{call} \cup \Sigma_{ret}$ *is a pushdown alphabet;*
- Σ_{proc} *is a procedural alphabet w.r.t.* $\widetilde{\Sigma}$;
- $\Lambda = \{\mathcal{A}^J \mid J \in \Sigma_{proc}\}$ *is a set of finite automata* over $\Sigma_{int} \cup \Sigma_{proc}$ *such that* $\mathcal{A}^J = \langle \Sigma_{int} \cup \Sigma_{proc}, Q^J, I^J, F^J, \delta^J \rangle$ *for each* J;
- $\mathcal{A}^S = \langle \Sigma_{int} \cup \Sigma_{proc}, Q^S, I^S, F^S, \delta^S \rangle$ *is a starting automaton.*

We write $Q_{\mathcal{A}} = \bigcup_{J \in \Sigma_{proc} \cup \{S\}} Q^J$ *(resp.* $\delta_{\mathcal{A}} = \bigcup_{J \in \Sigma_{proc} \cup \{S\}} \delta^J$*) the set of all states (resp. transitions) of a VRA* $\mathcal{A}$. *Its size, denoted by* $|\mathcal{A}|$, *is* $|Q_{\mathcal{A}}| + |\delta_{\mathcal{A}}|$.

In this definition, we assume that $\mathcal{A}^S \notin \Lambda$, $S \notin \Sigma_{proc}$, and the sets of states Q^J, with $J \in \Sigma_{proc} \cup \{S\}$, are pairwise disjoint. A transition in δ^J on an internal (resp. procedural) symbol is called an *internal* (resp. *procedural*) transition.

A VRA $\mathcal{A}$ accepts words over $\widetilde{\Sigma}$ as follows. The semantics of $\mathcal{A}$ use *configurations* $\langle q, \sigma \rangle$ where $q \in Q_{\mathcal{A}}$ is a state and $\sigma \in Q_{\mathcal{A}}^*$ is a *stack word* whose symbols are states of the VRA. A *recursive run* of $\mathcal{A}$ on a word $w = a_1 \ldots a_n \in \widetilde{\Sigma}^*$ is a sequence $\pi = \langle q_0, \sigma_0 \rangle \xrightarrow{a_1} \langle q_1, \sigma_1 \rangle \xrightarrow{a_2} \ldots \xrightarrow{a_n} \langle q_n, \sigma_n \rangle$, where for all $i \in [1, n]$:

- If $a_i \in \Sigma_{int}$, there exists a transition $q_{i-1} \xrightarrow{a_i} q_i \in \delta_{\mathcal{A}}$ and $\sigma_i = \sigma_{i-1}$;
- If $a_i \in \Sigma_{call}$, there exists a procedural symbol $J \in \Sigma_{proc}$ such that $f_{call}(J) = a_i$, $q_i \in I^J$, and there exists $q_{i-1} \xrightarrow{J} p \in \delta_{\mathcal{A}}$ such that $\sigma_i = p\sigma_{i-1}$;[4]
  Hence, when reading $a_i \in \Sigma_{call}$, an automaton $\mathcal{A}^J$ such that $f_{call}(J) = a_i$ is called and there is a jump to an initial state q_i of $\mathcal{A}^J$, while a state p such that $q_{i-1} \xrightarrow{J} p$ is pushed on the stack word.
- If $a_i \in \Sigma_{ret}$, there exists a procedural symbol $J \in \Sigma_{proc}$ such that $f_{ret}(J) = a_i$, $q_{i-1} \in F^J$, and $\sigma_{i-1} = q_i\sigma_i$.
  Hence, when reading $a_i \in \Sigma_{ret}$, if q_{i-1} is a final state of $\mathcal{A}^J$ and $f_{ret}(J) = a_i$, the call to $\mathcal{A}^J$ is completed and the state q_i is popped from the stack word.

---

[4] A symbol is pushed on the left of a stack word.

See Example 1 below to better understand the semantics.

We denote by $\Pi(\mathcal{A})$ the set of all recursive runs of $\mathcal{A}$. The *recursive language*[5] of an automaton $\mathcal{A}^J \in \Lambda \cup \{\mathcal{A}^S\}$, denoted by $\widetilde{L}(\mathcal{A}^J)$, is the set of words such that there exists an *accepting* recursive run on them, i.e., from an initial configuration $\langle q_i, \varepsilon \rangle$, with $q_i \in I^J$, to a final configuration $\langle q_f, \varepsilon \rangle$, with $q_f \in F^J$:

$$\widetilde{L}(\mathcal{A}^J) = \left\{ w \in \widetilde{\Sigma}^* \mid \exists q_i \in I^J, q_f \in F^J, \langle q_i, \varepsilon \rangle \xrightarrow{w} \langle q_f, \varepsilon \rangle \in \Pi(\mathcal{A}) \right\}.$$

Notice that $f_{call}(J)$ and $f_{ret}(J)$ do not appear in the definition of $\widetilde{L}(\mathcal{A}^J)$. That is, a word of the recursive language of J can start (resp. end) with a symbol that is not $f_{call}(J)$ (resp. $f_{ret}(J)$). The language of a VRA $\mathcal{A}$, denoted by $\widetilde{L}(\mathcal{A})$, is the recursive language of its starting automaton: $\widetilde{L}(\mathcal{A}) = \widetilde{L}(\mathcal{A}^S)$.

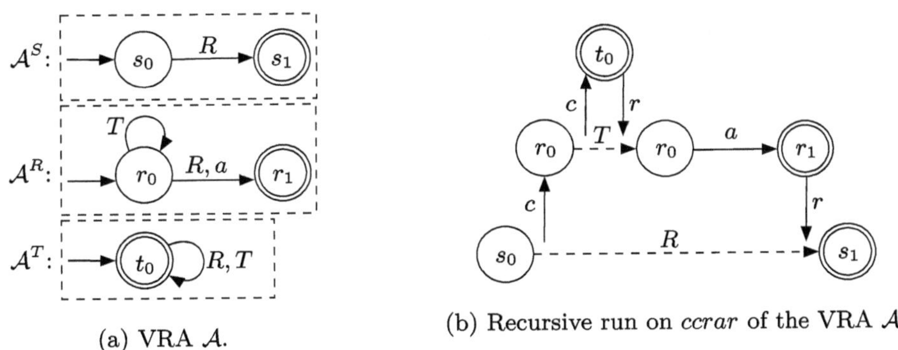

(a) VRA $\mathcal{A}$.

(b) Recursive run on *ccrar* of the VRA $\mathcal{A}$.

Fig. 1. Example of a VRA and a recursive run of it.

Example 1. Figure 1a shows an example of a VRA $\mathcal{A} = \langle \widetilde{\Sigma}, \Sigma_{proc}, \Lambda, \mathcal{A}^S \rangle$, with the pushdown alphabet $\widetilde{\Sigma} = \{a\} \cup \{c\} \cup \{r\}$, the procedural alphabet $\Sigma_{proc} = \{R, T\}$, and the linking function f such that $f(R) = f(T) = \langle c, r \rangle$. The VRA is composed of three DFAs $\mathcal{A}^S$, $\mathcal{A}^R$ and $\mathcal{A}^T$, where $\mathcal{A}^S$ is the starting one.

Let $w = ccrar \in WM(\widetilde{\Sigma})$. The following recursive run on w witnesses that $w \in \widetilde{L}(\mathcal{A})$ (it is also illustrated in Fig. 1b with the automata calls):

$$\langle s_0, \varepsilon \rangle \xrightarrow{c} \langle r_0, s_1 \rangle \xrightarrow{c} \langle t_0, r_0 s_1 \rangle \xrightarrow{r} \langle r_0, s_1 \rangle \xrightarrow{a} \langle r_1, s_1 \rangle \xrightarrow{r} \langle s_1, \varepsilon \rangle \in \Pi(\mathcal{A}).$$

We explain the first three transitions of the recursive run:

- $\langle s_0, \varepsilon \rangle \xrightarrow{c} \langle r_0, s_1 \rangle$ is possible since $s_0 \xrightarrow{R} s_1 \in \delta^S$ and $f_{call}(R) = c$. We call the FA $\mathcal{A}^R$, go to $r_0 \in I^R$, and push s_1 on top of the stack word ε.
- $\langle r_0, s_1 \rangle \xrightarrow{c} \langle t_0, r_0 s_1 \rangle$ is also possible, but with a call to the FA $\mathcal{A}^T$.

---

[5] This refers to a language defined in terms of a VRA and should not be confused with the class of recursive languages in theory of computation.

- $\langle t_0, r_0 s_1 \rangle \xrightarrow{r} \langle r_0, s_1 \rangle$ is possible since $t_0 \in F^T$ and $f_{ret}(T) = r$. The call to $\mathcal{A}^T$ is completed. We pop r_0 from the stack word and go to this state.

As this run starts in $s_0 \in I^S$ and ends in $s_1 \in F^S$, it follows that $w \in \widetilde{L}(\mathcal{A})$.

Given a VRA $\mathcal{A}$, each of its FAs $\mathcal{A}^J \in \Lambda$, can be seen as accepting either the recursive language $\widetilde{L}(\mathcal{A}^J) \subseteq \widetilde{\Sigma}^*$, or the language $L(\mathcal{A}^J) \subseteq (\Sigma_{proc} \cup \Sigma_{int})^*$. To avoid any confusion, a run of $\mathcal{A}^J$ on a word over $\Sigma_{int} \cup \Sigma_{proc}$ is called a *regular run*, and the language $L(\mathcal{A}^J)$ is called its *regular language*. Note that $\widetilde{L}(\mathcal{A}^J) \subseteq WM(\widetilde{\Sigma})$. Indeed, a recursive run on $w \in \widetilde{L}(\mathcal{A})$ begins and ends with an empty stack word, and we cannot pop a symbol from an empty stack word.

In order to better understand the VRA model, we state Proposition 1 below, which provides a recursive definition of the semantics of VRAs: to follow a procedural transition $q \xrightarrow{J} p$, a VRA must read a word cwr such that $f(J) = \langle c, r \rangle$ and w is accepted by $\mathcal{A}^J$. Proposition 1 is illustrated in Fig. 2 (see also the example of Fig. 1b). Given $c \in \Sigma_{call}$ and $r \in \Sigma_{ret}$, we write $\Sigma_{proc}^{\langle c,r \rangle}$ as the set of procedural symbols linked by f to $\langle c, r \rangle$: $\Sigma_{proc}^{\langle c,r \rangle} = \{J \in \Sigma_{proc} \mid f(J) = \langle c, r \rangle\}$.

Proposition 1. *Given a VRA $\mathcal{A}$, let $cwr \in \Sigma_{call} \cdot WM(\widetilde{\Sigma}) \cdot \Sigma_{ret}$ and $p, q \in Q_{\mathcal{A}}$:*

$$\langle q, \varepsilon \rangle \xrightarrow{cwr} \langle p, \varepsilon \rangle \in \Pi(\mathcal{A}) \iff \exists J \in \Sigma_{proc}^{\langle c,r \rangle} : q \xrightarrow{J} p \in \delta_{\mathcal{A}} \wedge w \in \widetilde{L}(\mathcal{A}^J).$$

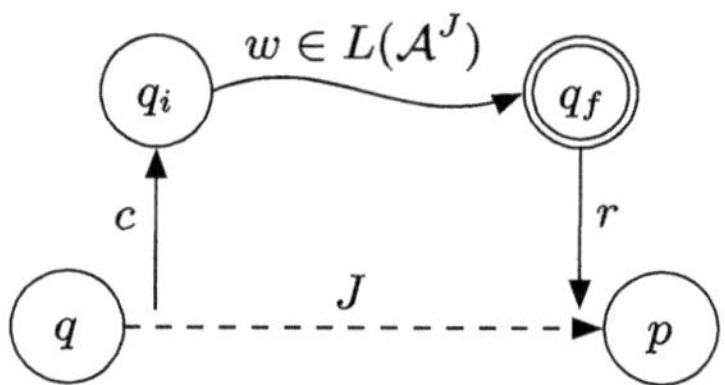

Fig. 2. Illustration of the semantics of a VRA on $cwr \in \Sigma_{call} \cdot WM(\widetilde{\Sigma}) \cdot \Sigma_{ret}$, with $q_i \in I^J$, $q_f \in F^J$ and $f(J) = \langle c, r \rangle$.

This proposition provides a link between the recursive language of an automaton and its regular language. Consider an accepting recursive run of $\mathcal{A}^J \in \Lambda \cup \{\mathcal{A}^S\}$ on a well-matched $w = u_0 c_1 w_1 r_1 \ldots c_n w_n r_n u_n \in \widetilde{L}(\mathcal{A}^J)$, with $n \in \mathbb{N}$, $u_i \in \Sigma_{int}^*$, $c_i \in \Sigma_{call}$, $r_i \in \Sigma_{ret}$ and $w_i \in WM(\widetilde{\Sigma})$. We can decompose the recursive run into $\langle q, \varepsilon \rangle \xrightarrow{u_0} \langle q_1, \varepsilon \rangle \xrightarrow{c_1 w_1 r_1} \langle p_1, \varepsilon \rangle \xrightarrow{u_1} \ldots \xrightarrow{u_{n-1}} \langle q_n, \varepsilon \rangle \xrightarrow{c_n w_n r_n} \langle p_n, \varepsilon \rangle \xrightarrow{u_n} \langle p, \varepsilon \rangle$, with $q \in I^J$, $p \in F^J$ and $q_i, p_i \in Q^J$ for all $i \in [1, n]$. By Proposition 1, we can replace each recursive run $\langle q_i, \varepsilon \rangle \xrightarrow{c_i w_i r_i} \langle p_i, \varepsilon \rangle$ by a regular run $q_i \xrightarrow{J_i} p_i \in \delta_{\mathcal{A}}$, with $J_i \in \Sigma_{proc}^{\langle c_i, r_i \rangle}$ such that $w_i \in \widetilde{L}(\mathcal{A}^{J_i})$. This results in an accepting regular run on $u_0 J_1 \ldots J_n u_n \in L(\mathcal{A}^J)$. Note that the converse also holds: from the word $u_0 J_1 \ldots J_n u_n$, we can replace each J_i by a word $c_i w_i' r_i$, with w_i' any word in $\widetilde{L}(\mathcal{A}^{J_i})$, to obtain a word in the recursive language of $\mathcal{A}^J$.

2.3   Comparison with Other Models

A VRA $\mathcal{A} = \langle \widetilde{\Sigma}, \Sigma_{proc}, \Lambda, \mathcal{A}^S \rangle$ is *deterministic* if all its automata in $\Lambda \cup \{\mathcal{A}^S\}$ are DFAs and, for all $q \in Q_{\mathcal{A}}$, if there exist two transitions $(q, J, p), (q, J', p') \in \delta_{\mathcal{A}}$ with distinct $J, J' \in \Sigma_{proc}$, then $f_{call}(J) \neq f_{call}(J')$. The VRA of Fig. 1a is not deterministic because $r_0 \xrightarrow{T} r_0, r_0 \xrightarrow{R} r_1 \in \delta^R$ and $f_{call}(R) = f_{call}(T)$.

In a deterministic VRA, for all configurations $\langle q, \sigma \rangle \in Q_{\mathcal{A}} \times Q_{\mathcal{A}}^*$ and symbols $a \in \widetilde{\Sigma}$, there exists at most one recursive run on a from $\langle q, \sigma \rangle$. This is clear when $a \in \Sigma_{ret}$, by the semantics of VRAs, and when $a \in \Sigma_{int}$, since all FAs are DFAs. When $a \in \Sigma_{call}$, since there exists at most one procedural symbol $J \in \Sigma_{proc}$ such that $f_{call}(J) = a$ and $(q, J, p) \in \delta_{\mathcal{A}}$, the only reachable configuration is $\langle q_i, p\sigma \rangle$, with $q_i \in I^J$ the unique initial state of $\mathcal{A}^J$. The next proposition states that deterministic VRAs are less expressive than VRAs.

Proposition 2. *There exists no deterministic VRA accepting the recursive language accepted by the VRA depicted in Fig. 1a.*

Proof (Sketch). The main idea is that a deterministic VRA cannot simulate the nondeterministic transitions $r_0 \xrightarrow{T} r_0, r_0 \xrightarrow{R} r_1 \in \delta^R$ without altering the recursive language of $\mathcal{A}^R$. $\qquad\qquad\square$

A particular class of VRAs, called systems of procedural automata, is studied in [18]. It consists of VRAs such that for all distinct $J, J' \in \Sigma_{proc}$, $f_{call}(J) \neq f_{call}(J')$. This class forms a strict subclass of the deterministic VRAs.

Visibly pushdown automata (VPAs) form a subclass of pushdown automata [4]. The next theorem states that VRAs and VPAs are equivalent models.

Theorem 1 (Equivalence of VRAs and VPAs). *Let $L \subseteq WM(\widetilde{\Sigma})$. There exists a VRA $\mathcal{A}$ accepting L iff there exists a VPA $\mathcal{B}$ accepting L. Moreover, there exists a logspace-computable construction for $\mathcal{B}$ with $|\mathcal{B}| = \mathcal{O}(|\mathcal{A}|)$ (resp. for $\mathcal{A}$ with $|\mathcal{A}| = \mathcal{O}(|\mathcal{B}|^4)$).*

3   Codeterministic and Complete VRAs

Proposition 2 states that not all VRAs have an equivalent deterministic VRA. We introduce in this section the notions of *codeterministic* VRA and *complete* VRA, and prove that any VRA can be transformed into a codeterministic and complete one. This property is notably useful to show that the class of VRAs is closed under complement (see Theorem 3 below). Our concept of codeterminism is inspired by the concept of *codeterministic grammars* introduced in [7].

Definition 5 (Codeterministic VRA). *A VRA $\mathcal{A}$ is* codeterministic *if all automata linked to the same call/return symbols have pairwise disjoint languages:*

$$\forall c \in \Sigma_{call}, \forall r \in \Sigma_{ret}, \forall J, J' \in \Sigma_{proc}^{\langle c,r \rangle} : J \neq J' \Rightarrow \widetilde{L}(\mathcal{A}^J) \cap \widetilde{L}(\mathcal{A}^{J'}) = \varnothing.$$

The notion of complete VRA requires two conditions. The first asks all the FAs of the VRA to be complete. The second is a universality condition on the recursive languages. These conditions guarantee that there always exists a recursive run on any well-matched word, regardless of the starting configuration.

Definition 6 (Complete VRA). *A VRA $\mathcal{A}$ is* complete *if*

- *for all $q \in Q_{\mathcal{A}}$ and $a \in \Sigma_{int} \cup \Sigma_{proc}$, there exists $(q, a, p) \in \delta_{\mathcal{A}}$;*
- *for all $c \in \Sigma_{call}$ and $r \in \Sigma_{ret}$: $\bigcup_{J \in \Sigma_{proc}^{\langle c,r \rangle}} \widetilde{L}(\mathcal{A}^J) = WM(\widetilde{\Sigma})$.*

We remark that the second equation in the definition above holds for the partition of Σ_{proc} into the $\Sigma_{proc}^{\langle c,r \rangle}$.

The conditions on recursive languages for a VRA to be codeterministic and complete can be replaced by a condition at the level of regular languages:

Proposition 3. *Let $\mathcal{A}$ be a VRA with all its automata being complete FAs. If, for all $c \in \Sigma_{call}$, $r \in \Sigma_{ret}$, the regular languages $L(\mathcal{A}^J)$, with $J \in \Sigma_{proc}^{\langle c,r \rangle}$, form a partition of $(\Sigma_{int} \cup \Sigma_{proc})^*$, then $\mathcal{A}$ is codeterministic and complete.*

Proof (Sketch). By Proposition 1, since all regular languages $L(\mathcal{A}^J)$, $J \in \Sigma_{proc}^{\langle c,r \rangle}$, are pairwise disjoint, we can see that the recursive languages are pairwise disjoint too, i.e., $\mathcal{A}$ is codeterministic. Additionally, with Proposition 1 again, as the union of all $L(\mathcal{A}^J)$'s is equal to $(\Sigma_{int} \cup \Sigma_{proc})^*$, we can check that the union of the recursive languages is $WM(\widetilde{\Sigma})$. Since all FAs of $\mathcal{A}$ are complete by hypothesis, it follows that $\mathcal{A}$ is complete. $\qquad\square$

We now show that, given any VRA, we can construct an equivalent codeterministic complete VRA with an exponential size in the size of the input VRA.

Theorem 2 (Power of Codeterministic Complete VRAs). *Given a VRA $\mathcal{A}$, one can construct an equivalent codeterministic complete VRA $\mathcal{B}$ such that $|\mathcal{B}| = 2^{\mathcal{O}(|\mathcal{A}|)}$. Moreover, the automata that compose the VRA $\mathcal{B}$ are all DFAs.*

Proof (Sketch). Given a VRA $\mathcal{A} = \langle \widetilde{\Sigma}, \Sigma_{proc}, \Lambda, \mathcal{A}^S \rangle$, we want to construct an equivalent VRA $\mathcal{B} = \langle \widetilde{\Sigma}, \Sigma'_{proc}, \Lambda', \mathcal{B}^S \rangle$ that is codeterministic and complete. Merging Definitions 5 and 6, $\mathcal{B}$ must respect the following conditions: all its automata must be complete FAs and, for all $\langle c, r \rangle \in \Sigma_{call} \times \Sigma_{ret}$, the recursive languages of all $\mathcal{B}^{\mathcal{J}}$, with $\mathcal{J} \in \Sigma'^{\langle c,r \rangle}_{proc}$, must be a partition of $WM(\widetilde{\Sigma})$. The main idea is the following. We define $\Sigma'^{\langle c,r \rangle}_{proc} = 2^{\Sigma^{\langle c,r \rangle}_{proc}}$, leading to the procedural alphabet of $\mathcal{B}$ equal to $\Sigma'_{proc} = \bigcup_{\langle c,r \rangle \in \Sigma_{call} \times \Sigma_{ret}} \Sigma'^{\langle c,r \rangle}_{proc}$. Then, for each $\langle c, r \rangle$, we want to obtain, for all $\mathcal{J} \in \Sigma'^{\langle c,r \rangle}_{proc}$:

$$\widetilde{L}(\mathcal{B}^{\mathcal{J}}) = \bigcap_{J \in \mathcal{J}} \widetilde{L}(\mathcal{A}^J) \setminus \bigcup_{J \in \overline{\mathcal{J}}} \widetilde{L}(\mathcal{A}^J). \tag{1}$$

Recall (see Subsect. 2.1) that $\overline{\mathcal{J}} = \Sigma_{proc}^{\langle c,r \rangle} \setminus \mathcal{J}$ and, when $\mathcal{J} = \varnothing$, $\bigcap_{J \in \mathcal{J}} \widetilde{L}(\mathcal{A}^J) = WM(\widetilde{\Sigma})$. In this way, the recursive languages of all $\mathcal{B}^{\mathcal{J}}$, $\mathcal{J} \in \Sigma_{proc}'^{\langle c,r \rangle}$, form a partition of $WM(\widetilde{\Sigma})$ (see Fig. 3). Note that for each $\langle c, r \rangle$, the set $\varnothing$ belongs to $\Sigma_{proc}'^{\langle c,r \rangle}$, each time corresponding to a distinct automaton.

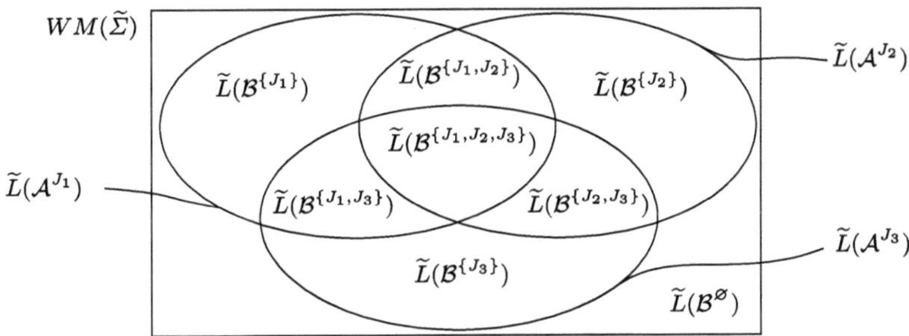

Fig. 3. Set of recursive languages of three automata $\mathcal{A}^{J_1}$, $\mathcal{A}^{J_2}$ and $\mathcal{A}^{J_3}$, and the corresponding set of languages of the automata $\mathcal{B}^{\mathcal{J}}$, for all $\mathcal{J} \subseteq \{J_1, J_2, J_3\}$.

Before detailing the construction of each $\mathcal{B}^{\mathcal{J}}$, we transform each $\mathcal{A}^J \in \Lambda \cup \{\mathcal{A}^S\}$ into a complete DFA $\mathcal{A}'^J$ over the alphabet $\Sigma_{int} \cup \Sigma'_{proc}$ as follows. For all $J \in \Sigma_{proc}$, we replace each procedural transition $q \xrightarrow{J} p \in \delta_{\mathcal{A}}$ by the transitions $q \xrightarrow{\mathcal{J}} p$, for all $\mathcal{J} \in \Sigma'_{proc}$ such that $\mathcal{J} \ni J$, and we then apply the subset construction to get a complete DFA [20]. This first step later helps obtain (1). With this construction, the regular language of $\mathcal{A}'^J$ is equal to the one of $\mathcal{A}^J$, up to the replacement of the procedural symbols appearing in the accepted words:

Property 1. Let $J \in \Sigma_{proc}$, $n \in \mathbb{N}$, $u_i \in \Sigma_{int}^*$ and $\mathcal{J}_i \in \Sigma'_{proc}$ for all i:

$$u_0 \mathcal{J}_1 \ldots \mathcal{J}_n u_n \in L(\mathcal{A}'^J) \iff \forall i \in [1, n], \exists J_i \in \mathcal{J}_i : u_0 J_1 \ldots J_n u_n \in L(\mathcal{A}^J).$$

Proof (of Property 1 - Sketch). This follows from the replacement of the procedural transitions $q \xrightarrow{J} p$ by $q \xrightarrow{\mathcal{J}} p$, for all $\mathcal{J} \in \Sigma'_{proc}$ such that $J \in \mathcal{J}$. ⌟

Since all $\mathcal{A}'^J$, $J \in \Sigma_{proc} \cup \{S\}$, are complete DFAs, they are closed under Boolean operations with well-known constructions [20,27]. For each $\langle c, r \rangle$, we can thus construct an automaton $\mathcal{B}^{\mathcal{J}}$, $\mathcal{J} \in \Sigma_{proc}'^{\langle c,r \rangle}$, such that its regular language respects a form similar to (1):

$$L(\mathcal{B}^{\mathcal{J}}) = \bigcap_{J \in \mathcal{J}} L(\mathcal{A}'^J) \setminus \bigcup_{J \in \overline{\mathcal{J}}} L(\mathcal{A}'^J) = \bigcap_{J \in \mathcal{J}} L(\mathcal{A}'^J) \cap \bigcap_{J \in \overline{\mathcal{J}}} \overline{L(\mathcal{A}'^J)}. \qquad (2)$$

The automaton $\mathcal{B}^{\mathcal{J}}$ is obtained with an appropriate Cartesian product of the FAs $\mathcal{A}'^J$, with $J \in \Sigma_{proc}^{\langle c,r \rangle}$, in a way that $\mathcal{B}^{\mathcal{J}}$ accepts the regular language of (2).

Finally, we construct the required codeterministic and complete VRA $\mathcal{B} = \langle \widetilde{\Sigma}, \Sigma'_{proc}, \Lambda', \mathcal{B}^S \rangle$ such that $\Lambda' = \{\mathcal{B}^{\mathcal{J}} \mid \mathcal{J} \in \Sigma'_{proc}\}$ where each $\mathcal{B}^{\mathcal{J}}$ is obtained as described before, and $\mathcal{B}^S = \mathcal{A}'^S$.

Let us prove that $\mathcal{B}$ is codeterministic and complete. For all $\langle c, r \rangle \in \Sigma_{call} \times \Sigma_{ret}$, according to (2), the regular languages of all DFAs $\mathcal{B}^{\mathcal{J}}$, with $\mathcal{J} \in \Sigma'^{\langle c,r \rangle}_{proc}$, form a partition of $(\Sigma_{int} \cup \Sigma'_{proc})^*$. By Proposition 3, since all automata are complete DFAs, it follows that $\mathcal{B}$ is codeterministic and complete.

We now prove that $\mathcal{B}$ accepts the same language as $\mathcal{A}$. We first prove the correctness of the recursive languages $\widetilde{L}(\mathcal{B}^{\mathcal{J}})$ as exposed in (1), which is a consequence of $\mathcal{B}$ being codeterministic and complete and the next property (see also Fig. 3).

Property 2. For all $J \in \Sigma_{proc}$, $\widetilde{L}(\mathcal{A}^J) = \bigcup_{\mathcal{J} \in \Sigma'_{proc}, \mathcal{J} \ni J} \widetilde{L}(\mathcal{B}^{\mathcal{J}})$.

Proof (of Property 2 - Sketch). This follows from (2), Proposition 1 and Property 1. ⌟

Finally, to show that $\mathcal{A}$ and $\mathcal{B}$ are equivalent, we must prove that for all $w \in WM(\widetilde{\Sigma})$, $w \in \widetilde{L}(\mathcal{A}^S) \Leftrightarrow w \in \widetilde{L}(\mathcal{B}^S)$. Suppose that $w = u_0 c_1 w_1 r_1 \ldots c_n w_n r_n u_n \in WM(\widetilde{\Sigma})$ with $n \in \mathbb{N}$, $u_i \in \Sigma^*_{int}$, $c_i \in \Sigma_{call}$, $r_i \in \Sigma_{ret}$, $w_i \in WM(\widetilde{\Sigma})$ for all i:

$\Rightarrow$ If $w \in \widetilde{L}(\mathcal{A}^S)$, by Proposition 1, we have $u_0 J_1 \ldots J_n u_n \in L(\mathcal{A}^S)$ for some $J_i \in \Sigma^{\langle c_i, r_i \rangle}_{proc}$ such that $w_i \in \widetilde{L}(\mathcal{A}^{J_i})$, for all $i \in [1, n]$. By Property 2, for all i, as $w_i \in \widetilde{L}(\mathcal{A}^{J_i})$, there exists $\mathcal{J}_i \ni J_i$ such that $w_i \in \widetilde{L}(\mathcal{B}^{\mathcal{J}_i})$. Then by Property 1, we have that $u_0 \mathcal{J}_1 \ldots \mathcal{J}_n u_n \in L(\mathcal{A}'^S)$. By Proposition 1, it follows that $w \in \widetilde{L}(\mathcal{A}'^S) = \widetilde{L}(\mathcal{B}^S)$.

$\Leftarrow$ The other implication is proved similarly.

To complete the proof, it remains to study the size of $\mathcal{B}$. The number of states of $\mathcal{B}^S$ is equal to $2^{|Q^S|}$. For each $\langle c, r \rangle \in \Sigma_{call} \times \Sigma_{ret}$, there are $2^{|\Sigma^{\langle c,r \rangle}_{proc}|}$ automata $\mathcal{B}^{\mathcal{J}}$, each with a number of states $\prod_{J \in \Sigma^{\langle c,r \rangle}_{proc}} 2^{|Q^J|} = 2^{\sum |Q^J|}$. Hence,

$$|Q_{\mathcal{B}}| = 2^{|Q^S|} + \sum_{\langle c,r \rangle \in \Sigma_{call} \times \Sigma_{ret}} 2^{|\Sigma^{\langle c,r \rangle}_{proc}|} \cdot 2^{\sum_{J \in \Sigma^{\langle c,r \rangle}_{proc}} |Q^J|} = 2^{\mathcal{O}(|Q_{\mathcal{A}}|)}.$$

Since the number of transitions $|\delta_{\mathcal{B}}|$ of $\mathcal{B}$ is in $\mathcal{O}(|Q_{\mathcal{B}}|^2 \cdot |\Sigma'_{proc}|) = 2^{\mathcal{O}(|\mathcal{A}|)}$ ($|\Sigma_{int}|$ is supposed constant and $|\Sigma'_{proc}| \leq |Q_{\mathcal{B}}|$), we conclude that $|\mathcal{B}| = 2^{\mathcal{O}(|\mathcal{A}|)}$. $\square$

4   Closure Properties and Decision Problems

4.1   Closure Properties of VRAs

As VRAs and VPAs are interreducible (Theorem 1), VRAs inherit the same closure properties as VPAs, which are closed under concatenation, Kleene-*, and Boolean operations [4]. Although correct, translating VRAs into VPAs in order to perform these closure operations leads to a polynomially larger automaton. We therefore provide direct constructions over VRAs, yielding automata with sizes as exposed in Table 1.

Theorem 3 (Closure properties). *Let $\mathcal{A}_1$ and $\mathcal{A}_2$ be VRAs with $\widetilde{L}(\mathcal{A}_1) = L_1$ and $\widetilde{L}(\mathcal{A}_2) = L_2$. One can construct a VRA accepting $L_1 \cdot L_2$, L_1^*, $L_1 \cup L_2$, $L_1 \cap L_2$, and $\overline{L_1}$ with respective sizes in $\mathcal{O}(|\mathcal{A}_1| + |\mathcal{A}_2|)$, $\mathcal{O}(|\mathcal{A}_1|)$, $\mathcal{O}(|\mathcal{A}_1| + |\mathcal{A}_2|)$, $\mathcal{O}(|\mathcal{A}_1| \cdot |\mathcal{A}_2|)$, and $2^{\mathcal{O}(|\mathcal{A}_1|)}$.*

Proof (Sketch). We only provide intuition of the constructions. Let $\mathcal{A}_1 = \langle \widetilde{\Sigma}, \Sigma_{proc1}, \Lambda_1, \mathcal{A}_1^S \rangle$ and $\mathcal{A}_2 = \langle \widetilde{\Sigma}, \Sigma_{proc2}, \Lambda_2, \mathcal{A}_2^S \rangle$. We assume, without loss of generality, that their input alphabets are the same, and that their procedural alphabets are disjoint.

In case of concatenation, Kleene-$*$ and union operations, the constructions are easy, as they only involve the starting automaton. For instance, to obtain a VRA $\mathcal{B}$ accepting $\widetilde{L}(\mathcal{B}) = L_1 \cdot L_2$, we simply copy all automata of Λ_1 and Λ_2 and define the starting automaton $\mathcal{B}^S$ such that it accepts the regular language $L(\mathcal{B}^S) = L(\mathcal{A}_1^S) \cdot L(\mathcal{A}_2^S)$, using the concatenation construction for FAs [20].

For the intersection, we need to compute the intersection of each pair of recursive languages $\widetilde{L}(\mathcal{A}_1^{J_1})$ and $\widetilde{L}(\mathcal{A}_2^{J_2})$ of $\mathcal{A}_1$ and $\mathcal{A}_2$. Intuitively, we define the new procedural symbols $\langle J_1, J_2 \rangle \in \Sigma_{proc1} \times \Sigma_{proc2}$ and we replace all transitions $q_1 \xrightarrow{J_1} p_1 \in \delta_{\mathcal{A}_1}$ and $q_2 \xrightarrow{J_2} p_2 \in \delta_{\mathcal{A}_2}$, respectively by $q_1 \xrightarrow{\langle J_1, J_2 \rangle} p_1$ and $q_2 \xrightarrow{\langle J_1, J_2 \rangle} p_2$. We then construct $\mathcal{B}^{\langle J_1, J_2 \rangle}$ equal to the Cartesian product of $\mathcal{A}_1^{J_1}$ and $\mathcal{A}_2^{J_2}$, such that $\widetilde{L}(\mathcal{B}^{\langle J_1, J_2 \rangle}) = \widetilde{L}(\mathcal{A}_1^{J_1}) \cap \widetilde{L}(\mathcal{A}_2^{J_2})$. As we must construct a Cartesian product for all pairs of procedural symbols, it follows that $|\mathcal{B}| = \mathcal{O}(|\mathcal{A}_1| \cdot |\mathcal{A}_2|)$.

Lastly, for the complementation, if $\mathcal{A}_1$ is codeterministic and complete, with all its automata being complete DFAs, then we construct $\mathcal{B} = \langle \widetilde{\Sigma}, \Sigma_{proc1}, \Lambda_1, \mathcal{B}^S \rangle$, with $\mathcal{B}^S$ accepting the regular language $L(\mathcal{B}^S) = \overline{L(\mathcal{A}_1^S)}$ (i.e., final states of $\mathcal{B}^S$ are the non final states of $\mathcal{A}_1^S$ [20]). If not, we first apply the construction of Theorem 2, and then apply the previous construction. Thus, $|\mathcal{B}| = 2^{\mathcal{O}(|\mathcal{A}_1|)}$. $\qquad\square$

4.2   Decision Problems for VRAs

We here study the complexity of the emptiness, universality, inclusion, and equivalence decision problems. They belong to the same complexity class as for VPAs [4,24], since Theorem 1 states that VRAs and VPAs are equivalent under a logspace reduction. However, using direct algorithms without translations into equivalent VPAs yields lower upper-bounds, as summarized in Table 1.

Theorem 4 (Decision problems for VRAs). *Let $\mathcal{A}_1, \mathcal{A}_2$ be two VRAs. The emptiness decision problem is PTIME-complete, with an upper-bound time complexity in $\mathcal{O}(|\mathcal{A}_1|)$. The universality, inclusion, and equivalence decision problems are EXPTIME-complete, with an upper-bound time complexity respectively in $2^{\mathcal{O}(|\mathcal{A}_1|)}$, $\mathcal{O}(|\mathcal{A}_1|) \cdot 2^{\mathcal{O}(|\mathcal{A}_2|)}$, and $2^{\mathcal{O}(|\mathcal{A}_1| + |\mathcal{A}_2|)}$.*

Proof (Sketch). The complexity class was discussed above. We here propose an algorithm that solves the emptiness decision problem for VRAs. The other decision problems are solved with classical methods and have better complexity than VPA due to the simpler complementation construction (see Theorem 3).

The main idea to solve the emptiness problem is to progressively compute the set of automata in $\Lambda \cup \{\mathcal{A}^S\}$ whose languages are not empty, using a reachability algorithm. Initially, the algorithm starts from all initial states and is limited to the internal transitions. When a final state of an automaton $\mathcal{A}^J \in \Lambda$ is reached, this automaton is marked as having a nonempty recursive language, and the algorithm is then allowed to take procedural transitions reading the symbol J.

The algorithm updates a set $Reach_i \subseteq Q_{\mathcal{A}}$, for $i \in \mathbb{N}$, which contains the states marked as reachable from an initial state. It also uses a set $\mathcal{J}_i \subseteq \Sigma_{proc}$ which contains symbols $J \in \Sigma_{proc}$ such that $Reach_i$ contains a final state of $\mathcal{A}^J$:

- **Initialization:** $Reach_0 = \bigcup_{J \in \Sigma_{proc} \cup \{S\}} I^J$;
- **Main loop:** Let $\mathcal{J}_i = \{J \in \Sigma_{proc} \mid F^J \cap Reach_i \neq \varnothing\}$, then $Reach_{i+1} = Reach_i \cup \{p \in Q_{\mathcal{A}} \mid \exists q \in Reach_i, a \in \Sigma_{int} \cup \mathcal{J}_i : q \xrightarrow{a} p \in \delta_{\mathcal{A}}\}$;
- **Output:** When $Reach_{i+1} = Reach_i$, $\widetilde{L}(\mathcal{A}) = \varnothing$ iff $F^S \cap Reach_i = \varnothing$.   □

5   Conclusion

We studied an extension of a class of modular automata proposed in [18]: visibly recursive automata, in which the modules are FAs. We showed that they are equivalent to VPAs, and provided complexity results about the classical language operations and automata decision problems. In line with this paper, we intend to study the problem of deciding the determinization problem for VRAs.

Our future main goal is the design of a learning algorithm for VRAs (in the Angluin's sense [6]), to obtain readable automata instead of VPAs. This paper is the first step in that direction. The second step is to study the existence of a canonical VRA as Angluin's algorithm requires such a canonical model. We believe that codeterministic and complete VRAs are a promising avenue for this task. Once the VRA learning algorithm is designed, we intend to continue our search for an effective and scalable inference algorithm for JSON schemas to enable efficient validation of JSON documents as initiated in [15].

References

1. Alur, R.: Marrying words and trees. In: Diekert, V., Volkov, M.V., Voronkov, A. (eds.) CSR 2007. LNCS, vol. 4649, pp. 5–5. Springer, Heidelberg (2007). https://doi.org/10.1007/978-3-540-74510-5_3
2. Alur, R., Benedikt, M., Etessami, K., Godefroid, P., Reps, T.W., Yannakakis, M.: Analysis of recursive state machines. ACM Trans. Program. Lang. Syst. **27**(4), 786–818 (2005). https://doi.org/10.1145/1075382.1075387
3. Alur, R., Kumar, V., Madhusudan, P., Viswanathan, M.: Congruences for Visibly Pushdown Languages. In: Caires, L., Italiano, G.F., Monteiro, L., Palamidessi, C., Yung, M. (eds.) ICALP 2005. LNCS, vol. 3580, pp. 1102–1114. Springer, Heidelberg (2005). https://doi.org/10.1007/11523468_89

4. Alur, R., Madhusudan, P.: Visibly pushdown languages. In: Babai, L., (ed.) Proceedings of the 36th Annual ACM Symposium on Theory of Computing, Chicago, IL, USA, 13–16 June 2004, pp. 202–211. ACM (2004). https://doi.org/10.1145/1007352.1007390

5. Alur, R., Madhusudan, P.: Adding nesting structure to words. J. ACM **56**(3), 16:1–16:43 (2009). https://doi.org/10.1145/1516512.1516518

6. Angluin, D.: Learning regular sets from queries and counterexamples. Inf. Comput. **75**(2), 87–106 (1987). https://doi.org/10.1016/0890-5401(87)90052-6

7. Berstel, J., Boasson, L.: Balanced grammars and their languages. In: Brauer, W., Ehrig, H., Karhumäki, J., Salomaa, A. (eds.) Formal and Natural Computing. LNCS, vol. 2300, pp. 3–25. Springer, Heidelberg (2002). https://doi.org/10.1007/3-540-45711-9_1

8. Bouajjani, A., Esparza, J., Maler, O.: Reachability analysis of pushdown automata: application to model-checking. In: Mazurkiewicz, A., Winkowski, J. (eds.) CONCUR 1997. LNCS, vol. 1243, pp. 135–150. Springer, Heidelberg (1997). https://doi.org/10.1007/3-540-63141-0_10

9. Bruyère, V., Pérez, G.A. , Staquet, G.: Validating streaming JSON documents with learned VPAs. In: Sankaranarayanan, S., Sharygina, N. (eds.) TACAS 2023, LNCS. vol. 13993, pp. 271–289. Springer, Cham (2023). https://doi.org/10.1007/978-3-031-30823-9_14

10. Chen, H., Wagner, D.A.: MOPS: an infrastructure for examining security properties of software. In: Atluri, V., (ed.) Proceedings of the 9th ACM Conference on Computer and Communications Security, CCS 2002, Washington, DC, USA, 18–22 November 2002, pp. 235–244. ACM (2002). https://doi.org/10.1145/586110.586142

11. Chiari, M., Mandrioli, D., Pontiggia, F., Pradella, M.: A model checker for operator precedence languages. ACM Trans. Program. Lang. Syst. **45**(3), 19:1–19:66 (2023). https://doi.org/10.1145/3608443

12. Chiari, M., Mandrioli, D., Pradella, M.: Model-Checking Structured Context-Free Languages. In: Silva, A., Leino, K.R.M. (eds.) CAV 2021. LNCS, vol. 12760, pp. 387–410. Springer, Cham (2021). https://doi.org/10.1007/978-3-030-81688-9_18

13. Dubrulle, K.: Visibly systems of procedural automata validating JSON documents. https://github.com/BlueTorche/VSPA

14. Dubrulle, K., Bruyère, V., Pérez, G.A., Staquet, G.: Visibly recursive automata. CoRR, abs/ arXiv: 2603.11648 (2026)

15. Dubrulle, K.: Mutually recursive procedural systems - streaming validation of JSON documents. Master's thesis, UMONS - Université de Mons [Faculté Polytechnique de Mons], Mons, Belgium (2025)

16. Esparza, J., Kučera, A., Schwoon, S.: Model-checking LTL with regular valuations for pushdown systems. In: Kobayashi, N., Pierce, B.C. (eds.) TACS 2001. LNCS, vol. 2215, pp. 316–339. Springer, Heidelberg (2001). https://doi.org/10.1007/3-540-45500-0_16

17. Fortz, S., et al.: A research agenda for active automata learning. Int. J. Softw. Tools Technol. Transfer (2026). https://doi.org/10.1007/s10009-026-00839-z

18. Frohme, M., Steffen, B.: Compositional learning of mutually recursive procedural systems. Int. J. Softw. Tools Technol. Transf. **23**(4), 521–543 (2021). https://doi.org/10.1007/S10009-021-00634-Y

19. Gallier, J.H., La Torre, S., Mukhopadhyay, S.: Deterministic finite automata with recursive calls and DPDAs. Inf. Process. Lett. **87**(4), 187–193 (2003). https://doi.org/10.1016/S0020-0190(03)00281-3

20. Hopcroft, J.E., Ullman, J.D.: Introduction to Automata Theory. Addison-Wesley, Languages and Computation (1979)
21. Isberner, M.: Foundations of active automata learning: an algorithmic perspective. PhD thesis, Technical University Dortmund, Germany (2015). https://hdl.handle.net/2003/34282
22. Jia, X., Tan, G.: V-Star: Learning visibly pushdown grammars from program inputs. Proc. ACM Program. Lang. **8**(PLDI), 2003–2026 (2024). https://doi.org/10.1145/3656458
23. Kumar, V., Madhusudan, P., Viswanathan, M.: Visibly pushdown automata for streaming XML. In: Williamson, C.L., Zurko, M.E., Patel-Schneider, P.F., Shenoy, P.J. (eds), Proceedings of the 16th International Conference on World Wide Web, WWW 2007, Banff, Alberta, Canada, 8–12 May 2007, pp. 1053–1062. ACM (2007). https://doi.org/10.1145/1242572.1242714
24. Lange, M.: P-hardness of the emptiness problem for visibly pushdown languages. Inf. Process. Lett. **111**(7), 338–341 (2011). https://doi.org/10.1016/J.IPL.2010.12.013
25. Le Glaunec, A., Li, A.W., Mamouras, K.: Streaming validation of JSON documents against schemas. Proc. VLDB Endow. **19**(3), 509–522 (2025)
26. Nguyen, H.-V., Touili, T.: CARET model checking for pushdown systems. In: Seffah, A., Penzenstadler, B., Alves, C., Peng, X., (eds.) Proceedings of the Symposium on Applied Computing, SAC 2017, Marrakech, Morocco, 3–7 April 2017, pp. 1393–1400. ACM (2017). https://doi.org/10.1145/3019612.3019829
27. Sipser, M.: Introduction to the theory of computation. PWS Publishing Company (1997)
28. Song, F., Touili, T.: Pushdown model checking for malware detection. Int. J. Softw. Tools Technol. Transf. **16**(2), 147–173 (2014). https://doi.org/10.1007/S10009-013-0290-1
29. Woods, W.A.: Transition network grammars for natural language analysis. Commun. ACM **13**(10), 591–606 (1970). https://doi.org/10.1145/355598.362773

On Languages Describing Large Graph Classes

Henning Fernau[1]($\boxtimes$) (iD), Pamela Fleischmann[2] (iD), Kevin Mann[1] (iD), and Silas Cato Sacher[1] (iD)

[1] Trier University, Trier, Germany
`fernau@uni-trier.de` , `mann@uni-trier.de` , `sacher@uni-trier.de`
[2] Kiel University, Kiel, Germany
`fpa@informatik.uni-kiel.de`

Abstract. In this work, we introduce a new notion for representing graph classes with formal languages. In contrast to the seminal work by Kitaev and Pyatkin to represent graphs by words, we use formal binary languages in order to have a set of patterns (given by the languages' words) defining the edges in the graph. In particular, we investigate famous languages like the palindromes, copy-words, Lyndon words and Dyck words to represent all graphs or specific graph classes by restricting these languages.

Keywords: Generalized word-representability · Graph classes · Word-representable graphs · Comparability graphs · k-colorable graphs · Palindromes · Dyck-language · Copy-language · Lyndon words

1 Introduction

Graphs are a fundamental theoretical notion for describing relations between anything imaginable. Thus, the question how to store efficiently formally given graphs in computers is as old as computer science itself, and the easiest way is the adjacency matrix which needs quadratic space in the number of the graph's nodes. In 2008 Kitaev and Pyatkin introduced the notion of word-representability [13] and even though it was meant to describe the Perkins semigroup, it also works as an elegant way to store a graph: the graph's nodes are the word's letters and there is an edge between the nodes u and v iff the letters u and v alternate in the representing word, e.g., the *hour-glass graph* on five nodes $V = \{1, 2, 3, 4, 5\}$ where $1, 2, 3$ and $3, 4, 5$ build two triangles can be represented by 124534512. Whereas every graph can be represented by an adjacency matrix, not all graphs are representable by words in this way (cf. the wheel graph W_5 [12, 13]). Moreover, it is known to be NP-hard to decide whether a graph is representable or not. Such results motivate research into several directions including the one of altering the definition in order to represent all graphs.

Several variations and generalizations of word-representability have been described in the literature. The alternation of letters a and b in a word's projection onto $\{\mathsf{a}, \mathsf{b}\}$ equivalent to the avoidance of the factors aa and bb. This leads to

M.-P. Béal and P. Caron (Eds.): DLT 2026, LNCS 16578, pp. 209–222, 2026.
https://doi.org/10.1007/978-3-032-28404-4_16

the generalization via pattern avoiding words in [11]. For any pattern $u \in \{1,2\}^*$ such that $|u| \geq 2$, a graph $G = (V, E)$ is u-representable iff $\mathrm{alph}(w) = V$ and $\{\mathsf{a},\mathsf{b}\} \in E$ iff w avoids the pattern u (cf. [11, Def. 2] for a formal definition). However, only in the case $u \in \{1\}^*$, their setting can be seen as defining abstract graphs and is hence comparable to classical word-representable graphs. Remarkably, for $|u| \geq 3$ every graph is u-representable. In particular, for $k \geq 3$, every abstract graph is 1^k-representable. The *k-11-representability* generalizes this notion: a graph $G = (V, E)$ is *k-11-represented* by a word $w \in V^*$ if, for all distinct vertices $\mathsf{a}, \mathsf{b} \in V$, at most k times the patterns (factors) aa and bb occur in w's projection onto $\{\mathsf{a}, \mathsf{b}\}$. By definition, the 0-11-representable graphs are the classical word-representable graphs, while it can be shown that the 2-11-representable graphs are the class of all graphs [3]. A different approach is the *permutation representability* as defined in [14]: here additional conditions on the words are given that can be used to represent graphs, not only on the patterns that define edges. Lozin [15] discussed several representations of graphs with finite automata. Another completely different way to connect the theory of regular languages with basic notions of graph theory, e.g., with bounded treewidth, was proposed in [5]. In [9] it is shown that the language containing all words representing a specific graph is regular. However, there only one graph is investigated rather than an entire graph class. Yet, there are other ways to connect words to graphs, and hence languages to graph classes. The notion of *letter graphs* introduced by Petkovšek in [16] gives rise to an infinite chain of graph classes where the words are simply restricted by the alphabet's size; in this context, this yields the graph parameter lettericity that recently obtained a certain popularity, cf. [1] and related papers. A generalization of lettericity is contained in [7].

Own Contributions. Our approach seeks to generalize classical word-representability by allowing for a broad variety of patterns in contrast to allowing or avoiding just one. As with classical word-representable graphs, the graph's nodes are the word's letters but there is an edge between the nodes u and v iff the pattern of u and v matches a word from a given binary language L, i.e., $h_{u,v}$ projects the word to $\{u, v\}$, maps u to 0, v to 1 and checks whether the resulting word is in L. For example, $h_{u,v}(uuxvx) = 001$. The language $L_{\mathrm{cl}} := (1 \cup \lambda) \cdot (01)^* \cdot (0 \cup \lambda)$ (λ is the empty word) is modeling classical word representability. Moreover, graph representation via pattern avoiding words, yields exactly the graphs characterized by $L_{\neg u} = \overline{\{0,1\}^*\{h_{1,2}(u)\}\{0,1\}^*} \cap \overline{\{0,1\}^*\{h_{2,1}(u)\}\{0,1\}^*}$ for a given pattern $u \in \{1,2\}^*$ with $|u| \geq 2$. Thus, language-representability can be understood as a generalization of graph representation via pattern-avoiding words. In particular, $L_{\neg 1^k}$ describes the class of all graphs for each $k \geq 3$. The k-11-representable graphs can be modeled in our approach as well.

In this work, we study the graph classes described by binary versions of famous languages such as the palindromes, the copy-language, the Dyck-language or Lyndon words. All of those, except for the Dyck-language, represent the class of all graphs and yield a representation that is as good as other implicit graph representations. The graphs described by the Dyck-language describe exactly the comparability graphs. We also show that we obtain specific graph classes if we intersect the aforementioned languages with other languages, e.g., the language for the classical word-representability.

2 Preliminaries: Fixing General Notions and Notation

In this section we introduce all notations and notions we need for our results and the framework in general. Let $\mathbb{N}$ denote the natural numbers including 0, let $[n] = \{i \in \mathbb{N} \mid 1 \leq i \leq n\}$ and $\mathbb{N}_{\geq n} = \{k \in \mathbb{N} \mid k \geq n\}$ for some $n \in \mathbb{N}$. For a set X and $k \in \mathbb{N}$, define $\binom{X}{k} = \{Y \subseteq X \mid |Y| = k\}$.

Language Theory and Combinatorics on Words. An *alphabet* Σ is a non-empty, finite set with *letters* as elements. A *word* over Σ is a finite *concatenation* of letters from Σ; Σ^* denotes the set of all words over Σ, including the *empty word* λ. The *length of a word* w, i.e., the number of its letters, is denoted by $|w|$. Let $\Sigma^+ = \Sigma^* \setminus \{\lambda\}$. For $w \in \Sigma^*$ and for all $i \in [|w|]$, $w[i]$ denotes the ith letter of w. The number of occurrences of $\mathsf{a} \in \Sigma$ in $w \in \Sigma^*$ is defined as $|w|_{\mathsf{a}} = |\{i \in [|w|] \mid w[i] = \mathsf{a}\}|$ and w's alphabet is given by $\mathrm{alph}(w) = \{\mathsf{a} \in \Sigma \mid \exists i \in [|w|] : w[i] = \mathsf{a}\}$ as the symbols that occur in w. Define $F(w) = \{|w|_a \mid a \in \Sigma\}$ as the *frequentnesses*. A word $w \in \Sigma^*$ is called *k-uniform* for some $k \in \mathbb{N}$ if $|w|_{\mathsf{a}} = k$ for all $\mathsf{a} \in \Sigma$.

Define the *reversal* w^R of $w \in \Sigma^*$ by $w^R = w[|w|] \cdot w[|w| - 1] \cdots w[1]$; w is a *palindrome* if $w = w^R$ holds. The word $w \in \Sigma^*$ is called a *repetition* if there exist $u \in \Sigma^*$ and $k \in \mathbb{N}_{\geq 2}$ such that $w = u^k$ where $u^0 = \lambda$ and $u^k = uu^{k-1}$. Repetitions with $k = 2$ are called *copy-words* and words that are not a repetition are called *primitive*. We also use the *shuffle product* to combine two words: for $u, v \in \Sigma^*$, define the *shuffle* by $u \sqcup\!\!\sqcup v := \{x_1 y_1 x_2 y_2 \cdots x_n y_n \mid \exists x_1, \ldots, x_n, y_1, \ldots, y_n \in \Sigma^* : u = x_1 \cdots x_n \wedge v = y_1 \cdots y_n\}$.

One very famous set of words are the *Lyndon words*. Given a linear ordering $\prec$ on Σ, we extend it to the linear *lexicographical order* on Σ^*, again denoted by $\prec$ and defined by $u \prec w$ iff either u is a *prefix* of w, i.e., $w = ux$ for some $x \in \Sigma^*$, or $w = x\mathsf{b}y_1$, $u = x\mathsf{a}y_2$ for letters $\mathsf{a} \prec \mathsf{b}$ and $x, y_1, y_2 \in \Sigma^*$. The second required basis for Lyndon words is the notion of *conjugates*: the words uv and vu (for $u, v \in \Sigma^*$) are known as *conjugates*. Thus, Σ^* is partitioned in *conjugacy classes*. Finally, a primitive word is a *Lyndon word* if it is the lexicographically smallest element in its conjugacy class. If we want to make the ordering explicit, we also write $\prec$-Lyndon word, or $\prec$-smaller, etc. A mapping f from the free monoid Σ^* into another monoid is called a *morphism* if $f(xy) = f(x)f(y)$ holds for all $x, y \in \Sigma^*$. Note that a morphism is uniquely defined by giving the images of all letters. For $A \subseteq \Sigma$, the *projective morphism* $h_A : \Sigma^* \to A^*$ is defined by $h_A(\mathsf{a}) = \mathsf{a}$ for $\mathsf{a} \in A$ and $h_A(\mathsf{a}) = \lambda$ otherwise. Let $\widetilde{} : \{0,1\}^* \to \{0,1\}^*$ be the *complement morphism* mapping 0 to 1 and 1 to 0 which is an involution, i.e., $\widetilde{\widetilde{w}} = w$.

Each $L \subseteq \Sigma^*$ is called a *language* over Σ. We extend the concatenation to languages, so that we can define powers of a language L and the *Kleene* *star* of L as $L^* = \bigcup_{n \in \mathbb{N}} L^n$ (analogously, we define other operations on words for entire languages). We define the *symmetric hull* operator $\langle L \rangle := L \cup \widetilde{L}$. A language $L \subseteq \{0,1\}^*$ is called 0-1-*symmetric* if $L = \widetilde{L}$ (i.e., $L = \langle L \rangle$). For instance, the language $L_{\mathrm{cl}} = (1 \cup \lambda)(01)^*(0 \cup \lambda)$ is 0-1-symmetric. This also holds for its complement $\overline{L_{\mathrm{cl}}}$ where $\overline{L}$ denotes $\Sigma^* \backslash L$ for some language $L \subseteq \Sigma^*$. Let $\mathrm{freq}(L) = \{n \in \mathbb{N}_{\geq 1} \mid \exists w \in L : |w|_0 = n\}$ denote the set of frequentnesses in a 0-1-symmetric language $L \subseteq \{0,1\}^*$. The language $\mathcal{D} = \{w \in \{0,1\}^* \mid |w|_0 = |w|_1 \wedge \forall i \in [|w|] : |w[1..i]|_0 \leq |w[1..i]|_1\}$ is the restricted Dyck-language with one pair of parentheses with 0 for the opening and 1 for the closing bracket (cf. $D_1'^*$ in [2]). More generally, $D_k'^*$ is the set of correctly parenthesized words formed with k different types of parentheses, forming an alphabet Σ of size $2k$. Identifying singleton sets with their elements, we can build *regular expressions* from letters by using concatenation $\cdot$, union $\cup$, and Kleene star. We also include set complementation $\bar{}$ and the *shuffle* $\sqcup\!\sqcup$ when building such expressions.

Graph Theory. Throughout this paper, we only consider undirected finite graphs. A graph G is a pair (V, E) with the finite, non-empty set of vertices V and the set of edges $E \subseteq \binom{V}{2}$. For a given graph G, let $V(G)$ and $E(G)$ denote its sets of vertices and edges respectively. The cardinality of V is called the *order* of G. If $\{u, v\} \in E$, u is called a *neighbor* of v, and $N(v) \subseteq V$ are the neighbors of v; $N[v] := N(v) \cup \{v\}$ is the *closed neighborhood* of v. The *degree* of $v \in V$ is $\deg(v) := |N(v)|$. The *complement* of the graph G, written $\overline{G}$, satisfies $\overline{G} = (V, \binom{V}{2} \setminus E)$. For a graph class $\mathcal{G}$, co-$\mathcal{G} := \{\overline{G} \mid G \in \mathcal{G}\}$. If $G_1 = (V_1, E_1)$ and $G_2 = (V_2, E_2)$ are graphs, then $\varphi : V_1 \to V_2$ is a *graph morphism* iff $\{u, v\} \in E_1$ implies $\{\varphi(u), \varphi(v)\} \in E_2$; a bijection φ is a *graph isomorphism* iff $\{u, v\} \in E_1 \iff \{\varphi(u), \varphi(v)\} \in E_2$; then we write $G_1 \simeq G_2$. We call a graph with $n \in \mathbb{N}$ vertices and an empty edge set a *null graph*, denoted by N_n, and a graph with n vertices and all possible edges a *complete graph*, denoted by K_n, i.e. $\overline{N_n} = K_n$. A graph with $n + m$ vertices, with $n, m \in \mathbb{N}_{\geq 1}$ is called a *complete bipartite graph* $K_{n,m} = (V, E)$ if $V = U \cup W$, $U \cap W = \emptyset$, with $|U| = n$ and $|W| = m$ and E contains all edges between U and W but no other edges. A graph with $n \in \mathbb{N}$ vertices is a *path* $P_n = (V, E)$ if there is a linear ordering $<$ on its vertex set such that $\{u, v\} \in E$ if either u is the immediate predecessor of v in $<$ or u is the immediate successor of v in $<$. A graph with $n \in \mathbb{N}$ vertices is a *cycle* $C_n = (V, E)$ if it contains a path $P_n = (V, E')$ and one additional edge e that connects the two vertices of this path that have degree one. A vertex in a graph is called *isolated* if $N(v) = \emptyset$. If $G_1 = (V_1, E_1)$ and $G_2 = (V_2, E_2)$ are two graphs with disjoint sets of vertices V_1 and V_2, then the (graph) *union* of G_1 and G_2 is $G_1 \cup G_2 = (V_1 \cup V_2, E_1 \cup E_2)$. Let $G = (V, E)$ be a graph with a partition $V = V_1 \cup V_2$. If $\binom{V_1}{2} \cap E = \emptyset$ and $\binom{V_2}{2} \cap E = \emptyset$, then G is called *bipartite*. If $\binom{V_1}{2} \cap E = \emptyset$ and $\binom{V_2}{2} \cap E = \binom{V_2}{2}$, then G is called a *split graph*. If $\binom{V_1}{2} \cap E = \binom{V_1}{2}$ and $\binom{V_2}{2} \cap E = \binom{V_2}{2}$, then G is called *cobipartite*. k-colorability generalizes bipartiteness by partitioning the vertex set into k independent sets instead of two. If there exists a partial order on the vertex set V of $G = (V, E)$

such that $u, v \in V$ are incomparable if and only if $\{u, v\} \in E$, then G is a *comparability graph*.

Our Main Notions. After establishing the basics from language and graph theory, we are now able to introduce the main notions for our setting. We start with a definition, which allows us to compare words over an alphabet of a graph's nodes with a binary language. This new morphism $h_{\mathsf{a,b}}$ can be viewed as the composition of the projection $h_{\{u,v\}}$ and the renaming isomorphism $\iota : \{u, v\}^* \to \{0,1\}^*$ with $u \mapsto 0$ and $v \mapsto 1$ for some given nodes u, v. For example, if we have $\mathsf{a, b, n} \in V$ and $\mathsf{banana} \in V^*$, we get $h_{\mathsf{a,b}}(\mathsf{banana}) = 1000$. If we have 0-1-symmetric languages, it is unimportant whether u or v is mapped to 0 or 1 resp.; it only influences later the names of the represented nodes.

Definition 1. *Let V be an alphabet and fix $u, v \in V$ with $u \neq v$. Define $h_{u,v} : V^* \to \{0,1\}^*$ by $u \mapsto 0$, $v \mapsto 1$ and $x \mapsto \lambda$ for $x \in V \setminus \{u, v\}$.*

Based on $h_{u,v}$, we introduce the notion of L-representability of graphs.

Definition 2. *Let $L \subseteq \{0,1\}^*$ be 0-1-symmetric. Let $w \in V^*$ with $V = \mathrm{alph}(w)$. We call a graph $G = (V, E)$ L-represented by $w \in V^*$ iff, for all $u, v \in V$ with $u \neq v$, we have $\{u, v\} \in E \Leftrightarrow h_{u,v}(w) \in L$. A graph $G = (V, E)$ is L-representable if it can be L-represented by some word $w \in V^*$. Let $G(L, w)$ denote the graph G that is L-represented by $w \in \Sigma^*$ and let $\mathcal{G}_L = \{G(L, w) \mid w \in \Sigma^*, \Sigma = \mathrm{alph}(w)\}$.*

As 0-1-symmetry is essential for $G(L, w)$ and $\mathcal{G}_L$ being well-defined, we will assume this condition for any binary language used for defining graphs and graph classes in the following. We finish this section with some immediate insights and examples. Since $h_{u,v} \in L$ iff $h_{u,v} \notin \overline{L}$ for every language L and vertices u, v, we have $G(L, w) = \overline{G(\overline{L}, w)}$ and we get:

Lemma 3. *For each $L \subseteq \{0,1\}^*$, $\mathcal{G}_{\overline{L}} = co\text{-}\mathcal{G}_L$.*

Lemma 4. *Let V be an alphabet and $w \in V^*$. If $0^k \sqcup\!\sqcup 1^\ell \subseteq L \subseteq \{0,1\}^*$ for all $k, \ell \in F(w)$, then $K_{|V|}$ is L-represented by w.*

Lemma 5. *A graph G can be L-represented iff it can be L^R-represented.*

Corollary 6. *If $L \subseteq \{0,1\}^*$ satisfies $L = L^R$, then $G(L, w) = G(L, w^R)$ for each word w.*

Intuitively, one may assume that G is L-representable iff G is $(L \cup L^R)$-representable but consider given $L = \overline{\langle 0001 \rangle}$ and $w = \mathsf{aaab}$, we have $G(L, w) = (\{\mathsf{a, b}\}, \emptyset)$, $L \cup L^R = \{0,1\}^*$, and $\mathcal{G}_{L \cup L^R}$ is the class of complete graphs and $G \notin \mathcal{G}_{L \cup L^R}$. We finish this section by briefly looking into 0-1-symmetric languages L that are closed under reversal ($L = L^R$), e.g., L_{cl} is closed under reversal. Note that L_{cl} equals $L_{\neg 1^k} = \overline{\{0,1\}^*\{0^k\}\{0,1\}^*} \cap \overline{\{0,1\}^*\{1^k\}\{0,1\}^*}$ for $k = 2$. By [11], we know that $\mathcal{G}_{L_{\neg 1^k}}$ corresponds to the family of graphs called 1^k-representable. There it is also shown that, for every $k \geq 3$, *every* graph is 1^k-representable., i.e., $\mathcal{G}_{L_{\neg 1^k}}$ is the class of all (undirected) graphs. For instance, we have for $k = 3$, $G(L_{\neg 1^3}, w) = ([4], \{\{1, 2\}, \{2, 3\}, \{3, 4\}, \{4, 1\}, \{2, 4\}\})$ with the word $w = 1124112341234$. More representations of the cycle C_4 can be found in Fig. 1.

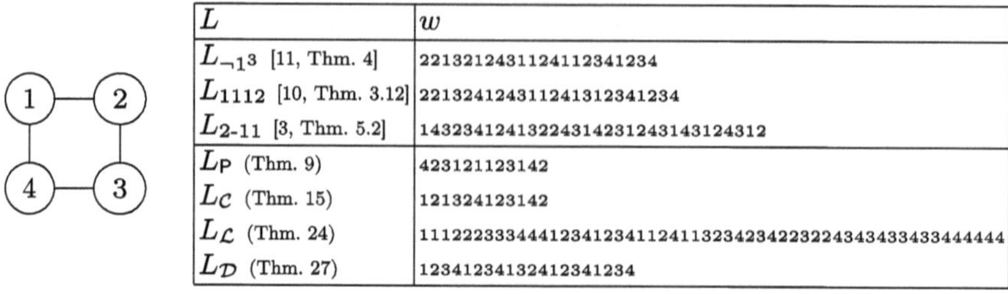

L	w
$L_{\neg 1}3$ [11, Thm. 4]	221321243112411234
L_{1112} [10, Thm. 3.12]	22132412431124131234
$L_{2\text{-}11}$ [3, Thm. 5.2]	1432341241322431423124314312
L_{P} (Thm. 9)	423121123142
$L_{\mathcal{C}}$ (Thm. 15)	121324123142
$L_{\mathcal{L}}$ (Thm. 24)	1112223334441234123411241132342342232243434334334444444
$L_{\mathcal{D}}$ (Thm. 27)	123412341324123441234

Fig. 1. Language L represents C_4 with the word w.

3 Representing Graphs with Famous Languages

In this section, we turn our attention to rather 'famous' (infinite, non-regular) languages as the set of palindromes, the-copy language, the Dyck-language or the set of Lyndon words and look at their associated graph classes. Thus, in all paragraphs we start with the associated binary 0-1-symmetric languages. We will see that most of these languages over the arbitrary alphabets allow to describe all graphs in a rather efficient way. Illustrations of these representations and the underlying constructions can be found in Fig. 1. Some proofs were omitted due to space restrictions; this is marked by $(*)$.

Palindromes. Set $L_{\mathcal{P}} := \{u \in \{0,1\}^+ \mid u = u^R\}$. This language can be restricted to the deterministic context-free version of the binary palindrome language $L_{\text{detP}} := \{h_{\text{double}}(w)01(h_{\text{double}}(w))^R, h_{\text{double}}(w)10(h_{\text{double}}(w))^R \mid w \in \{0,1\}^*\}$ with the morphism $h_{\text{double}} : \{0,1\} \to \{0,1\}$ such that $h(\mathsf{a}) = \mathsf{aa}$ for all $\mathsf{a} \in \{0,1\}$. Motivated by [9], we also define $\mathcal{G}_\star = \{K_{1,n} \cup N_m \mid n,m \in \mathbb{N}\}$ as the set of all extended star graphs. Regarding palindromes in general, notice that $h_{\mathsf{a},\mathsf{b}}(w^R) = h_{\mathsf{a},\mathsf{b}}(w)^R$ for each palindrome $w \in \Sigma^*$ and all $\mathsf{a}, \mathsf{b} \in \Sigma$. Our first lemma shows that being a palindrome can be characterized by the palindrome property of all projections to binary sub-alphabets. Besides being a nice word-combinatorial property, it has a graph-theoretic interpretation, see Corollary 8.

Lemma 7. $(*)$ *A word $w \in \Sigma^*$ is a palindrome iff, for all $\mathsf{a}, \mathsf{b} \in \Sigma$ with $\mathsf{a} \neq \mathsf{b}$, $h_{\mathsf{a},\mathsf{b}}(w)$ is a palindrome.*

The previous lemma links the set of all palindromes to $L_{\mathcal{P}}$ and we show that the (deterministic) palindrome languages represents all graphs.

Corollary 8. *A word w is a palindrome iff $G(L_{\mathcal{P}}, w) \simeq K_{|\operatorname{alph}(w)|}$.*

Theorem 9. *The binary palindrome language $L_{\mathcal{P}}$ represent every graph.*

Proof. Let $G = ([n], E)$ be a graph with $n \in \mathbb{N}$. Define, for $i \in [n]$, $u_i \in [i-1]^*$ as the word which enumerates $\overline{N(i)} \cap [i-1]$ in linear order as well as $w_1 := 11$ and $w_i := iu_iw_{i-1}iu_i^R$ if $i \in \mathbb{N}_{\geq 2}$. We want to show that $h_{j,k}(w_i) = h_{j,k}(w_i)^R$ iff $\{j,k\} \in E$ for all $i \in [n]$ and $j,k \in [i]$ with $j \neq k$. For $i = 1$ this

trivially holds. So let $i > 1$. Since $h_{j,k}$ is a morphism, $h_{j,k}(iu_iw_{i-1}iu_i^R)^R =$
$h_{j,k}(u_i^R)^R h_{j,k}(i)^R h_{j,k}(w_{i-1})^R h_{j,k}(u_i^R)h_{j,k}(i)^R = h_{j,k}(u_i iw_{i-1}^R u_i^R i)$.

First we assume $i \notin \{j,k\}$. Then $h_{j,k}(w_i) = h_{j,k}(u_i)h_{j,k}(w_{i-1})h_{j,k}(u_i^R)$. This
equals $h_{j,k}(w_i^R) = h_{j,k}(u_i)h_{j,k}(w_{i-1})^R h_{j,k}(u_i^R)$ iff $h_{j,k}(w_{i-1})^R = h_{j,k}(w_{i-1})$. By
induction, we know this is the case iff $\{j,k\} \in E$. Now we can assume $j = i$. If
$\{i,k\} \notin E$, then $h_{i,k}(w_i) = h_{i,k}(u_i) \cdot h_{i,k}(i) \cdot h_{i,k}(w_{i-1}) \cdot h_{i,k}(u_i^R) \cdot h_{i,k}(i) =$
$1 \cdot 0 \cdot h_{i,k}(w_{i-1}) \cdot 1 \cdot 0$ as $k < i$, which is no palindrome. For $\{i,k\} \in E$,
$h_{i,k}(w_i) = 0h_{i,k}(w_{i-1})0$. This is a palindrome as $h_{i,k}(w_{i-1}) \in 1^+$. By induction, $G \simeq G(L_{\mathcal{P}}, w_n)$ follows. $\qquad\square$

Theorem 10. $\mathcal{G}_{L_{\mathrm{detP}}}$ *is the class of all graphs.*

Proof. Let $G = ([n], E)$ be a graph with $n \in \mathbb{N}$. Define $\langle M \rangle$ as the words that
enumerate the vertices of the set $M \subseteq [n]$ in linear order. Define, for $i \in [n]$,
$u_i := h_{\mathrm{double}}(\langle \overline{N(i)} \cap [i-1] \rangle)$ and
$\qquad w_i := 11\langle [n] \rangle$ if $i = 1$ and $w_i := iiu_iw_{i-1}iiu_i^R$ if $i \neq 1$. We show $G(L_{\mathrm{detP}}, w_n)$
analogous to the proof of Theorem 9. $\qquad\square$

By [9, Theorem 28], the non-empty palindromes represent exactly
the extended star graphs in the following sense: $\mathcal{G}_\star = \{G(L_{\mathrm{cl}}, w) \mid$
w is a palindrome$\}$. This raises the question if we can combine the classical
word-representability with our setting in the sense of intersecting $L_{\mathcal{P}}$ and L_{cl} to
obtain the class of extended star graphs. The following theorem shows that in
fact, we have $\mathcal{G}_\star \subsetneq \mathcal{G}_{L_{\mathcal{P}} \cap L_{\mathrm{cl}}}$, since the latter one is the class of bipartite graphs.

Theorem 11. $(\ast)$ $\mathcal{G}_{L_{\mathcal{P}} \cap L_{\mathrm{cl}}}$ *is the class of bipartite graphs.*

Since Theorem 11 shows that intersecting the palindrome language with the
language for the classical word-representability may lead to interesting (even on
first sight unexpected) results, we finish the paragraph about palindromes with
two other combinations of languages.

Corollary 12. $(\ast)$ *1.* $\mathcal{G}_{(L_{\mathcal{P}} \cap L_{\mathrm{cl}}) \cup ((0^2)^+ \sqcup (1^2)^+)}$ *is the class of split graphs.*
2. $\mathcal{G}_{\overline{L_{\mathcal{P}}} \cup \overline{L_{\mathrm{cl}}}}$ *is the class of cobipartite graphs.*

Copy-Language. Let $L_{\mathcal{C}} = \{ww \mid w \in \{0,1\}^*\}$ denote the binary copy-
language which is 0-1-symmetric. Apart from the fact that the language of
palindromes is context-free, while the copy-language is not, both languages enjoy
quite similar properties. We exemplify this by an analogue to Lemma 7. We follow a presentation similar to the one on palindromes, starting with a morphic
characterization.

Lemma 13. $(\ast)$ *For every alphabet Σ such that $|\Sigma| \geq 2$, a word $w \in \Sigma^*$ is a*
copy-word iff, for all $\mathsf{a}, \mathsf{b} \in \Sigma$ with $\mathsf{a} \neq \mathsf{b}$, $h_{\mathsf{a},\mathsf{b}}(w)$ is a copy-word.

Similarly to the palindromes, we can infer from Lemma 13 that we again
obtain the complete graphs and in general that $_c$ represents all words.

Corollary 14. *A word w is a copy-word iff $G(L_{\mathcal{C}}, w) \simeq K_{|\operatorname{alph}(w)|}$.*

Theorem 15. *The binary copy-language $L_{\mathcal{C}}$ represent every graph.*

Proof. Let $G = ([n], E)$ be a graph. Define $w = u_1 1 u_2 2 \cdots u_n n 1 u_1 \cdots n u_n$ where u_i enumerates each vertex of $\overline{N(i)} \cap [i]$ exactly once in the natural order for $i \in [n]$. Consider $G' = ([n], E') = G(L_{\mathcal{C}}, w)$. Let $i, j \in [n]$ with $i < j$. By definition, $h_{i,j}(u_i) = \lambda$. Also, $h_{i,j}(u_j) \in \{\lambda, 0\}$ with $h_{i,j}(u_j) = \lambda$ being equivalent to $i \in N(j)$. For $k < i$, $h_{i,j}(u_k) = \lambda$. Hence, $h_{i,j}(w) = w_1 w_2$ with

$$w_1 := h_{i,j}(u_i) 0 h_{i,j}(u_{i+1}) \cdots h_{i,j}(u_j) 1 h_{i,j}(u_{j+1}) \cdots h_{i,j}(u_n) \text{ and}$$
$$w_2 := 0 h_{i,j}(u_i) h_{i,j}(u_{i+1}) \cdots 1 h_{i,j}(u_j) h_{i,j}(u_{j+1}) \cdots h_{i,j}(u_n).$$

It is easy to see that $|w_1| = |w_2|$. Furthermore, $w_1 = w_2$ holds iff $1 h_{i,j}(u_j) = h_{i,j}(u_j) 1$. Since $h_{i,j}(u_j) \in \{\lambda, 0\}$, $1 h_{i,j}(u_j) = h_{i,j}(u_j) 1$ iff $h_{i,j}(u_j) = \lambda$. As $h_{i,j}(u_j) = \lambda$ iff $i \in N(j)$, we can conclude $E = E'$ and $G = G'$. $\square$

Estimating the complexity of a language by the Chomsky hierarchy, the copy-language is quite complicated. Notice that by Lemma 3, we get another characterization of all graphs through complementation, and the complement of the copy-language is a one-counter language, a subclass of context-free.

Corollary 16. *The complement of $L_{\mathcal{C}}$ can represent every graph.*

Now we intersect $L_{\mathcal{C}}$ with the the classical word-representation. Moreover, we generalise this result: copy-words are only a special case of repetitions w^i for some $i \in \mathbb{N}_{\geq 2}$.

Theorem 17. $(*)$ *$\mathcal{G}_{L_{\mathcal{C}} \cap L_{\mathrm{cl}}}$ is the set of classically word-representable graphs.*

Lemma 18. $(*)$ *Let $w \in \Sigma^*$ and $i \in \mathbb{N}_{\geq 1}$. Then w^i and w L_{cl}-represent the same graph iff for all $\mathsf{a}, \mathsf{b} \in \Sigma$ we have $h_{\{\mathsf{a},\mathsf{b}\}}(w) \in \mathsf{a}\{\mathsf{a}, \mathsf{b}\}^* \mathsf{b} \cup \mathsf{b}\{\mathsf{a}, \mathsf{b}\}^* \mathsf{a}$.*

Proposition 19. $(*)$ *Let $L = \langle 0\{0, 1\}^* 1 \rangle$. If w L-represents G, then also w^i L-represents G for all $i \in \mathbb{N}_{\geq 1}$.*

We finish this paragraph with an – in our opinion – unexpected twist w.r.t. restricting the copy-language $L_{\mathcal{C}}$. Define for all $k \in \mathbb{N}$, the restriction $L_{\mathcal{C},k} := \{ww \mid |w|_0 \not\equiv_k |w|_1\}$ of $L_{\mathcal{C}}$ where the number of 0s and 1 resp. is not allowed to be congruent modulo k. This language represents exactly the k-colorable graphs.

Theorem 20. $(*)$ *$\mathcal{G}_{L_{\mathcal{C},k}}$ is the class of k-colorable graphs.*

The construction of the words representing $G = ([n], E)$ is similar to the idea of the copy-language: Let $f : V \to \{1, \ldots, k\}$ be the coloring. We use $x = 1^{f(1)} 2^{f(2)} \cdots n^{f(n)}$ and $w = x u_1 1^k u_2 2^k \cdots u_n n^k x 1^k u_1 \cdots n^k u_n$ where u_i writes t^k exactly once per $t \in \overline{N(i)} \cap [i]$ times in the natural order. Observe that x ensures the coloring. For all $i \in [n]$, $u_i i^k$ and $i^k u_i$ deletes the edges between i and the elements in $\overline{N(i)} \cap [i]$. This construction can be generalized to graph morphisms.

Theorem 21. (*) *For all graphs $H = (U, F)$, there is a language $L(H)$ such that $\mathcal{G}_{L(H)}$ is the set of all graphs $G = (V, E)$ for which there exists a graph homomorphism $f : V \to U$.*

Lyndon Words. Proposition 19 establishes that repetitions of a representant are not needed and even though this word is not necessarily primitive, this insight encourages to look at the Lyndon words. By [13, Prop. 5] we have that if a k-uniform word w represents a graph G in the classical sense, then every word in the conjugacy class of w also represents G, i.e., by [13, Thm. 7], for representing any word-representable graph, Lyndon words suffice. As we focus on 0-1-symmetric languages and as not both ab and ba can be Lyndon for any two different $\mathsf{a}, \mathsf{b} \in \Sigma$, we look at the symmetric closure of binary Lyndon words: set $L_{\mathcal{L}} := \langle L_{\mathrm{Lyndon},\{0,1\},<} \rangle$ with the ordering $0 < 1$. Observe that $L_{\mathcal{L}} = L_{\mathrm{Lyndon},\{0,1\},<} \cup L_{\mathrm{Lyndon},\{0,1\}>}$. The results for the copy-language imply that repetitions cannot occur in projections when dealing with Lyndon words. Therefore, we can, w.l.o.g., focus on the property of being the smallest conjugate in its class during the investigation of Lyndon words. N ot all renaming projections $h_{\mathsf{a},\mathsf{b}}(w)$ of a Lyndon word w have to be Lyndon: we have for the Lyndon word $w = \mathsf{abca}$ that $h_{\mathsf{a},\mathsf{b}}(w) = 010$, is not Lyndon. Hence, we only get:

Lemma 22. (*) *Let $w \in \Sigma^*$ such that $h_{\mathsf{a},\mathsf{b}}(w)$ is $<$-Lyndon for all $\mathsf{a}, \mathsf{b} \in \Sigma$ with $\mathsf{a} \prec \mathsf{b}$. Then w is $\prec$-Lyndon.*

Lemma 23. (*) *Let $\mathsf{a}, \mathsf{b} \in \Sigma$ with $\mathsf{a} \neq \mathsf{b}$ and $w \in \Sigma^*$.*
1. $h_{\mathsf{a},\mathsf{b}}(w)$ is a $<$-Lyndon word iff $h_{\mathsf{b},\mathsf{a}}(w)$ is a $>$-Lyndon word.
2. If $h_{\mathsf{a},\mathsf{b}}(w)$ is a $<$-Lyndon word then $h_{\mathsf{a},\mathsf{b}}(w)$ is no $>$-Lyndon word.

Theorem 24. $\mathcal{G}_{L_{\mathcal{L}}}$ *includes every graph.*

Proof. Let $G = ([n], E)$ be an graph with $n \in \mathbb{N}$. Define for $i \in [n]$ the words $v_i := i \cdots n$, $x_i := j_{i,1} \cdots j_{i,\ell_i}$, $y_i := k_{i,1} \cdots k_{i,\ell_i'}$ and $u_i := i^2 x_i i^2 y_i$ where $N(i) \cap [i..n] = \{j_{i,1}, \ldots, j_{i,\ell_i}\}$, $\overline{N(i)} \cap [i..n] = \{i, k_1, \ldots, k_{i,\ell_i'}\}$ and $j_{i,1}, \ldots, j_{i,\ell_i}$ as well as $k_{i,1}, \ldots, k_{i,\ell_i'}$ are strictly monotone increasing. We will show $G = G' = (V, E') := G(L_{\mathcal{L}}, w)$, $w := 1^3 2^3 \cdots n^3 v_1^2 u_1 v_2^2 u_2 \cdots v_n^2 u_n$.

Let $p, q, r \in [n]$ with $p < q$. By Lemma 23 only $h_{p,q}(w), h_{q,p}(w)$ with respect to $0 < 1$ needs to be considered. Since p appears before q in w, it is enough to consider $h_{p,q}(w)$. Observe that for $q < r$, $h_{p,q}(v_r u_r) = \lambda$. Hence, $h_{p,q}(v_{q+1} u_{q+1} \cdots v_n^2 u_n) = \lambda$. For $p < r < q$, $h_{p,q}(v_r^2 u_r) = 111$ and $h_{p,q}(v_q^2 u_q) = 111111$. Thus, $h_{p,q}(v_q u_q \cdots v_n^2 u_n) \in 111111(111)^*$.

Next, assume $r < p < q$. Then $h_{p,q}(v_i) = 01$. If $p, q \in N(r)$, $h_{p,q}(x_r) = 01$ and $h_{p,q}(y_r) = \lambda$. Analogously, $h_{p,q}(x_r) = \lambda$ and $h_{p,q}(y_r) = 01$ for $p, q \notin N(r)$. In the case of $p \in N(r)$ and $q \notin N(r)$, $h_{p,q}(x_r) = 0$ and $h_{p,q}(y_r) = 1$. For $p \notin N(r)$ and $q \in N(r)$, $h_{p,q}(x_r) = 1$ and $h_{p,q}(y_r) = 0$. Therefore, $h_{p,q}(v_r v_r u_r) \in \{010101, 010110\}$ and $h_{p,q}(v_1^2 u_1 \cdots v_{p-1}^2 u_{p-1}) \in \{010101, 010110\}^{p-1}$. Clearly, there does not exist an $i \in [6(p-1)]$ with $w[3n + i..3n + i + 2] = 000$. Finally, consider $r = p$. If $\{p, q\} \in E$ then $h_{p,q}(v_p^2, u_q) = 010100100$ and $h_{p,q}(w) \in$

$000111\{010101, 010110\}^{p-1}0101001001111111(111)^*$. Clearly, $i = 1$ is the only $i \in |h_{p,q}(w)|$ such that $h_{p,q}(w)[i..i+2] = 000$. Since $h_{p,q}(w)[|h_{p,q}(w)|] = 1$, $h_{p,q}(w)$ is Lyndon and $\{p, q\} \in E'$.

Assume $\{p, q\} \notin E$. Then $h_{p,q}(v_p p^2 x_p p^2 y_p) = 0100001$. As $v_p u_p = v_p p^2 x_p p^2 y_p$ is an infix of w, 0100001 is an infix of $h_{p,q}(w)$. As 000111 is a prefix of $h_{p,q}(w)$, $h_{p,q}(w)$ is no Lyndon word (a conjugate with prefix $h_{p,q}(u_p)$ is $<$-smaller). Thus, $\{p, q\} \notin E$. In total, $G = G'$. $\qquad\square$

We conclude this paragraph with some insights about graph classes when we only look at odd or even length Lyndon words with a similar result to Corollary 12.

Theorem 25. $(*)$ $\mathcal{G}_{\{w \in L_{\mathcal{L}} \,||w| \text{ is odd}\}}$ *is the class of bipartite graphs.*

Corollary 26. *Let* $L := \{w \in L_{\mathcal{L}} \mid |w| \text{ is odd}\}$, $L_{odd} := \{w \in \{0,1\}^* \mid |w|_0, |w|_1 \text{ are odd}\}$ *and* $L_{even} := \{w \in \{0,1\}^* \mid |w|_0, |w|_1 \text{ are even}\}$. *Then* $\mathcal{G}_{L \cup L_{odd}}$ *is the class of split graphs and* $\mathcal{G}_{L \cup L_{odd} \cup L_{even}}$ *is the class of cobipartite graphs.*

Dyck-language. Similar to the Lyndon words, we need here to define the symmetric closure of the Dyck-language as $L_{\mathcal{D}} := \langle \mathcal{D} \rangle$. Also similar, we have $h_{(,)}(w) \in L_{\mathcal{D}}$ for any pair of parentheses $(,) \in \Sigma$ if $w \in D_k'^*$, but the converse is not true, as $w = (|])$ shows. Our main result is that $\mathcal{G}_{L_{\mathcal{D}}}$ is the class of comparability graphs. To see that each graph in $\mathcal{G}_{L_{\mathcal{D}}}$ is a comparability graph, consider the order $\prec_V$ such that, for all $v, u \in V$, $v \prec_V u$ iff $v \neq u$, $|w|_v = |w|_u$ and, for all prefixes w' of w, $|w'|_v \leq |w'|_u$. For the other direction, let $G = (V, E)$ be a comparability graph with the order $\prec$ on V. For each $v \in V$, $z_v := y_v v x_v$ where x_v enumerates all $u \in V$ with $v \prec u$, and y_v the remaining vertices. For all $v, u \in V$, if $v \prec u$ then $h_{v,u}(z_v) = 01$. The final word enumerates first all vertices and then all z_v in the order of $\prec$.

Theorem 27. $\mathcal{G}_{L_{\mathcal{D}}}$ *is the class of comparability graphs.*

Proof. Let $G = (V, E)$ be a comparability graph w.r.t. the strict order $\prec$ on V, i.e., $\{u, v\} \in E$ iff u, v are comparable with respect to $\prec$. Define the upper set $U_{\prec}(v) := \{u \in V \mid v \prec u\}$ for $v \in V$. $\prec_{lin}$ denotes an arbitrary but fixed linear order on V that extends $\prec$. Define $z := v_1 \cdots v_n$ to be an enumeration of V in the order $\prec_{lin}$. For $v \in V$, let x_v, y_v be words such that x_v enumerates $U_{\prec}(v)$ and y_v enumerates $\overline{U_{\prec}(v)} \setminus \{v\}$ according to the order $\prec_{lin}$ (increasingly). Further, denote $z_v := y_v v x_v$, i.e., z_v first enumerates the elements of V that are in the upper set of v, then v, and then the remaining elements of V. Define $w := z z_{v_1} \cdots z_{v_n}$ and $G' = (V, E') = G(L_{\mathcal{D}}, w)$. Note that for each $v, u \in V$, $|z_v|_u = |z|_u = 1$. Thus, for all $v, a, b \in V$, $h_{a,b}(z_v), h_{a,b}(z) \in \{01, 10\}$. Let w.l.o.g. $a, b \in V$ with $a \prec_{lin} b$ in our following discussion. Thus, $h_{a,b}(z) = 01$ $(*)$.

Assume $\{a, b\} \in E$. As G is a comparability graph, this implies $a \prec b$ $[*]$, as $b \nprec a$ because $a \prec_{lin} b$. We show that $h_{a,b}(z_v) = 01$ for all $v \in V$. Then, with $(*)$, we can conclude $h_{a,b}(w) = (01)^{n+1}$. Hence, $h_{a,b}(w) \in L_{\mathcal{D}}$. Suppose there existed

$v \in V$ such that $h_{a,b}(z_v) = 10$. Hence, b precedes a in the enumeration z_v. If $v = a$ (or $v = b$, resp.), then $b \in \mathrm{alph}(y_v) = \overline{U_{\prec}(v)} \setminus \{v\}$ (or $a \in \mathrm{alph}(x_v) = U_{\prec}(v)$, resp.), contradicting $[*]$. Therefore, $v \notin \{a, b\}$. Since x_v, y_v enumerate in accordance with $\prec_{lin}$ and as $a \prec b$, b has to appear in y_v and a in x_v. This is a contradiction to the transitivity of $\prec$, as $a \prec b$ and $v \prec a$ (because $a \in U_{\prec}(v)$) would imply $b \in U_{\prec}(v) = \mathrm{alph}(x_v)$. Hence, $h_{a,b}(z_v) = 01$ for all $v \in V$. Now assume $\{a, b\} \notin E$. Then $|y_a|_b = 1$ and $h_{a,b}(z_a) = 10$. Let $a = v_i$. This implies $h_{a,b}(w) \in 01\{01, 10\}^{i-1}10\{01, 10\}^{n-i}$. Then $\{a, b\} \notin E'$, since $|h_{a,b}(w[1..2i + 1])|_0 = i < i + 1 = |h_{a,b}(w[1..2i + 1])|_1$ and $|h_{a,b}(w[1])|_1 = 0 < 1 = |h_{a,b}(w[1])|_0$.

Conversely, let $G = (V, E) \in \mathcal{G}_{L_{\mathcal{D}}}$. There exists $w \in V^*$ such that $G = G(L_{\mathcal{D}}, w)$. Define the order $\prec_V$ such that, for all $v, u \in V$, $v \prec_V u$ iff $v \neq u$, $|w|_v = |w|_u$ and, for all prefixes w' of w, $|w'|_v \leq |w'|_u$. We show that $v \prec_V u$ is a strict partial order and that $\{v, u\} \in E$ iff $v \prec_V u$, i.e., G is a comparability graph. By definition, $\prec_V$ is irreflexive. Let $a, b, c \in V$ with $a \prec_V b$ and $b \prec_V c$. Thus, $|w|_a = |w|_b = |w|_c$. For any prefix w' of w, $|w|_a \leq |w|_b \leq |w|_c$. Let $i \in [|w|]$ be the the smallest i with $w[i] = b$. This implies $|w[1..i - 1]|_a \leq 0 = |w[1..i - 1]|_b$ and $|w[1..i]|_a \leq 0 < 1 = |w[1..i]|_b \leq |w[1..i]|_c$. Thus, $a \neq c$ and $a \prec_V c$. Therefore, $\prec_V$ is a strict partial order.

Let $a, b \in V$. For $|w|_a \neq |w|_b$, $\{a, b\} \notin E$ and $a \not\prec_V b$ as well as $b \not\prec_V a$. Thus, we can assume $|w|_a = |w|_b$. Since $h_{a,b}$ is a morphism, if w' is a prefix of w, then $h_{a,b}(w')$ is a prefix of $h_{a,b}(w)$. Therefore, $|w'|_a = |h_{a,b}(w')|_0$ and $|w'|_b = |h_{a,b}(w')|_1$. By definition, $a \not\prec_V b$ and $b \not\prec_V a$ iff there are prefixes w', w'' such that $|h_{a,b}(w')|_1 = |w'|_b < |w'|_a = |h_{a,b}(w')|_0$ and $|h_{a,b}(w'')|_0 = |w''|_a < |w''|_b = |h_{a,b}(w'')|_1$. Thus, $\{a, b\} \notin E$. Conversely, assume $\{a, b\} \notin E$. Hence, there are prefixes h', h'' of $h_{a,b}(w)$ with $|h'|_1 < |h'|_0$ and $|h''|_0 < |h''|_1$. Since $|h_{a,b}(v)| \leq 1$ for all $v \in V$, we can show by an inductive argument that, for each prefix h of $h_{a,b}(w)$, there exists a prefix w' of w such that $h = h_{a,b}(w')$. Again by definition, this means that $a \not\prec_V b$ and $b \not\prec_V a$. $\qquad\square$

There is a rather unexpected connection between the number d of pairs of parentheses and the order dimension, a famous notion from order theory: Dyck words with d of pairs of parentheses describe exactly the comparability graphs of partial orders of dimension d, as shown in [6].

With Lemma 3, we get a characterization of co-comparability graphs in terms of complements of the symmetric hull of the Dyck language. We finish this paragraph with the symmetric hull of probably the most famous Dyck language $L_1 := \langle\{0^n 1^n \mid n \in \mathbb{N}\}\rangle$ and the non-regular language $L_2 := \{w \in \{0, 1\}^* \mid |w|_0 = |w|_1\}$ where the Dyck-property is dropped (i.e., we have 2-uniform words).

Proposition 28. $(*)$ $\mathcal{G}_{L_1}$ *is the class of graph unions of co-interval graphs.*[1] $\mathcal{G}_{L_2}$ *is the class of cluster graphs.*

By Lemma 3, we get a characterization of complete multipartite graphs (as complements of cluster graphs) as $\mathcal{G}_{\overline{L_2}}$.

---

[1] Co-interval graphs are not closed under graph union : $2K_2$ is not a co-interval graph.

Size of the Graph Representations. The languages that we just discussed, most of which enable us to describe any graph, are not the first ones observed: for instance, $L_{\overline{1^3}}$ is such a language, as described in [11], see our discussions above. However, there is one drawback of such an encoding: the word $w(G)$ describing a given graph $G = (V, E)$ will have length $\Theta(|V|!)$, so that we need an exponential number of bits to describe a graph, measured in its order. Even worse is the situation with 1112-avoidance, see [10, Theorem 3.12], where the recursively constructed words can have length $\mathcal{O}(2^{|V|^2})$. With 2-11-representable graphs, a more concise representation of every graph was proven in [3, Theorem 5.1], using only $\mathcal{O}(n^3 \log(n))$ many bits. This is still worse than using traditional adjacency lists or adjacency matrices. In particular, the latter data structure needs $\mathcal{O}(n^2)$ bits to describe any (un)directed graph of order n. The good news is that the graph representations that we obtain with the help of palindromes, copy-words or Lyndon words are better. Taking as an example the cycle C_4, palindromes and copy languages performs much better, as illustrated in Fig. 1.

Corollary 29. *With the help of palindromes, copy-words or Lyndon words, each graph of order n can be represented with $\mathcal{O}(n^2 \log(n))$ many bits. By Lemma 3, similar statements are true for complement languages.*

Keep in mind that the log-factor has to be added, as writing a single letter of the alphabet V needs $\mathcal{O}(\log(|V|))$ many bits. Notice that this matches other implicit graph representations, e.g., in terms of sum graphs, see [8], or also that of adjacency lists. For sparse graphs, it is also important to take the number of edges into account. For instance, for adjacency lists, one needs $\mathcal{O}((n + m) \log n)$ many bits. Next, we show that a slight modification of our constructions yields a graph representation that matches this bound.

Theorem 30. (∗) *There exists a language L such that any graph $G = (V, E)$, with $|V| = n$ and $|E| = m$, can be represented as $G(L, w)$ for a word $w \in V^*$ of length $\mathcal{O}(n + m)$, so that w can be encoded as a binary string of size $\mathcal{O}((n + m) \log n)$.*

This theorem is also interesting from a language-theoretic perspective. While the copy-language is not context-free, its complement is even a 1-counter language. Hence, there is no need to go very far in the Chomsky hierarchy to obtain such an efficient graph encoding. On the other hand, we neither have any arguments why regular languages have to lead to essentially bigger graph encodings. We leave this as an open question.

In contrast, it might be interesting to have encodings in hand that are suitable for storing dense graphs, as opposed to sparse graphs as in Theorem 30. By the proof of this theorem, the language of copy-words L_C seems to be very suitable. For a concrete example, reconsider Corollary 14: for a vertex set V of size n, ww encodes K_n if w enumerates V in some order.

4   Conclusions and Further Open Questions

In this paper, we have shown various examples of rather famous languages that can represent large classes of graphs, if not all graphs. For instance, we showed that both Lyndon words and palindromes are suitable to represent all graphs. This motivates the question what graph class Christoffel words (introduced in [4]) can represent. Notice that they are special Lyndon words and that for a Christoffel word $w = 0v1$, v is a palindrome.

One of the nice combinatorial properties of classical word representability is that each word-representable graph can be represented by a uniform word. This kind of property does not hold for the language representations that we studied in this paper. However, there are indeed languages that can represent all graphs and that allow for uniform word representations. In [6], it is shown that every graph of so-called boxicity b can be represented by a finite $2b$-uniform language L_b, so that their union $L_B := \bigcup_{b \geq 1} L_b$ represents all graphs as every graph has some finite boxicity [17]. However, L_B is no "nice" and well-known formal language, so that the question remains to find a "known" formal language representing all graphs by uniform words.

References

1. Alecu, B., Alekseev, V.E., Atminas, A., Lozin, V.V., Zamaraev, V.: Graph parameters, implicit representations and factorial properties. Discrete Mathematics **346**(10), 113573 (2023)
2. Berstel, J.: Transductions and Context-Free Languages, LAMM, vol. 38. Stuttgart: Teubner (1979)
3. Cheon, G.S., Kim, J., Kim, M., Kitaev, S., Pyatkin, A.: On k-11-representable graphs. Journal of Combinatorics **10**(3), 491–513 (2019)
4. Christoffel, E.B.: Observatio arithmetica. Annali di Matematica Pura ed Applicata **6**, 148–152 (1875)
5. Diekert, V., Fernau, H., Wolf, P.: Properties of graphs specified by a regular language. Acta Informatica **59**(4), 357–385 (2022)
6. Feng, Z., Fernau, H., Fleischmann, P., Kindermann, P., Sacher, S.C.: Determining factorial speed fast. Tech. Rep. 2602.24064, ArXiv, Cornell University (Feb 2026)
7. Feng, Z., Fernau, H., Mann, K., Raman, I., Sacher, S.C.: Generalized lettericity of graphs. In: Gaur, D.R., Mathew, R. (eds.) Algorithms and Discrete Applied Mathematics - 11th International Conference, CALDAM. LNCS, vol. 15536, pp. 134–146. Springer (2025)
8. Fernau, H., Gajjar, K.: The space complexity of sum labelling. Theory of Computing Systems **67**(5), 1026–1049 (2023)
9. Fleischmann, P., Haschke, L., Löck, T., Nowotka, D.: Word-representable graphs from a word's perspective. Acta Informatica **61**(4), 383–400 (2024)
10. Gaetz, M.R., Ji, C.: Enumeration and extensions of word-representants. Discrete Applied Mathematics **284**, 423–433 (2020)
11. Jones, M.E., Kitaev, S., Pyatkin, A.V., Remmel, J.B.: Representing graphs via pattern avoiding words. The Electronic Journal of Combinatorics **22**(2), (2015). P2.53

12. Kitaev, S., Lozin, V.V.: Words and Graphs. Monographs in Theoretical Computer Science. An EATCS Series, Springer (2015)
13. Kitaev, S., Pyatkin, A.V.: On representable graphs. Journal of Automata, Languages and Combinatorics **13**(1), 45–54 (2008)
14. Kitaev, S., Seif, S.: Word problem of the Perkins semigroup via directed acyclic graphs. Order **25**, 177–194 (2008)
15. Lozin, V.V.: Graph representation functions computable by finite automata. Journal of Automata, Languages and Combinatorics **13**(1), 73–90 (2008)
16. Petkovšek, M.: Letter graphs and well-quasi-order by induced subgraphs. Discrete Mathematics **244**(1), 375–388 (2002)
17. Roberts, F.S.: On the boxicity and cubicity of a graph. In: Recent Progress in Combinatorics, pp. 301–310. Academic Press (1969)

Efficient Computation of Discriminative Absent Words for String Collections

Giuseppa Castigione⬤, Sabrina Mantaci⬤, Antonio Restivo⬤,
Giuseppe Romana$^{(\boxtimes)}$⬤, and Marinella Sciortino⬤

Università degli Studi di Palermo, Palermo, Italy
{giuseppa.castiglione,sabrina.mantaci,antonio.restivo,giuseppe.romana01,
marinella.sciortino}@unipa.it

Abstract. Minimal absent words (MAWs) provide a compact and combinatorially meaningful representation of a string or of a collection of strings, and have been successfully used in alignment-free sequence comparison and classification methods. Among MAW-based approaches, the set of discriminative absent words (DAWs) for two string collections isolates those minimal absent words of one collection that actually occur as a factor in the other, thus capturing the shortest witnesses of structural differences.

In this paper, we present a linear-time algorithm for computing the set of discriminative absent words for two collections of strings. We detect DAWs through a single scan of LCP-based arrays, without explicitly constructing the full sets of minimal absent words of the two string collections. The algorithm naturally extends to the circular setting, preserving linear time and space complexity with respect to the total input size. This yields a conceptually simple and efficient framework for large-scale sequence comparison.

Keywords: Minimal Absent Words · Discriminative Absent Words · String comparison · Generalized LCP array

1 Introduction

Minimal absent words (MAWs) are a very well-known notion in combinatorics of words, introduced in [13] and further investigated in [3,26,27]. They have found applications in several contexts, such as symbolic dynamics [4] and text compression [14]. Given a word $x \in \Sigma^*$, the set $M(x)$ of its minimal absent words consists of those words that do not occur as factors of x while all their proper factors do. Equivalently, $M(x)$ forms the minimal generating set (or base) of the ideal $\Sigma^* M(x) \Sigma^*$, which coincides with the complement of the language $F(x)$ of the factors of x.

Beyond their intrinsic combinatorial interest, minimal absent words have been employed in the definition of alignment-free similarity measures between words, and more generally as an effective combinatorial tool for describing and

© The Author(s), under exclusive license to Springer Nature Switzerland AG 2026
M.-P. Béal and P. Caron (Eds.): DLT 2026, LNCS 16578, pp. 223–237, 2026.
https://doi.org/10.1007/978-3-032-28404-4_17

comparing strings. A first systematic use of MAWs for sequence comparison was proposed by Chairungsee and Crochemore [10], who introduced a similarity measure between two words x and y based on a length-weighted evaluation of the elements in the symmetric difference $M(x) \triangle M(y)$. Later, Charalampopoulos et al. [11] provided a linear-time algorithm for its computation. Their method is based on a merge-like traversal guided by the suffix array of the concatenation xy, allowing the linear construction of the words in $M(x) \triangle M(y)$ by starting from $M(x)$ and $M(y)$. They also extended the approach to the cyclic setting, preserving linear time and space complexity.

A related direction concerns the notion of *T-specific strings*. Given a target collection T and a reference collection R, a T-specific string is a substring of T that does not occur in R and is minimal with respect to this property. Bonizzoni et al. [5] introduced this notion in the context of comparative genome analysis and proposed an algorithm based on backward and forward extensions over an FMD-index, a bidirectional variant of the FM-index [23]. While effective in practice, their method has quadratic worst-case time, with a relaxed linear-time variant obtained by restricting overlaps. Subsequently, Béal and Crochemore [1,2] developed a combinatorial and an automata-theoretic perspective on specific strings. They showed how to construct in linear time a trie of all T-specific factors by marking states in the DAWG of $R \cup T$ and identifying transitions corresponding to minimal absent words.

While T-specific strings capture minimal asymmetric differences between two languages, Castiglione, Mantaci and Restivo [8] introduced a symmetric construction leading to a similarity measure that evaluates the lengths of words when the roles of target and reference are assumed symmetrically by both strings. They considered the set $\mathcal{D}(x, y)$ which consists of the factors of x that are minimal absent words of y together with the factors of y that are minimal absent words of x. In [8], it is shown that $\mathcal{D}(x, y)$ coincides with the base of the two-sided ideal generated by $M(x) \triangle M(y)$. In particular, every element of the symmetric difference contains an element of $\mathcal{D}(x, y)$ as a factor. This characterization clarifies the algebraic relation between the two sets and highlights a limitation of the symmetric difference: it may contain non-minimal elements that do not provide additional structural information. Experimental evidence [7] shows that $\mathcal{D}(x, y)$ is typically much smaller than $M(x) \triangle M(y)$, both in cardinality and in total length, while retaining the essential discriminative information. Indeed, the symmetric difference may contain absent words from both strings for unrelated structural reasons, and therefore do not necessarily capture actual structural differences.

In this paper, we consider the extension of this approach to collections of strings. Throughout the paper, the term *collection* means both sets and multisets. From now on, we will focus on sets of strings; however, all definitions and results extend naturally to multisets.

Given two sets X and Y, we consider the set of *discriminative absent words* (DAWs)

$$\mathcal{D}(X, Y) = (F(X) \cap M(Y)) \cup (F(Y) \cap M(X)),$$

which contains exactly those minimal absent words of one set that actually occur as factors in the other. These words represent the shortest witnesses of structural differences between the languages of the factors in the sets X and Y.

We focus on the algorithmic computation of DAWs. We aim to characterize and compute $\mathcal{D}(X,Y)$ without passing through the explicit construction of the full sets $M(X)$ and $M(Y)$. To this end, we introduce a linear-time framework that combines different data structures, typically employed for combinatorial pattern matching: the key idea is a local characterization of DAWs in terms of two LCP-like arrays, the *Colored LCP* (CLCP) and the *Merge LCP* (MLCP) arrays, together with the Ψ array which is closely related to the notion of *standard permutation* of a word [21,31]. This yields a conceptually simple procedure based on a single scan of numeric arrays, which reports each element of $\mathcal{D}(X,Y)$ in compact form as a triple encoding the string where the discriminative absent word occurs, together with its starting position and its length.

Our approach works in $O(N)$ time and space, where N is the total length of the input. The underlying characterization extends directly to T-specific words and to circular settings, yielding linear-time algorithms in these cases as well. Hence, we provide an algorithmically efficient combinatorial tool for computing discriminative absent words and the associated DAW-based distance measures.

2   Preliminaries

2.1   Words, Factors and Minimal Absent Words

Let $\Sigma = \{a_1, a_2, \ldots, a_\sigma\}$ be a finite set of ordered *letters* $a_1 < a_2 < \cdots < a_\sigma$, that we call an *alphabet*. A *word* (or *string*) $w = w[1]w[2]\cdots w[n]$ is a finite sequence of *length* $|w| = n$, where $w[i] \in \Sigma$ denotes the ith letter in w. The *empty word* ε is the only word such that $|\varepsilon| = 0$. The set of all finite words (resp. all non-empty words) over the alphabet Σ is denoted by Σ^* (resp. Σ^+). Throughout the paper, we assume the alphabet has polynomial size, that is, $|\Sigma| = n^{O(1)}$.

For any word $w \in \Sigma^*$ the n-th power of w is w^n, i.e. w is concatenated to itself n times. A word w is called *primitive* if $w = u^k$ implies $w = u$ and $k = 1$. For every word w, there exists a unique primitive word u and a unique integer k such that $w = u^k$. The word u is called $\mathrm{root}(w)$ and k is called $\exp(w)$; thus, $w = \mathrm{root}(w)^{\exp(w)}$.

The *standard permutation* of a word $w = w[1]w[2]\cdots w[n]$ over Σ is the permutation π_w of $\{1, 2, \ldots, n\}$ such that $\pi_w(i) < \pi_w(j)$ if and only if $w[i] < w[j]$ or $w[i] = w[j]$ and $i < j$. In other words, π_w orders distinct letters of w lexicographically, and equal letters by occurrence order.

A *factor* of w is any word $u \in \Sigma^*$ such that there exists i, j such that $u = w[i..j]$. If $i > j$, then we assume $w[i..j] = \varepsilon$. We let $F(w)$ denote the set of all factors of w, while if $X = \{w_1, w_2, \ldots, w_{|X|}\}$ is a set of words, then $F(X) = \bigcup_{i \in [1..|X|]} F(w_i)$. We denote by $\mathsf{alph}(X)$ the set of letters of Σ occurring in the words of X. The factor $w[i..j]$ is called a *prefix* when $i = 1$, and a *suffix* when $j = n$. The *longest common prefix* between two words u and v is the longest word that is a prefix of both words. The length of this word is denoted by $\ell cp(u, v)$.

A *rotation*, or *conjugate*, of the word $w = w[1]w[2]\cdots w[n]$ is the word $w[i..n]w[1..i-1]$, for some $1 \le i \le n$. We say that u is a cyclic factor of w if it is a factor of one of its rotations. We denote by $\mathring{F}(w)$ the set of the cyclic factors of w.

A word $w \in \Sigma^*$ of length m is a *minimal absent word* (or MAW, in short) of a word $x \in \Sigma^*$ if $w \notin F(x)$ and all its proper factors are in $F(x)$, or, equivalently, the prefix and the suffix of w of length $m-1$, namely $w[1..m-1]$ and $w[2..m]$, are in $F(x)$. We denote by $M(x)$ the set of all minimal absent words of x. Furthermore, if X is a set of words, $M(X)$ denotes the set of its minimal absent words, i.e., the set of words $w \notin F(X)$ whose proper factors are all in $F(X)$. Note that $M(X)$ is an antifactorial set.

A word w is a *circular minimal absent word* (cMAW, in short) of $x \in \Sigma^*$, if w is not a cyclic factor of x, and each of its proper factors does. This last statement is equivalent to saying each of the prefix and the suffix of w with length $|w| - 1$ (namely $w[1..n-1]$ and $w[2..n]$) belongs to $\mathring{F}(x)$. The set of circular minimal absent words of x is denoted by $\mathring{M}(x)$.

2.2 Sorting Suffixes in a Set of Words

Let $X = \{x_1, \ldots, x_m\}$ be a *finite set* of words over an ordered alphabet Σ, with $|x_d| = n_d$ and total length $\|X\| = \sum_{d=1}^{m} n_d$. In order to compare suffixes belonging to different words, we append to each x_d a distinct end-marker $\$_d$, with $\$_1 < \cdots < \$_m$ and $\$_d < c$ for every $c \in \Sigma$, and define $X_\$ = \{x_1\$_1, \ldots, x_m\$_m\}$. Notice that even if the input is given as a *multiset* (allowing repetitions), $X_\$$ is still a set: distinct end-markers make all strings $x_d\$_d$ pairwise different.

Let $S(X_\$)$ be the list of all suffixes of all strings in $X_\$$ sorted in lexicographic order and let $N = \|X_\$\|$. The *generalized suffix array* GSA_X is an array of length N that stores pairs (d, p), with $d \in [1..m]$ and $p \in [1..|x_d| + 1]$, such that $\mathsf{GSA}[i] = (d, p)$ iff the i-th suffix in lexicographic order is the suffix of $x_d\$_d$ starting at position p. Accordingly, $S(X_\$)[i] = (x_d\$_d)[p..|x_d| + 1]$ when $\mathsf{GSA}[i] = (d, p)$.

The *generalized longest common prefix array* GLCP is the array of length N defined by $\mathsf{GLCP}[1] = 0$ and, for $2 \le i \le N$, $\mathsf{GLCP}[i] = \ell cp\big(S(X_\$)[i], S(X_\$)[i-1]\big)$, that is, $\mathsf{GLCP}[i]$ is the length of the longest common prefix between two consecutive suffixes in the lexicographic order.

The *extended Burrows–Wheeler transform* (eBWT) is an extension of the Burrows–Wheeler transform to a set (or multiset) of words [16, 25]. When applied to $X_\$$, the transform eBWT produces the word obtained by concatenating the characters that immediately precede the suffixes in the lexicographically sorted list $S(X_\$)$. More formally, if $\mathsf{GSA}[i] = (d, p)$, then $\mathsf{eBWT}[i] = x_d\$_d[p - 1]$, if $p \ne 1$, or $\mathsf{eBWT}[i] = \$_d$ otherwise. More generally, the same transform can be defined without explicit end-markers by sorting all conjugates of the input words according to the ω-order, which compares the corresponding infinite periodic words. It can be thought as a bijection between words and multisets of conjugacy classes in Σ^* [12,21].

The function $\psi_X : [1..N] \to [1..N]$ associated to $X_\$$ is defined as follows: if $\mathsf{GSA}[i] = (d, k)$, then $\psi(i)$ is the unique index such that $\mathsf{GSA}[\psi(i)] = (d, 1 + k \bmod (|x_d|+1))$, that is, ψ_X maps the suffix of $x_d\$_d$ starting at k to the suffix of the same word starting at $k+1$, and maps the suffix starting at the end-marker back to the whole word. From a combinatorial viewpoint, ψ_X coincides with the inverse of the standard permutation π_U of the word $U = \mathsf{eBWT}(X_\$)$, that is, $\psi_X(i) = j$ if and only if $\pi_U(j) = i$ [9]. The permutation π_U is also known as the LF-mapping. We observe that both the standard permutation π_U and the function ψ_X can be evaluated by applying one rank and one select operation on $\mathsf{eBWT}(X_\$)$, respectively[1]. The operations rank and select are defined as follows: given an array T and a symbol c, $\mathsf{rank}_c(T, i)$ returns the number of occurrences of c in the prefix $T[1..i]$, while $\mathsf{select}_c(T, t)$ returns the position in T of the t-th occurrence of c (when it exists).

3   Discriminative Absent Words for Collections of Words

Let X and Y be two sets of words. The set of X-specific words against the reference set Y is defined as

$$D(X \leftarrow Y) = F(X) \cap M(Y).$$

We define the set of *discriminative absent words* (DAWs) for X and Y defined as:

$$\mathcal{D}(X, Y) = D(X \leftarrow Y) \cup D(Y \leftarrow X) = \big(F(X) \cap M(Y)\big) \cup \big(F(Y) \cap M(X)\big).$$

When X and Y are singletons, i.e., $X = \{x\}$ and $Y = \{y\}$, we simply write $\mathcal{D}(x, y)$. The set $\mathcal{D}(x, y)$ has been studied in [8] and further investigated in [7] as a combinatorial tool for comparing two words. Note that the definition of discriminative absent words extends naturally to the case where X and Y are multisets of words.

The following proposition gives some combinatorial properties of $\mathcal{D}(X, Y)$.

Proposition 1. *Let X, Y be two sets of words in Σ^*. Then,*

1. *if $F(X) \subseteq F(Y)$ then $\mathcal{D}(X, Y) = D(Y \leftarrow X)$;*
2. *$\mathcal{D}(X, Y) = \emptyset$ if and only if $F(X) = F(Y)$;*
3. *$\mathcal{D}(X, Y)$ is an antifactorial set;*
4. *$\mathcal{D}(X, Y)$ is the base of the two-sided ideal $\Sigma^*(M(X) \triangle M(Y))\Sigma^*$ of Σ^*.*

Example 1. Let us consider the sets $X = \{\mathtt{GTGT}\}$ and $Y = \{\mathtt{GTGT}, \mathtt{GTG}, \mathtt{TGT}\}$. It is easy to verify that $\mathcal{D}(X, Y) = \emptyset$ since $F(X) = F(Y)$. Moreover, $M(X) = M(Y) = \{\mathtt{GG}, \mathtt{TT}, \mathtt{TGTG}\}$.

---

[1] Let $C[c]$ be the number of symbols in eBWT lexicographically smaller than c. If $c = \mathsf{eBWT}(X_\$)[i]$, then $\pi_U[i] = C[c] + \mathsf{rank}_c(\mathsf{eBWT}(X_\$), i)$ and $\psi_X(i) = \mathsf{select}_c(\mathsf{eBWT}(X_\$), i - C[c])$.

Note that Proposition 1(4.) shows that $\mathcal{D}(X,Y)$ is not only a subset of $M(X)\triangle M(Y)$, but in fact represents the set of the shortest words among the ones witnessing a structural difference between the two sets X and Y.

Example 2. Consider the sets $X = \{\text{ACCCCCCCA, GT}\}$ and $Y = \{\text{GTTTTTTTG, AC}\}$. The symmetric difference between $M(X)$ and $M(Y)$ is $M(X)\triangle M(Y) = \{\text{ACA, ACCA, ACCCA, ACCCCA, ACCCCCA, ACCCCCCA, CA, CAC, CC, CCCCCCCC, GTG, GTTG, GTTTG, GTTTTG, GTTTTTG, GTTTTTTG, TG, TGT, TT, TTTTTTTT}\}$. Note that this set has cardinality 20 and size 96. The discriminative absent word set is $\mathcal{D}(X,Y) = \{\text{CA, CC, TG, TT}\}$ with cardinality 4 and size 8.

The following proposition considers the case when X and Y contain words on disjoint alphabets.

Proposition 2. *Let X, Y be two sets of words in Σ^*. Then, $\mathcal{D}(X,Y) = \mathsf{alph}(X) \cup \mathsf{alph}(Y)$ if and only if $\mathsf{alph}(X) \cap \mathsf{alph}(Y) = \emptyset$.*

Example 3. Consider the words $x = \text{ACCAC}$ and $y = \text{GGTGTG}$. The respective minimal absent words are $M(x) = \{\text{G, T, AA, ACA, CCC, CACC}\}$ and $M(y) = \{\text{A, C, TT, GGG, TGG, TGTGT}\}$. The symmetric difference is therefore $M(x)\triangle M(y) = \{\text{A, C, G, T, AA, TT, ACA, CCC, GGG, TGG, CACC, TGTGT}\}$. The DAWs are $\mathcal{D}(x,y) = \{\text{A, C, G, T}\}$.

In several applications, such as comparative genomics, sequences are naturally modeled as circular words. In this setting, circular factors are considered, and minimal absent words are computed accordingly.

The set of *circular discriminative absent words* (CDAWs) for X and Y is defined as

$$\mathring{\mathcal{D}}(X,Y) = \left(\mathring{F}(X)\cap\mathring{M}(Y)\right) \cup \left(\mathring{F}(Y)\cap\mathring{M}(X)\right).$$

When X and Y are singletons, we simply write $\mathring{\mathcal{D}}(x,y)$.

4   Linear-Time Computation of $\mathcal{D}(X,Y)$

In this section, we present our algorithm for computing $\mathcal{D}(X,Y)$, without constructing the full sets of factors and minimal absent words of X and Y. The algorithm consists of a linear-time scan of two LCP-like arrays.

Let $X = \{x_1,\ldots,x_m\}$ and $Y = \{y_1,\ldots,y_n\}$ be two set of words, for some integer $m, n > 0$. We denote by $X \uplus Y$ the set of words by taking all words in X and in Y and appending to each of them a distinct end-marker, i.e., $X \uplus Y = \{x_i\$_i \mid i \in [1..m]\} \cup \{y_j\$_{m+j} \mid j \in [1..n]\}$, with $\$_1 < \cdots < \$_{m+n}$. Let us denote by N the sum of the lengths of the words in $X \uplus Y$, i.e., $N = \|X \uplus Y\| = \|X\|+\|Y\|+m+n$. For instance, given $X = \{\text{AC, CG, TGT}\}$ and $Y = \{\text{AC, CG, ATA}\}$, one has $X \uplus Y = \{\text{AC}\$_1, \text{CG}\$_2, \text{TGT}\$_3, \text{AC}\$_4, \text{CG}\$_5, \text{ATA}\$_6\}$. In this case, $N = 20$.

We define a *color function* as a mapping $\mathrm{color} : X \uplus Y \rightarrow \{0,1\}$, that associates to each word a label, called its *color*, such that $\mathrm{color}(v\$_i) = 0$ if $i \in [1..m]$, and $color(v\$_i) = 1$ if $i \in [m + 1..m + n]$. We denote the *complementary color* of $w \in X \uplus Y$ by $\overline{color(w)}$.

Given the lexicographically sorted list of suffixes $S(X \uplus Y)$, the *color array* col is the array such that $\text{col}[i] = color(w)$ whenever the i-th suffix in $S(X \uplus Y)$ belongs to the word w.

Before describing our algorithm, we give the definition of Colored LCP and Merge LCP, similar to the notions previously introduced in [19] for the lightweight computation of matching statistics in external memory.

The *Colored Longest Common Prefix Array* CLCP of $X \uplus Y$, is the array that counts for each i the maximum length of the longest common prefix between the ith suffix in $S(X \uplus Y)$ and any other suffix of the complementary color. More formally, for each position $i \in [1..\|X \uplus Y\|]$, let $j_\ell = \max\{j < i \mid \text{col}[j] = \overline{\text{col}[i]}\}$, and $j_r = \min\{j > i \mid \text{col}[j] = \overline{\text{col}[i]}\}$, whenever such indices exist. Then,

$$\text{CLCP}[i] = \max\left(\ell cp(S(X \uplus Y)[i], S(X \uplus Y)[j_\ell]),\ \ell cp(S(X \uplus Y)[i], S(X \uplus Y)[j_r])\right),$$

where a term is ignored if the corresponding index does not exist.

The *Merge Longest Common Prefix Array* MLCP of two sets X and Y is obtained by interleaving the arrays GLCP_X and GLCP_Y according to the color array col, i.e.

$$\text{MLCP}[i] = \begin{cases} \text{GLCP}_X[\text{rank}_0(\text{col}, i)] & \text{if } \text{col}[i] = 0, \\ \text{GLCP}_Y[\text{rank}_1(\text{col}, i)] & \text{if } \text{col}[i] = 1. \end{cases}$$

For simplicity of exposition, in the following two subsections, we describe a new characterization of the DAWs and the algorithm for their computation, by assuming that each of the two sets contains a single word. However, all the combinatorial properties and all data structures and operations used can be naturally extended to the case in which two generic sets of words X and Y are given as input. This means that the algorithm immediately generalizes to computing $\mathcal{D}(X, Y)$ as defined in Sect. 3, without any change in the asymptotic complexity or data-structures requirements.

4.1   Discriminative Absent Words for Two Strings

Let us assume that $X = \{x\}$ and $Y = \{y\}$ and let us denote by $S(x, y)$ the list $S(X \uplus Y)$. Denoted $N = \|X \uplus Y\|$, we define map : $\mathcal{D}(x, y) \rightarrow [1, N]$ as the function which associates each $w \in \mathcal{D}(x, y)$ to the rank of the first suffix in lexicographic order with w as prefix. Observe that map is injective, since by Proposition 1(3.) the set $\mathcal{D}(x, y)$ is antifactorial. The ψ function is supposed to be associated to $X \uplus Y$.

Lemma 1. *Let $w \in \mathcal{D}(x, y)$. Then,*

$$\text{CLCP}[\psi(\text{map}(w))] \geq \text{CLCP}[\text{map}(w)] = |w| - 1.$$

Proof. If $w \in \mathcal{D}(x, y)$, then either $w \in D(x \leftarrow y)$ (and $w \notin D(y \leftarrow x)$) or $w \in D(y \leftarrow x)$ (and $w \notin D(x \leftarrow y)$). Without loss of generality, suppose $w \in D(x \leftarrow y)$. For w to be in $D(x \leftarrow y)$, the following properties hold:

1. $w \in F(x)$;
2. $w \notin F(y)$;
3. $w[1..|w| - 1] \in F(y)$;
4. $w[2..|w|] \in F(y)$.

By Properties 1. and 2.n, it follows that $\mathsf{map}(w) = i$ implies $\mathsf{col}[i] = color(x)$ and $\mathsf{CLCP}[i] < |w|$, while by Property 3. $\mathsf{CLCP}[i] \geq |w[1..|w| - 1]| = |w| - 1$; combining these, we obtain $\mathsf{CLCP}[i] = |w| - 1$. Moreover, since $\mathsf{col}[i] = \mathsf{col}[\psi(i)]$ for all i and $S(x, y)[\psi(i)]$ starts with $w[2..|w|]$, by Property 4. there exists i' such that $\mathsf{col}[i'] = color(y)$ and $S(x, y)[i']$ starts with $w[2..|w|]$ as well, implying that $\mathsf{CLCP}[\psi(\mathsf{map}(w))] \geq \ell cp(S(x, y)[\psi(\mathsf{map}(w))], S(x, y)[i']) \geq |w[2..|w|]| = |w| - 1$.

The case $w \in D(y \leftarrow x)$ is treated symmetrically, and the thesis follows. $\square$

Lemma 2. *Let $i \geq 3$ such that $i \neq \mathsf{map}(w)$ for all $w \in \mathcal{D}(x, y)$. Then, at least one of the following is true:*

1. $\mathsf{CLCP}[i] < \mathsf{MLCP}[i]$;
2. $\mathsf{CLCP}[i] > \mathsf{CLCP}[\psi(i)]$.

Proof. Without loss of generality, suppose $\mathsf{col}[i] = color(x)$. Let us suppose $S(x, y)[i]$ is prefixed by some $w \in \mathcal{D}(x, y)$, and by Lemma 1 $\mathsf{CLCP}[i] = |w| - 1$. From the hypothesis $\mathsf{map}(w) \neq i$, it follows that there exists $j < i$ such that $\mathsf{map}(w) = j$. Moreover, since w does not occur in y, we have that $\mathsf{col}[j] = color(x)$, and therefore $\mathsf{MLCP}[i] \geq \ell cp(S(x, y)[j], S(x, y)[i]) \geq |w| > \mathsf{CLCP}[i]$.

Let us suppose that there does not exist $w \in \mathcal{D}(x, y)$ such that $S(x, y)[i]$ is prefixed by w. Then, each prefix $\tilde{w} \notin \mathcal{D}(x, y)$ of $S(x, y)[i]$ is either a factor of both x and y or a factor of x and at least one between its prefix and its suffix of length $|\tilde{w}| - 1$ does not occur in y. Moreover, since $i \geq 3$, we also have that $\tilde{w} \notin \{\$_1, \$_2\}$. Based on the length of each prefix $\tilde{w}$, we can distinguish between three cases:

a. if $|\tilde{w}| \leq \mathsf{CLCP}[i]$, then $\tilde{w} \in F(x) \cap F(y)$;
b. if $|\tilde{w}| > \mathsf{CLCP}[i] + 1$, then $\tilde{w}, \tilde{w}[1, |\tilde{w}| - 1] \in F(x)$ and $\tilde{w}, \tilde{w}[1..|\tilde{w}| - 1] \notin F(y)$;
c. if $|\tilde{w}| = \mathsf{CLCP}[i] + 1$, observe that $\tilde{w}[1..|\tilde{w}| - 1] \in F(y)$, while $\tilde{w} \in F(x)$ and $\tilde{w} \notin F(y)$; Since $S(x, y)[\psi(i)]$ has $\tilde{w}[2..|\tilde{w}|]$ as prefix, to not contradict the hypothesis $\tilde{w} \notin \mathcal{D}(x, y)$, it follows that $\tilde{w}[2..|\tilde{w}|] \notin F(y)$, and therefore $\mathsf{CLCP}[\psi(i)] < |\tilde{w}[2..|\tilde{w}|]| = |\tilde{w}| - 1 = \mathsf{CLCP}[i]$, and the thesis follows. $\square$

By using previous lemmas, the following corollary gives a characterization of the DAWs of x and y.

Corollary 1. *Let $w \in \Sigma^*$. Then, $w \in \mathcal{D}(x, y)$ if and only if*

$$\mathsf{CLCP}[\psi(\mathsf{map}(w))] \geq \mathsf{CLCP}[\mathsf{map}(w)] \geq \mathsf{MLCP}[\mathsf{map}(w)].$$

Example 4. Let us consider $x = \mathtt{AGCAGT\$}_1$ and $y = \mathtt{ACACGT\$}_2$, where $\$_1 < \$_2 < \mathtt{A} < \mathtt{C} < \mathtt{G} < \mathtt{T}$, and $color(x) = 0$ and $color(y) = 1$. In Table 1, we show the sorted suffixes in $S(x, y)$ and the corresponding values of the generalized suffix array GSA of X, the col array, the MLCP array, and the CLCP array. We highlight in red (resp. blue) the DAWs in $D(x \leftarrow y)$ (resp. $D(y \leftarrow x)$). Observe that only for $i \in \{\mathsf{map(AC)}, \mathsf{map(AG)}, \mathsf{map(CG)}, \mathsf{map(GC)}\} = \{3, 5, 9, 10\}$, the conditions of Corollary 1 are verified, while for $j \in \{4, 6\}$ we have that $\mathsf{MLCP}[j] > \mathsf{CLCP}[j]$, and for $j' \in \{7, 8, 11, 12, 13, 14\}$ we have $\mathsf{CLCP}[\psi(j')] < \mathsf{CLCP}[j']$.

Table 1. Example for the computation of DAWs of $x = \mathtt{AGCAGT\$}_1$ and $y = \mathtt{ACACGT\$}_2$. The table reports the lexicographically sorted suffixes in $S(x, y)$ together with the corresponding entries of the generalized suffix array GSA, the eBWT, the standard permutation $\pi = \pi_{\mathsf{eBWT}}$, the ψ function, the color array col, and the values of MLCP and CLCP. The suffixes whose prefixes satisfy the conditions of Corollary 1 are highlighted: red entries (**AG** and **GC** at lines 5 and 10 resp.) correspond to elements in $D(x \leftarrow y)$, while blue entries (**AC** and **CG** at lines 3 and 9 resp.) correspond to elements in $D(y \leftarrow x)$.

i	$S(x,y)$	GSA($\{x, y\}$)	eBWT	π	ψ	col	MLCP	CLCP
1	$\$_1$	(1,7)	T	13	5	0	0	0
2	$\$_2$	(2,7)	T	14	3	1	0	0
3	ACACGT$\$_2$	(2,1)	$\$_2$	2	7	1	0	1
4	ACGT$\$_2$	(2,3)	C	7	9	1	2	1
5	AGCAGT$\$_1$	(1,1)	$\$_1$	1	10	0	0	1
6	AGT$\$_1$	(1,4)	C	8	11	0	2	1
7	CACGT$\$_2$	(2,2)	A	3	4	1	0	2
8	CAGT$\$_1$	(1,3)	G	10	6	0	0	2
9	CGT$\$_2$	(2,4)	A	4	12	1	1	1
10	GCAGT$\$_1$	(1,2)	A	5	8	0	0	1
11	GT$\$_1$	(1,5)	A	6	13	0	1	2
12	GT$\$_2$	(2,5)	C	9	14	1	0	2
13	T$\$_1$	(1,6)	G	11	1	0	0	1
14	T$\$_2$	(2,6)	G	12	2	1	0	1

4.2   Computing $\mathcal{D}(x, y)$ in Linear Time

Let us consider the Ψ array of length $N = |x| + |y| + 2$ that stores the values of the ψ function associated to $X \uplus Y$. Corollary 1 provides a local characterization of discriminative absent words. It suggests that, once the arrays CLCP, MLCP, and Ψ of length N have been computed, the set $\mathcal{D}(x, y)$ can be constructed by a single scan of these arrays, as described in Algorithm 1.

When the condition of Corollary 1 holds, the pair $(\mathsf{GSA}[i], \mathsf{CLCP}[i] + 1)$ is added to the set $\mathcal{D}(x, y)$ of DAWs. Recall that $\mathsf{GSA}[i] = (d, p)$ identifies the

---

Algorithm 1:   Input: Colored Longest Common Prefix Array CLCP, Merge Longest Common Prefix Array MLCP **Output:** $\mathcal{D}(x,y)$

---

1 $\mathcal{D} \leftarrow \emptyset$
2 $k \leftarrow 2$
3 $N \leftarrow |x| + |y| + k$
4 **for** $i \in [k+1, N]$ **do**
5 | **if** $\mathsf{CLCP}[\Psi[i]] \geq \mathsf{CLCP}[i] \geq \mathsf{MLCP}[i]$ **then**
6 | |_ $\mathcal{D} \leftarrow \mathcal{D} \cup \{(\mathsf{GSA}[i], \mathsf{CLCP}[i]+1)\}$

7 **return** $\mathcal{D}$

---

i-th suffix in lexicographic order, where d is the index of the word in the set (equivalently, its document identifier) and p is the starting position of the suffix within that word. Hence, the pair $(\mathsf{GSA}[i], \mathsf{CLCP}[i]+1)$ uniquely represents a DAW by specifying: (i) the word in which it occurs, (ii) its starting position in that word, and (iii) its length.

We need to preprocess the GLCP array for the set $\{x\} \uplus \{y\}$ with a Range Minimum Query (RMQ) structure, which requires linear time and space and supports constant-time queries [17]. Given an array T of length N, a Range Minimum Query $\mathsf{RMQ}_T(i,j)$ returns an index $k \in [i..j]$ such that $T[k]$ is minimum in the interval $T[i..j]$.

In the following lemmas, we show how to compute the values $\mathsf{CLCP}[i]$ and $\mathsf{MLCP}[i]$, for each position $i \geq 3$.

Lemma 3. *Given the* GLCP *for the set* $\{x\} \uplus \{y\}$ *and any integer* $0 \leq i \leq N$, *let* $j_l = \mathsf{select}_{\overline{col[i]}}(\mathsf{col}, \mathsf{rank}_{\overline{col[i]}}(\mathsf{col}, i))$ *and* $j_r = \mathsf{select}_{\overline{col[i]}}(\mathsf{col}, \mathsf{rank}_{\overline{col[i]}}(\mathsf{col}, i)+1)$. *Then,* $\mathsf{CLCP}[i] = \max\{\mathsf{GLCP}[m_l], \mathsf{GLCP}[m_r]\}$, *where* $m_l = \mathsf{RMQ}_{\mathsf{GLCP}}(j_l, i)$ *and* $m_r = \mathsf{RMQ}_{\mathsf{GLCP}}(i, j_r)$.

Lemma 4. *Given the* GLCP *for the set* $\{x\} \uplus \{y\}$ *and any integer* $0 \leq i \leq |x| + |y|$, *let* $j = \mathsf{select}_{col[i]}(\mathsf{col}, \mathsf{rank}_{col[i]}(\mathsf{col}, i) - 1)$. *Then,* $\mathsf{MLCP}[i] = \mathsf{GLCP}[\mathsf{RMQ}_{\mathsf{GLCP}}(j, i)]$.

Note that the previous lemma also shows that the array MLCP corresponds to a special case of the notion of Upper Longest Common Prefix array introduced in [19] for the computation of matching statistics in external memory.

Proposition 3. *Given two strings* x *and* y *of total length* N, *the set* $\mathcal{D}(x,y)$ *can be computed in* $O(N)$ *time and* $O(N)$ *words of space.*

Proof. The generalized suffix array GSA and the generalized LCP array GLCP of x and y takes $O(N)$ words of space and can be constructed in linear time [24]. The col array can be easily built at the construction time of the GSA, taking $O(N)$ bits of space. Then, in $O(N)$ words of space and $O(N)$ time, we can build a data structure supporting constant-time range minimum queries on GLCP [17]. Instead of computing the ψ function at each step, we first build the Ψ array of

$O(N)$ words of space in $O(N)$ time by augmenting eBWT with data structures supporting constant-time select queries and granting access in constant time [30]. We can build in $O(N)$ time a data structure supporting constant-time rank and select queries on the bit vector col taking $O(N)$ words of space [28]. Overall, all the data structures used can be preprocessed in $O(N)$ time and stored in $O(N)$ words of space. By Lemma 3, each value CLCP$[i]$ can be computed by a constant number of rank and select operations on the col array together with two RMQ queries, hence in $O(1)$ time per position. Similarly, by Lemma 4, each value MLCP$[i]$ is obtained by one rank, one select, and one RMQ query, thus also in $O(1)$ time per position. Thus, Algorithm 1 makes $O(N)$ comparisons, each in $O(1)$ time, and the thesis follows. $\square$

4.3   The General Case

Given two sets of words X and Y, in the set $X \uplus Y$, each word is terminated by a distinct end-of-string. Moreover, we assign a distinct color to each set of words.

The combinatorial characterization of Corollary 1 remains valid, since it depends only on local comparisons of auxiliary arrays and the Ψ array, which preserves the same properties. Therefore, Algorithm 1, where $k = |X| + |Y|$ and $N = ||X \uplus Y|| = ||X|| + ||Y|| + k$ computes $\mathcal{D}(X, Y)$ in linear time with respect to the total input size.

5   Variants and Algorithmic Solutions

Algorithm 1 can be immediately adapted to compute words specific to a target set against a reference set, as well as the circular version of DAWs.

Computation of X-specific Words Against Y. Given the sets of words X and Y, to compute the X-specific words against Y, that is $D(X \leftarrow Y) = F(X) \cap M(Y)$, we need to check, for each position $i \geq k + 1$ if the i-th element corresponds to a suffix of a word in X. Concretely, the test at line 5 is performed only when col$[i] = 0$ (assuming 0 labels words in X and 1 labels words in Y); if col$[i] = 1$ the iteration is skipped and the algorithm proceeds to the next index. Symmetrically, the computation of $D(Y \leftarrow X) = F(Y) \cap M(X)$ is obtained by applying the same procedure while requiring col$[i] = 1$. All data structures and operations (rank/select on eBWT, RMQ on GLCP, and the auxiliary LCP-based arrays) are unchanged, and the time and space complexities remain linear in the total input size.

Circular Discriminative Absent Words. Given two sets X and Y, we are interested in computing discriminative absent words when *circular factors* of the input words are considered, namely $\mathring{\mathcal{D}}(X, Y) = (\mathring{F}(X) \cap \mathring{M}(Y)) \cup (\mathring{F}(Y) \cap \mathring{M}(X))$. For simplicity, we assume that all words in X and Y are *primitive*; the general case will be discussed in a full version of this paper. In this circular setting, no end-of-string symbol is appended; then, here we use $X \uplus Y$ to denote the multiset given by the disjoint union of X and Y, and we let $S(X \uplus Y)$ be the list of all conjugates of the words in $X \uplus Y$ sorted according to the ω-order, i.e., the order induced by comparing the corresponding infinite periodic words [25]. Accordingly, instead of the generalized suffix array, we use the *generalized conjugate array* GCA, which stores at each position the identifier of the originating word in $X \uplus Y$ together with the starting position of the corresponding conjugate within that word. The algorithm then proceeds exactly as in the linear case, replacing GSA and GLCP with their circular counterparts induced by GCA and by the ω-sorted conjugates, while keeping the same local conditions on the corresponding CLCP and MLCP arrays and the same ψ function (now defined over the circular order). In this variant, $k = 0$ and N is the total length $\|X\| + \|Y\|$. Therefore, the circular computation preserves the $O(N)$ time and $O(N)$ space bounds.

6   Conclusions and Further Works

We presented a linear-time and linear-space algorithm for computing the set of DAWs between two sets of words. DAWs are detected through simple local conditions that can be evaluated during a single scan of auxiliary LCP-like arrays, thus avoiding the explicit construction of the full sets of minimal absent words. Each element of $\mathcal{D}(X, Y)$ is reported in compact form, as a triple encoding the originating word (document identifier), the starting position, and the length.

The proposed framework is flexible and admits natural extensions. First, the same methodology applies to circular words by replacing GSA with GCA. Second, the computation of one-sided specific sets $D(X \leftarrow Y)$ can be obtained with minor modifications of the selection phase.

An additional application of our results is related to the similarity measures based on absent words. In particular, the similarity measure introduced in [8] can be extended from pairs of words to pairs of string collections, yielding the function

$$\delta(X, Y) = \sum_{w \in \mathcal{D}(X,Y)} \frac{1}{|w|^2}.$$

that depends only on the lengths of the discriminative absent words of X and Y. Since our algorithm reports directly the lengths of the words in $\mathcal{D}(X, Y)$, it enables the comparison of two collections of words, avoiding the potentially larger symmetric difference $M(X) \triangle M(Y)$. It would be interesting to investigate the mathematical properties of $\delta(X, Y)$ for collections and to assess whether, as observed experimentally in the single-string setting [7], $\mathcal{D}(X, Y)$ is substantially smaller on average (about 30%) than $M(X) \triangle M(Y)$ while still producing coherent classifications with respect to standard benchmarks.

Furthermore, a promising direction for further research concerns the adaptation of our algorithm to compressed and highly repetitive collections. In particular, representations based on run-length encoded BWTs or other compressed suffix structures could reduce the space usage while preserving efficient rank, select, and RMQ operations [6,15,18,20,22,29]. Investigating whether DAWs can be computed in space proportional to measures of repetitiveness, such as the number of BWT-runs, remains an interesting open problem.

Acknowledgements. SM, GR, and MS are partially supported by the INdAM - GNCS Project CUP_E53C25002010001. GC, SM, GR, and MS are partially supported by the project "ACoMPA – Algorithmic and Combinatorial Methods for Pangenome Analysis" (CUP B73C24001050001) funded by the NextGeneration EU programme PNRR MUR M4 C2 Inv. 1.5 – Project ECS00000017 Tuscany Health Ecosystem (Spoke 6), CUP Master B63C22000680007.

Disclosure of Interests. The authors have no competing interests to declare that are relevant to the content of this article.

References

1. Béal, M., Crochemore, M.: Fast detection of specific fragments against a set of sequences. In: DLT, volume 13911 of Lecture Notes in Computer Science, pp. 51–60. Springer (2023). https://doi.org/10.1007/978-3-031-33264-7_5
2. Béal, M., Crochemore, M.: Specific patterns against reference sequences. In: From Strings to Graphs, and Back Again, volume 132 of OASIcs, pp. 14:1–14:12. Schloss Dagstuhl - Leibniz-Zentrum für Informatik (2025). https://doi.org/10.4230/OASICS.GROSSI.14
3. Béal, M.-P., Crochemore, M., Mignosi, F., Restivo, A., Sciortino, M.: Computing forbidden words of regular languages. Fund. Inform. **56**(1–2), 121–135 (2003)
4. Béal, M.-P., Mignosi, F., Restivo, A., Sciortino, M.: Forbidden words in symbolic dynamics. Adv. Appl. Math. **25**(2), 163–193 (2000). https://doi.org/10.1006/AAMA.2000.0682
5. Bonizzoni, P., De Felice, C., Pirola, Y., Rizzi, R., Zaccagnino, R., Zizza, R.: Can formal languages help pangenomics to represent and analyze multiple genomes? In: DLT, volume 13257 of Lecture Notes in Computer Science, pp. 3–12. Springer (2022). https://doi.org/10.1007/978-3-031-05578-2_1
6. Brown, N.K., Gagie, T., Manzini, G., Navarro, G., Sciortino, M.: Faster run-length compressed suffix arrays. In: From Strings to Graphs, and Back Again, volume 132 of OASIcs, pp. 10:1–10:15. Schloss Dagstuhl - Leibniz-Zentrum für Informatik (2025). https://doi.org/10.4230/OASICS.GROSSI.10
7. Castiglione, G., Mantaci, S., Pizzuto, S.L., Restivo, A.: A comparison between similarity measures based on minimal absent words: an experimental approach. In: ICTCS, volume 3811 of CEUR Workshop Proceedings, pp. 95–105. CEUR-WS.org (2024). https://ceur-ws.org/Vol-3811/paper140.pdf
8. Castiglione, G., Mantaci, S., Restivo, A.: Some investigations on similarity measures based on absent words. Fundam. Informaticae **171**(1–4), 97–112 (2020). https://doi.org/10.3233/FI-2020-1874
9. Cenzato, D., Lipták, Z.: A survey of BWT variants for string collections. Bioinform. **40**(6) (2024). https://doi.org/10.1093/BIOINFORMATICS/BTAE333
10. Chairungsee, S., Crochemore, M.: Using minimal absent words to build phylogeny. Theor. Comput. Sci. **450**, 109–116 (2012). https://doi.org/10.1016/j.tcs.2012.04.031

11. Charalampopoulos, P., Crochemore, M., Fici, G., Mercas, R., Pissis, S.P.: Alignment-free sequence comparison using absent words. Inf. Comput. **262**(Part), 57–68 (2018). https://doi.org/10.1016/j.ic.2018.06.002
12. Crochemore, M., Désarménien, J., Perrin, D.: A note on the burrows - wheeler transformation. Theor. Comput. Sci. **332**(1–3), 567–572 (2005). https://doi.org/10.1016/J.TCS.2004.11.014
13. Crochemore, M., Mignosi, F., Restivo, A.: Automata and forbidden words. Inf. Process. Lett. **67**(3), 111–117 (1998). https://doi.org/10.1016/S0020-0190(98)00104-5
14. Crochemore, M., Mignosi, F., Restivo, A., Salemi, S.: Data compression using antidictionaries. Proc. IEEE **88**(11), 1756–1768 (2000). https://doi.org/10.1109/5.892711
15. Ferragina, P., Lari, F.: FL-RMQ: a learned approach to range minimum queries. In: CPM, volume 331 of LIPIcs, pp. 7:1–7:23. Schloss Dagstuhl - Leibniz-Zentrum für Informatik (2025). https://doi.org/10.4230/LIPICS.CPM.2025.7
16. Fici, G., Mantaci, S., Restivo, A., Romana, G., Rosone, G., Sciortino, M.: BWT and combinatorics on words. In: The Expanding World of Compressed Data, volume 131 of OASIcs, pp. 1:1–1:23. Schloss Dagstuhl - Leibniz-Zentrum für Informatik (2025). https://doi.org/10.4230/OASICS.MANZINI.1
17. Fischer, J., Heun, V.: Space-efficient preprocessing schemes for range minimum queries on static arrays. SIAM J. Comput. **40**(2), 465–492 (2011). https://doi.org/10.1137/090779759
18. Gagie, T., Navarro, G., Prezza, N.: Fully functional suffix trees and optimal text searching in bwt-runs bounded space. J. ACM **67**(1), 2:1–2:54 (2020). https://doi.org/10.1145/3375890
19. Garofalo, F., Rosone, G., Sciortino, M., Verzotto, D.: The colored longest common prefix array computed via sequential scans. In: SPIRE, volume 11147 of Lecture Notes in Computer Science, pp. 153–167. Springer (2018). https://doi.org/10.1007/978-3-030-00479-8_13
20. Gawrychowski, P., Jo, S., Mozes, S., Weimann, O.: Compressed range minimum queries. Theor. Comput. Sci. **812**, 39–48 (2020). https://doi.org/10.1016/J.TCS.2019.07.002
21. Gessel, I.M., Reutenauer, C.: Counting permutations with given cycle structure and descent set. J. Comb. Theory A **64**(2), 189–215 (1993)
22. Jo, S., Rao Satti, S.: Encoding data structures for range queries on arrays. In: From Strings to Graphs, and Back Again, volume 132 of OASIcs, pp. 12:1–12:12. Schloss Dagstuhl - Leibniz-Zentrum für Informatik (2025). https://doi.org/10.4230/OASICS.GROSSI.12
23. Li, H.: Exploring single-sample SNP and indel calling with whole-genome de novo assembly. Bioinformatics **28**(14), 1838–1844 (2012). https://doi.org/10.1093/BIOINFORMATICS/BTS280
24. Louza, F.A., Telles, G.P., Gog, S., Prezza, N., Rosone, G.: GSUFSORT: constructing suffix arrays, LCP arrays and BWTs for string collections. Algorithms Mol. Biol. **15**(1), 18 (2020). https://doi.org/10.1186/S13015-020-00177-Y
25. Mantaci, S., Restivo, A., Rosone, G., Sciortino, M.: An extension of the burrows-wheeler transform. Theor. Comput. Sci. **387**(3), 298–312 (2007). https://doi.org/10.1016/J.TCS.2007.07.014
26. Mignosi, F., Restivo, A., Sciortino, M.: Forbidden factors in finite and infinite words. In: Karhumäki, J., Maurer, H.A., Păun, G., Rozenberg, G., (eds.), Jewels Are Forever: Contributions on Theoretical Computer Science in Honor of Arto Salomaa, pp. 339–350. Springer (1999)

27. Mignosi, F., Restivo, A., Sciortino, M.: Forbidden factors and fragment assembly. RAIRO Theor. Inf. Appl. **35**(6), 565–577 (2001). https://doi.org/10.1051/ITA:2001132
28. Navarro, G.: Compact Data Structures - A Practical Approach, Cambridge University Press (2016)
29. Nishimoto, T., Tabei, Y.: Optimal-time queries on bwt-runs compressed indexes. In ICALP, volume 198 of LIPIcs, pp. 101:1–101:15. Schloss Dagstuhl - Leibniz-Zentrum für Informatik (2021). https://doi.org/10.4230/LIPICS.ICALP.2021.101
30. Ohlebusch, E.: Bioinformatics Algorithms: Sequence Analysis, Genome Rearrangements, and Phylogenetic Reconstruction, Oldenbusch Verlag (2013)
31. Perrin, D., Restivo, A.: Words. In: Bóna, M. (ed.) Handbook of Enumerative Combinatorics. CRC Press, Boca Raton, FL (2015)

Passive Learning of Symbolic Automata over Monotonic Algebras

Peter Habermehl[1]([✉]) [iD] and Erwann Loulergue[2] [iD]

[1] Université Paris Cité, CNRS, IRIF, 75013 Paris, France
haberm@irif.fr
[2] Université Paris-Saclay, CNRS, ENS Paris-Saclay, Laboratoire Méthodes Formelles, 91190 Gif-sur-Yvette, France
ErwannL@lmf.cnrs.fr

Abstract. Symbolic automata extend classical finite-state automata to handle large or infinite alphabets by labeling transitions by predicates coming from a boolean algebra. Many results from automata theory have been lifted to this model, and it has proved its usefulness for example in multiple software verification applications. Here, we tackle the passive learning problem of identification in the limit, i.e. learning a model from a sample without access to an oracle to query. We provide an algorithm, SAI, that efficiently identifies in the limit symbolic automata over any monotonic algebra where predicates labeling transitions are of the form $a \preceq x \prec b$. The algorithm extends the RPNI framework for passive learning of finite-state automata to symbolic automata thanks to a new *splitting* operation inspired by RTI, a passive learning algorithm for deterministic real-time automata, a subclass of timed automata. The learning algorithm combines merging of states and splitting of states allowing to infer the predicates on transitions in a top-down fashion. We prove that SAI admits polynomial size characteristic samples.

Keywords: Symbolic Automata · Automata Learning · Identification In The Limit

1 Introduction

Symbolic finite-state automata (SFA) have attracted a lot of interest recently. In contrast to usual finite-state automata (FA) they can define languages over an infinite alphabet. Their transitions are predicates coming from a boolean algebra denoting sets of letters from a potentially infinite domain (natural numbers, reals, etc.). Even if the alphabet is finite, it is often useful to define predicates describing succinctly finite sets of letters like all characters of some kind in Unicode. There has been a flurry of papers lifting results from the setting of FA to SFA (overviewed in [5]). For example, determinization, minimization, ω-automata, etc. have been considered. SFA are used successfully in practice [5].

The problem of exact *learning* FA and extensions is fundamental [1,7,8]. There are roughly two types of learning. Passive learning considers inferring an

M.-P. Béal and P. Caron (Eds.): DLT 2026, LNCS 16578, pp. 238–251, 2026.
https://doi.org/10.1007/978-3-032-28404-4_18

automaton from a sample of (positive and negative) words from an unknown *target* automaton. The learned automaton should be consistent with the sample, i.e. accept *all* positive examples and reject *all* negative ones. Furthermore, it should *generalize*, i.e. accept more words than the positive examples. One would like to have a theoretical guarantee that the learning algorithm *identifies in the limit* the target automaton i.e. given more and more examples it converges to the target automaton at some point. A *characteristic sample* (CS) is a sample which guarantees that the algorithm infers the target automaton given a sample containing at least the CS. A class of languages (or automata representing them) is called *efficiently identifiable*, if such a (polynomial) learning algorithm exists, with a CS of polynomial size which can be computed efficiently. FA are efficiently identifiable [10] and a lot of passive learning algorithms have been proposed [10, 12]. These algorithms do not necessarily have CS but work well in practice. A lot of algorithms are based on starting with a *prefix-tree automaton* for the sample which only accepts the positive words in the sample and then *merging* states in some fixed or heuristically chosen order generalizing the language accepted by the automaton while ensuring that negative examples are all rejected.

In active learning pioneered by Angluin [1], a learner tries to infer a target automaton by asking membership and equivalence queries to a teacher which knows the target. Angluin's L* algorithm for FA has been extended to SFA [2] recently. The main difficulty is that the learner not only must infer the structure of the automaton but also the predicates labeling transitions. This is solved elegantly in [2] by using an active learning algorithm for predicates of the boolean algebra as subroutines in automata learning. Coming back to passive learning of SFA considered in this paper, here the difficulty is the same: How to combine inference of the structure of the automaton with inference of predicates of the boolean algebra? Passive learning for SFA has been considered very recently in [6]. Among others, they give a sufficient condition for efficient learnability of SFA which allows one to show that SFA over any *monotonic* algebra (predicates can be seen as intervals) are efficiently learnable. Their learning algorithm is based on the existence of a *generalization* function over the alphabet which given a set of domain values outputs predicates in the underlying boolean algebra generalizing them. Then, passive learning of an SFA from a sample S is done by first *decontaminating* S, then passively learning a DFA A_S over the concrete alphabet and then finally generalizing A_S to get an SFA. However, the obtained SFA might not be consistent with the sample. In that case, the algorithm just outputs the prefix-tree automaton. There is a CS of polynomial size whose presence in the sample guarantees identification, but it can be said that the algorithm is only of theoretical interest as in practice there is no guarantee that the sample contains a CS and one would like to obtain a generalization of the sample regardless.

In a more practically oriented paper [14] passive learning of deterministic real-time automata (DRTA) is considered. DRTA are a restricted form of timed-automata where only constraints on the time spent between two consecutive events can be expressed. Therefore transitions are labelled with (time) intervals. Therefore, DRTA can be seen as a special case of SFA. The problem of inferring predicates on the transitions is elegantly solved by starting with a prefix-tree automaton which is typically contradictory (it contains states which should be accepting and rejecting at the same time) with all transitions labeled by $\top$ denoting the whole underlying domain. Then, an operation allows to split states to solve some of the contradictions. This is mixed with merging as in classical passive learning algorithms. The order splitting and merging is done guarantees that the algorithm will stop with typically a generalization of the sample.

In this work, we present a novel passive learning algorithm for SFA called SAI (Symbolic Automaton Inference) which integrates the learning of the structure of the automaton with the learning of the predicates. The structure of the automaton is learned similarly to algorithms for FA, predicates are learned with a *top-down* approach by starting from an automaton where transitions accept all letters at first and splitting predicates only when necessary. To tackle the main issue in passively learning SFA which is generalizing the behavior of the model to letters not appearing in the sample, our approach starts from the most generic model (although contradictory) and then specifies just enough for it to become coherent. SAI is based on adapting the algorithm of [14] to SFA. We show that our algorithm efficiently identifies SFA over monotonic algebras. We exhibit a CS of quadratic size in the size of the automaton (two exponents smaller than the one from [6]). Contrary to the algorithm of [6] SAI generalizes in all situations (for infinite alphabets). The CS is defined directly from the symbolic automaton whereas [14] define the CS operationally on a run of the algorithm.

The rest of the paper is organized as follows. In Sect. 2 we give preliminary definitions. Section 3 presents our learning algorithm and analyzes it. We compare with related work in Sect. 4 before concluding in Sect. 5.

2   Preliminaries

2.1   Symbolic Automata

Symbolic automata theory (see [4] for a tutorial) extends classical automata theory to encompass more complex, or larger, alphabet structures. For example, a complete automaton to recognize some regular expression over a text encoded in UTF16 will need 2^{16} transitions from each state, which quickly becomes far too costly. Instead of using a concretely represented finite alphabet, symbolic automata employ an alphabet implicitly characterized by a Boolean algebra: this allows for handling of massive finite size or even infinite ($\mathbb{N}, \mathbb{Z}, \ldots$) alphabets.

Definition 1. *A **Boolean algebra** is a tuple* $(\mathbb{D}, \mathbb{P}, [\![\cdot]\!], \bot, \top, \wedge, \vee, \neg)$ *where* $\mathbb{D}$ *is a set called **domain**,* $\mathbb{P}$ *a set of **predicates** including* $\top$, $\bot$, *closed under boolean operations.* $[\![\cdot]\!] : \mathbb{P} \to 2^{\mathbb{D}}$ *is a **semantics function** verifying:* $[\![\bot]\!] = \emptyset$, $[\![\top]\!] = \mathbb{D}$ *and* $\forall \varphi, \psi \in \mathbb{P}:$ $[\![\varphi \wedge \psi]\!] = [\![\varphi]\!] \cap [\![\psi]\!]$, $[\![\varphi \vee \psi]\!] = [\![\varphi]\!] \cup [\![\psi]\!]$, *and* $[\![\neg\varphi]\!] = \mathbb{D} \setminus [\![\varphi]\!]$.

A Boolean algebra is said to be *effective* when its boolean operations are computable, and predicate satisfiability is decidable. All Boolean algebra we will consider in this document are assumed to be effective. There are two examples of Boolean algebras we will come back to:

The Interval Algebra, whose domain is $\mathbb{N}$, where predicates are intervals $[a, b[$ $(a \in \mathbb{N}, b \in \mathbb{N} \cup \{+\infty\})$ and their boolean combinations with natural semantics. Similar algebras can be obtained when using $\mathbb{Z}$ or $\mathbb{R}$ instead of $\mathbb{N}$.

The Propositional Algebra is defined relative to a set of atomic propositions $AP = \{p_1, p_2, \ldots, p_n\}$: in contains these, $\top, \bot$, and their boolean closure. Its domain is $\{0, 1\}^n$, with semantics given by $[\![p_i]\!] = \{v \in \{0, 1\}^n, v[i] = 1\}$. This algebra and SFA defined over it play a central role in *model checking* [3].

One way of defining a Boolean algebra is by taking a set of atomic formulae, $\mathbb{P}_0$, including $\top$ and $\bot$ and setting $\mathbb{P}$ as its closure under boolean operations. In the interval algebra, $\mathbb{P}_0$ is the set of intervals, and in the propositional algebra, it is the set of atomic propositions and their negations.

A Boolean algebra is said to be **monotonic**[1] when (1) there exists a total order $\prec$ on $\mathbb{D}$, (2) there exists symbols $d_{-\infty} \in \mathbb{D}$ and d_∞ (not in $\mathbb{D}$) such that for all $d \in \mathbb{D}$, $d_{-\infty} \preccurlyeq d \prec d_\infty$, (3) for all atomic predicates $\varphi \in \mathbb{P}_0$ there exists $a \in \mathbb{D}, b \in \mathbb{D} \cup \{d_\infty\}$ such that $[\![\varphi]\!] = \{d \in \mathbb{D}, a \preccurlyeq d \prec b\}$. In a monotonic algebra, atomic predicates will simply be denoted as $[a, b[$. The interval algebra is an example of a monotonic algebra. These algebras also appear in natural settings (think for example regular expressions `[a-z]`,`[0-9]`, or other ranges of Unicode characters).

Definition 2. *A* ***symbolic automaton*** *(SFA) is a tuple* $\mathcal{A} = (\mathbb{A}, Q, q_\iota, F, \Delta)$ *with* $\mathbb{A}$ *a Boolean algebra,* Q *a finite set of states,* $q_\iota \in Q$ *the initial state,* $F \subseteq Q$ *the set of final states, and* $\Delta \subseteq Q \times \mathbb{P}_\mathbb{A} \times Q$ *a finite set of transitions, where* $\mathbb{P}_\mathbb{A}$ *is the set of predicates of* $\mathbb{A}$*. In a transition* (q_1, φ, q_2)*,* φ *is called the* ***guard***.

We call letters the elements of $\mathbb{D}$, and words those of $\mathbb{D}^*$. An execution of $\mathcal{A}$ on a word $a_1 a_2 \ldots a_n$ is a sequence of states $q_0 q_1 \ldots q_n$ with $q_0 = q_\iota$ and, for $0 \leq i < n$, $a_i \in [\![\varphi_i]\!]$ for a certain φ_i such that $(q_i, \varphi_i, q_{i+1}) \in \Delta$. As usual, an execution is accepting when the last state of the sequence is in F, and the language $\mathcal{L}(\mathcal{A})$

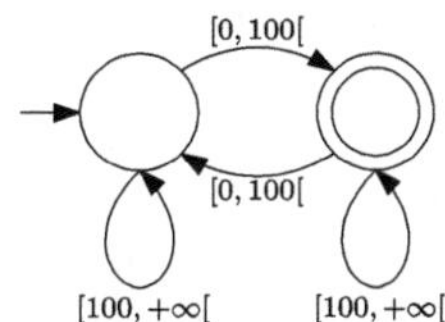

Fig. 1. An example SFA

of an automaton is the set of words that label an accepting execution of $\mathcal{A}$. A SFA is **complete** when at each state q, guards of outgoing transitions from q cover the whole of $\mathbb{D}$: that is, at each state, for each concrete letter there exists at least a transition. A SFA is **normalized** when, given states p and q there is at most one transition from p to q (there might be one from q to p). It is **neat** when every transition's guard is a conjunction of atomic predicates. Neat automata value simple transitions at the cost of having a lot of them, while normalized automata do the inverse. Figure 1 shows an SFA over the interval algebra which

---

[1] This definition slightly differs from the one in [6] for technical reasons related to the symbols $d_{-\infty}$ and d_∞.

is complete, neat, normalized, and recognizes the language of words over $\mathbb{N}$ with an odd number of letters below 100.

2.2 Automata Learning

Automata learning aims to construct a model, in the form of an automaton, that explains observed behavior (either sampled or queried). In essence, we aim to learn a class of languages: regular, context-free, . . . over an alphabet Σ, from examples, i.e. words over Σ with a label indicating whether they are member of the target language or not. In a theoretical setting, the examples come from an oracle, also called teacher. The algorithm that formulates hypotheses is often called the learner. In practice, the sample might come from observations, or a black-box software or hardware to which we provide inputs and observe outputs.

Multiple paradigms exist, some allowing errors and trying to minimize it and some disallowing errors entirely: this is called *exact* learning, which we will focus on. In exact learning, there are two main approaches: In *active learning*, the learner can query an oracle. Allowed queries depend on the paradigm. They can be for example membership and equivalence queries like in Angluin's seminal work [1]. In *passive learning*, the learner has no control over the data it receives. It might receive it in full as an input, or in a streaming setting. In this paper, we focus on the former case: the algorithms we discuss receive a **sample** as input, which consists of words together with a label indicating whether they are a positive or negative example (that is, a subset of $\Sigma^* \times \{+,-\})^2$, and output one automaton. Examples of passive learning algorithms include Gold's algorithm for DFA [8] and the RPNI framework [12]. Measuring the success of an algorithm in this setting is not trivial: only asking that the output automaton correctly accepts all words labeled as positive examples and rejects negative ones is not enough as the language recognized might agree on the sample and still be different from the target language. For example, returning a basic automaton that accepts exactly the positive examples of the sample (nothing more, nothing less) satisfies this condition but gives no *generalization* of the sample. One may therefore ask, given a language $\mathcal{L}$, what a sample should contain in order for a passive learner to identify $\mathcal{L}$. An answer to this question was proposed by Gold in the form of *identification in the limit*, and a more general definition was given by de la Higuera [9], which we follow. The idea of a *characteristic sample* is that it should describe the target automaton, so that it is not only correctly identified by chance, but that additional correct examples do not confuse the learner.

Definition 3. *A **characteristic sample** CS for a target language $\mathcal{L}$ and a learning algorithm A is an input sample $\{S_+ \subset \mathcal{L}, S_- \subset \mathbb{D}^* \setminus \mathcal{L}\}$ such that:*

- *Given CS as input, A returns an automaton $\mathcal{A}$ with $\mathcal{L}(\mathcal{A}) = \mathcal{L}$,*
- *Given a sample $S' \supset CS$ labeled in accordance with $\mathcal{L}$, A still returns an automaton with language $\mathcal{L}$.*

---

[2] This can also take the form of a sample S_+ and a sample S_-. We will use both conventions interchangeably. A sample is not contradictory, i.e. $S_+ \cap S_- = \emptyset$.

Definition 4. *A class of automata $\mathcal{C}$ is said to be **identifiable in the limit** when there exists an algorithm A such that for every automaton $\mathcal{A} \in \mathcal{C}$, there exists a characteristic sample of $\mathcal{L}(\mathcal{A})$. If these characteristic samples are of size polynomial in the size of the minimal automaton[3] recognizing $\mathcal{L}(\mathcal{A})$, computable in time polynomial in the size of this automaton, and the learning algorithm runs in polynomial time, the class is **efficiently identified**.*

3   Learning SFAs

In [6], the authors have shown that SFAs over a monotonic algebra are efficiently identifiable in the limit, but SFAs over the propositional algebra are not efficiently identifiable unless $P = NP$. Learning a symbolic automaton from sampled data runs into the problem of deciding how to identify letters that are not in the sample. For example, say the sample contains 0 as a positive example and 100 as a negative one. The automaton learned could accept letters from 0 to 99, or reject all letters from 1 to 100, or accept from 0 to 49 and reject above 50, and still be consistent with the sample. Note that this is inherent to this precise situation: in active learning, further queries could resolve this issue, and in learning DFAs, the alphabet is finite (and typically small) and no line has to be drawn as usually all letters appear in the sample.

We define an algorithm (see Algorithm 4), called SAI for Symbolic Automaton Inference, to learn symbolic automata over any *monotonic* algebra, inspired by RTI [14]. RTI was defined for a subclass of timed automata, namely *real-time automata*, that simultaneously recognize both a word and a sequence of numbers representing *timings* of the letters from that word. In our case, there are no letters, only timings, if we consider our intervals as windows in time.

The algorithm uses a state-merging framework, also used for DFA learning: it starts with a basic automaton, the *prefix-tree automaton*, then tries to merge the states. More precisely, it uses the red-blue framework [11], where red states are states we have finished identifying, and blue states (successors of red states) are candidates for becoming red (that might still need more work). During a run, there will be states that are neither blue nor red, and the algorithm will return when all remaining states are red. As our sample contains both positive and negative examples, some states will be explicitly rejecting (instead of simply not accepting). An automaton with explicitly rejecting states is called *augmented*, and all automata further on will be augmented with a set R of rejecting states. For brevity, coloring of states is stored implicitly in the following algorithms.

At first, only the initial state is red and all guards are $\top$, so the only thing separating examples is their length. Then, at each iteration, SAI tries to, in this order: (1) merge a blue and a red state, (2) promote a blue state to red, (3) split a transition from a red to a blue state. We check that no irreversible inconsistency (see below) is induced by such an operation : e.g., a state both accepting and rejecting cannot be colored red, and must be split into different states.

---

[3] This definition thus depends on the existence of a minimal representant in the class.

3.1   The Symbolic PTA

Usually, in state-merging algorithms, the first automaton constructed is the *prefix-tree automaton*, or PTA, a simple tree-shaped automaton that accepts (all and only) the positive samples. States of this PTA then get merged and colored, constructing the automaton in a *bottom-up* fashion: first all letters are separated and transitions are very specific, then merges allow for generalization. However, if we were to do the same here, all letters being separated would mean no generalization, which defeats the purpose of symbolic automata on huge or infinite alphabets. The introduction of splits allows us to use a *top-down* strategy.

Thus, to create the initial automaton from a sample, called the symbolic PTA (SPTA), all guards are set to $\top$ (i.e. $[d_{-\infty}, d_\infty[$) and the words from the sample are only distinguished by their lengths. Concretely, the SPTA function takes a sample as input, and outputs an augmented

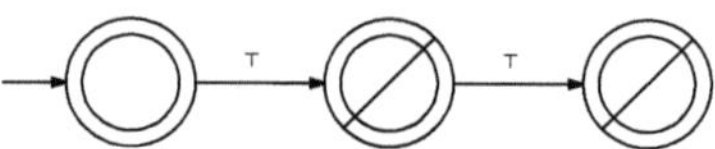

Fig. 2. The SPTA for a sample with accepting and rejecting samples of length 1 and 2, and containing ε as a positive example.

SFA $(\mathcal{A}, R)$ with as many states as the length of the longest word in the sample. The n-th state is accepting (resp. rejecting) when there is an n-letter word with positive (resp. Negative) label in the sample (Fig. 2). Intuitively, this amounts to construct the usual-sense PTA and then merge all states with the same depth in the tree.

3.2   Splitting

Splitting a transition $\delta = (q, [a, b[, q')$ at value $c \in [a, b[$ means creating two transitions from q with guards $[a, c[$ and $[c, b[$. This is the operation [14] brought to state-merging algorithms: in the usual *bottom-up* strategy, there is no need for it. For a split operation, we will remove the targeted state and its descendants from $\mathcal{A}$, then split the transition δ in two, classify samples that fired the removed transition in two sets, and generate the SPTA for these sets. Every string that fires δ, can be written as $\omega = \tau \cdot t \cdot \tau'$ with $a \preceq t \prec b$. We call $t \cdot \tau'$ the tail of ω for δ, and denote it ω^δ. Every transition being split in SAI is leading to an SPTA, whose tree structure ensures the uniqueness of this decomposition (which is not guaranteed in the general case). Denote the subsample of tails from S for δ as S^δ, that is $S^\delta = (\{\omega^\delta \mid \omega \in S^+ \text{ fires } \delta\}, \{\omega^\delta \mid \omega \in S^- \text{ fires } \delta\})$. For a split at value c, denote S^{δ_1} the subset of S^δ with the first letter $\prec c$, and S^{δ_2} those with first letter $\succeq c$.

Using these definitions, the splitting algorithm is given in Algorithm 1.

The computation of S^{δ_i} on line 4 can be done in polynomial time by running the automaton on the sample. However, in an actual implementation, these samples can just be stored in the states.

Algorithm 1: Splitting a transition: $\mathtt{Split}((\mathcal{A}, R), \delta, c, S)$

Data: An augmented SFA $(\mathcal{A} = (\mathbb{A}, Q, q_\iota, F, \Delta), R)$, a transition $\delta = (q, [a, b[, q')$ going to a SPTA, a value $c \in [a, b[$, a sample S

Result: Split δ at c and update $\mathcal{A}$

1 **begin**
2   Remove δ from $\mathcal{A}$
3   Remove q' and all its descendants from $\mathcal{A}$
4   Compute $S^{\delta 1}$ and $S^{\delta 2}$
5     $A_1 \leftarrow \mathtt{SPTA}(S^{\delta 1})$, $A_2 \leftarrow \mathtt{SPTA}(S^{\delta 2})$
6   Let q_1, q_2 be the roots of A_1, A_2 respectively in:
7     $\Delta \leftarrow \Delta \cup \{(q, [a, c[, q_1), (q, [c, b[, q_2)\}$

3.3 Merging

Although we are constructing guards in a top-down fashion, merging operations still occur: at the beginning, states are merged if and only if they are at the same depth in the (usual sense) PTA, thus other merges have not been considered.

Algorithm 2: Merging two states: $\mathtt{Merge}((\mathcal{A}, R), q, q', S)$

Data: An augmented SFA $(\mathcal{A} = (\mathbb{A}, Q, q_\iota, F, \Delta), R)$, two states $q, q' \in Q$ with q' not being red, a sample S

Result: Update $\mathcal{A}$ by merging q and q'

1 **begin**
2   Add a new state q_n to $\mathcal{A}$
3   **if** $q \in F$ *or* $q' \in F$ **then** $F \leftarrow F \cup \{q_n\}$
4   **if** $q \in R$ *or* $q' \in R$ **then** $R \leftarrow R \cup \{q_n\}$
5   **if** *there exists a transition* δ' *from* q' **then**
6   **forall** $\delta = (q, [a, b[, \_)$ *from* q, *by increasing* a **do**
7   **if** $b \neq d_\infty$ **then** $\mathtt{Split}((\mathcal{A}, R), \delta', b, S)$ and $\delta' \leftarrow (q', [b, d_\infty[, \_)$
8   **forall** $\delta = (q_1, \varphi, q_2) \in \Delta$ // (replace q and q' by q_n in transitions)
9   **do**
10   **if** $q_1 = q$ *or* $q_1 = q'$ **then** $\delta \leftarrow (q_n, \varphi, q_2)$
11   **if** $q_2 = q$ *or* $q_2 = q'$ **then** $\delta \leftarrow (q_1, \varphi, q_n)$
12   **while** Δ *contains two transitions* (q_n, φ, q_1) *and* (q_n, φ, q_2) *with the same guard such that* q_2 *is not red* **do**
13        $\mathtt{Merge}((\mathcal{A}, R), q_1, q_2, S)$
14   Remove q and q' from $\mathcal{A}, F, R$

Merging two states is recursive: successors of the two states (and their successors, etc.) are merged to maintain determinism in the automaton. When this function is called from the main algorithm, it is always with a blue state and a red one: in recursive calls, this might not be the case, but one of the two states will always be guaranteed to not be red. See Fig. 3 for an illustration.

Algorithm 2 creates a fresh state q_n replacing both q and q' in all appearances in Δ. It is rejecting (resp. Accepting) when one of the two is (lines 2–4), and takes the "highest" (red>blue>uncolored) color among the two states it replaces. The loop on lines 6 and 7 relies on the fact that q' is not red: when it has an outgoing transition, its guard is $\top$: repeatedly splitting it according to outgoing

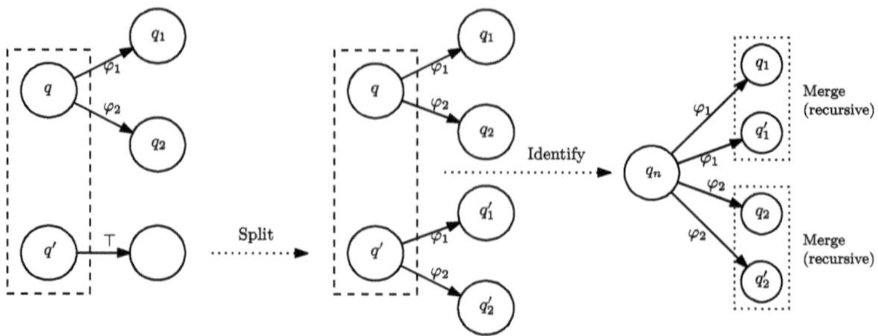

Fig. 3. The steps of a merge: splits, identification, then recursive calls.

transitions of q ensures that outgoing transitions from both states are the same (as the outgoing transitions from any state form a partition of $\mathbb{D}$). Then, the replacement and recursive merges occur on lines 8 through 12.

Checking Inconsistency. During the run of SAI, some states are *inconsistent*, i.e. both accepting and rejecting, or two words of the sample reach the same state with the same suffix remaining, but are labeled differently. This can be fixed by further splits. However, such a state cannot be colored red, as this would be an *irreversible inconsistency.* Thus, when SAI tries the different operations, it tests if the obtained automaton can still be made consistent with the sample without modifying red states. If not, the operation needs to be undone (see Algorithm 4).

Algorithm 3: Checking for irreversible inconsistency: $\texttt{Consistent}(\mathcal{A}, R, S)$

Data: An augmented SFA $(\mathcal{A} = (\mathbb{A}, Q, q_\iota, F, \Delta), R)$, a sample S
Result: true iff $(\mathcal{A}, R)$ can still be made consistent with S

1 **begin**
2   **forall** *red states* $q \in Q$ **do**
3   **if** $q \in F$ *and* $q \in R$ **then return** false
4   **forall** $(q, \varphi, q') \in \Delta$ *such that* q' *is not red* **do**
5   Let $S^\delta = (S^\delta_+, S^\delta_-)$ be the set of tails of S for δ
6   **if** $S^\delta_+ \cap S^\delta_- \neq \emptyset$ **then return** false
7   **return** true

3.4 The SAI Algorithm

Algorithm 4 below contains our passive learning algorithm for symbolic automata SAI. Let us define as the *shortlex blue state*, and denote as q_b, the blue state reached by the smallest string in shortlex order (shorter strings before longer strings, and strings of the same length ordered by lexicographic order). At each step, SAI first tries to merge it into each red state, then to color it, and if it can't do either, splits it. See Fig. 4 for an example run of the algorithm.

Algorithm 4: The SAI algorithm

Data: A sample S
Result: An SFA consistent with S

```
 1  begin
 2      (A = (𝔸, Q, qₗ, F, Δ), R) ← SPTA(S)
 3      Color the initial state qₗ of A red
 4      while A contains non-red states do
 5          forall (q, φ, q′) ∈ Δ such that q is red and q′ is not do Color q′ blue
 6          Let q_b be the shortlex blue state
 7          forall red states q ∈ Q do
 8              Merge((A, R), q, q_b, S)
 9              if Consistent(A, R, S) then skip else Undo the merge
10          Color q_b red
11          if Consistent(A, R, S) then skip else Undo the coloring
12          Let δ, S^δ be the transition leading to q_b and its set of tails
13          forall tails t · τ ∈ S^δ (picked by increasing t) do
14              Split((A, R), δ, t, S)
15              if Consistent(A with the new shotlex state colored red, R, S) then
16                  Undo the split
17              else
18                  Undo the split and Redo the previous split tried
19                  skip
20      return A
```

By skip, we mean go to the next iteration of the line 4 while loop. On line 3, as we assume the sample does not contain a word both positively and negatively, the initial state (corresponding to ε) cannot be both accepting and rejecting, so it is colored right away. In lines 13 and onward, our method of finding the right split is that we are trying to find the split with the largest value such that the resulting new shortlex blue state can be colored red. Thus, as long as splitting and coloring the new q_b red doesn't create an inconsistency, we keep trying bigger and bigger splits. The first one that fails tells us that the previous was the right one (line 18).

In the example run of Fig. 4, only the action that is actually performed is shown at each step. For example at the very first step, the only blue state is both accepting and rejecting, so coloring it red, or merging it into the red state, would create an irreversible inconsistency: thus, the only possible action is to split the transition leading to it. Splitting at value 0 works, but splitting at 100 does also, while any larger value would not: thus, the split at 100 is performed.

In the full paper, we show that SAI runs in polynomial time and returns a SFA consistent with its input sample and we prove the main result of this paper:

Theorem 1. *There exist polynomial characteristic samples for SAI, computable in polynomial time. That is, SAI efficiently identifies in the limit the class of SFAs over monotonic algebras.*

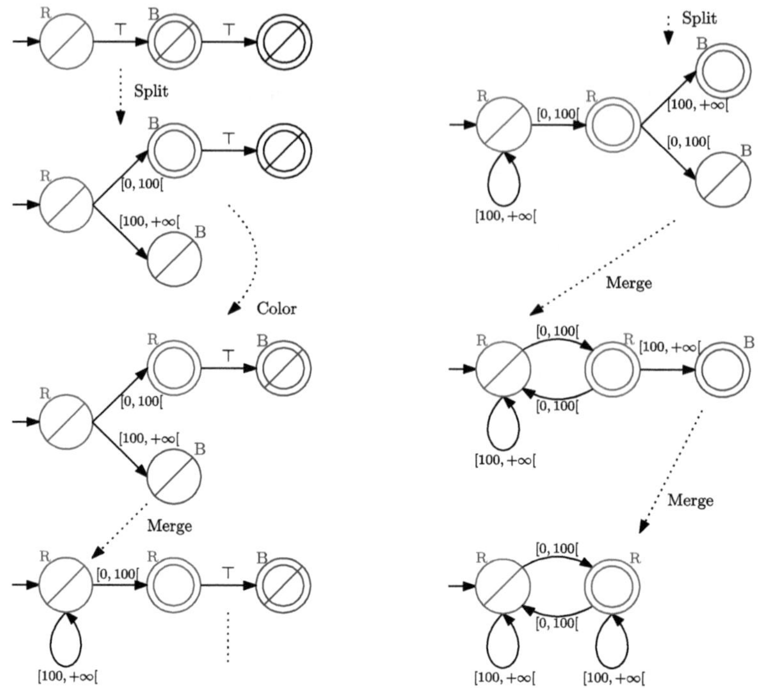

Fig. 4. An example run of SAI for the sample: $\{(\varepsilon, -), (0, +), (100, -), (0 \cdot 0, -),$ $(0 \cdot 100, +)\}$ over the interval algebra. Capital letters next to states denote their color (Blue, Red) when they have one. (Color figure online)

Let us detail the characteristic sample. Let $\mathcal{L}_t$ be any SFA language, and $\mathcal{A}_t$ the minimal-state neat SFA recognizing it (it exists and is unique [6, Lemma 5.7]). Given a state $q \in Q_t$, denote its shortlex access word by τ_q. Given q, q' two distinct states, denote as $\delta_{q,q'}$ a distinguishing word between the two (that is, a word accepted from q iff it is rejected from q'). These words' existence is guaranteed by the minimality of $\mathcal{A}_t$; they are all of length linear in $|Q_t|$, and can be computed in polynomial time. The characteristic sample is then defined as:

$$\bigcup_{q_1 \in Q_t} \bigcup_{\substack{a \in \mathbb{D} \text{ s.t.} \\ \exists (q_1, [a, b[, q_2) \in \Delta_t}} \bigcup_{q \neq q_2} \tau_{q_1} \cdot a \cdot \delta_{q_2, q} \cup \bigcup_{q, q' \in Q_t^2} \tau_q \delta_{q,q'}$$

4   Comparison to Previous-Known Results

We compare our results to Fisman *et al.* [6], the, to the best of our knowledge, only previously known algorithm for passive learning of SFA. With the same formalism as for SAI's the characteristic sample for Fisman *et al.*'s algorithm is:

$$\bigcup_{\substack{a\in\mathbb{D}\ \text{s.t.}\\ \exists(q_1,[a,b[,q_2)\in\Delta_t}} \ \bigcup_{\substack{q\in Q_t\\ q'\neq q''\in Q_t}} \tau_q\cdot a\cdot\delta_{q',q''} \ \cup\ \bigcup_{q,q'\in Q_t^2} \tau_q\delta_{q,q'}$$

Intuitively, if a appears as the lower bound of some transition, then there are words indicating at each state to which state reading a leads (recall τ_q is the access word to q, and $\delta_{q',q''}$ is a word distinguishing q' from q''); while we only ask that a letter appears after the access word for a state if it does appear as a lower bound of a transition from that state. In the worst case (if all lower bounds of transitions are distinct), Fisman *et al.*'s CS contains $|\Delta|\,(|Q|-1)\,|Q|^2+|Q|^2$ words. The one of SAI instead has always $|\Delta|\,(|Q|-1)+|Q|^2$ words. In both, the words' lengths are linear in $|Q|$. Furthermore, the CS for SAI is included in the one for Fisman *et al.*'s algorithm. Therefore, if the latter is guaranteed to learn an automaton, SAI also learns it, whereas the converse is false.

For example, consider the sample of Fig. 4 with one 100 replaced by 99: $\{(\varepsilon,-),(0,+),(100,-),(0\cdot0,-),(0\cdot99,+)\}$. SAI handles it like before, the only difference being that outgoing transitions from the right state are guarded by $[0,99[$ and $[99,+\infty[$. Yet, in Fisman *et al.*'s algorithm, that sole 99 is removed by their *decontaminating* function, so the DFA learner uses sample $\{(\varepsilon,-),(0,+),$ $(100,-),(0\cdot0,-)\}$ and learns an automaton accepting 0 and rejecting everything else. After being lifted back to symbolic automata, the automaton is not consistent with the sample, and thus, the algorithm returns the prefix-tree automaton for the sample accepting only two words, thus not generalizing the sample.

We also compare our results to the ones from Verwer *et al.* [14]: the *real-time automata* (RTA) their RTI algorithm learns are similar to SFA. They recognize words that have letters from a finite alphabet labeled with natural numbers corresponding to delays between letters. RTA over a single-letter alphabet are thus SFA over the natural interval algebra. SAI thus generalizes RTI by using any monotonic algebra and not only $\mathbb{N}$: it could be adapted to recognize words over a finite alphabet with delays from any monotonic algebra by redefining SPTAs for this case. Also, RTI relies on the structure of $\mathbb{N}$ by asking, when learning a transition with lower bound n, for both n and $n-1$ to appear in the sample. The final main difference is that characteristic samples are not defined explicitly but by giving an algorithm to generate them: running RTI with an empty sample, then each time a transition not in the target automaton is learned, adding words to the sample that forbid this operation, and restarting. This makes the CS smaller than the ones we defined —although SAI would also learn the target automaton given RTI's CS—, but practically impossible to express explicitly.

5   Conclusion and Future Work

A way to generalize SAI is to handle the non-monotonic case. As it is known [6] that SFAs over the propositional algebra cannot be efficiently identifiable unless $P=NP$, this can take various forms: (1) defining a polynomial-time algorithm without polynomial-size characteristic samples, (2) defining an algorithm that doesn't run in polynomial-time (with or without polynomial-size characteristic samples), (3) finding sufficient and/or necessary conditions for an algebra for corresponding SFA to be efficiently identifiable. We expect the third option to be the most promising one, as results for active learning are similar: Argyros and D'Antoni's MAT^* [2] showed that predicates of a boolean algebra being learnable in polynomial time using membership (of letters) and equivalence queries, is a sufficient condition for SFA over this algebra to be learnable in polynomial time using membership (of words) and equivalence queries by creating instances of the predicate learning algorithms. Similarly, we would like to define a framework that relies on passive predicate-learning algorithms to get automata-learning ones.

Another generalization would be using the *Evidence-Driven State Merging* (EDSM) framework, where operations do not happen in a fixed order, but a heuristic, called *evidence value*, leads us to do "the most productive" one at every iteration (see [11] for a presentation, and [13] for a survey of evidence values). Although it was defined in cases where the only operations were coloring and merges, results with splits are promising [14]. SAI is currently equivalent to an EDSM algorithm with evidence value such that consistent merges score higher than the coloring, who scores higher than splitting (all only of the shortlex blue state, any other operation scoring 0). The existence of characteristic samples for these evidence-based algorithms is not always guaranteed, but in practice they converge quickly towards small automata consistent with the sample.

Acknowledgments. This work was partially supported by the "France 2030" government investment plan managed by ANR, under the reference ANR-23-PEIA-0006.

References

1. Angluin, D.: Learning regular sets from queries and counterexamples. Inf. Comput. **75**(2), 87–106 (1987)
2. Argyros, G., D'Antoni, L.: The learnability of symbolic automata. In: Chockler, H., Weissenbacher, G. (eds.) CAV 2018. LNCS, vol. 10981, pp. 427–445. Springer, Cham (2018). https://doi.org/10.1007/978-3-319-96145-3_23
3. Baier, C., Katoen, J.P.: Principles of model checking. MIT press (2008)
4. D'Antoni, L., Veanes, M.: The power of symbolic automata and transducers. In: Computer Aided Verification, 29th International Conference (CAV'2017). Springer (2017). https://www.microsoft.com/en-us/research/publication/power-symbolic-automata-transducers-invited-tutorial/
5. D'Antoni, L., Veanes, M.: Automata modulo theories. Commun. ACM **64**(5), 86–95 (2021)
6. Fisman, D., Frenkel, H., Zilles, S.: Inferring symbolic automata. Logical Methods Comput. Sci. **19**(2), 5 (2023). https://doi.org/10.46298/lmcs-19(2:5)2023. https://lmcs.episciences.org/8899
7. Gold, E.M.: Language identification in the limit. Inf. Control **10**(5), 447–474 (1967)
8. Gold, E.M.: Complexity of automaton identification from given data. Inf. Control **37**(3), 302–320 (1978)

9. de la Higuera, C.: Characteristic sets for polynomial grammatical inference. Mach. Learn. **27**, 125–138 (1997)
10. de la Higuera, C.: Grammatical Inference: Learning Automata and Grammars. Cambridge University Press, USA (2010)
11. Lang, K.J., Pearlmutter, B.A., Price, R.A.: Results of the Abbadingo one DFA learning competition and a new evidence-driven state merging algorithm. In: International Colloquium on Grammatical Inference, pp. 1–12. Springer (1998)
12. Oncina, J., García, P.: Inferring regular languages in polynomial update time. World Scientific (1992). https://doi.org/10.1142/9789812797902_0004
13. Tîrnăucă, C.: A survey of state merging strategies for DFA identification in the limit. Triangle p. 121 (2018). https://doi.org/10.17345/triangle8.121-136
14. Verwer, S., Weerdt, M., Witteveen, C.: Efficiently identifying deterministic realtime automata from labeled data. Mach. Learn. **86**, 295–333 (2012). https://doi.org/10.1007/s10994-011-5265-4

Transformations between Minimally f-free Words

Marcella Anselmo[1] , Giuseppa Castiglione[2] , Manuela Flores[3] ,
Dora Giammarresi[4(✉)] , Maria Madonia[5] , and Sabrina Mantaci[2]

[1] Dipartimento di Informatica, Università di Salerno, Fisciano, SA, Italy
manselmo@unisa.it
[2] Dipartimento di Matematica e Informatica, Università di Palermo, Palermo, Italy
{giuseppa.castiglione,sabrina.mantaci}@unipa.it
[3] Dipartimento di Bioscienze e Territorio, Università del Molise, Pesche, IS, Italy
manuela.flores@unimol.it
[4] Dipartimento di Matematica, Università di Roma "Tor Vergata", Roma, Italy
giammarr@mat.uniroma2.it
[5] Dipartimento di Matematica e Informatica, Università di Catania, Catania, Italy
madonia@dmi.unict.it

Abstract. A *transformation* between a pair of binary words u and v of the same length and Hamming distance d is a sequence of words $w_0 = u, w_1, w_2, \ldots, w_d = v$, such that w_i differs from w_{i+1} only in one position. The transformation is said to be *f-free*, for a given word f, if f is *absent* in all w_i, i.e. if f is not a factor of any w_i. A word f is *good* if there exists an f-free transformation between any pair of f-free words of same length. Good words have been completely characterized using combinatorial methods on words. A word f is a *minimal absent word* for w if f is an absent word in w but all its proper factors occur in w. We consider f-free transformations between pairs of words u and v for which f is a minimal absent word and study how the property of including the proper factors of f can be maintained in the intermediate words of the transformation.

Keywords: Minimal Absent Words (MAW) · Good words · Transformations of words · Hamming distance.

1 Introduction

Given two binary words u and v of the same length, their Hamming distance is the number of positions at which they differ, also called errors. A transformation from u to v is a sequence of words starting with u, where each step corrects one error until v is obtained. Two subsequent words in such a transformation thus have a Hamming distance of 1.

An interesting general problem is to determine whether there exists a transformation between two words that share the same combinatorial property, where all intermediate words also satisfy that property. From another perspective:

M.-P. Béal and P. Caron (Eds.): DLT 2026, LNCS 16578, pp. 252–265, 2026.
https://doi.org/10.1007/978-3-032-28404-4_19

decide, for a word with a certain combinatorial property and a certain number of errors to correct, if there exists a transformation that corrects these errors step by step, such that all intermediate words still satisfy the given property.

An important combinatorial property of strings is related to the concept of absent word. An *absent word* (or forbidden factor) in a string is a word that does not appear as a factor anywhere in the string. This property is central in combinatorics on words, formal language theory, and bioinformatics, where it is used to study the structure and constraints of sequences. If a word f is absent in a word u, we say that u is *f-free*. Using this property, the above mentioned general problem can be specified as follows. Given two f-free binary strings of the same length, is it possible to correct the errors in the first one to get the second one, one bit at a time, without ever introducing factor f in any intermediate string? More formally, given a binary word f and a pair of f-free binary words u and v of equal length and Hamming distance d, a *transformation* from u to v consists of a sequence of words $w_0, w_1, w_2, \ldots, w_d, d \geq 1$, where $w_0 = u$, $w_d = v$ and each consecutive pair w_i, w_{i+1} differs exactly in one position. If every word in the sequence avoids f as a factor, the transformation $w_0, w_1, w_2, \ldots, w_d$ is referred to as an *f-free transformation*. It can happen that no f-free transformation does exist for a given pair. Consider, as a simple example, the case with $f = 101$, $u = 1001$ and $v = 1111$; it is not possible to avoid 101 when correcting the errors between u and v.

The problem of which strings can be always avoided was introduced in 2012 by Ilić *et al.* in [20] and, at the same time, by Klavžar and Shpectorov in [21], in the context of finding isometric subgraphs of the hypercubes. Specifically, a word f is called *good* (*isometric* in the hypergraph context) if for any pair of f-free words of the same length there exists an f-free transformation between them. Conversely, a word is *bad* if it does not satisfy this property.

The structure of good words has been extensively analyzed and completely characterized by a property based on error-overlaps [20, 21, 23, 25, 26]. Research in this area remains highly active [7, 10, 19, 24], with recent studies exploring various generalizations, including extensions to larger alphabets, sets of words, alternative edit distances, and even two-dimensional words [1–6, 8, 9, 12, 14]. Additionally, good words can be related to pattern matching with errors and in the study of words avoiding special factors.

In this paper, we exploit the notion of *Minimal Absent Word (MAW)* and study f-free transformations under this stricter hypothesis. They were first introduced by Béal *et al.* in 1996 in [13] and, since then, they have been the subject of extensive research due to their rich combinatorial and algorithmic properties. A word f is a minimal absent word for w if f is absent in w, but all its proper factors are present in w. Informally, although w does not contain f, it still retains most of the information about f, that is, all its proper factors. If f is a minimal absent word for w, then the word w is called *minimally f-free*.

Minimal absent words are particularly useful in applications such as data compression, musical information retrieval, and bioinformatics. As a result, several efficient algorithms have been developed for computing MAWs (see

[11,18,22]). More recently, MAWs have also been explored as a tool for measuring sequence similarity [15–17], with applications in comparative genomics.

Guided by these motivations, in this paper, we consider f-free transformations restricted to pairs of minimally f-free words, only. The paper focuses on answering several key questions. We first ask: Is the notion of "good" word given only on these pairs the same as the classical one? We prove that this question has an affirmative answer. This means that if a word f is such that for all pairs of minimally f-free words there exists an f-free transformation then it is so also for any pair of f-free words. In some sense, the pairs of minimally f-free words are relevant pairs to check the property.

The second question is: What happens in these transformations? Can the intermediate words also remain "close" to f? Namely: Are there good words such that for any pair of minimally f-free words there exists a transformation whose intermediate words are all minimally f-free? The answer now is negative, unless for the trivial cases of $f = 01$ or $f = 10$. The proof goes by showing that for each minimal absent word f, it is possible to identify some "isolated" minimally f-free words for which there is no other minimally f-free word at distance one. This negative answer implies that, in general, to correct the errors in a minimally f-free word, we are forced to pass through some words somehow "far away" from f, since we cannot keep all the factors of f in any intermediate words.

In the last part of the paper, we then relax the constraint and define a word w of length n to be *proximally f-free* if w is f-free, but it contains the prefix or suffix of length $n - 1$, respectively. This condition guaranties that not all but "many" factors of f are in w. Using this notion, the third question we pose is: Can we perform an f-free transformation between a pair of minimally f-free words in a way that all intermediate words are proximally f-free? In other words, each time we correct an error, the new string contains either the longest proper prefix or the longest proper suffix, or both? We address this third question by proving some conditions on the pairs of minimally f-free words that ensure or prevent the existence of proximally f-free transformation. We present some interesting examples of words with and without this property and claim a conjecture for their characterization.

2   Preliminaries

Let Σ be the binary alphabet $\{0,1\}$. A word (or string) w of length $|w| = n$ is $w = a_1 a_2 \cdots a_n$, where $a_1, a_2, \ldots, a_n$ are symbols in Σ. The set of all finite words over Σ is denoted Σ^* and the set of all words over Σ of length n is denoted Σ^n. Finally, ε denotes the *empty word* of length zero and $\Sigma^+ = \Sigma^* \backslash \{\varepsilon\}$. If $a \in \{0,1\}$, we denote by $\bar{a}$ the opposite of a, i.e., $\bar{a} = 1$ if $a = 0$ and vice versa. Then we define the *complement* of w as the word $\overline{w} = \bar{a}_1 \bar{a}_2 \cdots \bar{a}_n$. Moreover, for any $1 \leq i \leq n$, we denote by R_i the *replacement operation* at position i and by $R_i(w)$ the word obtained from w by complementing the bit at position i, i.e., $R_i(w) = a_1 \ldots a_{i-1} \bar{a}_i a_{i+1} \ldots a_n$. Let $w[i]$ denote the symbol of w in position i, i.e., $w[i] = a_i$. Then $w[i..j] = a_i \cdots a_j$, for $1 \leq i \leq j \leq n$, is the *factor* of w that occurs at position i; it is a *proper factor* if $i \neq 1$ or $j \neq n$. Given two words f and w, we say that f is *absent* for w, or that w is f-*free*, if f is not a factor of w. The *prefix* (resp. *suffix*) of w of length ℓ, with $1 \leq \ell \leq n-1$ is $pre_\ell(w) = w[1..\ell]$ (resp. $suf_\ell(w) = w[n-\ell+1..n]$). When $pre_\ell(w) = suf_\ell(w) = u$ then u is an *overlap* of w of length ℓ and *shift* $r = n - \ell$; it is also referred to as *border*. A word without any overlap is called *unbordered*. Finally, given two words w and z, if $pre_\ell(w) = suf_\ell(z) = u$ then u is an *overlap* of w on z and we say that w *overlaps* z (cf. [2]).

Given two equal-length words $u = a_1 \cdots a_n$ and $v = b_1 \cdots b_n$, they have a *mismatch* at position i if $a_i \neq b_i$; a mismatch is also called an *error* and i is referred to as an *error position*. Then, the *Hamming distance* $\mathrm{dist}_H(u, v)$ is the number of positions where u and v have a mismatch.

Definition 1. *Let $u, v \in \Sigma^*$, with $|u| = |v|$ and $\mathrm{dist}_H(u, v) = d$. A transformation τ from u to v is a sequence of $d+1$ words, $\tau = (w_0, w_1, \ldots, w_d)$, such that $w_0 = u$, $w_d = v$, and for any $k = 0, 1, \ldots, d-1$, $\mathrm{dist}_H(w_k, w_{k+1}) = 1$. Moreover, τ is f-free if for any $i = 0, 1, \ldots, d$, the word w_i is f-free.*

A trivial observation is that if $\tau = (w_0, w_1, \ldots, w_d)$ is a transformation from u to v, then $\tau^{-1} = (w_d, w_{d-1}, \ldots, w_1)$ is a transformation from v to u. Also τ is f-free if and only if τ^{-1} is f-free. Moreover, note that a transformation τ from u to v can be seen as a sequence of distinct replacement operations (error corrections) applied to u.

Example 2. Let $f = 1010$, $u = 010110$ and $v = 011100$: u and v are 1010-free. Moreover they have Hamming distance 2 and therefore there exist two different transformations from u to v, namely $\tau_1 = (010110, 011110, 011100)$ and $\tau_2 = (010110, 010100, 011100)$. Note that τ_1 is f-free, whereas τ_2 is not.

Definition 3. *A word $f \in \Sigma^*$ is* good *if for any pair of f-free words u and v, with $|u| = |v|$ and $\mathrm{dist}_H(u, v) \geq 2$, there exists an f-free transformation from u to v. If f is not good, it is called* bad *and a pair of words (u, v) for which no transformation from u to v is f-free is called a* pair of witnesses *for f.*

The word w has a *2-error overlap* (2-eo, for short) if, for some length ℓ, $pre_\ell(w)$ and $suf_\ell(w)$ have 2 mismatches [20]. In [23] it is proven that a word is good if and only if it does not have 2 error overlaps.

Example 4. The word $f = 11100$ is bad because it has a 2-error overlap of length $\ell = 2$. Moreover, (u, v), with $u = 11101100$ and $v = 11110100$, is a pair of witnesses for f, in fact all the transformations from u to v are not f-free. More generally, every word of the form $1^h 0^k$, with $h, k \geq 2$, has a 2-error overlap and therefore it is bad.

We conclude by recalling a notion that will be central in this paper. Given two words, f and w, the word f is a *minimal absent word* for w if w is f-free, yet all its proper factors are present in w. In other terms, f is minimal with respect to the property of being absent. In [18] the following remark is given, which highlights a way to verify the property.

Remark 5. Let $f \in \Sigma^n$ be a word of length $n > 1$. Then f is a minimal absent word for a word w if and only if w is f-free and both $pre_{n-1}(f)$ and $suf_{n-1}(f)$ are factors of w.

3   Transformations between Two Minimally f-free Words

In this section, we explore f-free transformations between words u and v for which f is minimal absent. Those are words that, although they do not include f, are quite close to f. The aim is to investigate whether and how the intermediate words in a transformation between two such words can maintain this property.

Definition 6. *Let $f, w \in \Sigma^*$, with $|f| \leq |w|$. The word w is* minimally f-free *if f is a minimal absent word for w.*

The next lemma gives a combinatorial property of minimally f-free words.

Lemma 7. *For each $f \in \Sigma^*$ there exist two words $x_f, y_f \in \Sigma^*$ such that for any f-free word u, the word $x_f u y_f$ is minimally f-free.*

Proof. We consider only words that start with 1; everything remains valid when applying the complement operation. If $|f| = 1$ any f-free word is minimally f-free too, hence the thesis follows with $x_f = y_f = \varepsilon$.
If $f = 1^h$, with $h > 1$, then $u01^{h-1}$ is minimally f-free for each f-free word u.
If $f = 1^h 0^k$, with $h, k > 0$, let $x_f = 1^{h-1} 0^k$ and $y_f = 1^h 0^{k-1}$. One can easily verify that for each f-free word u, the word $x_f u y_f$ is f free and contains both $pre_{|f|-1}(f)$ and $suf_{|f|-1}(f)$ as factors.
Finally, if $f \neq 1^h 0^k$, with $h, k \geq 0$, let $f = af'b$, with $a, b \in \Sigma$ and z_f be a minimally f-free word. Let $x_f = \varepsilon$ and $y_f = (\overline{b})^{|f|-1}(\overline{a})^{|f|-1} \cdot z_f$, and let us prove that for each f-free word u, the word $x_f u y_f$ is minimally f-free. In fact, $u y_f$ is f-free because f does not occur in u, nor in z_f; it cannot occur in $u(\overline{b})^{|f|-1}$ since this word ends with b; and it cannot occur in $(\overline{a})^{|f|-1} \cdot z_f$, because f begins with a. Finally, f does not occur in $(\overline{b})^{|f|-1}(\overline{a})^{|f|-1}$ because $f \neq 1^h 0^k$. Moreover, $pre_{|f|-1}(f)$ and $suf_{|f|-1}(f)$ are factors of $u y_f$ because they are factors of z_f. The thesis follows from Remark 5.   □

The next proposition takes advantage of Lemma 7 to show that it is possible to determine whether a word is good by simply checking the transformations between minimally f-free words. Note that it can be interpreted as meaning that minimal f-free words have a main role in determining the goodness of f.

Proposition 8. *A word $f \in \Sigma^*$ is good if and only if for any pair of minimally f-free words u and v, there exists an f-free transformation from u to v.*

Proof. The first implication holds by definition. Now, let f be such that for every pair of minimally f-free words there exists an f-free transformation from the first to the second, and assume that f is not good. Then, there exists a pair (u, v) of witnesses for f, i.e., u, v are f-free but there is no f-free transformation from u to v. By Lemma 7, there exist two words $x_f, y_f \in \Sigma^*$ such that $x_f u y_f$ and $x_f v y_f$ are minimally f-free. Then, there is no f-free transformation from $x_f u y_f$ to $x_f v y_f$, contradicting the hypothesis.   □

Our next step is to investigate whether the f-free transformations between minimally f-free words can also preserve the factors of f in the intermediate words. In a sense, this would imply that it is possible to "transform" an f-free word u into another f-free v, always staying close to f. In particular, we investigate whether there are special good words f such that for every pair of minimally f-free words there always exists a transformation consisting entirely of minimally f-free words.

Proposition 9. *Let $f \in \Sigma^2$. Then, for any pair of minimally f-free words there exists an f-free transformation composed of only minimally f-free words if and only if $f \in \{01, 10\}$.*

Proof. The proof is given only for words that start with 1. The same reasoning applies when complementing symbols. The word $f = 10$ verifies the thesis. In fact, the set of minimally f-free words is 0^+1^+ and a minimally f-free word of length n is given as $w_h = 0^h 1^{n-h}$, with $h > 0$. For any pair of words w_h and w_k, $\mathrm{dist}_H(w_h, w_k) = |h - k|$. Assume $h < k$; then $(w_h, w_{h+1}, ..., w_k)$ is a transformation in which the intermediate words are all minimally f-free.
The word 11 does not verify the thesis. In fact, in this case, the set of minimally f-free words is $0^* 1 (0^+ 1)^* 0^*$. Let $u = 0^h 10$ and $v = 0^h 01$, with $h > 0$, be minimally f-free words. One can verify that in any transformation from $0^h 10$ to $0^h 01$ either the factor 11 occurs or 1 is absent in some intermediate word. $\square$

Next, we will prove that the condition of the above proposition does not hold for words of length greater than two. Let us introduce the definition of an isolated word that will be essential for this proof.

Definition 10. *Let $f \in \Sigma^*$. A minimally f-free word w, with $|f| \le |w|$, is isolated with respect to f if every word of the same length and at the Hamming distance one from w is not minimally f-free.*

Example 11. Let $f = 10$. Trivially, 01 is isolated, with respect to f, because 11 and 00 are not minimally f-free. Let $f = 11$. Then the minimally f-free word 010 is isolated since $110, 000$ and 011 are not minimally f-free.

Proposition 12. *Let $f \in \Sigma^*$ be a good word. Then, there exists at least a minimally f-free word $i(f)$ that is isolated with respect to f.*

Proof. The proof is given for words starting with 1. The same reasoning applies to the complemented words.

The case of $|f| = 2$ is handled in Example 11. Let us assume that $|f| = n > 2$ and $f = af'b$, where $a, b \in \Sigma$, $p = pre_{n-1}(f) = af'$ and $s = suf_{n-1}(f) = f'b$. The proof is then divided into three cases.

CASE 1: $f = 1^h 0$ for some $h \ge 2$. Then, it can be easily shown that $i(f) = 01^{h-1} 01^h$ is a minimally f-free isolated word, by definition.

CASE 2: f is unbordered and $f \ne 1^h 0$. Let us show that $i(f) = sp\bar{b}$ is a minimally f-free isolated word.

First, $i(f)$ is minimally f-free because it contains p and s, and f does not occur in $i(f)$. In fact, f cannot occur as a suffix of $i(f)$ because it ends with b and cannot occur elsewhere because f is unbordered.

Now we will prove that $i(f)$ is isolated. Consider the word $R_k(i(f))$ obtained by changing the k-th bit of $i(f)$, $1 \leq k \leq 2n-1$. If $1 \leq k \leq n-1$, then $R_k(i(f))$ cannot contain s, otherwise such s would overlap p, against f unbordered. If $n \leq k \leq 2n-2$, then $R_k(i(f))$ cannot contain p. More exactly, p may occur in $R_k(i(f))$ at a position r with $1 \leq r \leq n+1$. The case $1 \leq r \leq n-1$ is not possible since f is unbordered, and $r \neq n$ by construction. If $r = n+1$, then $f = a^l \overline{a}^m$, for some $l, m \neq 0$ and therefore, if $l = 1$ or $m = 1$, then CASE 1 applies, otherwise f would have a 2-error overlap, against f good. Finally, if $k = 2n-1$, then $R_k(i(f))$ contains f.

CASE 3: f has a border. If $f = 1^n$ then $i(f) = 1^{n-1}0$ satisfies the statement. Assume now that $f \neq 1^n$ and then $s \neq p$. Let y be the longest border of f and let $s = xy$, $p = yz$ for some non-empty words x, y, z. Consider the word $q = xyz$. Let us show that $i(f) = q\overline{b}$ is a minimally f-free isolated word. It is a minimally f-free word because it contains p and s and an occurrence of f in $i(f)$ contradicts the maximality of y. Let us show that $i(f)$ is isolated.

Consider the word $R_k(i(f))$, for any $1 \leq k \leq |i(f)|$. If $1 \leq k \leq |x|$, then $R_k(i(f))$ cannot contain s, otherwise f should have a border longer than y. Similarly, if $|xy| + 1 \leq k \leq |q|$, then $R_k(i(f))$ cannot contain p. If p occurs as a suffix, then $f = b^l \overline{b}^m$, for some $l, m \neq 0$ and the proof goes similarly to CASE 2. If $|x| + 1 \leq k \leq |xy|$, then $R_k(i(f))$ cannot contain both p and s. Indeed, suppose that p and s occur in $R_k(i(f))$ starting at positions r_p and r_s, respectively. Trivially, $r_s > 1$ and $r_p \neq |x| + 1$. Moreover, $r_p \neq |x| + 2$; otherwise it should be $f = a^l \overline{a}^m$, for some $l, m \neq 0$ and the proof goes similarly to that above. It cannot be $r_s < r_p$, since f should have a border longer than y. It cannot be $r_s = r_p$, since $s \neq p$. Lastly, it cannot be $r_s > r_p$. Otherwise, given that $i(f)$ is f-free, the symbol $\overline{a}$ must precede the occurrence of s in $R_k(i(f))$ and the symbol $\overline{b}$ must follow the occurrence of p in $R_k(i(f))$. This implies that f has a 2-error overlap, with shift $r = r_s - r_p - 1$ and error positions 1 and $n - r + 1$, against f good. Finally, if $k = |q| + 1$, then $R_k(i(f))$ contains f. $\square$

Example 13. Let us provide an example of the construction of $i(f)$ from the proof of Proposition 12, addressing each of the 3 considered cases.

CASE 1. $f = 1110$. Note that a minimally f-free word has length 6 at least. The set of all minimally f-free words of length 6 is $\{110111\}$, whereas of length 7 is $\{0110111, 1100111, 1101111\}$. Following the construction, $i(f) = 0110111$.

CASE 2. $f = 100$ (the example can be extended to any word $f = 10^h$). A minimally f-free word has length 4 at least. Then, the set of all minimally f-free words of length 6 is $\{0010\}$, whereas The set of all minimally f-free words of length 5 is $\{00010, 00101, 00110\}$. Following the construction, $i(f) = sp\overline{b} = 00101$ is an isolated word.

CASE 3. $f = 1010$. The word f is bordered and its longest border has length 2. Hence, following the construction $s = 010$ and $p = 101$, $q = 0101$ and $i(f) = 01011$.

Proposition 14. *Let $f \in \Sigma^*$ be a good word and $|f| \geq 3$. Then, there exist two minimally f-free words $i(f)$ and $j(f)$ such that $i(f) \neq j(f)$, $|i(f)| = |j(f)|$, and $i(f)$ is isolated.*

Proof. The proof follows the same cases as in Proposition 12 and uses the isolated word $i(f)$ constructed there. Again, let us assume that $|f| = n$ and express $f = af'b$, for some $a, b \in \Sigma$, $p = pre_{n-1}(f) = af'$ and $s = suf_{n-1}(f) = f'b$. Let us exhibit another word $j(f)$, minimally f-free word, such that $|i(f)| = |j(f)|$.

CASE 1: $f = 1^h 0$, for some $h \geq 2$. It can be easily verified that $i(f)$ and $j(f)$ can be chosen as follows: $i(f) = 01^{h-1}01^h$ and $j(f) = 1^{h-1}001^h$.

CASE 2: f is unbordered and $f \neq 1^h 0$. Using the hypotheses that f is unbordered, it can be easily shown that $i(f)$ and $j(f)$ can be chosen as follows: $i(f) = sp\bar{b}$ and $j(f)$ as one of the two words $j_0(f) = s0p$ and $j_1(f) = s1p$.

CASE 3: f has a border. If $f = 1^n$ then $i(f) = 1^{n-1}0$ and $j(f) = 01^{n-1}$ fulfill the statement. Assume that $f \neq 1^n$ and then $s \neq p$. Let y be the longest border of f and let $s = xy$, $p = yz$ for some non-empty words x, y, z, $q = xyz$. Then, denote $i(f) = q\bar{b}$ and $j(f) = \bar{a}q$. Note that $j(f)$ is minimally f-free because it contains p and s, and f does not occur in $j(f)$. Specifically, f cannot occur as a prefix of $j(f)$ because it starts by a and f cannot occur elsewhere in $j(f)$ because y is the longest border. $\qquad\square$

A direct consequence of Proposition 14 is that, for any good word f, with $|f| \geq 3$, f-free transformations that go through minimally f-free words only, may not exist. This is stated and proved in the following corollary.

Corollary 15. *For any good word $f \in \Sigma^*$ with $|f| \geq 3$, there exists a pair of minimally f-free words that does not admit any f-free transformation composed of only minimally f-free words.*

Proof. The claim follows considering the pair $(i(f), j(f))$ where $i(f)$ and $j(f)$ are constructed as in Proposition 14. In fact, $dist_H(i(f), j(f)) \geq 2$ since $i(f)$ is isolated. $\qquad\square$

The following example presents several pairs of minimally f-free words constructed according to Proposition 14 and Corollary 15.

Example 16. This example continues Example 13.

CASE 1. The word $f = 1110$ is a good word and following the construction, $i(f) = 0110111$ and $j(f) = 1100111$.

CASE 2: The word $f = 100$ is an unbordered and good word. Following the construction, $i(f) = sp\bar{b} = 00101$ is an isolated word, $j_0(f) = s0p = 00010$, $j_1(f) = s1p = 00110$. Furthermore, $dist_H(i(f), j_0(f)) = 3$ and $dist_H(i(f), j_1(f)) = 2$.

CASE 3: The word $f = 1010$ is a bordered good word and its longest border has length 2. Hence, following the construction, $s = 010$, $p = 101$ and $q = 0101$. Then, $i(f) = 01011$ and $j(f) = 00101$.

4　Proximally f-free Transformations and P-good Words

We observe that, in practice, not many pairs of minimally f-free words admit transformations whose intermediate words are all minimally f-free. For this reason, we relax the condition, still keeping the property of being very close to f. Let us start with new definitions.

Definition 17. *Let $f, w \in \Sigma^*$ with $|f| \leq |w|$. The word f is proximal absent for w, if w is f-free and at least one between $pre_{n-1}(f)$ and $suf_{n-1}(f)$ is a factor of w. In this case, w is said proximally f-free.*

Definition 18. *Let $f, u, v \in \Sigma^*$, with $|f| \leq |u| = |v|$. A transformation $\tau = (w_0, w_1, \ldots, w_d)$, such that $w_0 = u$, $w_d = v$ is proximally f-free if each word in τ is proximally f-free.*

Let us show a combinatorial property of a minimally f-free word, in the case that f is good and unbordered. This will provide some pairs of words for which a proximally f-free transformation exists.

Lemma 19. *Let $f \in \Sigma^*$ be a good unbordered word and let w be a minimally f-free word. Then, no occurrence of $pre_{n-1}(f)$ ($suf_{n-1}(f)$, resp.) overlaps an occurrence of $suf_{n-1}(f)$ ($pre_{n-1}(f)$, resp.) in w.*

Proof. Let $p = pre_{n-1}(f)$ and $s = suf_{n-1}(f)$. No overlap of p on s is possible because f is unbordered. Let $f = af'b$ with $a, b \in \{0, 1\}$ and $f' \in \{0, 1\}^*$. Assume, by contradiction, that in a minimally f-free word w there is an overlap of s on p. This means that $p = u_p a_p v$ and $s = v a_s u_s$ with $a_p, a_s \in \{0, 1\}$ and $u_p, u_s, v \in \{0, 1\}^*$, $v \neq \varepsilon$. Observe that $a_p = b$ and $a_s = a$ since w should not contain f. But this implies that f has a 2-error overlap against the hypothesis of f being a good word. $\qquad\qquad\square$

Proposition 20. *Let $f \in \Sigma^*$ be a good and unbordered word and let (u, v) be a pair of minimally f-free words. If in u (or in v) there is an occurrence of $suf_{n-1}(f)$ that precedes an occurrence of $pre_{n-1}(f)$, then there exists a proximally f-free transformation from u to v.*

Proof. Let $p = pre_{n-1}(f)$ and $s = suf_{n-1}(f)$. Consider an occurrence of s that precedes an occurrence of p in u and let $u = x_u s y_u p z_u$ be the corresponding factorization of u. Let $i = |x_u s|$ and $j = |x_u s y_u| + 1$ be the positions where s ends and where p starts, respectively; note that $i < j$ by Lemma 19. The idea of the proof is to divide the strings u and v at a same cut position that lets s and p be apart in u. Then, a proximally f-free transformation from u to v will be realized in two phases. Firstly, an f-free transformation acts only on one part of the strings by correcting its errors, while keeping the occurrence of s or p in the other part. In the second phase, an f-free transformation corrects the errors in the other part. The cut position is chosen following that in v, p or s occurs at a position $> i$ (CASE 1) or p or s ends in a position $\leq i < j$ (CASE 2).

CASE 1: the cut position is i, that is u is divided in $u = u_L u_R$, where $u_L = x_u s$ and $u_R = y_u p z_u$, whereas $v = v_L v_R$, with $|v_L| = |u_L|$. Refer to Fig. 1, for an example of this case. Note that the word $u' = u_L v_R$ is f-free, because u_L and v_R are f-free and no occurrence of f can include position i because otherwise f overlaps s, against the hypothesis that f is unbordered. Then, the first phase consists in an f-free transformation τ_1 from u to u' (recall that f is good), whereas the second phase consists in an f-free transformation τ_2 from u' to v. The transformation $\tau_1 \tau_2$ is f-free. Moreover, in the first phase, every intermediate word contains s, while, in the second phase, every intermediate word contains the word, between s and p, that occurs in v_R by the hypothesis of case 1.

CASE 2: the cut position is chosen to be j, that is u is divided in $u = u_L u_R$, where $u_L = x_u s y_u$ and $u_R = p z_u$, whereas $v = v_L v_R$, with $|v_L| = |u_L|$. Then, the proof goes analogously to case 1. $\square$

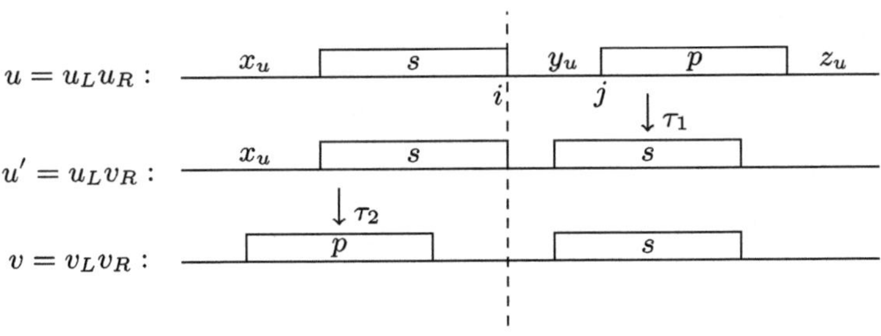

Fig. 1. A proximally f-free transformation from u to v as in Proposition 20.

In analogy with the definition of good words, let us introduce P-good words based on proximally f-free transformations.

Definition 21. *A word $f \in \Sigma^*$ is P-good if it is good and for all pairs of minimally f-free words (u, v), there exists a proximally f-free transformation τ from u to v. If a word is not P-good is P-bad and a pair of minimally f-free words (u, v) that does not satisfy the requirements is called a pair of P-witnesses.*

By using Proposition 20 we find a family of P-good words.

Example 22. The word $f = 10^k$ is a good and unbordered word, for any $k \in I\!N$. Let us show that 10^k is also P-good. In fact, all words that are minimally f-free must begin with $s = 0^k$, since such a block must be a factor and cannot follow a 1. An occurrence of $p = 10^{k-1}$ will appear in the subsequent part.

Next proposition shows another family of P-good words. The basic idea follows the one in the proof of Proposition 20. This time, the goal is reached using specific combinatorial properties of the words in the family.

Proposition 23. *The word $1^k 01^{k-1}0$ is P-good for any $k \geq 1$.*

Proof. Let $f = 1^k 01^{k-1}0$, for some $k \geq 1$, $p = 1^k 01^{k-1}$, $s = 1^{k-1}01^{k-1}0$ and (u,v) be a pair of minimally f-free words. Note that f is good and unbordered. If (u,v) fulfills the hypotheses of Proposition 20 then a proximally f-free transformation exists from u to v. Otherwise, all occurrences of p in u (and in v) precede any occurrences of s. Without loss of generality, assume that the leftmost occurrence of s in u occurs at position i and that i precedes the position of the leftmost occurrence of s in v. Position i is chosen as cut point, that is, u and v are factorized as $u = u_L u_R$, $v = v_L v_R$, with $|u_L| = |v_L| = i - 1$. Then, consider the word $u' = u_L v_R$. This word u' is f-free. In fact, u_L and v_R are f-free. Moreover, if f occurs across u_L and v_R then a proper prefix of f would occur as suffix of u_L. Note that the last symbol of u_L is 0, because otherwise f occurs in u, and that the unique proper prefix of f ending with 0 is $1^k 0$. This contradicts the f-freeness of $u = u_L v_L$ since v_L starts with $s = 1^{k-1}01^{k-1}0$.

Finally, a proximally f-free transformation τ from u to v can be obtained (as in the proof of Proposition 20) composing an f-free transformation τ_1 from u to u' with an f-free transformation τ_2 from u' to v. $\qquad\square$

We now exhibit examples of good words that are P-bad.

Example 24. The word $f = 1^n$ is P-bad, for any $n \geq 2$. In fact, the pair $(01^{n-1}, 1^{n-1}0)$ is a pair of P-witnesses for f.

Proposition 25. *Let $f \in \Sigma^*$ be a good and bordered word with $|f| \geq 2$, $f = af'b$ with $a,b \in \{0,1\}$, and let $\ell \geq 1$, be the length of the longest border of f. If*
  1. $\ell \geq |f|/2$ or
  2. $\ell < |f|/2$ and the length of the longest border of f' is less than ℓ
then f is P-bad.

Proof. (Sketch) For any f that satisfies the hypotheses, a pair of P-witnesses can be constructed as follows (refer to Example 26). Assume $|f| = n$. Let $p = pre_{n-1}(f)$ and $s = suf_{n-1}(f)$. Since f has a border of length ℓ, $f = af_1 f_2 = f_2 f_3 b$ where $|f_2| = \ell$. Define $q = f_1 f_2 f_3$ which is the word obtained by overlapping p on s with overlap length equal to ℓ. Then, let $u = q\bar{b}f_3 = f_1 f_2 f_3 \bar{b} f_3$ and $v = f_1 \bar{a} q = f_1 \bar{a} f_1 f_2 f_3$. Observe that both $p = f_2 f_3$ and $s = f_1 f_2$ occur in u and v. Moreover, $\mathrm{dist}_H(u,v) = 2$, because u and v differ at positions $i = |f_1| + 1$ and $j = |f_1 f_2 f_3| + 1$, because $u[i] = a$ while $v[i] = \bar{a}$, and $u[j] = \bar{b}$ while $v[j] = b$. To prove that (u,v) is a pair of P-witnesses for f, we have to show that: u and v are (minimally) f-free; p and s do not occur in the word $u_a = R_i(u)$ and $u_b = R_j(u)$ is not f-free. The fact that f occurs in u_b is immediate by construction.

If $\ell \geq |f|/2$ then it can be shown that $f = x^k x'$, for some primitive word x (i.e., x is not a proper power of another word) with x' prefix of x, and $k \geq 2$. Note that, since x is primitive, it can occur in xx only as a prefix or a suffix. It is possible to show that, assuming that f occurs in u (or v), or that p or s occurs in u_a, it either produces a contradiction to the maximality of ℓ, or implies that x is a factor (not prefix, nor suffix) of xx, or that $\bar{a}$ ($\bar{b}$, respectively) occurs instead of a (b, respectively).

Let us consider condition *2.* on f'. This implies that any border of s (p, resp.) has a length shorter than $\ell + 1$. It also implies that any overlap of s on p has a length shorter than ℓ. Similarly to case of condition *1.*, it is possible to show that, the assumptions that f occurs in u (or v), or that p or s occurs in u_a, yield a contradiction. $\qquad\qquad\square$

Example 26. The word $f = 100110$ is good, but it can be shown that it is P-bad, using Proposition 25. Indeed, f is bordered, the length of the longest border of f is 2 and, moreover, $f = af'b$ with $a = 1$, $b = 0$, $f' = 0011$ and f' unbordered. Therefore the structure of f falls into Case 2. Following the notation in the proof, $f_1 = 001$, $f_2 = 10$, $f_3 = 011$ and, hence, $q = 00110011$. Then, $u = q\bar{b}f_3 = 001100111011$ and $v = f_1\bar{a}q = 001000110011$, are f-free words that differ in position $i = 4$ and $j = 9$. The pair (u, v) is a pair of P-witnesses for f. Indeed, if the error in position j of u is corrected, f appears as factor in the resulting word; if the error in position i of u is corrected, neither the prefix 10011 nor the suffix 00110 of f occur in the resulting word.

5   Conclusions

Our results show that the existence of a border in f plays a crucial role in the property of being P-good or P-bad. In particular, Proposition 25 shows how the border of f is essential in the construction of P-witnesses. Then further work can be addressed to identify some additional property on the border that allows us to completely characterize the P-bad words.

Furthermore, Lemma 19 guaranties that, in the case of an unbordered word, the occurrences of the maximal prefix and suffix of f are separated, making it easy to construct a proximally transformation free of f under some conditions. This gives us confidence in proposing the following conjecture.

Conjecture 27. If f is good and unbordered then f is P-good.

Acknowledgements. MA, GC, DG and MM are partially supported by the INdAM-GNCS Project 2025 - CUP E53C24001950001, MA is partially supported by the FARB Project ORSA252274 of University of Salerno, MM is partially supported by the TEAMS Project and PNRR MUR Project PE0000013-FAIR University of Catania, GC and SM are partially supported by the FFR fund University of Palermo, DG is partially supported by the MUR Excellence Dept. Project CUP E83C23000330006, awarded to the Dept of Mathematics, University of Rome Tor Vergata. SM is partially supported by the Project MAGDA (CUP: B77G24000050001), funded by the European Union (PNRR) of NextGenerationEU under the "Future Artificial Intelligence - FAIR" program.

References

1. Anselmo, M., Castiglione, G., Flores, M., Giammarresi, D., Madonia, M., Mantaci, S.: Isometric words based on swap and mismatch distance. In: Developments in Language Theory. DLT23. Lecture Notes Computer Science, vol. 13911, pp. 23–35. Springer Nature (2023). https://doi.org/10.1007/978-3-031-33264-7.3
2. Anselmo, M., Castiglione, G., Flores, M., Giammarresi, D., Madonia, M., Mantaci, S.: Isometric sets of words and generalizations of the Fibonacci cubes. In: Computability in Europe. CIE24, volume LNCS 14773 of Lect. Notes Computer Science, pp. 1–14. Springer (2024). https://doi.org/10.1007/978-3-031-64309-5.35
3. Anselmo, M., Castiglione, G., Flores, M., Giammarresi, D., Madonia, M., Mantaci, S.: Hypercubes and isometric words based on swap and mismatch distance. In: DCFS23, volume 13918 of Lect. Notes Computer Science, pp. 21–35. Springer (2023). https://doi.org/10.1007/978-3-031-34326-1.2
4. Anselmo, M., Castiglione, G., Flores, M., Giammarresi, D., Madonia, M., Mantaci, S.: Characterization of isometric words based on swap and mismatch distance. IJFCS, $\mathbf{0}$(0), 1–25 (2025). https://doi.org/10.1142/S0129054125430051
5. Anselmo, M., Flores, M., Madonia, M.: Fun slot machines and transformations of words avoiding factors. In: Fun with Algorithms, volume 226 of LIPIcs, pp. 4:1–4:15 (2022). https://doi.org/10.4230/LIPIcs.FUN.2022.4
6. Anselmo, M., Flores, M., Madonia, M.: On K-ary n-cubes and isometric words. Theor. Comput. Sci. $\mathbf{938}$, 50–64 (2022). https://doi.org/10.1016/j.tcs.2022.10.007
7. Anselmo, M., Flores, M., Madonia, M.: Computing the index of non-isometric k-ary words with Hamming and Lee distance. Comput. IOS Press $\mathbf{13}$(3–4), 199–222 (2024). https://doi.org/10.3233/COM-230441
8. Anselmo, M., Flores, M., Madonia, M.: Density of k-ary words with 0, 1, 2 - error overlaps. Theor. Comput. Sci. $\mathbf{1025}$, 114958 (2025). https://doi.org/10.1016/j.tcs.2024.114958
9. Anselmo, M., Giammarresi, D., Madonia, M., Selmi, C.: Bad pictures: some structural properties related to overlaps. In: Jirásková, G., Pighizzini, G. (eds.) DCFS 2020. LNCS, vol. 12442, pp. 13–25. Springer, Cham (2020). https://doi.org/10.1007/978-3-030-62536-8_2
10. Azarija, J., Klavžar, S., Lee, J., Pantone, J., Rho, Y.: On isomorphism classes of generalized Fibonacci cubes. Eur. J. Comb. $\mathbf{51}$, 372–379 (2016). https://doi.org/10.1016/j.ejc.2015.05.011
11. Barton, C., Héliou, A., Mouchard, L., Pissis, S.P.: Linear-time computation of minimal absent words using suffix array. BMC Bioinform. $\mathbf{15}$, 388 (2014). https://doi.org/10.1186/s12859-014-0388-9
12. Béal, M.-P., Crochemore, M.: Checking whether a word is Hamming-isometric in linear time. Theor. Comput. Sci. $\mathbf{933}$, 55–59 (2022). https://doi.org/10.1016/j.tcs.2022.08.032
13. Béal, M.-P., Mignosi, F., Restivo, A.: Minimal forbidden words and symbolic dynamics. In: Puech, C., Reischuk, R. (eds.) STACS 1996. LNCS, vol. 1046, pp. 555–566. Springer, Heidelberg (1996). https://doi.org/10.1007/3-540-60922-9_45
14. Castiglione, G., Flores, M., Giammarresi, D.: Isometric words and edit distance: main notions and new variations. In: Cellular Automata and Discrete Complex Systems. AUTOMATA 2023, volume 14152 of Lecture Notes Computer Science, pp. 6–13. Springer (2023). https://doi.org/10.1007/978-3-031-42250-8.1

15. Castiglione, G., Mantaci, S., Pizzuto, S.L., Restivo, A.: A comparison between similarity measures based on minimal absent words: an experimental approach. In: Ugo de'Liguoro, Matteo Palazzo, and Luca Roversi, editors, Proceedings of the 25th Italian Conference on Theoretical Computer Science, Torino, Italy, September 11–13, 2024, CEUR Workshop Proceedings, pp. 95–105. CEUR-WS.org (2024). https://ceurws.org/Vol-3811/paper140.pdf
16. Castiglione, G., Mantaci, S., Restivo, A.: Some investigations on similarity measures based on absent words. Fundam. Informaticae **171**(1–4), 97–112 (2020). https://doi.org/10.3233/FI-2020-1874
17. Chairungsee, S., Crochemore, M.: Using minimal absent words to build phylogeny. Theor. Comput. Sci. **450**, 109–116 (2012). https://doi.org/10.1016/j.tcs.2012.04.031
18. Crochemore, M., Mignosi, F., Restivo, A.: Automata and forbidden words. Inf. Process. Lett. **67**(3), 111–117 (1998). https://doi.org/10.1016/S0020-0190(98)00104-5
19. Epifanio, C., Forlizzi, L., Marzi, F., Mignosi, F., Placidi, G., Spezialetti, M.: On the k-Hamming and k-edit distances. In: ICTCS'2023 Italian Conference on Theoretical Computer Science, volume 3587 of CEUR Workshop Proceedings, pp. 143–156 (2023). https://ceur-ws.org/Vol-3587/8136.pdf
20. Ilić, A., Klavžar, S., Rho, Y.: The index of a binary word. Theor. Comput. Sci. **452**, 100–106 (2012). https://doi.org/10.1016/j.tcs.2012.05.025
21. Klavžar, S., Shpectorov, S.V.: Asymptotic number of isometric generalized Fibonacci cubes. Eur. J. Comb. **33**(2), 220–226 (2012). https://doi.org/10.1016/j.ejc.2011.10.001
22. Pinho, A.J., Ferreira, P.J.S.G., Garcia, S.P., Rodrigues, J.M.O.S.: On finding minimal absent words. BMC Bioinform. **10** (2009). https://doi.org/10.1186/1471-2105-10-137
23. Wei, J.: The structures of bad words. Eur. J. Comb. **59**, 204–214 (2017). https://doi.org/10.1016/j.ejc.2016.05.003
24. Wei, J.: Proof of a conjecture on 2-isometric words. Theor. Comput. Sci. **855**, 68–73 (2021). https://doi.org/10.1016/j.tcs.2020.11.026
25. Wei, J., Yang, Y., Zhu, X.: A characterization of non-isometric binary words. Eur. J. Comb. **78**, 121–133 (2019). https://doi.org/10.1016/j.ejc.2019.02.001
26. Wei, J., Zhang, H.: Proofs of two conjectures on generalized Fibonacci cubes. Eur. J. Comb. **51**, 419–432 (2016). https://doi.org/10.1016/j.ejc.2015.07.018

Small Abelian Complexity
of Multidimensional Words

Olga Karmanova and Svetlana Puzynina[(✉)]

Saint Petersburg State University, Saint Petersburg, Russia
lgkrmnv8@gmail.com, s.puzynina@gmail.com

Abstract. A *complexity* of an infinite word is a function counting the number of its distinct blocks for each length. In a 1938 seminal paper on symbolic dynamics, Morse and Hedlund gave a link between periodicity and complexity of infinite words. Namely, they proved that if the complexity of an infinite word is bounded by n for some length n, then the word is periodic. In two dimensions, a similar assertion is known as Nivat's conjecture: it states that if the number of distinct $n \times m$ rectangular factors of a two-dimensional word is bounded by mn, then the word has a periodicity vector. We consider this problem in the abelian setting. Two finite words are called *abelian equivalent* if they are permutations of each other. An *abelian complexity* of a word counts the number of distinct abelian classes of its factors. In the paper we study relationships between small abelian complexity and periodicity in multidimensional words. We show that for $d \geq 2$, if a d-dimensional recurrent word contains at most two abelian classes of $n_1 \times \ldots \times n_d$-blocks for each d-tuple $(n_1, \ldots, n_d)$ of integers, then it has a periodicity vector. On the other hand, there exist recurrent aperiodic words with abelian complexity bounded by 3.

Keywords: Multidimensional words · Abelian complexity · Periodicity

1 Introduction

In this paper, we study the relationships between small abelian complexity and periodicity in multidimensional words. A classical theorem of Morse and Hendlund gives a link between periodicity and complexity in one-dimensional words [19]. A *complexity* of an infinite word is a function $p(n)$ counting for each length n the number of distinct blocks of consecutive letters in it. The theorem states that if $p(n) \leq n$ for some length n, then the word is periodic. Furthermore, this bound is optimal in the sense that there exist words with complexity $p(n) = n + 1$ for each n. These words correspond to the family of Sturmian words, an important and well-studied class of infinite words (see [20] and Chap. 2 in [18]).

In dimension 2, an analog of Morse and Hedlund theorem is given by Nivat's conjecture, introduced at ICALP 1997 [21]. A two-dimensinal word is understood as a function from $\mathbb{Z}^2$ to an alphabet, and, similarly to the one-dimensional case, its complexity is a function $p : \mathbb{N}^2 \mapsto \mathbb{N}$ counting the number of distinct

M.-P. Béal and P. Caron (Eds.): DLT 2026, LNCS 16578, pp. 266–279, 2026.
https://doi.org/10.1007/978-3-032-28404-4_20

rectangular $m \times n$-blocks in the word for each pair of integers m, n. Nivat's conjecture states that, given a two-dimensional word, if there exist two numbers m, n such that the complexity satisfies $p(n, m) \leq nm$, then w has a periodicity vector. Nivat's conjecture remains open despite of efforts of different scientists. It has been proven is some weak forms, for example, in an asymptotic form by Kari, Szabados [17], and for a stronger bound $\frac{mn}{2}$ [9]; see also [10,16].

Various modifications of the notion of a complexity of infinite words and generalizations of Morse and Hedlund theorem have been studied in the literature. This incudes maximal pattern complexity [15] and its abelian version [14], group complexity [4], arithmetic complexity [1] and others (see a survey [23]).

The goal of this paper is to study the generalizations of this question in the abelian setting. Two finite words u and v are said to be *abelian equivalent* if $|u|_a = |v|_a$ for all $a \in A$, where $|u|_a$ denotes the number of occurrences of the letter a in u. For recent survey on abelian properties of words we refer to [12]. The *abelian complexity* $a(n)$ of a (one-dimensional) word is the function counting number of distinct abelian classes of its blocks of length n. This definition can be extended to two or more dimensions in a natural way as a function counting the number of abelian classes of rectangular blocks.

In dimension 1, an abelian analogue of Morse and Hedlund theorem is straightforward: clearly, the condition $a(n) = 1$ implies that the word is n-periodic. Aperiodic words of abelian complexity 2 for each n exist and the set of words with this property coincides with the family of Sturmian words [7]. So, among aperiodic words, Sturmian words have minimal complexity both in the classical and in the abelian sense. The study of abelian complexity of one-dimensional infinite words has been developed, e.g., in [3,8,24,25]. For two-dimensional words, the abelian modifications of Nivat's conjecture have been studied in [22]. It has been shown that there exists an aperiodic word and integers m and n such that $a(m, n) = 1$. However, if the abelian complexity of a two-dimensional recurrent word is bounded by 2, then the word periodic in the sense that it has a periodicity vector. Moreover, for an aperiodic recurrent two-dimensional word there exist infinitely many pairs of numbers m, n such that $a(m, n) \geq 3$.

In this paper, we show that in the abelian setting, for dimension $d > 2$ the situation is similar to the two-dimensional case, generalizing the results from [22]. Remarkably, for factor complexity the situation is different: Nivat's conjecture does not hold in dimension higher than 2 [5]. In particular, we show that if the abelian complexity of a multidimensional recurrent word is bounded by 2, then the word has a periodicity vector (Theorem 3). This bound is optimal in the sense that there exist recurrent words with abelian complexity bounded by 3 without periodicity vectors (Proposition 2). In addition, we discuss multidimensional words with abelian complexity equal to 1 in some blocks (Proposition 4).

2   Preliminaries

We let Σ denote a finite non-empty set called an *alphabet*. Throughout the text we will denote vectors from $\mathbb{Z}^d$ by bold font as follows: $\mathbf{n} = (n_1, \ldots, n_d) \in \mathbb{Z}^d$. We write $\mathbf{nm}$ for a scalar product of $\mathbf{n}$ and $\mathbf{m}$: $\mathbf{nm} = n_1 m_1 + \ldots + n_d m_d$. A *d-dimensional infinite word* w is an element of $\Sigma^{\mathbb{Z}^d}$, and a d-dimensional finite word of size $\mathbf{k} \in \mathbb{N}^d$ is an element of $\Sigma^{\{0..k_1-1\} \times \ldots \times \{0..k_d-1\}}$. A *factor* of w of size $\mathbf{k} \in \mathbb{Z}^d$ is a finite word of the form

$$\{w_{\mathbf{m+i}} \mid \mathbf{i} \in \mathbb{Z}^d, 0 \le i_j < k_j, j = 1, \ldots, d\},$$

for some $\mathbf{m} \in \mathbb{Z}^d$. We denote it by $w_{\mathbf{m}..\mathbf{m+k-1}}$; here $\mathbf{1} = 1^d$ is the all-1 vector of length d. For each infinite d-dimensional word w, its complexity or *factor complexity* is a function $p_w : \mathbb{N}^d \to \mathbb{N}$ counting for each $\mathbf{k} \in \mathbb{N}^d$ the number of distinct factors of size $\mathbf{k}$ occurring in w.

An infinite word w is called *recurrent* if each its factor occurs in it infinitely many times. A d-dimensional infinite word w has a *periodicity vector* $\mathbf{v} \in \mathbb{Z}^d$ if $w_{\mathbf{x+v}} = w_{\mathbf{x}}$ for each $\mathbf{x} \in \mathbb{Z}^d$; otherwise w is *aperiodic*. A d-dimensional infinite word w is *fully periodic* if it has d linearly independent periodicity vectors.

The following theorem gives a link between periodicity and complexity in dimension 1 (we formulated it for two-way infinite words, although it is often formulated for one-way infinite words with ultimate periodicity property):

Theorem 1 (Morse and Hedlund, 1938). *Let w be a one-dimensional word. If there exists n such that $p_w(n) \le n$, then w is periodic.*

Aperiodic recurrent words satisfying $p_x(n) = n + 1$ for each $n \ge 0$ are called *Sturmian words*, and hence are regarded as the simplest aperiodic words. Sturmian words admit various types of characterizations of geometric and combinatorial nature (see Chap. 2 in [18]).

For a finite word v and a letter a, we let $|v|_a$ denote the number of occurrences of a in v. An infinite word w has *uniform frequency* of a letter a if the ratio $\frac{|w_{\mathbf{m}..\mathbf{m+k-1}}|_a}{k_1 \cdots k_d}$ has a limit $\mathrm{Freq}_a(w)$ when all $k_i \to \infty$, uniformly in $\mathbf{m}$. An infinite word is called *balanced* if for each $\mathbf{k}$ each its two factors u and v of size $\mathbf{k}$ satisfy $||u|_a - |v|_a| \le 1$ for each letter $a \in \Sigma$.

In this paper we are interested in extending the result of Morse and Hedlund to the abelian setting. Two finite words u and v of the same size $\mathbf{k} \in \mathbb{N}^d$ are *abelian equivalent* if $|u|_a) = |v|_a$ for each letter a. In other words, u and v are permutations of one another. It is straightforward that abelian equivalence is indeed an equivalence relation on the set of finite words. For each infinite d-dimensional word $w \in \Sigma^{\mathbb{Z}^d}$, the *abelian complexity* $a_w(\mathbf{k})$ counts the number of distinct abelian classes of factors of size $\mathbf{k} \in \mathbb{N}^d$ occurring in w. Ordering the alphabet $\Sigma = \{a_1, \ldots, a_k\}$, we define *Parikh vector* of a finite word v over Σ as $PV(v) = (|v|_{a_1}, \ldots, |v|_{a_k})$. The set of Parikh vectors of factors of size $\mathbf{k}$ of a d-dimensional infinite word w is then denoted by $PV_w(\mathbf{k})$ (the index w is omitted when no ambiguity arises).

In dimension 1, the following fact is straightforward and well-known:

Lemma 1. *Let w be a one-dimensional word. If there exists n such that $a_w(n) = 1$, then w is periodic.*

On the other hand, Sturmian words are aperiodic and have abelian complexity 2 for each n, and moreover this is a characterization:

Theorem 2 ([20]). *Sturmian words are recurrent aperiodic one-dimensional words with abelian complexity $a_w(n) = 2$ for each integer n.*

We remark that in dimension 1, abelian complexity bounded by 2 is equivalent to balance (see Lemma 6). For multidimensional words this is not the case. For example, consider a binary two-dimensional word such that $w(x, y) = 1$ if and only if x is odd. Although its abelian complexity is bounded by 2, it is not balanced. We will make use of the following fact about multidimensional balanced words ([2], see, e.g., a corollary stated in the introduction):

Proposition 1 ([2]). *A two-dimensional balanced infinite word has uniform frequencies of letters, and is moreover fully periodic, unless the frequency is 0 or 1.*

3   One-Dimensional Words

For the proof of the main result we will use several auxiliary lemmas about one-dimensional words.

Lemma 2 (Lemma 8 in [22]). *Let w be a one-dimensional binary word with uniform frequency of 1 equal to $\alpha, 0 < \alpha < 1$. Then for each N there exists $n \geq N$ such that $a_w(n) \geq 2$.*

Lemma 3 (Lemma 9 in [22]). *Let w and w' be one-dimensional binary words with uniform frequencies of 1 equal to α_0 and α_1, respectively. If $\alpha_0 \neq \alpha_1$, then there exists N such that $PV_w(n) \cap PV_{w'}(n) = \emptyset$ for each $n \geq N$.*

Lemma 4. *Let w be a one-dimensional word with frequency of 1 equal to 1. If for each n we have $a_w(n) \leq 2$, then w contains a finite number of 0's.*

Proof. Assume the converse. Then there exists an infinite prefix or an infinite suffix of w containing infinitely many 0's; without loss of generality we suppose that it is a suffix.

If the distance between consecutive occurrences of 0's is bounded (say by an integer k), then each factor u of length $l(k + 1)$, where $l \in \mathbb{N}$, of the suffix of w contains at least l occurrences of 0. Therefore, for each integer N there exists a factor u such that $|u| > N$, $1 - \frac{|u|_1}{|u|} = \frac{|u|_0}{|u|} \geq \frac{1}{k+1}$, which contradicts the fact that the frequency of 1 is 1.

If the distance between consecutive 0's is unbounded, then there exists a pair of 0's at some distance $x \geq 0$, and another pair of 0's at distance at least $x + 2$. Then we have $a_w(x + 2) \geq 3$, since there are factors of length $x + 2$ containing no 0's and containing one and two occurrences of 0; a contradiction.

The following is well known and actually holds for abelian complexity bounded by any constant (see, e.g., Theorem 16 in [6]):

Lemma 5. *Let w be a one-dimensional word. If for each integer n we have $a_w(n) \leq 2$, then w has uniform frequencies of letters.*

The following two lemmas are straightforward:

Lemma 6. *Let w be a one-dimensional word such that for each integer n we have $a_w(n) \leq 2$. Then w is balanced.*

Lemma 7. *Let w be a one-dimensional infinite word such that $a_w(n) \leq 2$ for each integer n. If w contains a finite number of occurrences of 1, then it has at most one occurrence of 1.*

Lemma 8 ([18], **Lemmas 2.1.14 and 2.1.15**). *Let w be a biinfinite one-dimensional balanced periodic word with frequency $\frac{p}{q}$, where $(p,q) = 1$. Then w has a period of length q.*

4 Periodicity of Multidimensional Words of Abelian Complexity Bounded by 2

The main result of this paper is the following theorem. Together with the example from Proposition 4, it gives a precise bound for abelian complexity of an aperiodic multidimensional word.

Theorem 3. *Let w be a d-dimensional recurrent infinite word. If for each $\mathbf{k} \in \mathbb{N}^d$ we have $a_w(\mathbf{k}) \leq 2$, then w has a periodicity vector.*

Remark 1. We set the condition on recurrence to exclude trivial examples which do not satisfy the theorem. Consider a d-dimensional word containing only one occurrence of 1. Clearly, its abelian complexity is equal to 2 for all block sizes, and it is aperiodic by definition. However, the word is not recurrent. Note that if $a_w(\mathbf{k}) = 1$ for each $\mathbf{k}$, then the word is unary and hence trivially fully periodic.

Proof. First note that w is binary since $a_w(1) \leq 2$. A *1-line along X* is a one-dimensional subword of w along axis X. Similarly, for $1 \leq k < d$, we will consider k-dimensional subwords along X, i.e., subwords in which we fix $d - k$ coordinates, all of them distinct from X. By definition, the abelian complexity of a 1-dimenional word along axis X is also at most 2. By Lemma 5, for each 1-line in w along each axis X there exist frequencies of letters. However, in different 1-lines frequencies could be different. We claim that we cannot have more than two distinct values of frequencies:

Claim. Let w be a d-dimensional infinite word such that for each its one-dimensional subword there exist frequencies of letters, and that for each $\mathbf{k} \in \mathbb{N}^d$ we have $a_w(\mathbf{k}) \leq 2$. Then we cannot have more than two distinct values for letter frequencies in 1-lines along each axis.

Proof of the Claim: Assume the converse: suppose that there exist three 1-dimensional subwords $w_\alpha, w_\beta, w_\gamma$ with distinct frequencies α, β, γ. Then by the definition of a frequency for each ε there exist factors u_α (resp., u_β, u_γ) of w_α (resp., w_β, w_γ) of length n such that $\left| \frac{|u_\alpha|_1}{|u_\alpha|} - \alpha \right| < \varepsilon$, $\left| \frac{|u_\beta|_1}{|u_\beta|} - \beta \right| < \varepsilon$, $\left| \frac{|u_\gamma|_1}{|u_\gamma|} - \gamma \right| < \varepsilon$. Consider $\varepsilon = \frac{1}{2} \cdot \min(|\alpha - \beta|, |\beta - \gamma|, |\gamma - \alpha|)$. Then the values $\frac{|u_\alpha|_1}{|u_\alpha|}, \frac{|u_\beta|_1}{|u_\beta|}, \frac{|u_\gamma|_1}{|u_\gamma|}$ are distinct. Since the denominators are equal to n, the numerators are also distinct, which contradicts the condition that $a_w(\mathbf{k}) \leq 2$ for each $\mathbf{k} \in \mathbb{N}^d$. The claim is proved.

We continue with the proof of the theorem:

Case 1. Suppose that there exists an axis X such that we have exactly two values for frequencies of the letter 1, i.e., in some 1-lines along the axis X the frequency is α, and in other 1-lines it is $\beta \neq \alpha$. Lemma 3 implies that for a sufficiently large n each 1-line has exactly one Parikh vector corresponding to a block size $n \times 1 \times \ldots \times 1$, where n is placed on the index corresponding to the axis X. If at least one of the numbers α and β is not equal to 0 and 1, then by Lemma 2 we can choose sufficiently large n such that the corresponding 1-line has at least two distinct Parikh vectors; a contradiction. So, the only possibility is that some 1-lines along the axis X have frequency 1, and others have frequency 0. By Lemma 4, along each 1-line along the axis X one of the symbols occurs only finitely many times.

Consider a factorization of w into 1-lines along axis X. As there are lines with frequencies 1 and 0, there are all-0 and all-1 factors for each length. If there exists a line with finitely many occurrences of 0 (or a line with finitely many occurrences of 1), then for each length there exist factors containing one occurrence of 0, e.g., starting with the last 0 (resp., one occurrence of 1). So, we obtain at least three Parikh vectors, which is impossible, since the abelian complexity is bounded by 2. So, all 1-lines along axis X are all-0 or all-1 lines, which means that the unit vector along the axis X is a periodicity vector.

Case 2. Suppose now that we have one value for frequency for each 1-line along each axis.

Case 2.1: The frequency is equal to 0 (or 1) in each 1-line along the axis X (without loss of generality we may assume it is 0). By the condition of Case 2, we also have frequency 0 along any other axis Y. Then the number of occurrences of the letter 1 in each 1-line is finite, and moreover is at most 1 by Lemma 7.

Claim. Let w be an aperiodic d-dimensional infinite word with abelian complexity bounded by 2 and such that all its 1-lines contain at most one occurrence of 1. Then for each $k \leq d$ in each k-dimensional subword of w we have at most one occurrence of 1.

Proof of the Claim. The proof is by induction on k. Lemma 7 gives the base case. Suppose that the statement holds for k-dimensional subwords; we will prove it for $(k + 1)$-dimensional subwords. Consider a $(k + 1)$-dimensional subword w_0; we split it into k-dimensional subwords along some axis X (we will call these subwords k-lines) and index them by integers. Suppose that w_0

contains at least two occurrences of 1; then these occurrences are in distinct k-lines by the induction hypothesis. Consider two k-lines such that their indices are consecutive in the sequence of indices of k-lines containing 1's.

We will prove that these occurrences of 1 have the same X-coordinate. If this is not the case, consider a minimal $(k + 1)$-dimensional rectangle f, containing these two 1's, and consider its shift along X by 1 (denote this rectangle by f_1) and by the length of the side of the rectangle which is parallel to X (denote this rectangle by f_2); see Fig. 1 for an illustration. Then f contains two occurrences of 1, f_1 contains one occurrence of 1, and f_2 does not contain 1's. Therefore, we found three distinct Parikh vectors for factors of the size of f; a contradiction.

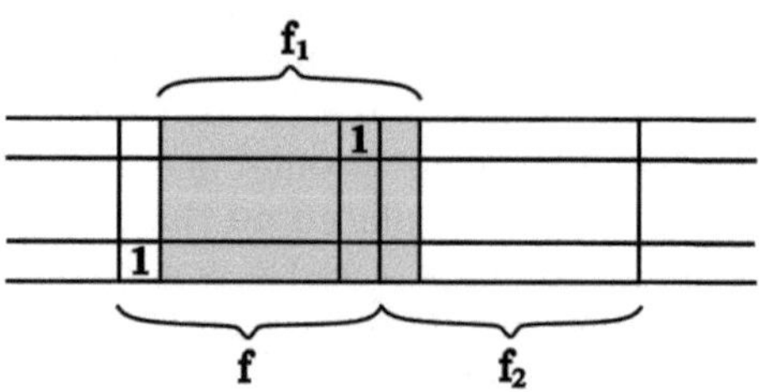

Fig. 1. Illustration to the proof of Theorem 3: three rectangular factors with different Parikh vectors (axis X is horizontal).

So, all 1's of w_0 have the same X-coordinate. This means that they are in the same k-line; it has as fixed coordinates the X-coordinate and all coordinates fixed in w_0. By induction hypothesis, we have one such 1; a contradiction. So, we proved the induction step, and hence the claim.

We now continue with the proof of the theorem. Applying the latter claim for $k = d$, we obtain that the word contains at most one occurrence of 1, and hence either it is not recurrent (if it has one occurrence of 1) or periodic (if it is all-0 word).

Case 2.2: the frequency in all 1-lines parallel to X is α, and $0 < \alpha < 1$. We split w into 2-lines, i.e., planes, parallel to X (and to some other axis, we can choose any). Each such plane consists of 1-lines parallel to X with the same frequencies of letters, which are not 0 or 1. We will prove that each such plane is balanced.

First note that for each integer k, the set of $1 \times k$-factors in the plane is balanced. Indeed, in one such line it is balanced by Lemma 6. Balance and the frequency equal to α imply that the Parikh vectors of $1 \times k$-factors are $(\lfloor \alpha k \rfloor, k - \lfloor \alpha k \rfloor)$ and $(\lceil \alpha k \rceil, k - \lceil \alpha k \rceil)$ if αk is not integer. If αk is integer, then one of the Parikh vectors is $(\alpha k, k - \alpha k)$, and there could be one more, either $(\alpha k - 1, k - \alpha k + 1)$ or $(\alpha k + 1, k - \alpha k - 1)$. If there are two distinct Parikh vectors, in any other 1-line along X they must be the same due to the condition $a_w \le 2$. If there is only one Parikh vector, then it is also a Parikh vector of any other 1-line with this frequency.

This implies that in each horizontal stripe of height k of the plane the word is balanced (i.e., in a stripe $\mathbb{Z} \times k$ for each integer l the numbers of occurrences of a letter a in any two $l \times k$-rectangles of this stripe differ by at most 1). With a symmetric argument we obtain that in each vertical stripe of each width l of the plane the word is also balanced in $l \times k$-rectangles for each k.

Suppose that the word in the plane however is not balanced, i.e., there exist two rectangles r_1 and r_2 of the same size $k \times l$ such that the numbers of 1's in them differ by at least 2: $|r_1|_1 = y$, $|r_2|_1 = y + z$, for some integers y and z, where $z \geq 2$). Consider a vertical stripe of width l containing the first rectangle, and a horizontal stripe of height k, the second rectangle. These stripes intersect by a rectangle r_3 of the same size, and we have balance inside each stripe. From the balance of the first stripe we have that $|r_3|_1 \in \{y - 1, y, y + 1\}$, and from the balance in the second one that $|r_3|_1 \in \{y + z - 1, y + z, y + z + 1\}$. These sets intersect if and only if $y + 1 = y + z - 1$, whence $z = 2$. However, in this case we have three rectangles r_1, r_2, r_3 of the same size with pairwise distinct Parikh vectors, which contradicts $a_w \leq 2$.

So, we proved that each such plane is balanced. Since we are in the case when the frequency is not 0 or 1, we can apply Proposition 1 to two-dimensional subwords of w. It implies that each such plane is fully periodic, and hence each 1-line along X is periodic (in particular, has rational frequency).

Since the word is doubly periodic, the frequencies of letters are rational, i.e., there exist integers p and q such that $\alpha = \frac{p}{q}$. By Lemma 8, each of 1-lines along X has period q. So, the vector $(q, 0, 0, \ldots, 0)$ (where the first coordinate corresponds to the axis X) is a periodicity vector for the word w. This completes the proof.

5 Aperiodic Multidimensional Words with Abelian Complexity Bounded by 3

Proposition 2. *For each $d \geq 2$, there exists an infinite aperiodic recurrent d-dimensional word w such that for each $\mathbf{n} \in \mathbb{N}^d$ we have $a_w(\mathbf{n}) = 2$ or $a_w(\mathbf{n}) = 3$.*

Proof. To build such an example, we start with any binary word $x \in \{0, 1\}^{\mathbb{Z}}$. We build a d-dimensional word w in the following way. For each $\mathbf{i} \in \{0, 1\}^d$ and $\mathbf{m} \in \mathbb{Z}^d$, we set

$$w_{2\mathbf{m}+\mathbf{i}} = \left(\sum_{j=1}^{d} (i_j + x_{m_j}) \right) \quad \mod 2.$$

For the proof we define a "domino pair" to be a pair of adjacent elements of the form $(w_{(m_1, \ldots, 2m_j, \ldots, m_d)}, w_{(m_1, \ldots, 2m_j+1, \ldots, m_d)})$. Note that each "domino pair" contains exactly one occurrence of 0 and one occurrence of 1 by construction.

We prove by induction on d that $a_w(\mathbf{n}) \leq 3$ for each $\mathbf{n} \in \mathbb{N}^d$. For that, we first note that fixing one coordinate, we obtain a $(d-1)$-dimensional word obtained by the same construction. We now consider two cases:

Case 1: At least one of the numbers $n_1, \ldots, n_d$ is odd. Without loss of generality we assume that n_d is odd. For each $\mathbf{k} \in \mathbb{Z}^d$, the rectangular factor $w_{\mathbf{k}..\mathbf{k}+\mathbf{n}-1}$ can be split into a factor of size $(n_1, \ldots, n_d-1)$ consisting of integral domino pairs and a factor of size $(n_1, \ldots, n_{d-1}, 1)$. In the first one we have equal numbers of occurrences of 0's and 1's, and the second one is a factor of a $(d-1)$-dimensional word, which has complexity at most 3 (see Fig. 2).

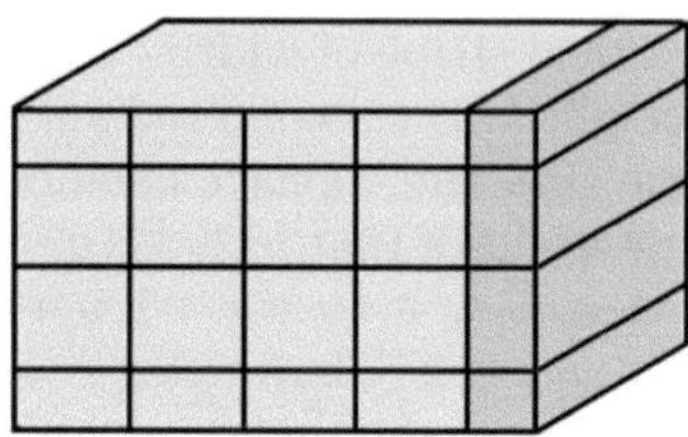

Fig. 2. Illustration to the proof of Proposition 2 in the case when there is at least one side of odd length. Here d-th coordinate is the horizontal axis, and the two partition classes are marked by colors.

Case 2: All the numbers $n_1, \ldots, n_d$ are even. In this case, we claim that the Parikh vector of a factor of this size is one of the following (here $N = \frac{n_1 \cdots n_d}{2}$):

$$PV(\mathbf{n}) = \left\{ (N, N), (N - 2^{d-1}, N + 2^{d-1}), (N + 2^{d-1}, N - 2^{d-1}) \right\}, \qquad (1)$$

and so $a_w(\mathbf{n}) = 3$. Here we have two subcases: either a $(n_1 \times \cdots \times n_d)$-block is situated at a position with at least one even coordinate, or all coordinates are odd. In the first subcase the $(n_1 \times \ldots \times n_d)$-block contains N domino pairs and hence the Parikh vector is (N, N) (see Fig. 3, left).

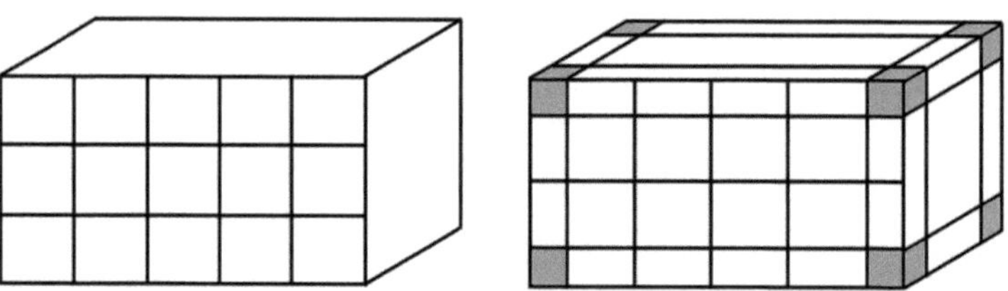

Fig. 3. Illustration to the proof of Proposition 2 in the case when all the sides are of even length. The set Cor_d is marked with a color.

In the other subcase a $(n_1 \times \cdots \times n_d)$-block is at a position of the form $2\mathbf{k} + 1$ for some $\mathbf{k} \in \mathbb{Z}^d$ and it contains $N - 2^{d-1}$ domino pairs plus 2^d points which are corners of the parallelepiped

$$Cor_d = \{ 2\mathbf{k} + 1 + \mathbf{i}(\mathbf{n} - 1) \mid \mathbf{i} \in \{0, 1\}^d \}$$

(see Fig. 3, right). We will prove that for such set Cor_d we have

$$PV(Cor_d) \in \{(2^{d-1}, 2^{d-1}), (2^d, 0), (0, 2^d)\}. \tag{2}$$

By construction, if $x_s = x_t$ for some $s, t \in \mathbb{Z}$, then for each $\mathbf{j} \in \mathbb{Z}^{d-1}$ we have

$$w_{j_1,\ldots,j_{d-1},2\,s} = w_{j_1,\ldots,j_{d-1},2t} = 1 - w_{j_1,\ldots,j_{d-1},2\,s+1} = 1 - w_{j_1,\ldots,j_{d-1},2t+1}.$$

If $x_s \neq x_t$, then for each $\mathbf{j} \in \mathbb{Z}^{d-1}$ we have

$$w_{j_1,\ldots,j_{d-1},2\,s} = 1 - w_{j_1,\ldots,j_{d-1},2t} = 1 - w_{j_1,\ldots,j_{d-1},2\,s+1} = w_{j_1,\ldots,j_{d-1},2t+1}.$$

Applying this observation to $s = k_d$, $t = k_d + n_d/2$ and

$$\mathbf{j} \in \{2\mathbf{k} + 1 + \mathbf{i}(\mathbf{n} - 1) \mid \mathbf{i} \in \{0,1\}^{d-1}\}$$

we obtain in the case $x_s = x_t$ for each $\mathbf{i} \in \{0,1\}^d$ the following:

$$w_{2\mathbf{k}+1+\mathbf{n}(i_1,\ldots,i_{d-1},0)} = 1 - w_{2\mathbf{k}+1+\mathbf{n}(i_1,\ldots,i_{d-1},1)}.$$

So,

$$PV(\{w_{2\mathbf{k}+1+\mathbf{i}(\mathbf{n}-1)} \mid \mathbf{i} \in \{0,1\}^d, i_d = 0\})+$$
$$+PV(\{w_{2\mathbf{k}+1+\mathbf{i}(\mathbf{n}-1)} \mid \mathbf{i} \in \{0,1\}^d, i_d = 1\}) = (2^{d-1}, 2^{d-1}),$$

and therefore

$$PV(Cor_d) = PV(\{2\mathbf{k} + 1 + \mathbf{i}(\mathbf{n} - 1)\} \mid \mathbf{i} \in \{0,1\}^d\}) = (2^{d-1}, 2^{d-1}).$$

In the case $x_s \neq x_t$ we have for each $\mathbf{i} \in \{0,1\}^d$

$$w_{2\mathbf{k}+1+\mathbf{n}(i_1,\ldots,i_{d-1},0)} = w_{2\mathbf{k}+1+\mathbf{n}(i_1,\ldots,i_{d-1},1)}.$$

So,

$$PV(\{2\mathbf{k} + 1 + \mathbf{i}(\mathbf{n} - 1)\} \mid \mathbf{i} \in \{0,1\}^d, i_d = 0\}) =$$
$$= PV(\{2\mathbf{k} + 1 + \mathbf{i}(\mathbf{n} - 1)\} \mid \mathbf{i} \in \{0,1\}^d, i_d = 1\}).$$

Since on both sides of the equality we have a set Cor_{d-1}, by induction on d we get (2) and hence (1).

Summing up, the abelian complexity of this word is given by:

$$a_w(n_1, \ldots, n_d) = \begin{cases} 2, & \text{if all the numbers } n_1, \ldots, n_d \text{ are odd,} \\ 3, & \text{otherwise.} \end{cases}$$

It remains to notice that choosing x we can make w recurrent and aperiodic. First, if x is recurrent, then so is w. Aperiodicity is achieved for example by breaking all periods as follows: to get that $(p_1, \ldots, p_d)$ is not a period, it is enough to choose a value x_{n_1} so that $\sum_{j=1}^d x_{n_j} \bmod 2 \neq \sum_{j=1}^d x_{n_j+2p_j} \bmod 2$ for some $n_1, \ldots, n_d$. Indeed, we then have $w_{2n_1,\ldots,2n_d} \neq w_{2n_1+2p_1,\ldots,2n_d+2p_d}$ by the definition of w; hence $(p_1, \ldots, p_d)$ is not a period.

6   Abelian Complexity Equal to 1 for Some Block Sizes

The following proposition states that for each $d \geq 2$ there exists an aperiodic d-dimensional word with abelian complexity 1 for some block size. This fact is a generalization of a similar statement from [22] for $d = 2$; see also [13].

Proposition 3. *For each $d \geq 2$, there exists an infinite aperiodic d-dimensional word w and infinitely many vectors $\mathbf{n} \in \mathbb{N}^d$ such that $a_w(\mathbf{n}) = 1$.*

Proof. We start with d binary words x^j, $j = 1, \ldots, d$. We build from them a d-dimensional word w as follows. For each $\mathbf{i} \in \{0,1\}^d$ and $\mathbf{m} \in \mathbb{Z}^d$, we set

$$w_{2\mathbf{m}+\mathbf{i}} = \begin{cases} \left(\sum_{s=1}^{d} i_s + x_{m_j}^j\right) \bmod 2, & \text{if } j \geq 2 \text{ is the largest index with odd } m_j, \\ \left(\sum_{s=1}^{d} i_s + x_{m_1}^1\right) \bmod 2, & \text{if for each } j \geq 2 \text{ the number } m_j \text{ is even.} \end{cases}$$

We will prove that in each $2 \times 4 \times 1 \times \ldots \times 1$ block we have equal numbers of 1's and 0's, i.e., abelian complexity is equal to 1 for this size of block (and hence in all blocks of size $2k_1 \times 4k_2 \times k_3 \times \ldots \times k_d$). We prove this by induction on d. For each d, if we fix the last coordinate, we get a word of a smaller dimension, in which we have abelian complexity 1 by induction. The base case for $d = 2$ is given by the example built in Proposition 6 in [22].

Multidimensional words with abelian complexity 1 in some block size satisfy the following property:

Proposition 4. *Let w be a d-dimensional word and $\mathbf{n} \in \mathbb{N}^d$ be d-dimensional integer vector such that $a_w(\mathbf{n}) = 1$. Then for each $\mathbf{x} \in \mathbb{Z}^d$ the following holds:*

$$PV(\{w_{\mathbf{x}+\mathbf{i}\mathbf{n}} | \mathbf{i} \in \{0,1\}^d, \sum_{j=1}^{d} i_j \equiv_2 0\}) = PV(\{w_{\mathbf{x}+\mathbf{i}\mathbf{n}} | \mathbf{i} \in \{0,1\}^d, \sum_{j=1}^{d} i_j \equiv_2 1\}).$$

Informally, the proposition says that for each d-dimensional parallelepiped of size $(n_1 + 1) \times \cdots \times (n_d + 1)$, if we color its corners in black and white such that vertices sharing an edge have distinct colors, then the Parikh vector of the set of black corners is the same as the Parikh vector of the set of white corners.

Proof. Consider a shift of a rectangular factor $w_{\mathbf{x}..\mathbf{x}+\mathbf{n}-1}$ by a vector $\mathbf{s} \in \{0,1\}^d$. Since $a_w(\mathbf{n}) = 1$, we have for each $\mathbf{s}' \in \{0,1\}^d$ that

$$PV(w_{\mathbf{x}+\mathbf{s}..\mathbf{x}+\mathbf{s}+\mathbf{n}-1}) = PV(w_{\mathbf{x}+\mathbf{s}'..\mathbf{x}+\mathbf{s}'+\mathbf{n}-1}).$$

All such shifts are situated inside a rectangular factor $w_{\mathbf{x}..\mathbf{x}+\mathbf{n}}$. This factor consists of the intersection of all such shifts, which is given by a rectangular factor $w_{\mathbf{x}+1..\mathbf{x}+\mathbf{n}-1}$, and the boundary of $w_{\mathbf{x}..\mathbf{x}+\mathbf{n}}$. The boundary can be further split into rectangular factors of smaller dimension: factors of dimension $d - 1$ (faces), factors of dimension $d - 2$, $\ldots$, factors of dimension 1 (edges) and finally factors of dimension 0 (corners) (see Fig. 4).

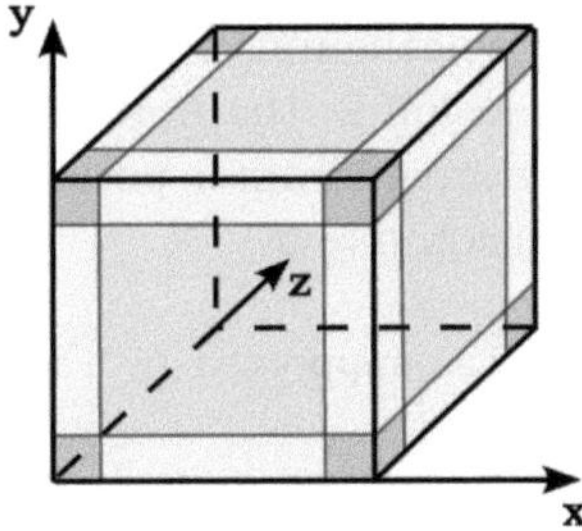

Fig. 4. Illustration to the proof of Proposition 4 for $d = 3$.

Consider 2^{d-1} such equalities, where on the left side we have all subsets of $\{0,1\}^d$ with even numbers of 1's, and on the right side all subsets of $\{0,1\}^d$ with odd numbers of 1's. Take their sum:

$$\sum_{\mathbf{s}:s_1+\ldots s_d\equiv_2 0} PV(w_{\mathbf{x+s}..\mathbf{x+s+n-1}}) = \sum_{\mathbf{s}:s_1+\ldots s_d\equiv_2 1} PV(w_{\mathbf{x+s}..\mathbf{x+s+n-1}}).$$

We now count, for each integer l, $0 \leq l \leq d$, and for each $\mathbf{i} \in \{0,1\}^d, i_1+\ldots+i_d = l$, how many times an l-dimensional rectangular factor $w_{\mathbf{x+i}..\mathbf{x+in-1}}$ has been counted, on the left and on the right. On the left, it is contained in all factors $w_{\mathbf{x+s}..\mathbf{x+s+n-1}}$ such that $\mathbf{s}$ is component-wise smaller than or equal to $\mathbf{i}$ (i.e., the set of indices with 1's in $\mathbf{s}$ is a subset of indices with 1's in $\mathbf{i}$). So, we count it $\sum_{0\leq k\leq l,k \text{ odd}} \binom{l}{k} = 2^{l-1}$ times. On the righthand side, with a similar argument, we count it $\sum_{0\leq k\leq l,k \text{ even}} \binom{l}{k} = 2^{l-1}$ times. The same holds for any other l-dimensional rectangular factor of the boundary of $w_{\mathbf{x}..\mathbf{x+n}}$ of the form $w_{\mathbf{x+i}+(1-\mathbf{i})\mathbf{nj}..\mathbf{x+in-1}+(1-\mathbf{i})\mathbf{nj}}$ for each $\mathbf{j} \in \{0,1\}^d$ due to symmetry. So, everything is counted the same number of times on both sides of the equality, except for the corners $\mathbf{x+tn}$, $\mathbf{t} \in \{0,1\}^d$ (corresponding to $l = 0$), which are counted once, on the left or on the right, depending on the parity of the number of 1's in $\mathbf{i}$.

Remark 2. It is easy to see that in dimension 2, the condition from Proposition 4 corresponds to periodicity in (m,n)-lattices described in [22], Theorem 5.

7   Conclusions and Open Questions

In this paper, we show that a recurrent d-dimensional word with abelian complexity bounded by 2 has a periodicity vector for $d \geq 2$. Moreover, the bound 2 is optimal in the sense that there exist d-dimensional words with abelian complexity bounded by 3 that are aperiodic in any direction. In dimension 1, the bound is different: aperiodic recurrent words of minimal abelian complexity have abelian complexity 2 for each length, and they correspond to the family of Sturmian words. In dimension $d \geq 2$, the structure of aperiodic recurrent words of minimal abelian complexity (i.e., bounded by 3) does not seem to be related to Sturmian words. An interesting question is to characterize such words for $d \geq 2$.

Another open question is strengthening Theorem 3 in the following sense. In [22] it was proved that in dimension 2, the bound 2 for abelian complexity for sufficiently large blocks is enough to guarantee periodicity. It would be interesting to generalize this fact to dimension $d \geq 3$.

Acknowledgments. This work was supported by the Russian Science Foundation, project 25-21-00535.

References

1. Avgustinovich, S.V., Fon-Der-Flaass, D.G., Frid, A.E.: Arithmetical Complexity of Infinite Words, Words, Languages & Combinatorics III, pp. 51–62 (2003)
2. Berthé, V., Tijdeman, R.: Balance properties of multi-dimensional words. Theor. Comput. Sci. **273**, 197–224 (2002)
3. Cassaigne, J., Richomme, G., Saari, K., Zamboni, L.Q.: Avoiding abelian powers in binary words with bounded abelian complexity. Int. J. Found. Comput. Sci. **22**(4), 905–920 (2011)
4. Charlier, É., Puzynina, S., Zamboni, L.: On a group theoretic generalization of the Morse-Hedlund theorem. In: Proceedings of the American Mathematical Society, vol. 145, issue (8), pp. 3381–3394 (2017)
5. Cassaigne, J.: Subword complexity and periodicity in two or more dimensions. Dev. Language Theory, 14–21 (1999)
6. Cassaigne, J., Kaboré, I.: Abelian complexity and frequencies of letters in infinite words. Int. J. Found. Comput. Sci. **27**(5), 631–650 (2016)
7. Coven, E., Hedlund, G.: Sequences with minimal block growth. Math. Syst. Theory **7**, 138–153 (1973)
8. Currie, J., Rampersad, N.: Recurrent words with constant Abelian complexity. Adv. Appl. Math. **47**, 116–124 (2011)
9. Cyr, V., Kra, B.: Nonexpansive $\mathbb{Z}^2$-subdynamics and Nivat's conjecture. Trans. Amer. Math. Soc. **367**, 6487–6537 (2015)
10. Cyr, V., Kra, B.: Complexity of short rectangles and periodicity. Eur. J. Comb. **52**, 146–173 (2016)
11. Durand, F., Rigo, M.: Multidimensional extension of the Morse-Hedlund theorem. Eur. J. Comb. **34**(2), 391–409 (2013)
12. Fici, G., Puzynina, S.: Abelian combinatorics on words: a survey. Comput. Sci. Rev. **47**, 100532 (2023)
13. Herva, P., Kari, J.: On forced periodicity of perfect colorings. Theory Comput. Syst. **67**(4), 732–759 (2023)
14. Kamae, T., Widmer, S., Zamboni, L.Q.: Abelian maximal pattern complexity of words. Ergod. Theory Dyn. Syst. **35**(1), 142–151 (2015)
15. Kamae, T., Zamboni, L.Q.: Sequence entropy and the maximal pattern complexity of infinite words. Ergod. Theory Dyn. Syst. **22**(4), 1191–1199 (2002)
16. Kari, J., Moutot, E.: Decidability and periodicity of low complexity tilings. Theory Comput. Syst. **67**(1), 125–148 (2023)
17. Kari, J., Szabados, M.: An algebraic geometric approach to Nivat's conjecture. Inf. Comput. **271**, 104481 (2020)
18. Lothaire, M.: Algebraic Combinatorics on Words. Cambridge University Press, Cambridge (2002)

19. Morse, M., Hedlund, G.: Symbolic dynamics. Amer. J. Math. **60**, 815–866 (1938)
20. Morse, M., Hedlund, G.: Symbolic dynamics II: Sturmian sequences. Amer. J. Math. **62**, 1–42 (1940)
21. Nivat, M.: Invited talk at ICALP (1997)
22. Puzynina, S.: Aperiodic two-dimensional words of small abelian complexity. Electron. J. Comb. (2019)
23. Puzynina, S.: Complexity of infinite words. In: LNCS 15270 (MCU 2024), pp. 1–16 (2024)
24. Richomme, G., Saari, K., Zamboni, L.Q.: Abelian complexity of minimal subshifts. J. London Math. Soc. **83**, 79–95 (2011)
25. Saarela, A.: Ultimately constant abelian complexity of infinite words. J. Autom. Lang. Comb. **14**, 255–258 (2009)

Tree Representations of Infinite Words and Their Logical Properties

Arnaud Carayol[(✉)] and Lucien Charamond

Univ Gustave Eiffel, CNRS, LIGM, 77454 Marne-la-Vallée, France
`{arnaud.carayol,lucien.charamond2}@univ-eiffel.fr`

Abstract. In this article, we investigate the relationship between the decidability of the MSO-theory of the linear representation and the decidability of the MSO-theory of the base-k tree representation of an infinite word. The base-k tree representation is a representation of the word by an infinite tree, inspired by the theory of automatic sequences. We show that decidability of MSO on the linear representation and on the base-k tree representation are independent properties. In particular, we show that if one fixes a recursive set of positions in an infinite word whose gaps sizes are increasing sufficiently fast, then any recursive infinite word can be placed on those positions and the remaining positions can be filled to obtain a recursive infinite word with a decidable MSO-theory.

Keywords: Infinite words · Monadic second order logic · Automatic sequences

1 Introduction

Infinite words over a finite alphabet are usually represented by an infinite line whose nodes are labeled by symbols of the alphabet. In this article, we consider a different representation, namely its base-k tree representation, which is a tree whose edges are labeled by the digits $0, \ldots, k-1$ and whose nodes are labeled by symbols of the alphabet. More formally, fix a base $k \geq 2$ and let $D_k := \{0, \ldots, k-1\}$. For an infinite word $w = w_0 w_1 w_2 \cdots \in \Sigma^\omega$, the *tree representation* of w in base k, denoted by $\mathrm{TRep}_k(w)$, is the tree with directions in D_k whose domain consists of all words in D_k^* that do not start with 0. For each node u in its domain, we write $[u]_k$ for the integer encoded by u in base k, with $[\varepsilon]_k = 0$, and we define the label of u to be $w_{[u]_k}$ (i.e., $\mathrm{TRep}_k(w)(u) = w_{[u]_k}$).

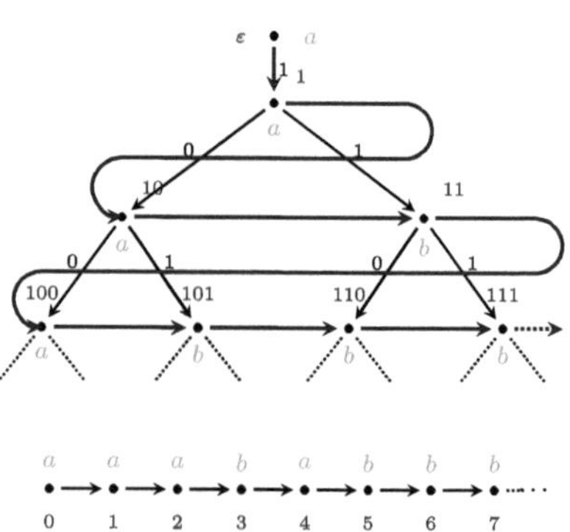

Fig. 1. The linear and base-2 tree representation of $aaababbb\cdots$.

M.-P. Béal and P. Caron (Eds.): DLT 2026, LNCS 16578, pp. 280–292, 2026.
https://doi.org/10.1007/978-3-032-28404-4_21

Thus the root ε carries the symbol w_0, the node 1 carries w_1, and, more generally, reading the nodes in length-lexicographic order produces the word w. This representation is illustrated in Fig. 1 for base 2.

One natural motivation for considering the base-k tree representation comes from the theory of automatic sequences (see, for instance, [3]). Recall that an infinite word w is k-automatic if there exists a deterministic finite automaton that, after reading the base-k expansion of n, outputs the symbol w_n. The tree $\mathrm{TRep}_k(w)$ is nothing but the unfolding of the finite automaton from its initial state. Thus, the automaticity of the infinite word is reflected in the regularity of its tree representation. More precisely, an infinite word is k-automatic if and only if its base-k tree representation is regular, that is, it has only finitely many non-isomorphic subtrees.

Another motivation comes from number theory. A real number $\alpha \in [0, 1]$ can be identified with its expansion in some base, and hence with an infinite word. In [2], the authors showed that the base-k tree representation[1] of the expansion of an irrational algebraic number carries structural information that is much less transparent in the linear presentation. More precisely, they showed that if the tree representation associated with such an expansion contains two isomorphic subtrees, then α cannot be irrational algebraic.

This implies that if the expansion is k-automatic, then α is either rational or transcendental (which was first established in [1]). This result extends beyond automatic sequences: replacing finite automata by deterministic pushdown automata in their definition leads to the context-free sequences of [9], and [2] shows that if the expansion is context-free, then α is again either rational or transcendental. This was later pushed further to extensions of the notion of automatic sequences based on deterministic higher-order pushdown automata [5].

A common feature of all these extended notions of automaticity is that their base-k tree representations have decidable MSO-theories: in the automatic case, by Rabin's theorem [12], and in the higher-order pushdown setting, by the decidability of MSO throughout the pushdown hierarchy [11]. This raises the question whether the expansion of an irrational algebraic number could also have a decidable MSO-theory when viewed through its tree representation. The intuition is that decidability of the MSO-theory of the tree representation could impose structural restrictions preventing the underlying word from being the expansion of an irrational algebraic number. The emphasis on tree representations is crucial here. In the linear setting, it was recently shown in [4] that the expansions of irrational algebraic numbers have decidable MSO-theories, assuming the long-standing conjecture that such expansions are weakly normal[2] in all bases. As a first step towards the tree case, we investigate here the relationship between decidability of the MSO-theory of the base-k tree representation and decidability of the MSO-theory of the corresponding linear representation.

---

[1] The base k of the tree representation has no connection with the base of the expansion of α.

[2] A base-b expansion is weakly normal if every finite block of digits occurs infinitely often.

Our first main result shows that no implication holds from the tree representation to the linear one. For every base $k \geq 2$, we construct an infinite word whose base-k tree representation has a decidable MSO-theory, while the MSO-theory of the word in its standard linear representation is undecidable. The construction proceeds through a deterministic tree t_{PCP} with a decidable MSO-theory whose linearization (the word obtained by reading the node labels in length-lexicographic order) has an undecidable MSO-theory, and then we embed it into genuine base-k tree representations.

Our second main result goes in the opposite direction. For every base $k \geq 2$, we construct an infinite word with a decidable MSO-theory whose base-k tree representation has an undecidable MSO-theory. Hence decidability of MSO on the linear representation and on the base-k tree representation are independent properties.

Beyond these separation results, the proof of the second direction yields a statement that may be useful in a wider setting. We show that if one fixes a recursive set of positions in an infinite word whose gaps sizes are increasing sufficiently fast, then any recursive infinite word can be placed on those positions and the remaining positions can be filled so as to obtain a recursive infinite word with decidable MSO-theory. In other terms, words with decidable MSO-theories can hide arbitrarily complex recursive subwords, provided that these subwords are read along sufficiently sparse positions. This gap-sequence construction is, in our view, the main reusable technical contribution of the paper.

2   Preliminaries

Finite and infinite words. For a finite alphabet Σ, a *finite word* u is a (possibly empty) sequence of letters in Σ, that is, an element of Σ^*. The *empty word* is denoted by ε. We write $|u|$ for the *length* of u. If $u, v \in \Sigma^*$, then their *concatenation* is denoted by uv.

We write $\sqsubseteq$ for the *prefix relation* on Σ^*: for $u, v \in \Sigma^*$, we have $u \sqsubseteq v$ if there exists $w \in \Sigma^*$ such that $v = uw$. We write $u \sqsubset v$ if $u \sqsubseteq v$ and $u \neq v$.

Assume that Σ comes with a total order $<_\Sigma$. The *lexicographic order* on Σ^* is defined for $u, v \in \Sigma^*$ by $u \prec_{\mathrm{lex}} v$ if either $u \sqsubset v$ or there exist $w \in \Sigma^*$ and $a, b \in \Sigma$ such that $u = wau'$ and $v = wbv'$ for some $u', v' \in \Sigma^*$, and $a <_\Sigma b$. The *length-lexicographic order* on Σ^* is defined for $u, v \in \Sigma^*$ by $u \prec_{\mathrm{llex}} v$ if either $|u| < |v|$, or $|u| = |v|$ and $u \prec_{\mathrm{lex}} v$. We write $u \preceq_{\mathrm{llex}} v$ if $u = v$ or $u \prec_{\mathrm{llex}} v$.

Let $u = u_1 \cdots u_{|u|}$, $v = v_1 \cdots v_{|v|}$, and $w = w_1 \cdots w_{|w|}$ be finite words, and let $r = \max(|u|, |v|, |w|)$. The *interleaving* of u, v, w is the word $[\![u, v, w]\!] := \tilde{u}_1 \tilde{v}_1 \tilde{w}_1 \ \tilde{u}_2 \tilde{v}_2 \tilde{w}_2 \ \cdots \ \tilde{u}_r \tilde{v}_r \tilde{w}_r$, where $\tilde{u}_k = u_k$ if $k \leq |u|$ and $\tilde{u}_k = \varepsilon$ otherwise, and similarly for $\tilde{v}_k$ and $\tilde{w}_k$.

Infinite deterministic trees. A *tree* t with directions in a finite alphabet Σ and colors in a finite alphabet C is a mapping from a prefix-closed subset of Σ^* to C. An element of $\mathrm{Dom}(t)$ is called a *node*. The *root* of t is the empty word $\varepsilon \in \mathrm{Dom}(t)$. For $u, v \in \mathrm{Dom}(t)$, we say that v is a *direct descendant* of u if there exists $a \in \Sigma$ such that $v = ua$; in this case, v is the *a-child* of u. More generally, v is a *descendant* of u if $u \sqsubseteq v$. Nodes without children are called leaves.

In other words, children are labeled by letters in Σ: from a node u and a direction $a \in \Sigma$, we reach the child ua (when it belongs to $\mathrm{Dom}(t)$). Nodes themselves are labeled (or "colored") by $t(u) \in C$.

Given a total ordering on the set of directions, the linearization of a tree t whose nodes are colored in C is the infinite word $\mathrm{Lin}(t) := t(u_0)t(u_1)t(u_2)\cdots \in C^\omega$ where $u_0 \prec_{\mathrm{llex}} u_1 \prec_{\mathrm{llex}} u_2 \prec_{\mathrm{llex}} \cdots$ is the enumeration of the nodes of t in length-lexicographic order. By definition, for every infinite word w and every base $k \geq 2$, $\mathrm{Lin}(\mathrm{TRep}_k(w)) = w$.

A (one-hole) *context* $C[\bullet]$ is a tree with a distinguished leaf, called the *hole*, which is the only leaf colored by $\bullet$. For two contexts $C_1[\bullet]$ and $C_2[\bullet]$, we denote by $C_1[C_2[\bullet]]$ the context obtained by gluing a copy of $C_2[\bullet]$ into the hole of $C_1[\bullet]$. Let $C_1[\bullet], C_2[\bullet], C_3[\bullet], \ldots$ be a sequence of non-trivial contexts, we denote by $C_1[C_2[C_3[\cdots]]]$ the infinite tree obtained by concatenating these contexts in sequence.

Monadic second-order logic (MSO) over graphs. Monadic second-order logic over graphs whose edges are labeled by letters from a finite alphabet and whose vertices are colored by letters from a finite alphabet is defined as usual, with quantification over both nodes and sets of nodes. We use lowercase letters for node variables and uppercase letters for sets of nodes. For a formula $\varphi(x_1, \ldots, x_n, X_1, \ldots, X_m)$, we write $G \models \varphi[u_1, \ldots, u_n, U_1, \ldots, U_m]$ when φ holds in the graph G under the assignment sending the variables $x_1, \ldots, x_n$ to the nodes $u_1, \ldots, u_n$ and the variables $X_1, \ldots, X_m$ to the sets of nodes $U_1, \ldots, U_m$. For a closed formula φ, we simply write $G \models \varphi$. The MSO-theory of a graph is the set of closed MSO sentences that it satisfies.

A relation on nodes is said to be *MSO-definable* if it is described by some MSO formula in this way. In particular, a vertex is MSO-definable if it is uniquely characterized by an MSO formula. On deterministic trees, the ancestor relation and the lexicographic ordering on nodes are well known to be MSO-definable. By contrast, the length-lexicographic order is not MSO-definable.

Graph transformations preserving the decidability of the MSO-theory. We say that a graph operation preserves the decidability of the MSO-theory if, whenever the input graph has a decidable MSO-theory, the output graph also has a decidable MSO-theory. Furthermore, a Turing machine deciding the MSO-theory of the output can be effectively constructed from a Turing machine deciding the MSO-theory of the input. We briefly present several such transformations that will be used to construct trees with decidable MSO-theories (we refer to [6] for a detailed presentation).

An *MSO-interpretation* transforms a graph into another graph by using MSO formulas to select which vertices are kept and which labeled edges and colors are present in the output graph. An *MSO-transduction* is obtained by first making finitely many copies of the graph: we fix a fresh set of edge labels, and for each label $\#_i$ in this set, every vertex u gets a fresh outgoing $\#_i$-edge to a new copy of u, and one then applies an MSO-interpretation.[3]

Both MSO-interpretations and MSO-transductions preserve the decidability of the MSO-theory, and both classes of operations are closed under composition.

Finally, given a graph G and a vertex s, the *unfolding* of G from s is the tree whose nodes are the finite paths of G starting from s. Its edges correspond to one-step extensions of paths, and its labels and colors are inherited from the underlying graph. This operation also preserves the decidability of the MSO-theory, provided that the vertex s from which one unfolds is MSO-definable.

3   Linearization Does Not Preserve the Decidability of MSO

In this section, we show that the linearization operation does not preserve the decidability of MSO-theories.

Theorem 3.1. *There exists a deterministic tree t_{PCP} with a decidable MSO-theory whose linearization has an undecidable MSO-theory.*

Before presenting the construction of t_{PCP}, we introduce a variant of the Post correspondence problem (PCP) used in the reduction. This variant, called the generalized Post correspondence problem (gPCP), is presented in Sect. 3.1. In Sect. 3.2, we define the tree t_{PCP} and show in Sect. 3.3 that the linearization of t_{PCP} has an undecidable MSO-theory by reduction from gPCP (Theorem 3.2). Finally, in Sect. 3.4, we show that t_{PCP} has a decidable MSO-theory. In Sect. 3.5, we construct for all bases $k \geq 2$ an infinite word w_k with an undecidable MSO-theory but whose base-k tree representation has a decidable MSO-theory.

3.1   Generalized Post Correspondence Problem (GPCP)

Let $\mathbb{B}$ be the binary alphabet $\{0, 1\}$ and let $\overline{\mathbb{B}}$ be the alphabet $\{0_L, 1_L, 0_R, 1_R\}$. Define two morphisms, π_L and π_R, from $\overline{\mathbb{B}}^*$ to $\mathbb{B}^*$. The morphism π_L maps every occurrence of 0_L to 0, every 1_L to 1, and erases all 0_R and 1_R. Dually, π_R maps every occurrence of 0_R to 0, every 1_R to 1, and erases all 0_L and 1_L.

The generalized Post correspondence problem[4] (gPCP) is the problem of deciding, given a regular language $S \subseteq \overline{\mathbb{B}}^*$, whether there exists a non-empty word $w \in S$ such that $\pi_L(w) = \pi_R(w)$.

Theorem 3.2. *The generalized Post correspondence problem is undecidable.*

---

[3] The MSO-interpretations and MSO-transductions we consider correspond to 1-dimensional parameterless interpretations in some of the litterature [10].

[4] The gPCP is equivalent to the problem of deciding whether a given rational relation over the binary alphabet $\mathbb{B}$ has a non-empty intersection with the identity relation on $\mathbb{B}^+$. It is very likely that this problem has already been observed to be undecidable, but we could not find a reference.

Proof (Sketch). The proof is by reduction from the Post correspondence problem (PCP). Given an instance $\sigma = (a_1, b_1) \cdots (a_n, b_n)$ of PCP which does not contain $(\varepsilon, \varepsilon)$, we consider the regular language $S_\sigma = \{a_i^L b_i^R \mid i \in [n]\}^+$, where u^L and u^R mark the letters of a binary word u with L and R, respectively. Every word $w \in S_\sigma$ encodes a sequence of pairs from σ, and its two projections are exactly the two concatenations produced by that sequence of pairs. Hence σ has a solution if and only if there exists $w \neq \varepsilon \in S_\sigma$ such that $\pi_L(w) = \pi_R(w)$. ∎

3.2 Construction of the Tree t_{PCP}

The tree t_{PCP} has directions in $\{0, 1, 2\}$ (equipped with the natural ordering) and node labels in the alphabet $\Sigma := \mathbb{B} \cup \overline{\mathbb{B}} \cup \{\triangleleft_u, \triangleleft_v, \$\}$.

The building blocks of our construction are the contexts $C_w[\bullet]$ depicted in Fig. 2 for $w \in \overline{\mathbb{B}}^*$. The context $C_w[\bullet]$ is a tree with three branches: the first is labeled by $\pi_L(w)\triangleleft_u$, the second by $\pi_R(w)\triangleleft_v$, and the third by w. The hole is placed at the end of the w-branch, and the root is colored by the special symbol $\$$. The key feature of the contexts $C_w[\bullet]$ is that their linearization allows us to determine, via an MSO formula, whether $\pi_L(w) = \pi_R(w)$. Indeed, the linearization of $C_w[\bullet]$ consists of the special symbol $\$$ followed by the first symbols

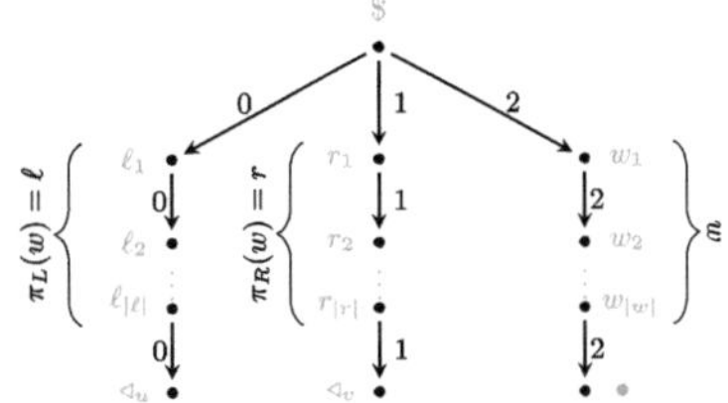

Fig. 2. The context $C_w[\bullet]$ represented as a tree.

of $\pi_L(w)$, $\pi_R(w)$, and w, then by their second symbols, and so on. The symbols $\triangleleft_u$ and $\triangleleft_v$ are used to signal, in the linearization, the ends of the $\pi_L(w)$-branch and the $\pi_R(w)$-branch, respectively. In other words, the linearization $\mathrm{Lin}(C_w[\bullet])$ is equal to $\$ \cdot [\![\pi_L(w)\triangleleft_u, \pi_R(w)\triangleleft_v, w]\!] \cdot \bullet$.

Hence, an MSO formula over the linearization of $C_w[\bullet]$ can test whether $\pi_L(w) = \pi_R(w)$ by checking that their first symbols are equal, their second symbols are equal, and so on until the pair $\triangleleft_u\triangleleft_v$ is reached.

Lemma 3.3. *There exists an MSO formula φ_{eq} over $\Sigma \cup \{\bullet\}$ such that for all words $w \in \overline{\mathbb{B}}^*$ we have $\mathrm{Lin}(C_w[\bullet]) \models \varphi_{eq}$ if and only if $\pi_L(w) = \pi_R(w)$.*

Definition 3.4. *Fix an arbitrary total order on $\overline{\mathbb{B}}$. Let $w_1 \prec_{\mathrm{llex}} w_2 \prec_{\mathrm{llex}} \cdots$ be the enumeration of $\overline{\mathbb{B}}^+$ in the induced length-lexicographic ordering. The infinite tree t_{PCP} is the tree with directions in $\{0, 1, 2\}$ (equipped with the natural ordering) and labels in the alphabet $\Sigma := \mathbb{B} \cup \overline{\mathbb{B}} \cup \{\triangleleft_u, \triangleleft_v, \$\}$ defined as the limit*

$$t_{\mathrm{PCP}} = C_{w_1}[C_{w_2}[C_{w_3}[\cdots]]]$$

3.3 Undecidability of the MSO-Theory of the Linearization of t_{PCP}

In this section, we show that the linearization of t_{PCP} has an undecidable MSO-theory. Let $w_1 \prec_{\mathrm{llex}} w_2 \prec_{\mathrm{llex}} \cdots$ be the enumeration of $\overline{\mathbb{B}}^+$ used to define t_{PCP} in Definition 3.4.

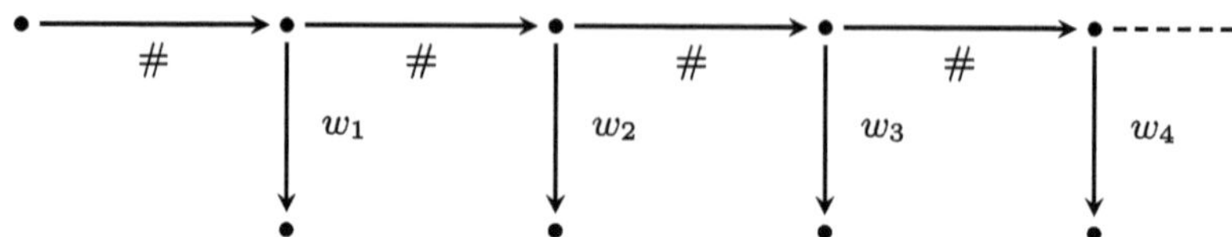

Fig. 3. The tree t_{llex}^{Σ}.

The linearization of t_{PCP} is $\text{Lin}(t_{\text{PCP}}) = \$\,\text{Lin}_1\,\$\,\text{Lin}_2\,\$\,\text{Lin}_3\,\$\cdots$, where $\text{Lin}_i = [\![\pi_L(w_i)\triangleleft_u, \pi_R(w_i)\triangleleft_v, w_i]\!]$ for all $i \geq 1$. Indeed, the linearization of $C_{w_i}[\bullet]$ is $\$\,\text{Lin}_i\,\bullet$, and the last node in the length-lexicographic order is the hole.

Lemma 3.5. *The linearization of the tree t_{PCP} defined in Definition 3.4 has an undecidable MSO-theory.*

Proof. Given a regular language S over $\overline{\mathbb{B}}$, we construct a formula φ_S over Σ^ω such that $\text{Lin}(t_{\text{PCP}}) \models \varphi_S$ if and only if there exists $w \neq \varepsilon \in S$ such that $\pi_L(w) = \pi_R(w)$.

The formula φ_S expresses that between two successive occurrences of the symbol $\$$ there is a factor ℓ such that the restriction of ℓ to $\overline{\mathbb{B}}$ is a word $w \in S$ and $\pi_L(w) = \pi_R(w)$.

The first condition is easily expressed by an MSO formula using the equivalence between MSO formulas and regular languages. The second condition can be expressed by the formula φ_{eq} from Lemma 3.3.

Clearly, if $\text{Lin}(t_{\text{PCP}}) \models \varphi_S$, then there exists a non-empty word $w \in S$ such that $\pi_L(w) = \pi_R(w)$, so the instance S of gPCP has a solution. Conversely, if there exists a non-empty word $w \in S$ such that $\pi_L(w) = \pi_R(w)$, then this word corresponds to some w_i in the enumeration of $\overline{\mathbb{B}}^+$ used to define t_{PCP} in Definition 3.4. Then Lin_i witnesses that the formula φ_S is satisfied. ∎

3.4 Decidability of the MSO-Theory of the Tree t_{PCP}

In this section, we show that the tree t_{PCP} has a decidable MSO-theory. To this end, we construct t_{PCP} from a graph with a decidable MSO-theory using the graph transformations recalled in Sect. 2, namely graph unfolding and MSO-transductions. This approach was developed in [6,8]; see also [13] for a survey.

As a stepping stone towards the tree t_{PCP}, we construct, using these operations, a comb-shaped tree that encodes the enumeration of all non-empty words in the length-lexicographic ordering over Σ^+. This tree, denoted t_{llex}^{Σ}, has directions in $\Sigma \cup \{\#\}$ and is depicted in Fig. 3.

Lemma 3.6. *Let $(\Sigma, <)$ be a finite alphabet with a total order. Let $w_1 \prec_{\text{llex}} w_2 \prec_{\text{llex}} \cdots$ be the enumeration of Σ^+ in the length-lexicographic ordering. The tree t_{llex}^{Σ} with directions in $\Sigma \cup \{\#\}$ whose domain consists of all prefixes of words of the form $\#^i w_i$ for $i \geq 1$ has a decidable MSO-theory.*

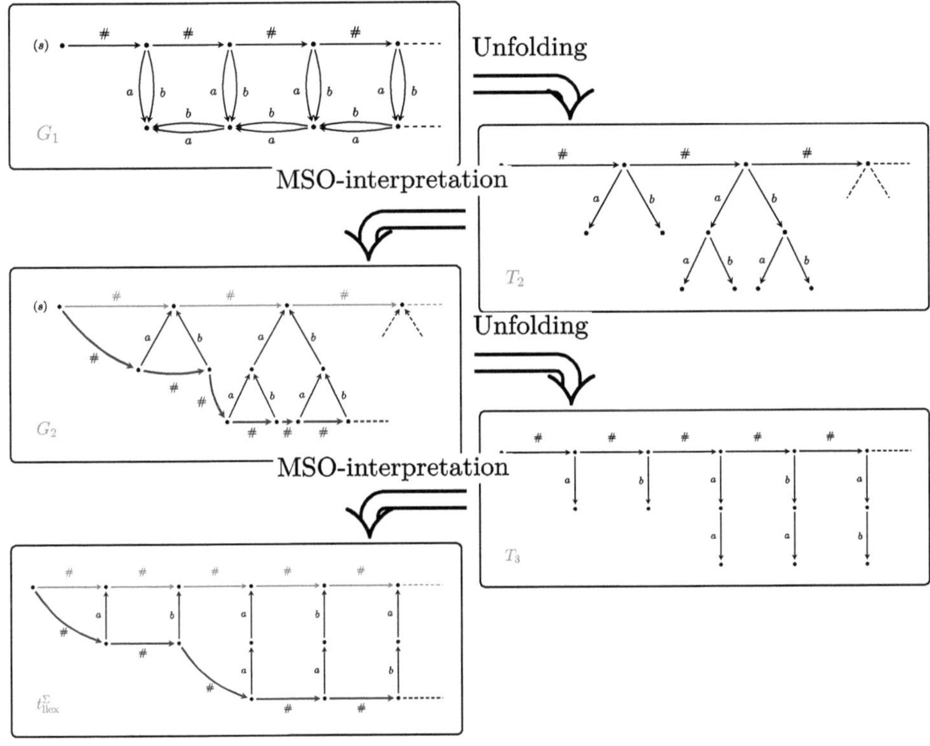

Fig. 4. Graph transformations constructing the tree t_{llex}^{Σ}.

Proof. To simplify the exposition, we consider only the case where $\Sigma = \{a, b\}$ with $a < b$. The general case follows from the same reasoning. The construction of the tree t_{llex}^{Σ} is depicted in Fig. 4. We start from the graph G_1, which can be MSO-interpreted in the full binary tree and therefore has a decidable MSO-theory [12]. By unfolding G_1 from the vertex s, we obtain the tree T_2. Below the node $\#^i$, this tree contains a complete tree of depth i, whose leaves correspond to the words of length i over Σ. Using an MSO-interpretation, we add a $\#$-labeled edge between two leaves of the same complete tree whenever they are consecutive in lexicographic order, and also between the last leaf of a complete tree and the first leaf of the next one. These edges are depicted in red in Fig. 4; they connect the leaves of the different complete trees according to the length-lexicographic order. This transformation is MSO-definable because the lexicographic order on the leaves of a complete tree is itself MSO-definable. Let G_2 denote the resulting graph. By unfolding G_2 from the vertex r, we obtain a tree T_3 that is very close to t_{llex}^{Σ}: it is comb-shaped and, below the node $\#^i$, contains the mirror image of the i-th word in the length-lexicographic order. The tree t_{llex}^{Σ} is then obtained by an MSO-interpretation that links the ends of consecutive teeths by a $\#$-labeled edge, reverses the directions of all a- and b-labeled edges, and removes the original $\#$-labeled edges. ∎

The tree t_{PCP} can be constructed from $t_{\mathrm{llex}}^{\overline{\mathbb{B}}}$ by an MSO-transduction and therefore has a decidable MSO-theory.

Lemma 3.7. *The tree t_{PCP} defined in Definition 3.4 has a decidable MSO-theory.*

Proof (Sketch). The tree t_{PCP} is obtained from $t_{\mathrm{llex}}^{\overline{\mathbb{B}}}$ by an MSO-transduction that replaces each branch encoding a word $w \in \overline{\mathbb{B}}^+$ by the context $C_w[\bullet]$, followed by an MSO-interpretation that links consecutive contexts together. ∎

3.5 Infinite Words with Undecidable MSO-Theories but Decidable Base-k Tree Representations

The tree t_{PCP} is not the base-3 tree representation of an infinite word, because its domain is not the set U_3 of all words over $\{0, 1, 2\}$ that do not start with 0.

However, we can readily define a tree t_3 with domain U_3 that contains a copy of t_{PCP} and in which all nodes not belonging to the copy of t_{PCP} are labeled by a fresh symbol $\diamond$.

Consider the tree t_3 defined by $t_3(1u) := t_{\mathrm{PCP}}(u)$ for $u \in \mathrm{Dom}(t_{\mathrm{PCP}})$ and $t_3(w) = \diamond$ for all other words $w \in U_3$. By removing the nodes labeled by $\diamond$ from t_3, we obtain a copy of t_{PCP}. It follows that the linearization w_3 of t_3 has an undecidable MSO-theory. Indeed, by erasing the symbols $\diamond$ from w_3, we obtain the linearization of t_{PCP}, which has an undecidable MSO-theory. The tree t_3 can be constructed from t_{PCP} using an MSO-transduction followed by an unfolding and hence has a decidable MSO-theory.

Corollary 3.8. *For all $k \geq 2$, there exists an infinite word w_k whose linear representation has an undecidable MSO-theory, while its base-k tree representation has a decidable MSO-theory.*

4 Infinite Words with Decidable Linear Representations but Undecidable Base-k Tree Representations

In this section, we prove Theorem 4.1.

Theorem 4.1. *For all $k \geq 2$, there exists an infinite word w_k with a decidable MSO-theory, while its base-k tree representation has an undecidable MSO-theory.*

The key observation is that for an infinite word w, it is possible to define particular subwords of w within $\mathrm{TRep}_k(w)$ using an MSO-interpretation. For instance, using the infinite branch $(k-1)^\omega$ of $\mathrm{TRep}_k(w)$, we can define the subword of w consisting of the letters occurring at positions $(k^{2n} - 1)_{n \geq 0}$.

Our main technical result, stated below, implies that it is possible to construct infinite words with decidable MSO-theories that contain any recursive infinite word as a subword at the positions $(k^{2n} - 1)_{n \geq 0}$. More precisely, Theorem 4.2 shows that this can be achieved for any set of positions, provided that the gaps between the positions increase sufficently fast.

A set of positions $p_0 < p_1 < \ldots$ in an infinite word can be encoded by the lengths of the successive gaps between its positions. The *gap sequence* $(s_i)_{i \geq 0}$ of the set of positions $p_0 < p_1 < \ldots$ is defined by $s_0 = p_0$ and $s_{i+1} = p_{i+1} - p_i - 1$ for all $i \geq 0$.

Theorem 4.2. *Let $\pi \in \{0,1\}^\omega$ be a recursive infinite word, and let $(s_i)_{i \geq 0}$ be a recursive gap sequence such that $s_{i+1} > s_i + 2$ for all $i \geq 0$. Then there exists a word $\sigma \in \{0,1\}^\omega$ with a decidable MSO-theory such that π is the subword of σ consisting of the symbols occurring at the positions encoded by the gap sequence $(s_i)_{i \geq 0}$. Furthermore, a Turing machines computing σ and its MSO-theory can be constructed from Turing machines computing π and $(s_i)_{i \geq 0}$.*

If we admit Theorem 4.2, we can prove Theorem 4.1 as follows. Let $k \geq 2$. We construct an infinite word w_k such that $\mathrm{TRep}_k(w_k)$ has an undecidable MSO-theory. Consider any recursive infinite word π_{und} over the binary alphabet $\mathbb{B}$ with an undecidable MSO-theory. By Theorem 4.2, we can construct an infinite word w_k with a decidable MSO-theory (and a Turing machine deciding it) such that the subword of w_k corresponding to the positions[5] $(k^{2n} - 1)_{n \geq 0}$ is precisely π_{und}. As already remarked, this directly implies that the base-k tree representation of w_k has an undecidable MSO-theory (as π_{und} labels the even positions in the branch $(k-1)^\omega$).

The reminder of this section is devoted to the proof of Theorem 4.2. We start by reducing the problem to a special case.

4.1   Proof of a Special Case of Theorem 4.2

We start by proving the theorem in a special case. To state this special case, we need to introduce the notion of gap word.

Given a gap sequence $s = (s_i)_{i \geq 0}$, the *gap word* $\langle s \rangle$ is the infinite word over $\{0,1\}$ which contains a 1 at every position prescribed by the gap sequence s and a 0 at all other positions. We say that a word σ *refines* a word σ' if for all $i \geq 0$, $\sigma'(i) = 1$ implies $\sigma(i) = 1$.

Consider the following special case of Theorem 4.2, restricted to the word 1^ω and with the requirement that the word produced does not contain two consecutive[6] 1s.

Lemma 4.3. *Let $s = (s_i)_{i \geq 0}$ be a recursive and strictly increasing gap sequence with $s_0 \geq 4$. There exists a recursive word σ without two consecutive 1s which refines the gap word $\langle s \rangle$ and has a decidable MSO-theory. Furthermore, Turing machines computing σ and its MSO-theory can be constructed from a Turing machine computing s.*

The main advantage of focusing on this special case is that there are very simple sufficient conditions ensuring that the gap word has a decidable MSO-theory. We use a simplified version of [7, Theorem 5.1].

Theorem 4.4 ([7]). *Let $s = (s_i)_{i \geq 0}$ be a recursive gap sequence. Assume that for all $k \geq 2$, we can effectively compute an index N_k such that for all $i \geq N_k$, we have $s_i \geq k$ and $s_i \mod k = 0$.*

The gap word $\langle s \rangle$ has a decidable MSO-theory and a Turing machine deciding the MSO-theory of $\langle s \rangle$ can be constructed from the Turing machines computing N_k.

The remainder of this section is devoted to the proof of Theorem 4.3. Let $s = (s_i)_{i \geq 0}$ be a recursive and strictly increasing gap sequence with $s_0 \geq 4$. Let $(p_i)_{i \geq 0}$ be the set of positions encoded by the gap sequence $(s_i)_{i \geq 0}$. Our aim is to add 1s to the word $\langle s \rangle$ to ensure that, for all $n \geq 2$, all the gaps in the resulting word are ultimately multiples of $n!$.

Consider the gap s_i between the positions p_i and p_{i+1} of the gap word $\langle s \rangle$. Let n be the unique integer such that $(n!)^2 \leq s_i < ((n+1)!)^2$. The integer s_i can be uniquely written as $s_i = n! \cdot a + b$ with $0 \leq b < n!$ and $a \geq n!$. If $b = 0$, then s_i is a multiple of $n!$ and we do nothing. Otherwise, we add b 1s at the positions $p_i + n! + 1$, $p_i + 2n! + 2$, ..., $p_i + bn! + b$. This replaces the gap s_i by $b+1$ gaps: the first b gaps have length $n!$ while the last gap has length $n!(a - b)$. Let σ be the word obtained by applying this procedure to all the gaps in $\langle s \rangle$. This construction is illustrated in Fig. 5.

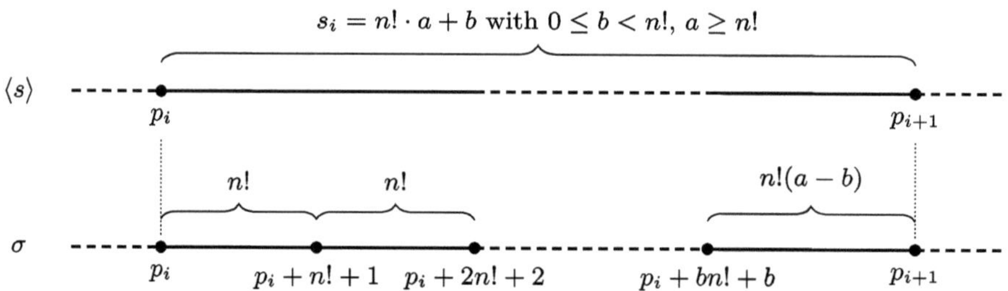

Fig. 5. Splitting a gap in the construction of the word σ.

The word σ is recursive and hence its gap sequence $(h_i)_{i \geq 0}$ is also recursive. By construction, the subword of σ consisting of the symbols occurring at the positions $(p_i)_{i \geq 0}$ is precisely the word 1^ω, and σ does not contain two consecutive 1s (i.e., it has no empty gap).

To show that the MSO-theory of σ is decidable, we give, for all $n \geq 2$, an effective bound N_n such that, for all $i \geq N_n$, h_i is a multiple of $n!$. Using Theorem 4.4, we can immediately conclude that the MSO-theory of σ is decidable and that a Turing machine deciding it can be constructed.

For the bound N_n, we take $(n!)^3$. Assume toward a contradiction that there exists an integer $i \geq N_n$ such that h_i is not a multiple of $n!$. Then it must be the case that h_i is produced by a gap s_ℓ for some ℓ such that $s_\ell < (n!)^2$. By strictness of the gap sequence, we must have $\ell < (n!)^2$. As each gap before s_ℓ produces at most $n!$ gaps, we have $i < \ell \cdot n! \leq (n!)^3$, which contradicts the assumption that $i \geq N_n = (n!)^3$. This concludes the proof of Lemma 4.3.

4.2   Proof of Theorem 4.2

We now prove Theorem 4.2. Consider a general word $\pi \in \{0,1\}^\omega$ and a gap sequence $(s_i)_{i \geq 0}$ such that $s_{i+1} > s_i + 2$ for all $i \geq 0$ and encoding a set of positions $p_0 < p_1 < \dots$. For now, assume that $s_0 \geq 4$; the case $s_0 < 4$ is handled at the end of the proof.

Consider the set of positions $(\overline{p_i})_{i \geq 0}$ defined by $\overline{p_i} = p_i$ if $\pi(i) = 1$ and $p_i - 1$ otherwise. As π and $(p_i)_{i \geq 0}$ are recursive, the set of positions $(\overline{p_i})_{i \geq 0}$ (and hence the associated gap sequence $(\overline{s_i})_{i \geq 0}$) is also recursive. As for all $i \geq 0$ $\overline{s_i}$ differs from s_i by at most 1, the gap sequence $(\overline{s_i})_{i \geq 0}$ is strictly increasing.

Assume we can construct a word $\overline{\sigma}$ without two consecutive 1s satisfying the conclusion of Lemma 4.3 for the gap sequence $(\overline{s_i})_{i \geq 0}$. It is easy to see that the subword of $\overline{\sigma}$ consisting of the symbols occurring at the positions $(p_i)_{i \geq 0}$ is precisely the word π. Indeed, if $\pi(i) = 1$, then $\overline{p_i} = p_i$ and $\overline{\sigma}(p_i) = 1$. If $\pi(i) = 0$, then $\overline{p_i} = p_i - 1$ and $\overline{\sigma}(p_i - 1) = 1$, and therefore $\overline{\sigma}(p_i) = 0$ because $\overline{\sigma}$ does not contain two consecutive 1s.

Now consider the case where $s_0 < 4$. We simply remove the first four symbols of π and the associated gaps. As the sequence is strictly increasing, all gaps starting from s_4 are at least 4. We obtain a word π' and a gap sequence $(s_i')_{i \geq 0}$. We construct the word σ' corresponding to π' and $(s_i')_{i \geq 0}$ as described above. We then take σ to be $0^{s_0}\pi(0)0^{s_1}\pi(1)\cdots 0^{s_3}\sigma'$. The word σ satisfies the conditions of Theorem 4.2 for the word π and the positions $(p_i)_{i \geq 0}$. Indeed, as σ can be MSO-transduced from σ', the MSO-theory of σ is decidable and recursive in the MSO-theory of σ'.

5   Conclusion

We have shown that decidability of the MSO-theory of the base-k tree representation of an infinite word is independent of decidability of the MSO-theory of its linear representation. Beyond this separation result, the proof of the second direction yields a construction showing that words with decidable MSO-theory can still encode arbitrary recursive subwords along sufficiently sparse positions. Two natural directions for further work are to characterize which MSO-definable properties are robust across tree representations in two multiplicatively independent bases, and to determine whether the expansion of an irrational algebraic real number can have a tree representation with decidable MSO-theory.

Acknowledgments. We thank the anonymous reviewers for their helpful comments and suggestions.

References

1. Adamczewski, B., Bugeaud, Y.: On the complexity of algebraic numbers I. expansions in integer bases. Ann. Math. **165**(2), 547–565 (2007)
2. Adamczewski, B., Cassaigne, J., Gonidec, M.L.: On the computational complexity of algebraic numbers: the Hartmanis-Stearns problem revisited. Trans. Am. Math. Soc. **373**, 3085–3115 (2020)
3. Allouche, J., Shallit, J.O.: Automatic Sequences - Theory, Applications, Generalizations. Cambridge University Press, Cambridge (2003)
4. Berthé, V., Karimov, T., Nieuwveld, J., Ouaknine, J., Vahanwala, M., Worrell, J.: On the decidability of monadic second-order logic with arithmetic predicates. In: Proceedings of the 39th Annual ACM/IEEE Symposium on Logic in Computer Science (LICS 2024), pp. 11:1–11:14 (2024). https://doi.org/10.1145/3661814.3662119
5. Carayol, A., Charamond, L.: The structure of trees in the pushdown hierarchy. In: 51st International Colloquium on Automata, Languages, and Programming (ICALP 2024). LIPIcs, vol. 297, pp. 143:1–143:18. Schloss Dagstuhl - Leibniz-Zentrum für Informatik (2024). https://doi.org/10.4230/LIPIcs.ICALP.2024.143
6. Carayol, A., Wöhrle, S.: The caucal hierarchy of infinite graphs in terms of logic and higher-order pushdown automata. In: Pandya, P.K., Radhakrishnan, J. (eds.) FSTTCS 2003. LNCS, vol. 2914, pp. 112–123. Springer, Heidelberg (2003). https://doi.org/10.1007/978-3-540-24597-1_10
7. Carton, O., Thomas, W.: The monadic theory of morphic infinite words and generalizations. Inf. Comput. **176**(1), 51–65 (2002). https://doi.org/10.1006/inco.2001.3139
8. Caucal, D.: On infinite terms having a decidable monadic theory. In: Diks, K., Rytter, W. (eds.) Mathematical Foundations of Computer Science 2002, 27th International Symposium, MFCS 2002, Warsaw, Poland, August 26–30, 2002, Proceedings. Lecture Notes in Computer Science, vol. 2420, pp. 165–176. Springer, Berlin, Heidelberg (2002). https://doi.org/10.1007/3-540-45687-2_13
9. Caucal, D., Gonidec, M.L.: Context-free sequences. In: Theoretical Aspects of Computing - ICTAC 2014–11th International Colloquium, Bucharest, Romania, September 17–19, 2014. Proceedings. Lecture Notes in Computer Science, vol. 8687, pp. 259–276. Springer, Cham (2014). https://doi.org/10.1007/978-3-319-10882-7_16
10. Ebbinghaus, H., Flum, J.: Finite Model Theory. Springer Monographs in Mathematics, 2 edn. Springer, Berlin, Heidelberg (2005). https://doi.org/10.1007/3-540-28788-4
11. Knapik, T., Niwinski, D., Urzyczyn, P.: Higher-order pushdown trees are easy. In: Nielsen, M., Engberg, U. (eds.) Foundations of Software Science and Computation Structures, 5th International Conference, FOSSACS 2002. Held as Part of the Joint European Conferences on Theory and Practice of Software, ETAPS 2002 Grenoble, France, April 8–12, 2002, Proceedings. Lecture Notes in Computer Science, vol. 2303, pp. 205–222. Springer, Berlin, Heidelberg (2002). https://doi.org/10.1007/3-540-45931-6_15
12. Rabin, M.: Decidability of second-order theories and automata on infinite trees. Trans. Amer. Math. Soc. **141**, 1–35 (1969)
13. Thomas, W.: Constructing infinite graphs with a decidable MSO-theory. In: Rovan, B., Vojtás, P. (eds.) Mathematical Foundations of Computer Science 2003, 28th International Symposium, MFCS 2003, Bratislava, Slovakia, August 25–29, 2003, Proceedings. Lecture Notes in Computer Science, vol. 2747, pp. 113–124. Springer, Berlin, Heidelberg (2003). https://doi.org/10.1007/978-3-540-45138-9_6

The Four Corners Problem Over Larger Alphabets

Daniel Průša[1(✉)], Michael Wehar[2], and Chen Xu[3]

[1] Faculty of Electrical Engineering, Czech Technical University, Karlovo nám. 13,
12135 Prague 2, Czech Republic
`daniel.prusa@fel.cvut.cz`
[2] Department of Computer Science, Bryn Mawr College, Bryn Mawr, PA, USA
`mwehar@brynmawr.edu`
[3] Department of Computer Science and Engineering, University at Buffalo, Buffalo,
NY, USA
`chenxu@buffalo.edu`

Abstract. We investigate algorithmic problems for two-dimensional pattern matching. We improve upon known complexity results by introducing fine-grained reductions that lead to conditional lower bounds and new algorithms that have modestly more efficient runtimes. In terms of conditional lower bounds, we establish triangle finding (i.e. 3-clique) and 4-clique hardness results for previously unexplored cases of the Four Corners Problem over larger than binary alphabets. In terms of runtime improvements, we introduce a $\sqrt{\log(n)}$ factor improvement for detecting if a matrix contains a square block whose four corners are all 1's.

Keywords: 2D Pattern Matching · Local Picture Languages · Fine-Grained Reductions · Matrix Algorithms

1 Background

1.1 Pattern Matching

Pattern matching is the task of searching a dataset to identify a subset that matches a specified pattern. The kinds of data and patterns can be varied to create different pattern matching problems. For example, if the data is text and the patterns are regular expressions, then we get the regular expression matching problem which has become a popular topic in recent years because of new improvements to the classical algorithmic solutions and novel applications to various areas of computer science [12,14,15].

For higher-dimensional data, efficient pattern matching algorithms are well-established when searching for one or a finite number of equally sized, fixed patterns [4,5,10]. However, in more flexible scenarios–where patterns are defined by varying constraints or templates–the problem becomes more complex. In these cases, existing research is limited, and the few studies available indicate that such generalized tasks often require algorithms with higher time complexity compared

M.-P. Béal and P. Caron (Eds.): DLT 2026, LNCS 16578, pp. 293–306, 2026.
https://doi.org/10.1007/978-3-032-28404-4_22

to simpler cases. For instance, this is observed by Mráz et al. [11] when patterns are described using local picture languages as well as picture languages accepted by finite-state two-dimensional (2D) automata, which are basic and well-studied families of picture languages.

1.2 Local Picture Language Matching

Picture languages are a classical topic in formal language theory, as surveyed in [9]. They naturally extend one-dimensional pattern matching concepts to two dimensions. Pictures can be viewed as matrices where the entries are restricted to characters from a fixed alphabet. A set of pictures is referred to as a picture language. A picture language L is called a local picture language if there exists a set of 2 by 2 tiles T such that a matrix M is in L if and only if every 2 by 2 block within M' (obtained by adding a border around M) is equal to a tile in T. The border addition allows the tile set T to restrict the appearance of the corner and border entries of M.

We next define the local picture language matching problem, as presented in [11]. Given a picture (i.e. an $n \times n$ matrix) M and a local picture language L specified by a tile set T, the problem asks whether there exists a subpicture (i.e. a contiguous block) of M that belongs to L. The general problem can be solved in $O(n^3)$ (i.e. cubic) time. However, for certain specific local picture languages, more efficient algorithms are known. This raises a natural and important open question: for a binary alphabet, can the matching problem always be solved in $\tilde{O}(n^2)$ (i.e. nearly quadratic) time?

Question 1. *Can the local picture language matching problem be solved in nearly quadratic time for all local picture languages over a binary alphabet?*

The Four Corners Problem is a subclass of specific cases of the local picture language matching problem. In particular, it is most simply formulated as follows. The Four Corners Problem (denoted by $\mathtt{Rect}\text{-}\Sigma^{m \times n}\text{-}\left[\begin{smallmatrix} a & b \\ c & d \end{smallmatrix}\right]$, see Section 2.2) asks if a given matrix M over alphabet Σ contains a non-trivial rectangular subblock whose top left corner is a, top right corner is b, bottom left corner is c, and bottom right corner is d. It has been established that if Σ is the binary alphabet, then for any selection of corner entries $a, b, c, d \in \{0, 1\}$, the matching problem is solvable in nearly quadratic time [13]. However, when the alphabet is increased beyond binary, it has been shown that the Four Corners Problem becomes triangle finding-hard [11,13]. In Sect. 3.1, we introduce previously uninvestigated cases of the Four Corners Problem over larger alphabets and prove new hardness results for these cases.

1.3 Triangle Finding and 4-Clique

The triangle finding problem (also known as 3-clique) asks whether a simple graph contains three distinct vertices that are pairwise adjacent. A problem is said to be triangle finding-hard if the triangle finding problem is efficiently reducible to it. More specifically, we have the following.

Definition 1. *We say that problem $\mathcal{P}$ is triangle finding-hard if we can reduce from any triangle finding instance of n vertices to an instance of $\mathcal{P}$ of size $\tilde{O}(n^2)$ in $\tilde{O}(n^2)$ time.*

The inputs for the problems that we investigate are matrices so the resulting instance for $\mathcal{P}$ could, for example, be an $\tilde{O}(n)$ by $\tilde{O}(n)$ matrix. We define hardness in this way so that the existence of a nearly linear time algorithm for $\mathcal{P}$ would imply the existence of an $\tilde{O}(n^2)$ time algorithm for triangle finding. We refer the reader to [6,17] for surveys on fine-grained reductions, related problems, and hardness results. It is also worth noting that triangle finding is closely related to matrix multiplication. The current best-known runtime for triangle finding is $O(n^\omega)$, where n is the number of vertices and ω is the matrix multiplication exponent. Additionally, an even closer relationship has been established between Boolean matrix multiplication (BMM) and triangle finding [20].

The 4-clique problem asks whether a simple graph contains four distinct vertices that are pairwise adjacent. In this work, we consider an additional kind of hardness, which we refer to as 4-clique hardness.

Definition 2. *We say that problem $\mathcal{P}$ is 4-clique-hard if we can reduce from any 4-clique instance with n vertices and m edges to an instance of $\mathcal{P}$ of size $\tilde{O}(mn)$ in $\tilde{O}(mn)$ time.*

The inputs for the problems that we investigate are matrices so the resulting instance for $\mathcal{P}$ could, for example, be an $\tilde{O}(m)$ by $\tilde{O}(n)$ matrix. We define hardness in this way so that the existence of a nearly linear time algorithm for $\mathcal{P}$ would imply the existence of an $\tilde{O}(n^3)$ time algorithm for 4-clique. As far as the authors are aware, the current best-known runtime for 4-clique is $O(n^{3.257})$ [19] where n is the number of vertices.

Notice that 4-clique-hard problems may be easier than those that are triangle finding-hard because the resulting instance sizes depend on both m and n. That being said, more efficient algorithms for 4-clique-hard problems would still result in faster algorithms for the k-clique problems, improving upon the best-known runtime upper bounds.

1.4   Finding Axis-Aligned Rectangles and Squares

There are natural connections between classical problems in computational geometry and two-dimensional pattern matching problems. In particular, consider the two-dimensional axis-aligned rectangle problem which is defined as follows. Given a set of 2D points P of size n, does P contain a subset of size 4 that forms an axis-aligned rectangle. In other words, do there exist four distinct points p_1, p_2, p_3, and $p_4 \in P$ such that:

- p_1 and p_2 form a line parallel to the x-axis
- p_3 and p_4 form a line parallel to the x-axis
- p_1 and p_3 form a line parallel to the y-axis
- p_2 and p_4 form a line parallel to the y-axis.

We denote this problem by `Aligned-Rect`. The axis-aligned square problem (denoted by `Aligned-Square`) is defined similarly for axis-aligned squares, instead of rectangles. In [16], it was shown that the axis-aligned rectangle problem is closely related to the classical four-cycle detection problem (4-cycle). That is, the problem of determining whether there exist four distinct vertices p_1, p_2, p_3, and p_4 in a simple graph such that $\{p_1, p_2\}$, $\{p_2, p_3\}$, $\{p_3, p_4\}$, and $\{p_4, p_1\}$ are edges. By transforming the axis-aligned rectangle problem into the four-cycle detection problem, we can apply the efficient algorithm for four-cycle detection from [2], which runs in $O(E^{\frac{4}{3}})$ time where E is the number of edges in the input graph. This leads to an $O(n^{\frac{4}{3}})$ time algorithm for `Aligned-Rect`. We refer the reader to [21] for more details on this approach. Also, an $O(n^{\frac{3}{2}})$ time algorithm is known for the `Aligned-Square` problem [16].

2 Our Contribution

2.1 Summary

In this work, we introduce conditional lower bounds, introduce new algorithms, and analyze existing algorithmic approaches for two-dimensional pattern matching problems. We specifically focus on pattern matching problems that we refer to as Four Corners Problems[1] which were previously studied in [13].

2.2 Problem Statements

To define the problems that we investigate in this work, we first introduce terminology for discussing matrices. A *subblock* (or subrectangle) within a matrix is the matrix that consists of all entries within a given rectangular range. A *submatrix* of a matrix is the matrix that is achieved by deleting any number of rows and columns of the original matrix. Every submatrix induces a subblock by adding back the removed elements that are between the entries in the submatrix. We use the term *square submatrix* to refer to a submatrix whose induced subblock has equal width and height.

The Four Corners Problem for an alphabet Σ and corner matrix $A = \begin{bmatrix} a & b \\ c & d \end{bmatrix}$ asks whether a given m by n matrix M contains a two by two submatrix that equals A (i.e. a non-trivial subblock with top left corner a, top right corner b, bottom left corner c, and bottom right corner d). We denote this problem by `Rect-`$\Sigma^{m \times n}$`-A` . Similarly, the Square Four Corners Problem asks whether a given m by n matrix M contains a square submatrix that equals A. We denote this problem by `Square-`$\Sigma^{m \times n}$`-A` .

Notice that the axis-aligned rectangle problem (denoted by `Aligned-Rect`) is a sparse version of the `Rect-`$\{0,1\}^{n \times n}$`-`$\begin{bmatrix} 1 & 1 \\ 1 & 1 \end{bmatrix}$ problem. Similarly, the axis-aligned square problem (denoted by `Aligned-Square`) is a sparse version of `Square-`$\{0,1\}^{n \times n}$`-`$\begin{bmatrix} 1 & 1 \\ 1 & 1 \end{bmatrix}$.

---

[1] The four corners problem and its variants have appeared as coding challenge problems on coding platforms (e.g. GeeksforGeeks) and online forums (e.g. StackExchange).

2.3 Runtime and Hardness Results

Here we list the primary conditional lower bounds and runtime improvements that we present in this work. In the following, A will be used to denote any corner matrix pattern over the implied alphabet.

- **Four Corners Problem over larger than binary alphabets—Rect-$\Sigma^{m \times n}$-A**

  Over a ternary alphabet, the Four Corners Problem Rect-$\{0, 1, 2\}^{m \times n}$-$\left[\begin{smallmatrix} 1 & 0 \\ 1 & 1 \end{smallmatrix}\right]$ has been shown to be triangle finding-hard [11]. We show computational hardness for solving several remaining cases of the Four Corners Problem over larger than binary alphabets which were left open in prior works. Most of these cases are shown to be triangle finding-hard while there is one special case Rect-$\{0, 1, 2\}^{m \times n}$-$\left[\begin{smallmatrix} 1 & 0 \\ 1 & 0 \end{smallmatrix}\right]$ that is shown to be 4-clique-hard in Theorem 2.

- **Square Four Corners Problem—Square-$\Sigma^{m \times n}$-A**

  Unlike the rectangular four corners problem from [13], the square problem is not known to be solvable in subcubic time [18] for n by n matrices. We achieve a speed up that improves on cubic time by a factor of $O\left(\sqrt{\log(n)}\right)$ in Theorem 3.

3 Complexity of Four Corners Problems

3.1 Hardness of Rectangular Case Over Larger Alphabets

The Rect-$\Sigma^{m \times n}$-A problem (also known as the Four Corners Problem) is to determine whether a given matrix with entries over an alphabet Σ contains a 2×2 submatrix A. In [13], it was demonstrated that when the alphabet is binary, all cases of Rect-$\{0, 1\}^{m \times n}$-A are solvable in $\tilde{O}(mn)$ time. Furthermore, all cases of the Rect-$\{0, 1\}^{m \times n}$-A problems can be reduced to four distinct corner pattern cases by applying matrix rotations and flipping 0's and 1's.

Lemma 1 ([13]). *Any given Rect-$\{0, 1\}^{m \times n}$-A problem can be reduced to one of four problems of the form Rect-$\{0, 1\}^{m \times n}$-B where B can be:*

$$\left[\begin{smallmatrix} 1 & 1 \\ 1 & 1 \end{smallmatrix}\right], \left[\begin{smallmatrix} 1 & 0 \\ 1 & 1 \end{smallmatrix}\right], \left[\begin{smallmatrix} 1 & 0 \\ 0 & 1 \end{smallmatrix}\right], \textit{ or } \left[\begin{smallmatrix} 1 & 0 \\ 1 & 0 \end{smallmatrix}\right].$$

It is worth noting that two of these cases, $\left[\begin{smallmatrix} 1 & 1 \\ 1 & 1 \end{smallmatrix}\right]$ and $\left[\begin{smallmatrix} 1 & 0 \\ 1 & 1 \end{smallmatrix}\right]$, each achieve a tight $\Theta(mn)$ linear time complexity with unique algorithms. However, the cases $\left[\begin{smallmatrix} 1 & 0 \\ 0 & 1 \end{smallmatrix}\right]$ and $\left[\begin{smallmatrix} 1 & 0 \\ 1 & 0 \end{smallmatrix}\right]$ appear to be more challenging and their tight bounds have not been achieved due to logarithmic factors. This suggests that there may be increased hardness for these two cases.

We now turn our attention to pattern matching in matrices over larger alphabets. Specifically, we investigate the time complexity of all cases of the Four Corners Problem over larger alphabets. We begin by demonstrating that any Four Corners matching problem over larger alphabets can be reduced to a small number of cases.

Lemma 2. *Any given* $\texttt{Rect-}\{0, 1, ..., C\}^{m \times n}\texttt{-}A$ *problem can be reduced to one of seven problems of the form* $\texttt{Rect-}\{0, 1, ..., C\}^{m \times n}\texttt{-}B$ *where* B *can be:*

$$\begin{bmatrix} 1 & 1 \\ 1 & 1 \end{bmatrix}, \begin{bmatrix} 1 & 0 \\ 1 & 1 \end{bmatrix}, \begin{bmatrix} 1 & 0 \\ 0 & 1 \end{bmatrix}, \begin{bmatrix} 1 & 0 \\ 1 & 0 \end{bmatrix}, \begin{bmatrix} 0 & 1 \\ 2 & 0 \end{bmatrix}, \begin{bmatrix} 0 & 0 \\ 1 & 2 \end{bmatrix}, \textit{ or } \begin{bmatrix} 0 & 1 \\ 2 & 3 \end{bmatrix}.$$

Proof. If the submatrix pattern has four identical corners, for instance $\begin{bmatrix} 2 & 2 \\ 2 & 2 \end{bmatrix}$, then we can use the same algorithm as matching the submatrix pattern $\begin{bmatrix} 1 & 1 \\ 1 & 1 \end{bmatrix}$.

If the corners consist of two distinct values, there are two possible cases. In the first, a single corner differs from the other three. Such cases can be reduced via rotation to the form $\begin{bmatrix} 1 & 0 \\ 1 & 1 \end{bmatrix}$. In the second, two corners differ from the other two. This can be further divided into diagonal and horizontal cases which reduce to $\begin{bmatrix} 1 & 0 \\ 0 & 1 \end{bmatrix}$ and $\begin{bmatrix} 1 & 0 \\ 1 & 0 \end{bmatrix}$, respectively.

If the corners consist of three different values, then there must be a pair of equal corners. In this case, the problem can again be divided into two subcases: diagonal and horizontal. The horizontal case reduces to $\begin{bmatrix} 0 & 0 \\ 1 & 2 \end{bmatrix}$ and the diagonal case reduces to $\begin{bmatrix} 0 & 1 \\ 2 & 0 \end{bmatrix}$.

Finally, we have the case of four corners with distinct values. This reduces to $\begin{bmatrix} 0 & 1 \\ 2 & 3 \end{bmatrix}$. $\qquad\square$

Adding a single additional alphabet character, not present in the pattern, to the input can potentially increase the problem's complexity. This is because a buffer character can prevent matches from occurring, adding an extra layer of difficulty. However, adding more than one buffer character does not affect the complexity, since each buffer character is treated identically when it comes to matching (i.e. all buffer characters serve the same function of preventing a match).

For example, consider the Four Corners Problem for the pattern $\begin{bmatrix} 1 & 1 \\ 1 & 1 \end{bmatrix}$ over a binary alphabet. This problem becomes no more difficult when we switch to a ternary alphabet $\{0, 1, 2\}$ because whether a character is 0 or 2 does not make a difference. They are both buffer characters that prevent a match from occurring. Therefore, we can reuse the $O(mn)$ time algorithm for $\texttt{Rect-}\{0, 1\}^{m \times n}\texttt{-}\begin{bmatrix} 1 & 1 \\ 1 & 1 \end{bmatrix}$ from [11, 13] to obtain the following.

Proposition 1. $\texttt{Rect-}\mathbb{N}^{m \times n}\texttt{-}\begin{bmatrix} 1 & 1 \\ 1 & 1 \end{bmatrix}$ *is solvable in* $O(mn)$ *time.*

In the subsequent theorem, we address the complexity of almost all of the remaining cases of the Four Corners Problem over larger alphabets. Specifically, we demonstrate triangle finding hardness for all cases except the pattern $\begin{bmatrix} 1 & 0 \\ 1 & 0 \end{bmatrix}$ over ternary matrices. This particular case is hard, but appears to be easier than the others. For this case, we use the concept of 4-clique hardness (defined in Sect. 1.3). The 4-clique and triangle finding problems are key graph problems in parameterized complexity [7] and are used to characterize the hardness for a variety of polynomial-time problems.

Theorem 1. $\texttt{Rect-}\{0, 1, 2\}^{m \times n}\texttt{-}A$ *is triangle finding-hard for any*

$$A \in \left\{ \begin{bmatrix} 1 & 0 \\ 1 & 1 \end{bmatrix}, \begin{bmatrix} 1 & 0 \\ 0 & 1 \end{bmatrix}, \begin{bmatrix} 0 & 1 \\ 2 & 0 \end{bmatrix}, \begin{bmatrix} 0 & 0 \\ 1 & 2 \end{bmatrix} \right\}.$$

Also, $\texttt{Rect-}\{0, 1, 2, 3\}^{m \times n}\texttt{-}\begin{bmatrix} 0 & 1 \\ 2 & 3 \end{bmatrix}$ *is triangle finding-hard.*

Now, consider $\text{Rect-}\{0,1,2\}^{m\times n}\text{-}\left[\begin{smallmatrix}1 & 0\\ 1 & 1\end{smallmatrix}\right]$. Recall that $\text{Rect-}\{0,1\}^{m\times n}\text{-}\left[\begin{smallmatrix}1 & 0\\ 1 & 1\end{smallmatrix}\right]$ is known to be solvable in $O(mn)$ time (Lemma 2 from [13]). However, when we switch to a ternary alphabet, the problem becomes triangle finding-hard (Theorem 4 from [11]).

Lemma 3. $\text{Rect-}\{0,1,2\}^{m\times n}\text{-}\left[\begin{smallmatrix}1 & 0\\ 1 & 1\end{smallmatrix}\right]$ *is triangle finding-hard.*

Proof. We refer the reader to [11] for the proof of this lemma.   □

We consider the remaining triangle finding-hard cases in the following lemmas.

Lemma 4. $\text{Rect-}\{0,1,2\}^{m\times n}\text{-}\left[\begin{smallmatrix}1 & 0\\ 0 & 1\end{smallmatrix}\right]$ *is triangle finding-hard.*

Proof. Given a graph $G = (V, E)$ where $n = |V|$, we construct a $2n \times 2n$ matrix M through the following steps:

1. Initialize four ternary matrices A, B, C, and D, each of size $n \times n$, with all entries set to 2. The matrices are indexed by the vertex indices.
2. For each edge[2] $e = \{i, j\} \in E$ where $i \neq j$, set $A[i][j] = 1$ and $B[i][j] = C[i][j] = 0$.
3. For all $k \in V$, set $D[k][k] = 1$.
4. Combine the four subblocks to form M such that $M = \left[\begin{smallmatrix}A & B\\ C & D\end{smallmatrix}\right]$.

Claim: There is a triangle in G if and only if M has a $\left[\begin{smallmatrix}1 & 0\\ 0 & 1\end{smallmatrix}\right]$ submatrix.

$\Rightarrow$: If there is a triangle in G formed by vertices $\{v_1, v_2, v_3\}$, then, by construction, we have $A[v_1][v_2] = 1$, $B[v_1][v_3] = C[v_3][v_2] = 0$, and $D[v_3][v_3] = 1$. These four entries in M form a submatrix $\left[\begin{smallmatrix}1 & 0\\ 0 & 1\end{smallmatrix}\right]$.

$\Leftarrow$: If there is a $\left[\begin{smallmatrix}1 & 0\\ 0 & 1\end{smallmatrix}\right]$ submatrix in M, then we know that each of its four corners is located in one of the four subblocks of M because, by construction, there are no 0 entries in the top-left and bottom-right parts and no 1 entries in the top-right and bottom-left parts. By step 3 of the construction, we know that the bottom-right 1 must be on the diagonal of M. Suppose its location is (k, k), then the location of the top-right 0 is (i, k) for some $i \neq k$, the location of the bottom-left 0 is (k, j) for some $j \neq k$, and the location of the top-left 1 is (i, j) where, by construction, we have $i \neq j$. Therefore, edges $\{i, j\}, \{i, k-n\}, \{k-n, j\}$ are elements of E that form a triangle in G.   □

Lemma 5. $\text{Rect-}\{0,1,2\}^{m\times n}\text{-}\left[\begin{smallmatrix}0 & 1\\ 2 & 0\end{smallmatrix}\right]$ *is triangle finding-hard.*

Proof. For case $\left[\begin{smallmatrix}0 & 1\\ 2 & 0\end{smallmatrix}\right]$, suppose M is divided into four subblocks $\left[\begin{smallmatrix}A & B\\ C & D\end{smallmatrix}\right]$. We proceed with the same construction as in Lemma 4, but A, B, D are initialized with 2's and C is initialized with 1's. For each edge $e = \{i, j\} \in E$, $A[i][j] = 0$, $B[i][j] = 1$, and $C[i][j] = 2$. For all $k \in V$, $D[k][k] = 0$.

If a submatrix $\left[\begin{smallmatrix}0 & 1\\ 2 & 0\end{smallmatrix}\right]$ exists in M, then:

- Similarly, it cannot be formed within any standalone A, B, C, D part.
- It cannot just be in the AB part because there are no 0's in B.

---

[2] We treat G as a simple graph, meaning it does not contain self-loop edges.

- It cannot just be in the AC part because there are no 0's in C.
- It cannot just be in the CD or BD part because there are no 1's in D.

Therefore, each of its corners must be in a different subblock. We can use the same argument as in Lemma 4 to show that there exists a triangle in G if and only if M has a $\left[\begin{smallmatrix} 0 & 1 \\ 2 & 0 \end{smallmatrix}\right]$ submatrix.   □

Lemma 6. $Rect\text{-}\{0,1,2\}^{m \times n}\text{-}\left[\begin{smallmatrix} 0 & 0 \\ 1 & 2 \end{smallmatrix}\right]$ *is triangle finding-hard.*

Proof. For case $\left[\begin{smallmatrix} 0 & 0 \\ 1 & 2 \end{smallmatrix}\right]$, suppose M is divided into four subblocks $\left[\begin{smallmatrix} A & B \\ C & D \end{smallmatrix}\right]$. We proceed with a similar construction to Lemma 4, but A, B are initialized with 1's and C is initialized with 0's. D has 2's on the diagonal, 0's below the diagonal and 1's above the diagonal. For each $e = \{i,j\} \in E$, $A[i][j] = B[i][j] = 0$ and $C[i][j] = 1$.

If a submatrix $\left[\begin{smallmatrix} 0 & 0 \\ 1 & 2 \end{smallmatrix}\right]$ exists in M, then:

- Similarly, it cannot be formed within any standalone A, B, C, D part.
- It cannot just be in the AB part because there are no 2's in B.
- It cannot just be in the AC part because there are no 2's in C.
- It cannot just be in the BD part because 2's are always left of 1's in D.
- It cannot just be in the CD part because 2's are always above 0's in D.

Therefore, each of its corners must be in a different subblock. We can use the same argument as in Lemma 4 to show that there exists a triangle in G if and only if M has a $\left[\begin{smallmatrix} 0 & 0 \\ 1 & 2 \end{smallmatrix}\right]$ submatrix.   □

Lemma 7. $Rect\text{-}\{0,1,2,3\}^{m \times n}\text{-}\left[\begin{smallmatrix} 0 & 1 \\ 2 & 3 \end{smallmatrix}\right]$ *is triangle finding-hard.*

Proof. For case $\left[\begin{smallmatrix} 0 & 1 \\ 2 & 3 \end{smallmatrix}\right]$, suppose M is divided into four subblocks $\left[\begin{smallmatrix} A & B \\ C & D \end{smallmatrix}\right]$. We proceed with a similar construction to Lemma 4, but A, C, D are initialized with 1's and B is initialized with 0's. D has 3's on the diagonal. For each $e = \{i,j\} \in E$, $A[i][j] = 0$, $B[i][j] = 1$, and $C[i][j] = 2$.

If a submatrix $\left[\begin{smallmatrix} 0 & 1 \\ 2 & 3 \end{smallmatrix}\right]$ exists in M, then:

- Similarly, it cannot be formed within any standalone A, B, C, D part.
- It cannot just be in the AB part because there are no 3's in B.
- It cannot just be in the AC part because there are no 3's in C.
- It cannot just be in the BD part because there are no 2's in D.
- It cannot just be in the CD part because there are no 0's in C.

Therefore, each of its corners must be in a different subblock. We can use the same argument as in Lemma 4 to show that there exists a triangle in G if and only if M has a $\left[\begin{smallmatrix} 0 & 1 \\ 2 & 3 \end{smallmatrix}\right]$ submatrix.   □

The pattern matching of $\left[\begin{smallmatrix} 1 & 0 \\ 1 & 0 \end{smallmatrix}\right]$ in ternary matrices presents a significant distinction. In particular, we demonstrate that $Rect\text{-}\{0,1,2\}^{m \times n}\text{-}\left[\begin{smallmatrix} 1 & 0 \\ 1 & 0 \end{smallmatrix}\right]$ is 4-clique-hard. One real-world application of this problem is in High-Frequency Trading (HFT) [1]. Consider a ternary matrix where the rows represent items and the columns represent traders. An entry at (i,j) is 1 if trader i wants to sell item

j, is 0 if trader i wants to buy item j, and is 2 if trader i is neither buying nor selling item j. The problem of determining whether there can be a transaction involving more than one item being sold from one trader to another reduces to solving Rect-$\{0,1,2\}^{m \times n}$-$\left[\begin{smallmatrix} 1 & 0 \\ 1 & 0 \end{smallmatrix}\right]$ and Rect-$\{0,1,2\}^{m \times n}$-$\left[\begin{smallmatrix} 0 & 1 \\ 0 & 1 \end{smallmatrix}\right]$. A faster detection algorithm for this pattern would lead to a more efficient transaction system.

Theorem 2. *Rect-$\{0,1,2\}^{m \times n}$-$\left[\begin{smallmatrix} 1 & 0 \\ 1 & 0 \end{smallmatrix}\right]$ is 4-clique-hard.*

Proof. Given a graph $G = (V, E)$ with n vertices and m edges. We build a $2n \times 2m$ matrix M over the alphabet $\{0, 1, 2\}$ by the following steps:

1. Initialize four matrices $M_{11}, M_{12}, M_{21}, M_{22}$ of size $n \times m$ each. All entries of the matrices are set to 2. For each matrix, we index the rows by vertices and the columns by edges.
2. For all $e = \{v_i, v_j\} \in E$ with $i < j$, set $M_{11}[v_i][e] = 1$ and $M_{21}[v_j][e] = 1$.
3. For all $e = \{v_i, v_j\} \in E$ and for all $v_k \neq v_i, v_j$ such that $\{v_i, v_k\}, \{v_j, v_k\} \in E$, set $M_{12}[v_k][e] = M_{22}[v_k][e] = 0$.
4. Combine the four matrices to form M such that $M = \left[\begin{smallmatrix} M_{11} & M_{12} \\ M_{21} & M_{22} \end{smallmatrix}\right]$.

Claim: There is a 4-clique in G if and only if M has a $\left[\begin{smallmatrix} 1 & 0 \\ 1 & 0 \end{smallmatrix}\right]$ submatrix.

$\Rightarrow$: If there is a 4-clique formed by vertices $\{v_1, v_2, v_3, v_4\}$ in G, then, by construction, entries $M_{11}[v_1][\{v_1, v_2\}] = M_{21}[v_2][\{v_1, v_2\}] = 1$ and entries $M_{12}[v_1][\{v_3, v_4\}] = M_{22}[v_2][\{v_3, v_4\}] = 0$. These four entries form a $\left[\begin{smallmatrix} 1 & 0 \\ 1 & 0 \end{smallmatrix}\right]$ submatrix.

$\Leftarrow$: If there exists a $\left[\begin{smallmatrix} 1 & 0 \\ 1 & 0 \end{smallmatrix}\right]$ submatrix in M, then take the row indexes of the two horizontal edges of the submatrix to be v_1 and v_2. By construction, we have $\{v_1, v_2\} \in E$. Take the column index of the right vertical edge to be $\{v_3, v_4\}$. By step 3 of the construction, we have $v_3 \neq v_1, v_2$ and $v_4 \neq v_1, v_2$ where $\{v_3, v_4\}, \{v_1, v_3\}, \{v_1, v_4\}, \{v_2, v_3\}, \{v_2, v_4\} \in E$ implying that there is a 4-clique in G. $\square$

For a 4-clique instance of n vertices and m edges, a $2n$ by $2m$ instance of the problem Rect-$\{0,1,2\}^{m \times n}$-$\left[\begin{smallmatrix} 1 & 0 \\ 1 & 0 \end{smallmatrix}\right]$ can be constructed in $O(mn)$ steps. Thus, if we can solve the problem Rect-$\{0,1,2\}^{m \times n}$-$\left[\begin{smallmatrix} 1 & 0 \\ 1 & 0 \end{smallmatrix}\right]$ in $O(mn)$ time, then we can solve the 4-clique problem in $O(mn)$ time. Furthermore, since $m = O(n^2)$, we could solve 4-clique in $O(n^3)$ time.

As far as we are aware, the best-known runtime for the 4-clique problem is $O(n^{3.257})$ [19]. For m by n instances (m now represents rows and n represents columns), we can achieve $O(mn + n^2)$ runtime for Rect-$\mathbb{N}^{m \times n}$-$\left[\begin{smallmatrix} 1 & 0 \\ 1 & 0 \end{smallmatrix}\right]$ by detecting the reoccurrence of pairwise indices of 1's and 0's for each row. Preprocessing rows to have constant time access to the next 1 and 0 entry in the sequence is also required. However, this approach does not extend to the column-wise case, meaning we are unable to achieve $O(mn + m^2)$ time complexity through this method. If it were possible to achieve $O(mn + m^2)$ time, then by Theorem 2, there would exist faster algorithms for 4-clique.

Here is a summary of the complexity and hardness results obtained for all cases:

- $\left[\begin{smallmatrix} 1 & 1 \\ 1 & 1 \end{smallmatrix}\right]$ in matrices of alphabet size 2 and above are of the same complexity.
- $\left[\begin{smallmatrix} 1 & 0 \\ 1 & 1 \end{smallmatrix}\right]$, $\left[\begin{smallmatrix} 1 & 0 \\ 0 & 1 \end{smallmatrix}\right]$ in matrices of alphabet size 3 and above are triangle finding-hard.
- $\left[\begin{smallmatrix} 1 & 0 \\ 1 & 0 \end{smallmatrix}\right]$ in matrices of alphabet size 3 and above are 4-clique-hard.
- $\left[\begin{smallmatrix} 0 & 0 \\ 1 & 2 \end{smallmatrix}\right]$, $\left[\begin{smallmatrix} 0 & 1 \\ 2 & 0 \end{smallmatrix}\right]$ in matrices of alphabet size 3 and above are triangle finding-hard.
- $\left[\begin{smallmatrix} 0 & 1 \\ 2 & 3 \end{smallmatrix}\right]$ in matrices of alphabet size 4 and above are triangle finding-hard.

From the preceding summary, we observe a trichotomy where one case has a tight $\Theta(mn)$ running time, one case is 4-clique-hard, and all other cases are triangle finding-hard.

3.2 Faster Algorithm for Square Case

The `Square-`$\Sigma^{m \times n}$`-A` problem is to determine whether a given matrix with entries over an alphabet Σ contains a 2×2 submatrix A where the induced subblock is a square. Although the rectangular variants of this problem over a binary alphabet (`Rect-`$\{0,1\}^{m \times n}$`-A`) are known to be solvable in quadratic or near-quadratic time [11,13], the square case has faced a cubic time barrier [18]. In this subsection, we present a $\sqrt{\log(m+n)}$ factor improvement for the square case.[3] This improvement loosely resembles a classic algorithmic approach from [3] where they obtain a $\log(n)$ factor improvement for constructing the transitive closure of an acyclic directed graph.

Theorem 3. *Square-*$\{0,1,...,C\}^{m \times n}$*-A   is solvable in* $O\left(\frac{mn \, \min(m,n)}{\sqrt{\log(m+n)}} \right)$ *time for every* $C \in \mathbb{N}$ *and 2 by 2 matrix* A.

Proof. Let $C \in \mathbb{N}$ and a 2 by 2 corner pattern A be given. Let $b = C + 1$. We first partition the matrix into blocks of size $\left\lceil \frac{1}{\sqrt{8}} \sqrt{\log_b(m+n)} \right\rceil$ by $\left\lceil \frac{1}{\sqrt{8}} \sqrt{\log_b(m+n)} \right\rceil$. This gives us a total of $\left\lceil \frac{8mn}{\log_b(m+n)} \right\rceil$ blocks. We can index the blocks just like we index the entries.

Notice that if there is a matching square submatrix whose top-left corner is in a block (i,j), then the possible blocks in which the four corners can appear are:

---

[3] A $\sqrt{\log(n)}$ factor improvement for `Aligned-Square`'s runtime was introduced in [16] as well, but their improvement only offsets the added log factors that result from their use of a primitive model of computation that lacks constant time random access and hash table lookups.

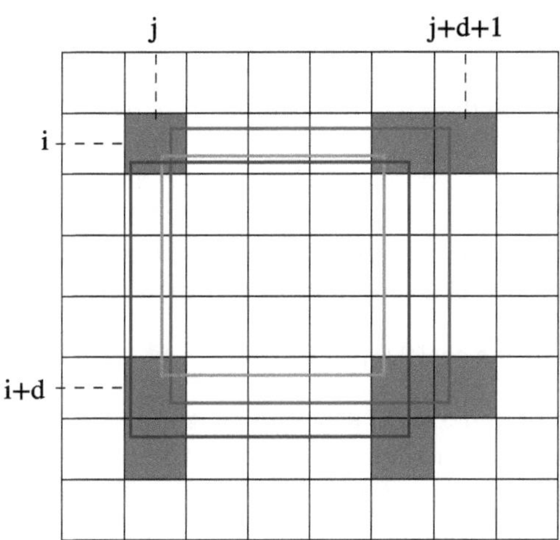

Fig. 1. The four corners of a square are located within these eight blocks.

- All in block (i, j).
- Top edge in block (i, j), bottom edge in block $(i + 1, j)$.
- Left edge in block (i, j), right edge in block $(i, j + 1)$.
- In blocks $(i, j), (i + d, j), (i, j + d), (i + d, j + d)$ for some $d \geq 1$.
- In blocks $(i, j), (i + d + 1, j), (i, j + d), (i + d + 1, j + d)$ for some $d \geq 1$.
- In blocks $(i, j), (i + d, j), (i, j + d + 1), (i + d, j + d + 1)$ for some $d \geq 1$.

Therefore, the four corners are always located in the union of at most eight blocks parameterized by i, j, and d (illustrated in Fig. 1). For a square whose top-left corner is in block (i, j), there are at most $\left\lceil \frac{\min(m,n)}{\frac{1}{\sqrt{8}}\sqrt{\log_b(m+n)}} \right\rceil$ possible 8-block-groups that could contain this square. Each group has at most $\left\lceil 8 \cdot \left(\frac{1}{\sqrt{8}}\sqrt{\log_b(m+n)} \right)^2 \right\rceil = \lceil \log_b(m+n) \rceil$ entries over $\{0, 1, ..., C\}$. Therefore, we can treat these entries from each group as a string over the alphabet $\{0, 1, ..., C\}$ and precompute a lookup table of query size $\lceil \log_b(m+n) \rceil$ to fast-check (with constant time lookup) if there exists a square in these blocks. The table returns true if and only if the query string of the corresponding group of blocks contains a square. The total size of this table is at most $b^{\lceil \log_b(m+n) \rceil} = O(m + n)$. Thus, the precomputation takes $\tilde{O}(m + n)$ time. The total number of lookup queries is at most

$$\left\lceil \frac{8mn}{\log_b(m+n)} \cdot \frac{\min(m,n)}{\frac{1}{\sqrt{8}}\sqrt{\log_b(m+n)}} \right\rceil = \left\lceil \frac{16\sqrt{2}\,mn\min(m,n)}{\log_b^{1.5}(m+n)} \right\rceil.$$

Also, it takes $\log_b(m + n)$ time to construct a query string for an 8-block-group. Hence, the total running time is $\left\lceil \frac{16\sqrt{2}\,mn\min(m,n)}{\log_b^{1.5}(m+n)} \cdot \log_b(m + n) \right\rceil = O\left(\frac{mn\min(m,n)}{\sqrt{\log(m+n)}} \right).$ $\qquad\square$

The preceding theorem is a slight improvement to the straightforward upper bound of $O(mn \min(m, n))$. We suspect that the presented approach can be applied to obtain a $\sqrt{\log(m + n)}$ factor improvement for other pattern matching problems as well.

Another variant of the problem (denoted by $\texttt{NotSquare-}\Sigma^{m \times n}\text{-}\left[\begin{smallmatrix}1 & 1 \\ 1 & 1\end{smallmatrix}\right]$) asks whether a given matrix contains a non-square 2×2 submatrix pattern. That is, a rectangular subblock with differing width and height. In contrast to the square case, we can readily solve this problem in $O(mn)$ time.

Proposition 2. $\texttt{NotSquare-}\{0, 1, ..., C\}^{m \times n}\text{-}\left[\begin{smallmatrix}1 & 1 \\ 1 & 1\end{smallmatrix}\right]$ *is solvable in $O(mn)$ time.*

Proof. Without loss of generality, suppose that $n \leq m$. To determine whether a non-square $\left[\begin{smallmatrix}1 & 1 \\ 1 & 1\end{smallmatrix}\right]$ submatrix exists, we process the matrix row by row from top to bottom. For each row, we efficiently examine all pairs of ones in that row. That is, pairs of column indices (i, j) such that $M(r, i) = 1$ and $M(r, j) = 1$. For each such pair, we record the row index where it first appears in an n by n integer matrix. If the same pair (i, j) appears again in a later row, we compute the vertical distance (height) as the difference between the current row and the first occurrence, and compare it to the horizontal distance (width), given by $j - i$. If these two values differ, we have found a non-square 2×2 block of ones. Each pair of columns is checked at most twice. Otherwise, we would have formed three rectangles where at least one is non-square. Because each pair of columns is checked at most twice we obtain the desired runtime bound similar to Lemma 1 from [13]. $\qquad\square$

Finally, the difficulty of finding faster algorithms for the $\texttt{Square-}\{0, 1\}^{n \times n}\text{-}\left[\begin{smallmatrix}1 & 1 \\ 1 & 1\end{smallmatrix}\right]$ problem can be observed by the existence of input matrices that are dense with 1's and do not contain a square submatrix matching $\left[\begin{smallmatrix}1 & 1 \\ 1 & 1\end{smallmatrix}\right]$. In particular, by using dense progression-free sets with no arithmetic triple (a lower bound was shown in [8]) we obtain $O(n)$ by $O(n)$ matrices that do not contain any $\left[\begin{smallmatrix}1 & 1 \\ 1 & 1\end{smallmatrix}\right]$ square submatrix with

$$\Omega\left(\frac{n^2 \cdot \sqrt{\log(n)}}{2^{4\sqrt{2 \log_2(n)}}} \right)$$

non-zero entries. This result was suggested to us by D. Eppstein [18]. Furthermore, we cannot achieve an $O(n^{3-\varepsilon})$ time algorithm for $\texttt{Square-}\{0, 1\}^{n \times n}\text{-}\left[\begin{smallmatrix}1 & 1 \\ 1 & 1\end{smallmatrix}\right]$ for any $\varepsilon > 0$ by simply checking if the matrix is dense enough to compel the existence of a square submatrix or if sparse, by brute-force checking whether each non-zero entry is the upper left corner of a square submatrix.

4   Conclusion

We introduced conditional lower bounds, introduced new algorithms, and analyzed existing algorithmic approaches for the Four Corners Problems. In particular, in Theorems 1 and 2 we presented triangle finding and 4-clique hardness results, respectively, for the $\texttt{Rect-}\Sigma^{m \times n}\text{-A}$ problems. In Theorem 3 we presented an improved algorithm for $\mathsf{Square}\text{-}\{0, 1, ..., C\}^{m \times n}\text{-}A$ and we observed potential challenges for finding a strongly subcubic time algorithm for $\mathsf{Square}\text{-}\{0, 1\}^{n \times n}\text{-}\left[\begin{smallmatrix} 1 & 1 \\ 1 & 1 \end{smallmatrix}\right]$.

These results have demonstrated a close relationship between the computational complexity of foundational algorithmic problems in 2D pattern matching, 2D computational geometry, and graph theory. Although this work has focused specifically on corner patterns and rectangles, we suggest that the presented techniques and ideas could be applied to more complex two-dimensional patterns. As future work, we hope to make progress towards efficiently solving the more general local picture language matching problem by developing 2D matching algorithms for border and interior patterns.

Acknowledgments. Preliminary research was pursued by the 2nd and 3rd authors in collaboration with the Algorithms Research Group at Swarthmore College during Spring and Fall 2022. We would especially like to thank and recognize the helpful discussions with A. Abdelmonsef, B. Xiang, D. Yang, E. Brickner, J. Mancini, J. Swernofsky, N. Almadbooh, X. Dong, and all 19 individuals who participated in the Algorithms Research Group. In addition, we would like to thank K. Regan as well as anonymous reviewers for helpful suggestions.

Disclosure of Interests. The authors have no competing interests to declare that are relevant to the content of this article.

References

1. Aldridge, I.: High-frequency trading: a practical guide to algorithmic strategies and trading systems, vol. 604. John Wiley & Sons (2013)
2. Alon, N., Yuster, R., Zwick, U.: Finding and counting given length cycles. Algorithmica **17**(3), 209–223 (1997)
3. Arlazarov, V.L., Dinic, E., Kronrod, M., Faradzev, I.: On economical construction of the transitive closure of a directed graph. In: Dokl. Akad. Nauk SSSR, vol. 194, pp. 1209–1210 (1970)
4. Baker, T.P.: A technique for extending rapid exact-match string matching to arrays of more than one dimension. SIAM J. Comput. **7**(4), 533–541 (1978). https://doi.org/10.1137/0207043
5. Bird, R.S.: Two dimensional pattern matching. Inf. Process. Lett. **6**(5), 168–170 (1977). https://doi.org/10.1016/0020-0190(77)90017-5
6. Bringmann, K.: Fine-grained complexity theory (tutorial). In: 36th International Symposium on Theoretical Aspects of Computer Science (STACS 2019). Schloss-Dagstuhl-Leibniz Zentrum für Informatik (2019)
7. Eisenbrand, F., Grandoni, F.: On the complexity of fixed parameter clique and dominating set. Theoret. Comput. Sci. **326**(1–3), 57–67 (2004)
8. Elkin, M.: An improved construction of progression-free sets. Israel J. Math. **184**(1), 93–128 (2011)
9. Giammarresi, D., Restivo, A.: Two-dimensional languages. In: Rozenberg, G., Salomaa, A. (eds.) Handbook of Formal Languages, pp. 215–267. Springer, Heidelberg (1997). https://doi.org/10.1007/978-3-642-59126-6_4

10. Kärkkäinen, J., Ukkonen, E.: Two- and higher-dimensional pattern matching in optimal expected time. SIAM J. Comput. **29**(2), 571–589 (1999). https://doi.org/10.1137/S0097539794275872
11. Mráz, F., Průša, D., Wehar, M.: Two-dimensional pattern matching against local and regular-like picture languages. Theoret. Comput. Sci. **870**, 137–152 (2021)
12. Murala, S., Wu, Q.J.: Local ternary co-occurrence patterns: a new feature descriptor for MRI and CT image retrieval. Neurocomputing **119**, 399–412 (2013)
13. Průša, D., Wehar, M.: Complexity of searching for 2 by 2 submatrices in boolean matrices. In: International Conference on Developments in Language Theory, pp. 266–279. Springer (2020)
14. Sun, X., Nobel, A.: On the size and recovery of submatrices of ones in a random binary matrix. J. Mach. Learn. Res.-JMLR **9**, 2431–2453 (2008)
15. Uguz, S., Akin, H., Siap, I., Sahin, U.: On the irreversibility of moore cellular automata over the ternary field and image application. Appl. Math. Model. **40**(17–18), 8017–8032 (2016)
16. Van Kreveld, M.J., De Berg, M.T.: Finding squares and rectangles in sets of points. BIT Numer. Math. **31**(2), 202–219 (1991)
17. Vassilevska Williams, V.: Hardness of easy problems: Basing hardness on popular conjectures such as the strong exponential time hypothesis (invited talk). In: 10th International Symposium on Parameterized and Exact Computation (IPEC 2015). Schloss-Dagstuhl-Leibniz Zentrum für Informatik (2015)
18. Wehar, M.: In an m by n boolean matrix, can you find a square block whose four corners are ones in $O(m \cdot n)$ time? Theoretical Computer Science Stack Exchange. https://cstheory.stackexchange.com/q/47588
19. Williams, V.V., Wang, J.R., Williams, R., Yu, H.: Finding four-node subgraphs in triangle time. In: Proceedings of the Twenty-Sixth Annual ACM-SIAM Symposium on Discrete Algorithms, pp. 1671–1680. SIAM (2014)
20. Williams, V.V., Williams, R.: Triangle detection versus matrix multiplication: A study of truly subcubic reducibility (2009)
21. Xu, C.: Rectangles in point set (2015). https://chaoxu.prof/posts/2015-02-02-rectangle-in-point-set.html

On the Descriptional Complexity
of Literal Shuffle

Guilherme Duarte[1]([✉]) [iD], Nelma Moreira[1] [iD], Luca Prigioniero[2] [iD],
and Rogério Reis[1] [iD]

[1] CMUP & DCC, Faculdade de Ciências, Universidade do Porto, Rua do Campo
Alegre, 4169-007 Porto, Portugal
{guilherme.duarte,nelma.moreira,rogerio.reis}@fc.up.pt
[2] Department of Computer Science, Loughborough University, Epinal Way,
LE11 3TU, Loughborough, UK
L.Prigioniero@lboro.ac.uk

Abstract. In this paper, we study the descriptional complexity of several variants of the shuffle operation on regular languages. In the perfect shuffle, words of equal length strictly alternate their symbols. If the words have different lengths, the initial literal shuffle performs the perfect shuffle while both words have symbols and then appends the remaining suffix of the longest word. Lastly, the literal shuffle is an extension of the previous operation, in which the initial shuffle may start at an arbitrary position in one of the two words. We analyze the number of states sufficient and necessary for nondeterministic and deterministic automata to recognize the resulting languages. For nondeterministic finite automata, the cost of the simulation is polynomial. In the worst case, a deterministic finite automaton requires exponentially many states to recognize the literal shuffle of two languages given also by deterministic finite automata, whereas a deterministic 1-limited automaton—namely, a one-tape Turing machine in which each cell may be rewritten only upon its first visit—requires only polynomial many states.

1 Introduction

The *shuffle* of two words u and v is the set of words obtained from interleaving the symbols u and v while preserving the internal order of each word. This operation extends naturally to languages, and the class of regular languages is known to be closed under shuffle. Berard [1] introduced several restricted variations on the shuffle operation. The *perfect shuffle* corresponds to the strict interleaving of symbols from the two words. The *initial literal shuffle* extends this operation to words of different lengths. The *general literal shuffle* further allows the initial literal shuffle to begin after consuming a prefix of one of the words.

Research partially supported by CMUP through FCT under the project with DOI 10.54499/UID/00144/2025 and by FCT project AVATAR 2023.12169.PEX. Guilherme Duarte is also supported by the FCT grant 2024.01528.BD.

M.-P. Béal and P. Caron (Eds.): DLT 2026, LNCS 16578, pp. 307–320, 2026.
https://doi.org/10.1007/978-3-032-28404-4_23

Unlike the general shuffle, these three restrictions are neither commutative nor associative. All of them extend naturally to languages. Moreover, the closure properties of these operations were studied for standard language classes, in particular establishing closure for regular languages. The closure properties of the n-ary versions of the initial literal shuffle and the general literal shuffle were later investigated by Hoffman [8].

In this work, we consider languages represented not only by nondeterministic finite automata (NFAs) and deterministic finite automata (DFAs), but also deterministic 1-limited automata (1-LAs). Limited automata are Turing machines with restrictions on their rewritings and were introduced by Hibbard [7]. These machines operate on a single tape, which initially contains the input. At the left and at the right of the input there are two special endmarkers $\triangleright$ and $\triangleleft$, respectively. The machine is restricted to move its head only between them. In particular, it cannot move to the left of the cell containing $\triangleright$ and to the right of the cell containing $\triangleleft$. Also, limited automata may overwrite the content of each cell only during its first d visits, for a fixed $d \geq 0$. We write d-limited automaton to make the parameter d explicit. For $d = 0$, no rewritings are allowed, hence these automata are standard two-way finite automata, thus they have the same expressive power as finite automata. For $d = 1$, the rewritings in each cell are restricted only to the first visit, and no additional expressive power is obtained [10]. For $d \geq 2$, nondeterministic d-limited automata recognize context-free languages.

In descriptional complexity of operations on languages, the goal is to determine the size of the target devices recognizing the languages resulting from an operation, as a function of the sizes of the source devices representing the input languages.

For the general shuffle, given two NFAs with m and n states, respectively, there exists an NFA with mn states that recognizes the shuffle of their languages [3]. In contrast, when the input automata are deterministic, every DFA recognizing the shuffle may require exponentially many states in the worst case. Domaratzki and Salomaa showed that a DFA with $2mn$ states suffices to recognize the perfect shuffle of the languages of two DFAs with m and n states, respectively, and that this bound is tight [4,5].

In this work, we present constructions for automata recognizing the perfect shuffle, the initial literal shuffle, and the general literal shuffle of two languages. In the nondeterministic case the cost in size of the simulation is polynomial. When DFAs are used both as source and target devices, the cost of the simulation of the general literal shuffle is exponential. On the other hand, when taking 1-limited automata as target devices, the cost becomes polynomial.

We remark that the idea of considering *extensions* of finite automata—that do not increase the recognizing power of the model—to obtain polynomial-size simulations for operations on regular languages is not new. In these cases, the extended devices are used to avoid exponential blow-ups in size. For instance, deterministic 1-LAs have been used to obtain polynomial simulations for the con-

catenation and the star of languages accepted by DFAs [9], and nondeterministic
1-LAs have been used to simulate the complementation of two-way NFAs [6].

2   Preliminaries

In this section, we establish the notation used. Given two integers i, j with
$i < j$, let $[i, j]$ denote the range from i to j, including both i and j, namely
$\{i, i + 1, \ldots, j\}$. Moreover, we shall omit the left bound if it is equal to 0, thus
$[j] = \{0, 1, \ldots, j\}$. Given a set S, we denote its cardinality by $|S|$ and 2^S the set
of all its subsets.

An *alphabet* Σ is a finite set of *symbols*. In this paper, we shall consider
alphabets Σ as ordered sets, for some fixed total order $<$ on the symbols of Σ.
A *word* is a sequence of symbols. A *language* is a set of words over an alphabet,
and $\Sigma^\star$ denotes the language containing all words over Σ, including the empty-
word ε. Given a word $w = \sigma_1 \sigma_2 \ldots \sigma_n$, with $\sigma_i \in \Sigma$, let $|w|$ denote its *length*. We
use $w_{[i,j]}$ to denote the *infix* of w from σ_i to σ_j, for $j > i$ and define $w_{[i,j]} = \varepsilon$ if
$i > j$, $i \notin [1, n]$ or $j \notin [1, n]$. Given two words $u, v \in \Sigma^\star$ we say that u is
lexicographically less than v, denoted as $u < v$ if either $|u| < |v|$, or $|u| = |v|$ and
the words can be written as $u = x\sigma_1 y$, $v = x\sigma_2 z$ with $\sigma_1 < \sigma_2$, for $\sigma_1, \sigma_2 \in \Sigma$
and $x, y, z \in \Sigma^\star$. The set Σ^ℓ corresponds to the set of all words on Σ of length
$\ell \geq 0$. The number of occurrences of a symbol $\sigma \in \Sigma$ in a word w is denoted
by $|w|_\sigma$. Given two words $u, v \in \Sigma^\star$ and a language $L \subseteq \Sigma^\star$, we say that they
are *indistinguishable* w.r.t. L, denoted as $u \equiv_L v$, if $uw \in L \Leftrightarrow vw \in L$, for all
$w \in \Sigma^\star$.

A *one-way nondeterministic finite automaton* (NFA) is defined as a tuple
$\mathscr{A} = \langle Q, \Sigma, \delta, I, F \rangle$, where Q is a finite set of *states*, $\delta : Q \times \Sigma \to 2^Q$ is the
transition function, $I \subseteq Q$ is the set of *initial states*, and $F \subseteq Q$ is the set
of *final states*. When $I = \{q_0\}$, we use $I = q_0$. As customary, we extend δ to
words in the natural way. The *right language* of a state $q \in Q$ is $\mathscr{L}_q(\mathscr{A}) = \{w \in
\Sigma^\star \mid \delta(q, w) \cap F \neq \emptyset\}$. The *language accepted* by $\mathscr{A}$ is $\mathscr{L}(\mathscr{A}) = \bigcup_{q \in I} \mathscr{L}_q(\mathscr{A})$.
An NFA is *minimal* if no NFA with fewer states recognizes the same language.
An NFA operates on a *read-only tape* containing the input word, with an *input
head* that moves strictly from left to right and points to the current symbol
being processed. An NFA is *deterministic* (DFA) if $|I| = 1$ and $|\delta(q, \sigma)| \leq 1$, for
all $q \in Q$ and $\sigma \in \Sigma$. Two states $q_1, q_2 \in Q$ are *equivalent* (or *indistinguishable*)
if $\mathscr{L}_{q_1}(\mathscr{A}) = \mathscr{L}_{q_2}(\mathscr{A})$. A *minimal* DFA has no equivalent states and it is unique
up to isomorphisms.

To prove lower bounds on the number of states of NFAs, we shall use the
fooling set technique [2], which we briefly recall. A fooling set for a language L
is a set of pairs of words $\mathscr{F} = \{\langle x_i, y_i \rangle \mid i \in [1, n]\}$ of size $n \geq 1$ where, for each
$i, j \in [1, n]$, $x_i y_i \in L$ and if $i \neq j$ then $x_i y_j \notin L$ or $x_j y_i \notin L$. It follows that every
NFA for L has at least n states.

A *deterministic 1-limited automaton* (1-LA) is a one-tape Turing machine
with restrictions on its rewritings. It has the ability to revisit the input symbols
already processed and also to rewrite the contents of each tape cell only upon

its first visit, but its head is not allowed to fall out of the input. To that end, we consider two new symbols $\triangleright$ and $\triangleleft$ not belonging to Σ, called the *left* and *right end-marker*, respectively, surrounding each input word on the tape. By $\Sigma_{\triangleright,\triangleleft}$ we denote the set $\Sigma \cup \{\triangleright, \triangleleft\}$. Formally, a deterministic 1-limited automaton is a tuple $\mathscr{A} = \langle Q, \Sigma, \Gamma, \delta, q_0, F \rangle$, where Q, q_0, F are defined as before, Γ is a finite *work alphabet* such that $\Sigma_{\triangleright,\triangleleft} \subseteq \Gamma$, and $\delta : Q \times \Gamma \to Q \times \Gamma \times \{-1, +1\}$ is the transition function. According to δ and the current state, in one move $\mathscr{A}$ reads a symbol from the tape, changes its state, replaces the symbol read from the tape with a new symbol, and moves its head one position either forward or backward. In particular, $\delta(q_1, \sigma) = \langle q_2, X, d \rangle$ means that when the automaton in the state q_1 scans a cell containing the symbol σ, it enters the state q_2, rewrites the cell contents by X, and moves the head to the left or right, according to the direction $d \in \{-1, +1\}$. However, replacing symbols is only allowed during the first visit, with the exception of the cells containing the end-markers, which are *never* modified. Moreover, the end-marker symbols cannot be used to replace the contents of any of the cells. Formally, if $\delta(q_1, \sigma) = \langle q_2, X, d \rangle$ then either $\sigma \in \Sigma$ and $X \in \Gamma \setminus \Sigma_{\triangleright,\triangleleft}$, or $X = \sigma$. After the first visit, the content of a tape cell can no longer be modified. The machine accepts an input if, starting from the initial state with the input head on the left end-marker, it ends the computation in a final state after passing the right end-marker.

The *size* of a machine is the number of symbols required to describe it. For deterministic finite automata, the size is linear in the size of the transition function and in the number of states. Since the transition function is bounded by a polynomial in the number of states and input symbols, the overall size is $\Theta(|Q| \cdot |\Sigma| \cdot \log(|Q|))$. For nondeterministic finite automata, each state and input symbol has a set of possible next states, whose size can be as large as $|Q|$. Hence, the size of the transition function may be quadratic in the number of states, yielding $\Theta(|Q|^2 \cdot |\Sigma|)$. For 1-LAs, the size is bounded by a polynomial in the number of states and work symbols, namely, $\Theta(|Q| \cdot |\Gamma| \cdot \log(|Q| \cdot |\Gamma|))$.

3   Shuffle Operations

Given $u, v \in \Sigma^\star$, the *(general) shuffle* operation between u and v, denoted as $u \amalg v$, corresponds to the set

$$u \amalg v = \{\, u_1 v_1 u_2 v_2 \cdots u_n v_n \mid u_i, v_i \in \Sigma^\star, i \in [1, n], n \geq 1,$$
$$u = u_1 u_2 \cdots u_n, v = v_1 v_2 \cdots v_n \,\}.$$

Now, let $u = u_1 u_2 \cdots u_m$ and $v = v_1 v_2 \cdots v_n$, such that $u_i, v_j \in \Sigma$, for $i \in [1, m]$ and $j \in [1, n]$. The *perfect shuffle* of u and v, denoted as $u \amalg_\pi v$, results in one word obtained by strictly alternating symbols from u and v, starting with the first symbol from u, and defined only when the two words are of equal length.

Formally,

$$u \sqcup\!\sqcup_\pi v = \begin{cases} \emptyset & if \ m \neq n; \\ \{\varepsilon\} & if \ m = n = 0; \\ \{u_1 v_1 u_2 v_2 \cdots u_n v_n\} & otherwise. \end{cases}$$

Example 1. Let $u = aba$ and $v = bab$. Then, $u \sqcup\!\sqcup_\pi v = \{abbaab\}$.

When the words u and v have different lengths, the perfect shuffle can be extended to the *initial literal shuffle*, denoted as $u \sqcup\!\sqcup_\iota v$ [1]. In this case, symbols from u and v are strictly interleaved while both words have symbols remaining, and once the shorter word is exhausted the remaining suffix of the longer word is appended to the result. It is formally defined as

$$u \sqcup\!\sqcup_\iota v = \begin{cases} (u \sqcup\!\sqcup_\pi v_{[1,m]}) \cdot v_{[m+1,n]} & if \ m \leq n; \\ (u_{[1,n]} \sqcup\!\sqcup_\pi v) \cdot u_{[n+1,m]} & otherwise, \end{cases}$$

where, given a set S and a word y, $S \cdot y = \{ xy \mid x \in S \}$ and $y \cdot S = \{ yx \mid x \in S \}$.

Example 2. Let $u = baaab$ and $v = abb$. Then, $u \sqcup\!\sqcup_\iota v = \{baababab\}$.

Finally, the operation most relevant to our study is the *general literal shuffle* of two words u and v, denoted as $u \sqcup\!\sqcup_\lambda v$. In this operation, the initial literal shuffle may start at an arbitrary position in one of the two words. Accordingly, the words can be partitioned into three (possibly empty) subwords: a prefix of one of the two words, followed by the perfect shuffle of symbols from both words, and a suffix of the word with symbols remaining. Formally, it is defined as

$$u \sqcup\!\sqcup_\lambda v = \{ u_{[1,i]} \cdot (u_{[i+1,m]} \sqcup\!\sqcup_\iota v) \mid i \in [m] \} \cup \{ v_{[1,j]} \cdot (u \sqcup\!\sqcup_\iota v_{[j+1,n]}) \mid j \in [n] \}.$$

Example 3. Let $u = bab$ and $v = ababa$, and let

$$L = u \sqcup\!\sqcup_\lambda v = \{abababab, ababbaab, abbaabba, baabbaba, babababa\}.$$

We have that $ababbaab \in L$ as

$$ababbaab = v_{[1,3]} \cdot (u \sqcup\!\sqcup_\iota v_{[4,5]}) = v_{[1,3]} \cdot (u_{[1,2]} \sqcup\!\sqcup_\pi v_{[4,5]}) \cdot u_{[3,3]},$$

and $babababa \in L$ as

$$babababa = u_{[1,2]} \cdot (u_{[3,3]} \sqcup\!\sqcup_\iota v) = u_{[1,2]} \cdot (u_{[3,3]} \sqcup\!\sqcup_\pi v_{[1,1]}) \cdot v_{[2,5]}.$$

These operations on words are extended to languages in the natural way, that is, by shuffling words from each language in all possible combinations. Formally, let $L', L'' \subseteq \Sigma^\star$ and $\sqcup\!\sqcup_o \in \{\sqcup\!\sqcup, \sqcup\!\sqcup_\pi, \sqcup\!\sqcup_\iota, \sqcup\!\sqcup_\lambda\}$. Then,

$$L' \sqcup\!\sqcup_o L'' = \bigcup_{u \in L', v \in L''} u \sqcup\!\sqcup_o v.$$

Example 4. Let $L' = a^\star$ and $L'' = b^\star$. Then,

$$L' \sqcup\!\sqcup L'' = (a+b)^\star, \qquad L' \sqcup\!\sqcup_\pi L'' = (ab)^\star,$$
$$L' \sqcup\!\sqcup_\iota L'' = (ab)^\star(a^\star + b^\star), \quad L' \sqcup\!\sqcup_\lambda L'' = (a^\star + b^\star)(ab)^\star(a^\star + b^\star).$$

For convenience, we refer to the set of operators $\{\sqcup\!\sqcup_\pi, \sqcup\!\sqcup_\iota, \sqcup\!\sqcup_\lambda\}$ as *literal shuffles*.

4   Finite Automata for Literal Shuffle

It is known that the class of regular languages is closed under the general shuffle operation. We shall now construct an NFA that recognizes the general literal shuffle of the languages accepted by the given operand NFAs. This construction shows that regular languages are also closed under the literal shuffles.

Let $\mathscr{A}' = \langle Q', \Sigma, \delta', I', F' \rangle$ and $\mathscr{A}'' = \langle Q'', \Sigma, \delta'', I'', F'' \rangle$ be the two operand NFAs. We describe $\mathscr{A}_\lambda = \langle Q, \Sigma, \delta, I, F \rangle$, such that $\mathscr{L}(\mathscr{A}) = \mathscr{L}(\mathscr{A}') \sqcup\!\sqcup_\lambda \mathscr{L}(\mathscr{A}'')$. First, consider the set of states

$$Q = (Q' \cup Q'' \cup (Q' \times Q'')) \times \{0, 1\}$$

and the initial and the set of final states, respectively,

$$I = (I' \cup I'' \cup (I' \times I'')) \times \{0\} \text{ and } F = ((F' \cup F'') \times \{1\}) \cup (F' \times F'' \times \{0\}).$$

Then, let the transition function δ be defined, for all $q' \in Q'$, $q'' \in Q''$, and $\sigma \in \Sigma$, as:

$$\delta(\langle q', 0 \rangle, \sigma) = \{ \langle p', 0 \rangle \mid p' \in \delta'(q', \sigma) \} \cup \{ \langle p', q_0'', 0 \rangle \mid p' \in \delta'(q', \sigma), q_0'' \in I'' \},$$

where $\langle q', 0 \rangle$ represents the state where, *prior* to performing the perfect shuffle, the input is being simulated on $\mathscr{A}'$ at the current state q'. From such a state, the simulation either continues on $\mathscr{A}'$ or initiates the perfect shuffle;

$$\delta(\langle q'', 0 \rangle, \sigma) = \{ \langle p'', 0 \rangle \mid p'' \in \delta''(q'', \sigma) \} \cup \{ \langle q_0', p'', 0 \rangle \mid p'' \in \delta''(q'', \sigma), q_0' \in I' \},$$

analogous to $\delta(\langle q', 0 \rangle, \sigma)$ where the input is being simulated on $\mathscr{A}''$;

$$\delta(\langle q', 1 \rangle, \sigma) = \{ \langle p', 1 \rangle \mid p' \in \delta'(q', \sigma) \},$$

where $\langle q', 1\rangle$ represents the state where, *after* performing the perfect shuffle, the input is being only simulated on $\mathscr{A}'$ at the current state q'. In this case, the simulation continues on $\mathscr{A}'$;

$$\delta(\langle q'', 1\rangle, \sigma) = \{\, \langle p'', 1\rangle \mid p'' \in \delta''(q'', \sigma) \,\},$$

analogous to $\delta(\langle q', 1\rangle, \sigma)$, where the simulation continues on $\mathscr{A}''$;

$$\begin{aligned}
\delta(\langle q', q'', 0\rangle, \sigma) = {} & \{\, \langle p', q'', 1\rangle \mid p' \in \delta'(q', \sigma) \,\} \cup \\
& \{\, \langle p'', 1\rangle \mid q' \in F', p'' \in \delta''(q'', \sigma) \,\} \cup \\
& \{\, \langle p', 1\rangle \mid q'' \in F'', p' \in \delta'(q', \sigma) \,\};
\end{aligned}$$

where $\langle q', q'', 0\rangle$ represents the state where the perfect shuffle is being performed, while on the state q' of $\mathscr{A}'$ and q'' of $\mathscr{A}''$. In this state, the simulation may proceed in one of three ways: continue the perfect shuffle, with the next input symbol simulated on $\mathscr{A}'$; if $q' \in F'$, switch to simulating only on $\mathscr{A}''$; if $q'' \in F''$, switch to $\mathscr{A}'$;

$$\delta(\langle q', q'', 1\rangle, \sigma) = \{\, \langle q', p'', 0\rangle \mid p'' \in \delta''(q'', \sigma) \,\},$$

analogous to $\delta(\langle q', q'', 0\rangle, \sigma)$, but now the next input symbol must be simulated on $\mathscr{A}''$.

Consider the following results:

Lemma 1. *Let $\mathscr{A}'$ and $\mathscr{A}''$ be two NFAs that recognize $L', L'' \subseteq \Sigma^\star$, respectively. Then, the automata for $L' \amalg_\pi L''$, $L' \amalg_\iota L''$, and $L' \amalg_\lambda L''$ can be effectively constructed from the construction given above.*

Sketch of Proof. Let $\mathscr{A}' = \langle Q', \Sigma, \delta', I', F'\rangle$ and $\mathscr{A}'' = \langle Q'', \Sigma, \delta'', I'', F''\rangle$. Moreover, let $\mathscr{A}_\lambda = \langle Q, \Sigma, \delta, I, F\rangle$ be the automaton resulting from the construction above when provided with the automata $\mathscr{A}'$ and $\mathscr{A}''$. Then,

1. $\mathscr{A}_\lambda$ recognizes $L' \amalg_\lambda L''$:
   The initial states correspond to the three different possible starting scenarios: perform the perfect shuffle from the beginning ($I' \times I'' \times \{0\}$); start with a simulation on $\mathscr{A}'$ alone ($I' \times \{0\}$); or on $\mathscr{A}''$ alone ($I'' \times \{0\}$). The final states correspond to the three different possible ending scenarios: end with the perfect shuffle ($F' \times F'' \times \{0\}$); end with the rest of input simulated on $\mathscr{A}'$ only ($F' \times \{1\}$); or on $\mathscr{A}''$ ($F'' \times \{1\}$).
2. $\mathscr{A}_\iota = \langle Q_\iota, \Sigma, \delta, I_\iota, F\rangle$, where

$$Q_\iota = Q \setminus ((Q' \cup Q'') \times \{0\}) \quad and \quad I_\iota = I \setminus ((I' \cup I'') \times \{0\})$$

   recognizes $L' \amalg_\iota L''$:
   In the initial literal shuffle, the simulation cannot begin solely on $\mathscr{A}'$ or $\mathscr{A}''$ prior to the perfect shuffle phase. Hence, we remove those states from Q and adjust the initial states accordingly, while the final states remain unchanged.
3. $\mathscr{A}_\pi = \langle Q_\pi, \Sigma, \delta, I_\iota, F_\pi\rangle$, where

$$Q_\pi = Q \setminus ((Q' \cup Q'') \times \{0, 1\}) \quad and \quad F_\pi = F \setminus ((F' \cup F'') \times \{1\})$$

recognizes $L' \sqcup\!\sqcup_\pi L''$:

For the perfect shuffle, only the states simulating the shuffle itself are required. Thus, the initial states are defined as in the case of the initial literal shuffle, but the final states are updated.

Note that $|Q| = 2(mn + m + n)$, $|Q_\iota| = 2mn + m + n$, and $|Q_\pi| = 2mn$. $\square$

4.1 Number of States of NFAs Recognizing the Literal Shuffles

Regarding the perfect shuffle, it is know that given two languages $L', L'' \subseteq \Sigma^\star$ recognized by two DFAs of m and n states, respectively, then $2mn$ states are sufficient for a DFA to recognize $L' \sqcup\!\sqcup_\pi L''$, and the bound is tight [4,5]. For every $\sigma \in \Sigma$ and $n \geq 0$, consider the language $L_{\sigma,n} = \{ w \in \Sigma^\star \mid |w|_\sigma \equiv 0 \bmod n \}$, over a k-symbol alphabet Σ.

Lemma 2. *Let $m, n > 0$ and $\Sigma = \{a, b\}$. Then, $2mn$ states are necessary for an NFA to recognize $L_{a,m} \sqcup\!\sqcup_\pi L_{b,n}$.*

Proof. Consider the set $\mathcal{F} = \{ \langle a^{2i}b^j, a^{2(m-i)}b^{2n-j} \rangle \mid i \in [m-1], j \in [2n-1] \}$. By adapting the proof for the tightness of the number of states required for a DFA [4,5], it follows that $\mathcal{F}$ forms a fooling set for $L_{a,m} \sqcup\!\sqcup_\pi L_{b,n}$. $\square$

Since the number of states of NFAs is bounded by that of their equivalent deterministic counterparts and by Lemma 2, we have the following result.

Theorem 1. *Let $\mathcal{A}'$ and $\mathcal{A}''$ be two NFAs with m and n states, respectively. Then, $2mn$ states are sufficient to recognize $\mathscr{L}(\mathcal{A}') \sqcup\!\sqcup_\pi \mathscr{L}(\mathcal{A}'')$, and the bound is tight.*

We shall now consider the remaining shuffle operations: the initial literal shuffle and the general literal shuffle. Again, from the construction presented above, we have an upper bound on the number of states of the NFAs recognizing the resulting languages.

Lemma 3. *Let $\mathcal{A}'$ and $\mathcal{A}''$ be two NFAs with m and n states, respectively. Then, $2mn + m + n$ and $2(mn + m + n)$ states are sufficient for an NFA to recognize $\mathscr{L}(\mathcal{A}') \sqcup\!\sqcup_\iota \mathscr{L}(\mathcal{A}'')$ and $\mathscr{L}(\mathcal{A}') \sqcup\!\sqcup_\lambda \mathscr{L}(\mathcal{A}'')$, respectively.*

Adapting the fooling set argument used for the perfect shuffle, it is possible to prove that the bound in Lemma 3 is tight. For $m, n > 0$, let $L_{a,m}$ be defined over the alphabet $\{a, b\}$ and $L_{c,n}$ over $\{c, d\}$.

Lemma 4. *Let $m, n > 0$ and $\Sigma = \{a, b, c, d\}$. Then, $2mn + m + n$ and $2(mn + m + n)$ states are necessary for an NFA to recognize $L_{a,m} \sqcup\!\sqcup_\iota L_{c,n}$ and $L_{a,m} \sqcup\!\sqcup_\lambda L_{c,n}$, respectively.*

We obtain the following result from Lemmas 3 and 4.

Theorem 2. *Let $\mathcal{A}'$ and $\mathcal{A}''$ be two NFAs with m and n states, respectively. Then, $2mn + m + n$ and $2(mn + m + n)$ states are sufficient for an NFA to recognize $\mathscr{L}(\mathcal{A}') \sqcup\!\sqcup_\iota \mathscr{L}(\mathcal{A}'')$ and $\mathscr{L}(\mathcal{A}') \sqcup\!\sqcup_\lambda \mathscr{L}(\mathcal{A}'')$, respectively, and the bound is tight for $|\Sigma| \geq 4$.*

4.2   Number of States of DFAs Recognizing the Literal Shuffles

We now turn to the size of DFAs for the various shuffle operators. Let $\mathscr{A}' = \langle Q', \Sigma, \delta', q_0', F' \rangle$ and $\mathscr{A}'' = \langle Q'', \Sigma, \delta'', q_0'', F'' \rangle$ be DFAs with m and n states, respectively, for two regular languages L' and L'', respectively. Moreover, let $\mathscr{A}_\iota = \langle Q, \Sigma, \delta, I, F \rangle$ denote the NFA resulting from the construction presented in the previous section recognizing $L' \sqcup\!\sqcup_\iota L''$.

The number of states for a DFA to recognize $L' \sqcup\!\sqcup_\iota L''$ has as upper bound the size of the resulting DFA from the well-known subset construction applied to $\mathscr{A}_\iota$, namely, $\mathscr{D}(\mathscr{A}_\iota)$. That is, $2^{|Q|} = 2^{2mn+m+n}$ states. However, one can notice that some of the subsets can be saved. Let $s \in 2^Q$ be a state from $\mathscr{D}(\mathscr{A}_\iota)$, which is a subset of states of $\mathscr{A}_\iota$. Then, s can be decomposed into two disjoint sets $s = s_1 \cup s_2$ such that $s_1 \subseteq Q' \times Q'' \times \{0, 1\}$ and $s_2 \subseteq (Q' \cup Q'') \times \{1\}$. Observe that if the state s is reachable in $\mathscr{D}(\mathscr{A}_\iota)$, then $|s_1| = 1$. This follows from the fact that δ' and δ'' are deterministic. Thus, we obtain the following upper bound on the number of states of $\mathscr{D}(\mathscr{A}_\iota)$.

Theorem 3. *Given two DFAs $\mathscr{A}'$ and $\mathscr{A}''$ with m and n states, respectively, $2^{m+n+1}mn$ states are sufficient for a DFA to recognize $\mathscr{L}(\mathscr{A}') \sqcup\!\sqcup_\iota \mathscr{L}(\mathscr{A}'')$.*

Now, let us consider the general literal shuffle operation between two regular languages $L', L'' \subseteq \Sigma^\star$. Once again, let $\mathscr{A}' = \langle Q', \Sigma, \delta', q_0', F' \rangle$ and $\mathscr{A}'' = \langle Q'', \Sigma, \delta'', q_0'', F'' \rangle$ be DFAs for L' and L'' with m and n states, respectively. Now, let $\mathscr{A}_\lambda = \langle Q, \Sigma, \delta, I, F \rangle$ denote the NFA resulting from the construction presented in the previous section recognizing $L' \sqcup\!\sqcup_\lambda L''$. Of course, $2^{|Q|} = 2^{2(mn+m+n)}$ is an upper bound on the number of states of $\mathscr{D}(\mathscr{A}_\lambda)$. However, again, some of the subsets can be saved. Let $s \in 2^Q$ be a state from $\mathscr{D}(\mathscr{A}_\lambda)$, which is a subset of states of $\mathscr{A}_\lambda$. Then, s can be decomposed into four disjoint sets $s = s_1 \cup s_2 \cup s_3 \cup s_4$ such that $s_1 \subseteq Q' \times \{0\}$, $s_2 \subseteq Q'' \times \{0\}$, $s_3 \subseteq (Q' \cup Q'') \times \{1\}$, and $s_4 \subseteq Q' \times Q'' \times \{0, 1\}$. Observe that if the state s is reachable in $\mathscr{D}(\mathscr{A}_\lambda)$, then $|s_1| = |s_2| = 1$. This follows again from the fact that δ' and δ'' are deterministic. Hence, the image of the states $Q' \times \{0\}$ (resp., $Q'' \times \{0\}$) under δ — that is, the states where the automaton simulates the input only on $\mathscr{A}'$ (resp., $\mathscr{A}''$) *prior* to the perfect shuffle — contains a single state within $Q' \times \{0\}$ (resp., $Q'' \times \{0\}$). Thus, we obtain the following upper bound on the number of states of $\mathscr{D}(\mathscr{A}_\lambda)$.

Theorem 4. *Given two DFAs $\mathscr{A}'$ and $\mathscr{A}''$ with m and n states, respectively, $2^{mn+m+n}mn$ states are sufficient for a DFA to recognize $\mathscr{L}(\mathscr{A}') \sqcup\!\sqcup_\lambda \mathscr{L}(\mathscr{A}'')$.*

We now introduce a family of languages that leads to an exponential blow-up in the size of the minimal DFA recognizing their literal shuffle. This result relies on the fact that the concatenation of two DFAs can result in an exponential increase in size. Since the general literal shuffle of two languages includes, in particular, their concatenation, we shall show how this complexity carries over to the shuffle. For $n \geq 0$, consider the language $L_n = \{\, \#w(\#\Sigma^n)^\star \#w \mid w \in \Sigma^n \,\}$ defined over a k-letter alphabet Σ. Of course, every automaton recognizing L_n must *remember* the prefix of length $n + 1$ of the input word, which implies that

$O(k^n)$ states are needed. It can be noticed that, given two words $x, y \in L_n$, where $x = \#x'(\#\Sigma^n)^\star \#x'$, words of the form $xy \in L_n L_n \subset L_n \sqcup\!\sqcup_\lambda L_n$ are the *hardest* to recognize. The reason is that, when concatenating these two strings, the automaton has no deterministic way to tell where x ends and y begins. Because of this ambiguity, the DFA is forced to keep track of every length-n factor that follows each occurrence of x', starting from the second one. As a consequence, the number of states necessary to keep this information is $\Omega(2^{k^n})$.

Lemma 5. *For $n \geq 0$ and $k \geq 2$, every DFA recognizing $L_n \sqcup\!\sqcup_\lambda L_n$ over a k-letter alphabet needs $\Omega(2^{k^n})$ states.*

5   Polynomial Simulations with 1-Limited Automata

In this section, we study the cost of simulating the initial and the general literal shuffle of two languages using deterministic 1-LAs, assuming the input languages are given by DFAs. By leveraging the ability of 1-LAs to rewrite the tape and perform two-way scanning, we show that the literal shuffle of two languages can be recognized by deterministic 1-LAs of polynomial size in the number of states of their DFAs.

Let $\mathscr{A}' = \langle Q', \Sigma, \delta', q'_0, F' \rangle$ and $\mathscr{A}'' = \langle Q'', \Sigma, \delta'', q''_0, F'' \rangle$ be two DFAs with m and n states, respectively, recognizing the languages $L', L'' \subseteq \Sigma^\star$. We shall describe how to obtain a deterministic 1-LA $\mathscr{A} = \langle Q, \Sigma, \Gamma, \delta, q_0, F \rangle$ accepting $L = L' \sqcup\!\sqcup_\lambda L''$ in such a way that the size of $\mathscr{A}$ is polynomial in the sizes of $\mathscr{A}'$ and $\mathscr{A}''$.

Our objective is to avoid the exponential blow-up in size by exploiting the rewriting capabilities of 1-LAs. Instead of storing the set of states in the input automata reached by the different literal shuffle scenarios within the states of the machine (as is required in a DFA), $\mathscr{A}$ encodes it on the tape. This information is then accessed using the ability of 1-LAs of scanning the tape in a two-way fashion, to access and resume the simulation corresponding to each scenario.

First, the tape is logically partitioned into blocks of $2mn$ cells, with a possibly shorter final block, where each cell in a block represents a state in the product space $Q' \times Q'' \times \{0, 1\}$. Given a state $\langle q', q'', \mathsf{b} \rangle \in Q' \times Q'' \times \{0, 1\}$, we define its index in an ordered enumeration of $Q' \times Q'' \times \{0, 1\}$ as $\mathsf{ind}(\langle q', q'', \mathsf{b} \rangle) = c$.

To enable the writing and reading of the information necessary to simulate the general literal shuffle of the languages recognized by $\mathscr{A}'$ and $\mathscr{A}''$, while ensuring the simulation cost is polynomial in the size of the input automata, the simulating 1-LA operates within sliding windows spanning two consecutive blocks of cells, i.e., virtual windows of size $4mn$. During the overwriting of a window, the *left block* has been fully overwritten, while the *right block* contains, at some position, the leftmost cell that has not yet been overwritten, referred to as the *relative frontier*. We refer to the positions of the current window as pairs in $[2mn-1] \times \{\mathsf{L}, \mathsf{R}\}$, where the second component indicates whether the position given in the first component belongs to the left (L) or right (R) block of the window. As the 1-LA operates in windows, we consider the infix $w = \sigma_1\sigma_2\cdots\sigma_{2mn}$ of the original input as the word read left-to-right of the left block of a window. As a special case, at the beginning of the simulation, the left block of the window has length 1 and only contains the left end-marker and thus $w = \varepsilon$. Moreover, the frontier is initially at position $(0, \mathsf{R})$, and the head of the machine is on the left end-marker.

The written information is organized into *four* tracks. In particular, consider the c-th cell of a given block, with $c \in [2mn - 1]$, and the state $\langle q', q'', \mathsf{b}\rangle$ for which $\mathsf{ind}(\langle q', q'', \mathsf{b}\rangle) = c$. Then,

- The *first track* retains a copy of the input symbol originally contained in the cell prior to any rewriting, ensuring that the symbol remains accessible throughout all possible simulations of the shuffle;
- The *second track* contains a marker indicating whether (✓) or not (✗), *prior* to executing the perfect shuffle, the state q' is reachable in $\mathscr{A}'$ (resp., q'' is reachable in $\mathscr{A}''$) depending on whether $\mathsf{b} = 0$ (resp., $\mathsf{b} = 1$), when simulating the input prefix up to the end of the previous window;
- The *third track* contains a marker indicating whether (✓) or not (✗), after processing symbols from either $\mathscr{A}'$ or $\mathscr{A}''$, the states q' and q'' are reachable on $\mathscr{A}'$ and $\mathscr{A}''$, respectively, while performing the perfect shuffle, when simulating the input prefix as before;
- The *fourth track* contains a marker indicating whether (✓) or not (✗), *after* executing the perfect shuffle, the state q' is reachable in $\mathscr{A}'$ (resp., q'' is reachable in $\mathscr{A}''$) depending on whether $\mathsf{b} = 0$ (resp., $\mathsf{b} = 1$), when simulating the input prefix as before.

We now present in detail on how $\mathscr{A}$ recognizes $L' \sqcup\!\sqcup_\lambda L''$ (see Procedure 1). The deterministic 1-LA stores in its finite control the position of the frontier within the right block of the current window (relativeFrontier), which is initially set to 0. At every step of the simulation, *before* entering the cell at the frontier, the 1-LA gathers the information regarding the reachability of the state encoded in that cell, which is given by ind(state). To do so, it uses the procedures `markSndTrack`, `markTrdTrack`, and `markFthTrack` to determine the values to be written in the second, third, and fourth tracks of the cell at the frontier, respectively. Then, it moves the input head to the right until reaching the cell at the frontier (this is possible since the position of the frontier is stored in the finite control), and, using the procedure `write`, it overwrites the cell with the information gathered. Finally, the frontier advances to the next cell updating the value of relativeFrontier accordingly, that is, relativeFrontier is incremented by one modulo $2mn$. This process is repeated until the right end-marker is reached, at which point the 1-LA performs a final check to determine whether to accept or reject the input word. The contents written in the *right block* of the final window, which may be shorter than $2mn$ in size, are going to be ignored during this check. Instead, for every state $q = \langle q', q'', \mathsf{b}\rangle \in F' \times F'' \times \{0, 1\}$, the 1-LA checks whether q can be reached by any of the states encoded in the *left block* of the final window, continuing (or starting) the simulation of the shuffles on the

---

Procedure 1 `literalShuffle`($\mathscr{A}', \mathscr{A}''$)

Given two DFAs $\mathscr{A}' = \langle Q', \Sigma, \delta', q_0', F' \rangle$ and $\mathscr{A}'' = \langle Q'', \Sigma, \delta'', q_0'', F'' \rangle$, where $m = |Q'|$ and $n = |Q''|$, the deterministic 1-LA implementing this procedure recognizes the language $\mathscr{L}(\mathscr{A}') \sqcup\!\sqcup_\lambda \mathscr{L}(\mathscr{A}'')$.

```
 1: relativeFrontier ← 0
 2: while TRUE do
 3:     ⟨q′, q″, b⟩ ← ind⁻¹(relativeFrontier)
 4:     sndTrack ← markSndTrack(⟨q′, q″, b⟩, 𝒜′, 𝒜″)
 5:     trdTrack ← markTrdTrack(⟨q′, q″, b⟩, 𝒜′, 𝒜″)
 6:     fthTrack ← markFthTrack(⟨q′, q″, b⟩, 𝒜′, 𝒜″)
 7:     move the head rightward until reaching position (relativeFrontier, R)
 8:     if current symbol ≠ ◁ then
 9:         write(current symbol, sndTrack, trdTrack, fthTrack)
10:         relativeFrontier ← (relativeFrontier + 1) mod 2mn
11:     else
12:         if checkReachability(𝒜′, 𝒜″) then ACCEPT else REJECT
```

---

suffix of the input word contained in the last window. This is done by using the procedure `checkReachability`, that reuses adapted versions of the functions `markSndTrack`, `markTrdTrack`, and `markFthTrack` and accepts according to the values returned by them.

Theorem 5. *Let $\mathscr{A}'$ and $\mathscr{A}''$ be two DFAs with m and n states, respectively, over a k-letter alphabet Σ. Then, there exists a deterministic 1-LA recognizing $\mathscr{L}(\mathscr{A}') \sqcup\!\sqcup_\lambda \mathscr{L}(\mathscr{A}'')$ with $O(m^5 n^5)$ states and $9k + 2$ work symbols.*

Proof. Let $\mathscr{A} = \langle Q, \Sigma, \Gamma, \delta, q_0, F \rangle$ be the deterministic 1-LA implementing Procedure 1 described above. The work alphabet Γ of $\mathscr{A}$ is $\Gamma = \Sigma_{\triangleright,\triangleleft} \cup (\Sigma \times \{\checkmark, ✗\} \times \{\checkmark, ✗\} \times \{\checkmark, ✗\})$ of size $|\Sigma_{\triangleright,\triangleleft}| + 8|\Sigma| = 9k + 2$. Moreover, the automaton stores the following components in its states:

- The position of the head within the window of size $4mn$. This is required to know the position of the head with respect to the frontier, and not to pass it before gathering the information to be written at the frontier cell. Moreover, it is also needed to navigate through the left block of the window to retrieve the information stored in the second, third, and fourth tracks;
- The position of the relative frontier (relativeFrontier in Procedure 1) within the right block of the window, whose size is $2mn$;
- Three bits for the values returned by `markSndTrack`, `markTrdTrack`, and `markFthTrack`;
- The procedure `markSndTrack` requires a component of size $2mn$ to loop through the cells of the left block of the window and retrieve the information stored in the second track, and a component of size $2mn$ to keep track of the states reached during the simulation of the prefix of the input word on either $\mathscr{A}'$ or $\mathscr{A}''$. This gives a total of $4m^2 n^2$ states for this part of the simulation;

- Similarly, the procedure `markTrdTrack` requires a component of size $2mn$ to loop through the cells of the left block of the window and retrieve the information stored in the second and third tracks, plus a component of size $2mn$ to keep track of the states reached during the simulations of the shuffles. Moreover, the procedure needs an extra component of size $2mn$ to consider all possible splits of the word w into two parts: one to continue the head on either $\mathscr{A}'$ or $\mathscr{A}''$, and another to simulate the shuffle (see Case 2). This gives a total of $8m^3n^3$ states for this part of the simulation;
- The procedure `markFthTrack` requires a component of size $2mn$ to loop through the cells of the left block of the window and retrieve the information stored in the second, third, and fourth tracks. It also needs a component of size $2mn$ to keep track of the reached states. Moreover, the procedure needs two extra components of size $2mn$ to consider all possible splits of the word w into three parts: one to continue the head on either $\mathscr{A}'$ or $\mathscr{A}''$, a second to perform the shuffle, and a third to start the tail on either $\mathscr{A}'$ or $\mathscr{A}''$ (see Case 3). This gives a total of $16m^4n^4$ states for this part of the simulation;
- Some additional state components of constant size used to keep track of the execution flow of the procedure.

Therefore, the number of states of $\mathscr{A}$ is $16mn(4m^2n^2 + 8m^3n^3 + 16m^4n^4) \in O(m^5n^5)$. $\qquad\square$

It can be observed that the 1-LA implementing Procedure 1, when executed without the second track, recognizes the initial literal shuffle of the languages accepted by the operand automata. We then have the following result:

Corollary 1. *Let $\mathscr{A}'$ and $\mathscr{A}''$ be two DFAs with m and n states, respectively, over a k-letter alphabet Σ. Then, there exists a deterministic 1-LA recognizing $\mathscr{L}(A') \sqcup_\iota \mathscr{L}(\mathscr{A}'')$ with $O(m^5n^5)$ states and $5k + 2$ work symbols.*

References

1. Berard, B.: Literal shuffle. Theoret. Comput. Sci. **51**(3), 281–299 (1987). https://doi.org/10.1016/0304-3975(87)90037-5
2. Birget, J.C.: Intersection and union of regular languages and state complexity. Inf. Process. Lett. **43**(4), 185–190 (1992). https://doi.org/10.1016/0020-0190(92)90198-5
3. Brzozowski, J.A., Jirásková, G., Liu, B., Rajasekaran, A., Szykula, M.: On the state complexity of the shuffle of regular languages. In: Câmpeanu, C., Manea, F., Shallit, J.O. (eds.) 18th DCFS, 2016. LNCS, vol. 9777, pp. 73–86. Springer, Cham (2016). https://doi.org/10.1007/978-3-319-41114-9_6
4. Domaratzki, M.: Trajectory-Based Operations, Ph.D. thesis. Kingston, Ontario, Canada (2004)
5. Domaratzki, M., Salomaa, K.: State complexity of shuffle on trajectories. J. Autom. Lang. Comb. **9**(2/3), 217–232 (2004). https://doi.org/10.25596/JALC-2004-217

6. Guillon, B., Prigioniero, L., Taheri, J.: Polynomial complementation of nondeterministic two-way finite automata by 1-limited automata. In: Mahajan, M., Manea, F., McIver, A., Nguyen, K.T. (eds.) STACS 2026. LIPIcs, vol. 364, pp. 48:1–48:18. Schloss Dagstuhl - Leibniz-Zentrum für Informatik (2026). https://doi.org/10.4230/LIPICS.STACS.2026.48
7. Hibbard, T.N.: A generalization of context-free determinism. Inf. Control **11**(1), 196–238 (1967). https://doi.org/10.1016/S0019-9958(67)90513-X
8. Hoffmann, S.: The n-ary initial literal and literal shuffle. Discret. Appl. Math. **376**, 112–132 (2025). https://doi.org/10.1016/j.dam.2025.04.023
9. Pighizzini, G., Prigioniero, L., Sádovský, S.: Performing regular operations with 1-limited automata. Theory Comput. Syst. **68**(3), 465–486 (2024). https://doi.org/10.1007/S00224-024-10163-1
10. Wagner, K.W., Wechsung, G.: Computational complexity. D. Reidel Publishing Company, Dordrecht (1986)

On Word Representations and Embeddings in Complex Matrices

Paul C. Bell[1], George Kenison[2], Reino Niskanen[1]([✉]), Igor Potapov[3],
and Pavel Semukhin[1]

[1] Liverpool John Moores University, Liverpool, UK
`{p.c.bell,r.niskanen,p.semukhin}@ljmu.ac.uk`
[2] KU Leuven, Leuven, Belgium
`george.kenison@kuleuven.be`
[3] University of Liverpool, Liverpool, UK
`potapov@liverpool.ac.uk`

Abstract. Embeddings of word structures into matrix semigroups provide a natural bridge between combinatorics on words and linear algebra. However, low-dimensional matrix semigroups impose strong structural restrictions on possible embeddings. Certain finitely generated groups admit faithful representations in $\mathrm{SL}(2,\mathbb{C})$ and other similar matrix groups. On the other hand, it is known that the product of two free semigroups on two generators cannot be embedded into the 2×2 complex matrices. In this paper we study embeddings of word structures into low-dimensional matrix semigroups over the complex numbers and develop new techniques for constructing word representations of the Euclidean Bianchi groups. These representations provide a symbolic framework and a natural first step towards analysing fundamental decision problems in 2×2 matrix semigroups.

1 Introduction

Matrix products are one of the most fundamental operations in mathematics. Originally introduced as a compact way to represent and solve systems of linear equations, matrix products later came to play an essential role in many fields, with applications ranging from engineering and control theory to modern data science and large-scale information systems such as search ranking algorithms.

The study of embeddings of word structures into matrices connects combinatorics on words with linear algebra. By representing words or generators as matrices, questions about concatenation of words and computational models operating on words can be translated into questions about matrix multiplication. Thus, solutions to problems in combinatorics on words and formal language theory can employ algebraic and analytic tools from matrix theory. Moreover, embeddings into integer or complex matrices provide a concrete and well-understood algebraic framework in which structural properties of (semi)groups can be analysed, while also enabling the transfer of results between the theory of words and the theory of matrix semigroups.

M.-P. Béal and P. Caron (Eds.): DLT 2026, LNCS 16578, pp. 321–334, 2026.
https://doi.org/10.1007/978-3-032-28404-4_24

Observe that it is not always possible to embed word structures within certain matrix classes or dimensions. This observation clarifies the expressive power of small matrix semigroups, showing that some algebraic structures generated by words cannot be faithfully represented in restricted matrix domains. Further, it helps determine which techniques from linear algebra can be applied to word problems, and which problems require different approaches.

The origins of the connection between combinatorial group theory and linear representations go back to the work of Nielsen in the 1920 s on free groups [19]. These techniques provided an early structural understanding of free groups and influenced approaches to representing their elements by algebraic objects. Later, Magnus showed that free groups admit faithful representations into groups of matrices over certain rings of formal power series. Interest in such embeddings grew further with the development of algorithmic problems in algebra and theoretical computer science, when researchers began investigating whether problems concerning words and formal languages could be translated into algebraic problems involving matrices and vice versa [8,9,11,18,22,23].

Recent work has also highlighted the importance of decision problems for matrix semigroups over complex numbers. In particular, the first decidability results for the identity and group problems in the complex Heisenberg group $H(3,\mathbb{C})$ demonstrate that matrix semigroups with non-trivial structural constraints require new number theoretical techniques [3] and new group theoretic techniques for algorithmic analysis [12]. At the same time, many fundamental questions for low-dimensional matrices over the complex numbers remain open. A central example concerns embeddings of word semigroups into the 2×2 complex matrices. It is known that the product of two free semigroups on two generators cannot be embedded into $\mathbb{C}^{2\times 2}$ [8]. On the other hand, deciding whether a finitely generated torsion-free group embeds in $SL(2,\mathbb{C})$ (see [6]) has deep connections to questions in geometric group theory. These results indicate that the expressive power of 2×2 complex matrices lies at a delicate boundary—while they allow rich algebraic representations, they also impose strong structural limitations on possible embeddings.

Table 1. The state of the art for the non-existence of embeddings from pairs of words into different matrix semigroups. Entries in red are new to the present paper.

$\sharp$	$S(\Sigma_1)$	$S(\Sigma_2)$	$F(\Sigma_1)$	$F(\Sigma_2)$
$\{\varepsilon\}$	?	?	?	$U(n,\mathbb{C}),\ n \geq 1$ (Prop. 7)
$S(\Sigma_1)$	?	?	?	$SL(2,\mathcal{O}_d)$ (Prop. 9)
$S(\Sigma_2)$		$\mathbb{C}^{2\times 2}[8],\ SL(3,\mathbb{Z})[18]$	$\mathcal{O}_d^{2\times 2}$ (Prop. 9)	$\mathcal{O}_d^{3\times 3}$ (Prop. 9)
$F(\Sigma_1)$			?	$\mathcal{O}_d^{2\times 2}$ (Prop. 9)
$F(\Sigma_2)$				$\mathbb{Z}^{3\times 3}[18]$

This paper explores the boundary between combinatorics on words and matrix semigroups, aiming to characterise which word structures admit faithful low-dimensional matrix representations over complex numbers and develops new techniques to find word representations for the Euclidean Bianchi groups (Theorem 3), an essential first step to study fundamental decision problems in matrix semigroups within their symbolic representations, like membership, freeness, and vector reachability in 2×2 matrix semigroups. Table 1 summarises non-existence results for embeddings in different settings, where $S(\Sigma_1)$ and $S(\Sigma_2)$ denote free semigroups over unary and binary alphabets, while $F(\Sigma_1)$ and $F(\Sigma_2)$ denote free groups over unary and binary group alphabets. The entries are read as "column" $\times$ "row". The grey background indicates symmetrical cases that do not require consideration, e.g., $S(\Sigma_2) \times S(\Sigma_1) = S(\Sigma_1) \times S(\Sigma_2)$. Table 2 details existing embeddings.

Table 2. The state of the art for the existence of embeddings from pairs of words into different matrix semigroups. Entries in blue are straightforward extensions of results in the literature.

$\exists$	$S(\Sigma_1)$	$S(\Sigma_2)$	$F(\Sigma_1)$	$F(\Sigma_2)$
$\{\varepsilon\}$		$U(2, \mathbb{N})$ (folklore)	$U(2, \mathbb{Z})$ (folklore), $\mathbb{Q}$ (folklore)	$\mathbb{C}^{2 \times 2}$ (folklore), $\mathbb{Z}^{2 \times 2}$ (folklore)
$S(\Sigma_1)$	$U(2, \mathbb{N})$ (Prop. 8)	$\mathbb{N}^{2 \times 2}$ (Prop. 8)	?	$\mathbb{Z}^{2 \times 2}$ (Prop. 8)
$S(\Sigma_2)$		SL$(3, \mathbb{Q})$ [18], $U(3, \mathbb{N})$ [22]	$\mathbb{Q}^{2 \times 2}, \mathbb{Z}^{3 \times 3}$ (Prop. 8)	?
$F(\Sigma_1)$			$\mathbb{Q}$ (folklore)	$\mathbb{Q}^{2 \times 2}$ (Prop. 8)
$F(\Sigma_2)$				$\mathbb{Z}^{4 \times 4}, \mathbb{H}(\mathbb{Q})^{2 \times 2}$ [4]

The paper is structured as follows. In the next section, we introduce necessary notations and preliminaries. In Sect. 3, we prove that an element of a Euclidean Bianchi group admits a canonical word representation. In Sect. 4, we investigate the border between the existence and non-existence of injective morphisms from pairs of word (semi)groups to various matrix semigroups. The full version of this paper with additional proof details and appendices is available at [2].

2 Preliminaries

2.1 Words, Semigroups, and Groups

Given an *alphabet* $\Sigma = \{a_1, a_2, \ldots, a_m\}$, a finite *word* u is a finite sequence of letters, $u = u_1 u_2 \cdots u_n$, where $u_i \in \Sigma$. We denote the free semigroup of finite words over Σ by $S(\Sigma)$ where $S(\Sigma) = \langle a_1, \ldots, a_m \rangle$, with $\langle a_1, \ldots, a_m \rangle$ denoting the semigroup *generated* by $a_1, \ldots a_m$. The *empty word* is denoted by ε. The length of a finite word u is denoted by $|u|$ and $|\varepsilon| = 0$. We denote by $\mathrm{F}(\Sigma)$ the free group $\langle a_1, a_1^{-1}, \ldots, a_m, a_m^{-1} \rangle$, i.e., the free group generated by the letters a_i and their inverses a_i^{-1}. Naturally, $a_i a_i^{-1} = a_i^{-1} a_i = \varepsilon$. The elements of $\mathrm{F}(\Sigma)$ are all *reduced* words over Σ, i.e., words not containing $a_i a_i^{-1}$ or $a_i^{-1} a_i$ as a subword. In this context, we call Σ a finite *group alphabet*, i.e., an alphabet with an involution. The multiplication of two elements (reduced words) $u, v \in \mathrm{F}(\Sigma)$ corresponds to the unique reduced word of the concatenation uv. The length of an element of $\mathrm{F}(\Sigma)$ is given by the length of its reduced word representation.

Let φ be a mapping from $\mathrm{S}(\Sigma)$ into $\mathbb{K}^{n \times n}$, where $\mathbb{K}$ is an algebraic structure (e.g., $\mathbb{N}, \mathbb{Z}, \mathbb{Q}, \mathbb{C}$) and $n \geq 1$. That is, a mapping of semigroup words into n-by-n matrices with elements from $\mathbb{K}$. We say that φ is an *injective morphism* or an *embedding* if $\varphi(u)\varphi(v) = \varphi(uv)$ for every $u, v \in \mathrm{S}(\Sigma)$ and if $\varphi(u) = \varphi(v)$ implies that $u = v$. We define *group embeddings* over the free group $\mathrm{F}(\Sigma)$ analogously.

The following proposition is folklore. It allows us to focus on small alphabets, namely unary or binary alphabets, in the sequel.

Proposition 1. *Let $\Sigma_k = \{a_1, a_2, \ldots, a_k\}$ be an alphabet for any $k \geq 2$.*

- *If there exists an embedding $\sigma : \mathrm{S}(\Sigma_2) \to \mathbb{F}^{n \times n}$, then there exists an embedding $\sigma : \mathrm{S}(\Sigma_k) \to \mathbb{F}^{n \times n}$ for any $k \geq 2$.*
- *If there exists a group embedding $\sigma : \mathrm{F}(\Sigma_2) \to \mathbb{F}^{n \times n}$, then there exists a group embedding $\sigma : \mathrm{F}(\Sigma_k) \to \mathbb{F}^{n \times n}$ for any $k \geq 2$.*

2.2 Integer Rings, Euclidean Domains, and Quadratic Fields

We shall assume some familiarity with concepts from algebraic number theory (cf. [25]). We refer the reader to the background material on properties of certain imaginary quadratic fields in [2, Appendix A].

Let $\overline{\mathbb{Q}}$ denote the field of algebraic numbers. Recall that a number is *algebraic* if it is the root of a non-zero polynomial $p \in \mathbb{Z}[x]$ with integer coefficients. Further, a number is an *algebraic integer* if it is the root of a non-zero monic polynomial $p \in \mathbb{Z}[x]$. We can effectively represent and compute algebraic numbers [10].

An *integral domain* is a non-zero commutative ring such that the product of any two non-zero elements is itself non-zero. An integral domain E is Euclidean (or a *Euclidean domain*) if there exists a function $g \colon E \setminus \{0\} \to \mathbb{Z}_{\geq 0}$ such that for every $x, y \in E \setminus \{0\}$,

- $g(xy) \geq g(x)$;
- if $x, y \in E$ and $y \neq 0$. The there exist $q, r \in E$ with $a = qb + r$ and either $r = 0$ or $g(r) < g(b)$.

A function g with such properties is a *Euclidean function* [13,16].

The group of *units* $R^\times$ of a ring R is the subset of elements that possess a multiplicative inverse element in the ring. Two elements $a, b \in R$ of an integral domain are *associates* if there exists a unit $u \in R$ such that $a = bu$.

Recall that a number field $\mathbb{K}$ is *quadratic* (or a *quadratic field*) if there is a square-free integer $-d \in \mathbb{Z}$ such that $\mathbb{K} = \mathbb{Q}(\sqrt{-d})$. If, in addition, $d \in \mathbb{N}$, then the field $\mathbb{Q}(\sqrt{-d})$ is an *imaginary quadratic field*. In the work that follows, we denote by $\mathcal{O}_d$ the ring of algebraic integers in $\mathbb{Q}(\sqrt{-d})$. For $d \in \mathbb{Z}$ square-free, the quadratic field $\mathbb{Q}(\sqrt{-d})$ admits a *field norm* $N \colon \mathbb{Q}(\sqrt{d}) \to \mathbb{Q}$ [25]. This norm is multiplicative, so that $N(\alpha\beta) = N(\alpha)N(\beta)$, for algebraic integers α we have $N(\alpha) \in \mathbb{Z}$, and for imaginary quadratic fields we have $N(\alpha) = |\alpha|^2$. For the ring of algebraic integers in a number field, the units are precisely those integers u for which $N(u) = 1$. The next result precisely characterises imaginary quadratic fields $\mathbb{Q}(\sqrt{-d})$ whose integer rings $\mathcal{O}_d$ are Euclidean domains.

Lemma 2 ([17, **Proposition 8.9**]). *The ring of integers $\mathcal{O}_d$ for the imaginary quadratic field $\mathbb{Q}(\sqrt{-d})$ is a Euclidean domain if and only if $d \in \{1, 2, 3, 7, 11\}$. For each such d, the field norm on $\mathbb{Q}(\sqrt{-d})$ is Euclidean.*

2.3   Matrix Groups

We shall assume some familiarity with (semi)groups of square matrices such as $\mathrm{GL}(2, R) = \{M \in R^{2 \times 2} : \det(M) \in R^{\times}\}$, the *General Linear group* of 2×2 matrices with entries in the ring R whose inverses also have entries in the ring R, the *Special Linear group* $\mathrm{SL}(2, R) = \{M \in \mathrm{GL}(2, R) : \det(M) = 1\}$, and the quotient group $\mathrm{SL}(2, R)/\{\pm \mathrm{Id}_2\} =: \mathrm{PSL}(2, R)$ called the *Projective Special Linear group*.

By definition, we associate to each element $m \in \mathrm{PSL}(2, R)$ a set $\{M, -M\}$ of two matrices in $\mathrm{SL}(2, R)$. Henceforth we employ a slight abuse of notation and write $m = \pm M$, or choose either matrix M or $-M$ to represent m. Intuitively, one can take $\mathrm{PSL}(2, R)$ as $\mathrm{SL}(2, R)$ by ignoring the sign.

In the literature, the *Bianchi groups* are the groups $\mathrm{PSL}(2, \mathcal{O}_d)$ where $\mathcal{O}_d$ is a ring of integers of an imaginary quadratic field $\mathbb{Q}(\sqrt{-d})$ [5]. In the sequel, we are interested in the *Euclidean Bianchi groups* $\mathrm{PSL}(2, \mathcal{O}_d)$ where $d = 1, 2, 3, 7, 11$. The naming convention follows from the result in Lemma 2: of the imaginary quadratic number rings, only $\mathcal{O}_1, \mathcal{O}_2, \mathcal{O}_3, \mathcal{O}_7$, and $\mathcal{O}_{11}$ are Euclidean domains. Observe that $\mathcal{O}_1$ is the Gaussian ring $\mathbb{Z}[\mathrm{i}]$. The group $\mathrm{PSL}(2, \mathbb{Z}[\mathrm{i}])$ is commonly called the Picard group [14, 15].

3   Word Representation Procedure

It is well-known that each of the Euclidean Bianchi groups is finitely generated [15, Theorem 4.3.1]. The groups $\mathrm{PSL}(2, \mathcal{O}_1)$ and $\mathrm{PSL}(2, \mathcal{O}_3)$ admit group presentations with four generators each, whilst $\mathrm{PSL}(2, \mathcal{O}_2)$, $\mathrm{PSL}(2, \mathcal{O}_7)$, and $\mathrm{PSL}(2, \mathcal{O}_{11})$ are each generated by three elements. The matrices associated to each of these generators (commonly denoted by a, t, u, and ℓ) are:

$$A = \begin{pmatrix} 0 & -1 \\ 1 & 0 \end{pmatrix}, \quad T = \begin{pmatrix} 1 & 1 \\ 0 & 1 \end{pmatrix}, \quad U = \begin{pmatrix} 1 & \zeta \\ 0 & 1 \end{pmatrix}.$$

The top-right entry ζ of U depends on the specific matrix group $\mathrm{PSL}(2, \mathcal{O}_d)$; we give further details in [2, Appendix A]. The fourth matrix L is given as follows,

$$L = \begin{cases} \begin{pmatrix} \mathrm{i} & 0 \\ 0 & -\mathrm{i} \end{pmatrix} & \text{if } d = 1, \text{ and} \\[2ex] \begin{pmatrix} \omega^2 & 0 \\ 0 & \omega \end{pmatrix} & \text{if } d = 3 \text{ where } \omega = -\tfrac{1}{2} + \tfrac{\sqrt{3}\mathrm{i}}{2}. \end{cases}$$

The following theorem generalises the procedure in [1, Lemma 3.1] that generates word representations for elements of $\mathrm{PSL}(2, \mathbb{Z})$.

Theorem 3. *There is a procedure that, given an element M in a Euclidean Bianchi group $\mathrm{PSL}(2, \mathcal{O}_d)$, outputs a word representation for M of the form*

$$(L^\epsilon T^{p_0} U^{q_0}) \cdot A T^{p_k} U^{q_k} \cdot A T^{p_{k-1}} U^{q_{k-1}} \cdots A T^{p_1} U^{q_1} \ \text{if } d \in \{1,3\}, \ \text{and}$$
$$(T^{p_0} U^{q_0}) \cdot A T^{p_k} U^{q_k} \cdot A T^{p_{k-1}} U^{q_{k-1}} \cdots A T^{p_1} U^{q_1} \ \text{if } d \in \{2,7,11\}.$$

Here $\epsilon \in \{0,1,2\}$, the exponent pairs $p_\ell, q_\ell \in \mathbb{Z}$ each satisfy $|p_\ell + q_\ell \omega|^2 \le \|M\|$ where $\|M\| := \max_{1 \le i,j \le 2} |M_{ij}|^2$, and $k < 1 - \log_{\kappa(d)} \|M\|$ where $\kappa(d)$ is the Euclidean minimum of $\mathcal{O}_d$ (see [2, Appendix A]). Moreover, this procedure runs in time polynomial in $-\log_{\kappa(d)} \|M\|$.

Proof. For the ease of presentation, we give the procedure for elements of the Picard group $\mathrm{PSL}(2, \mathcal{O}_1)$ $(=\mathrm{PSL}(2, \mathbb{Z}[\mathrm{i}]))$ here. *Mutatis mutandis*, analogous arguments for the remaining Euclidean Bianchi groups are given in [2, Appendix B]. There are two parts to our proof: the construction of the word representation and the polynomial runtime.

Construction of the Word Representation. Suppose that $M = \begin{pmatrix} \alpha & \beta \\ \gamma & \delta \end{pmatrix} \in \mathrm{PSL}(2, \mathbb{Z}[\mathrm{i}])$. Our first step is to construct an element $H \in \mathrm{PSL}(2, \mathbb{Z}[\mathrm{i}])$ such that MH is upper-triangular (thus reducing the representation problem to that of representing an upper-triangular element). The second step constructs the word representation of an upper-triangular element.

We begin with the first step. In the case that $\gamma = 0$, we choose $H = \mathrm{Id}_2$. We continue under the assumption that $\gamma \ne 0$. We claim that there is an $H_1 \in \mathrm{PSL}(2, \mathbb{Z}[\mathrm{i}])$ such that $MH_1 = \left(\begin{smallmatrix} \alpha_1 & \beta_1 \\ \gamma_1 & \delta_1 \end{smallmatrix} \right)$ where $N(\gamma_1) < N(\gamma)$. (Here N is the field norm on $\mathbb{Q}(\mathrm{i})$ with $N(x+y\mathrm{i}) = x^2 + y^2$ for $x+y\mathrm{i} \in \mathbb{Z}[\mathrm{i}]$.) Since $\mathbb{Z}[\mathrm{i}]$ equipped with the field norm N is a Euclidean domain, we can write $\delta = -\theta_1 \gamma + \gamma_1$ with $\theta_1, \gamma_1 \in \mathbb{Z}[\mathrm{i}]$ and so the inequality $N(\gamma_1) = N(\theta_1 \gamma + \delta) < N(\gamma)$ follows. Let us write $\theta_1 = -p_1 - q_1 \mathrm{i} \in \mathbb{Z}[\mathrm{i}]$ in terms of the integral basis $\{1, \mathrm{i}\}$, then

$$MU^{-q_1} T^{-p_1} A = \begin{pmatrix} \alpha & \beta \\ \gamma & \delta \end{pmatrix} \begin{pmatrix} 1 & \mathrm{i} \\ 0 & 1 \end{pmatrix}^{-q_1} \begin{pmatrix} 1 & 1 \\ 0 & 1 \end{pmatrix}^{-p_1} \begin{pmatrix} 0 & -1 \\ 1 & 0 \end{pmatrix}$$
$$= \begin{pmatrix} \alpha & \theta_1 \alpha + \beta \\ \gamma & \theta_1 \gamma + \delta \end{pmatrix} \begin{pmatrix} 0 & -1 \\ 1 & 0 \end{pmatrix} = \begin{pmatrix} \theta_1 \alpha + \beta & -\alpha \\ \theta_1 \gamma + \delta & -\gamma \end{pmatrix} =: \begin{pmatrix} \alpha_1 & \beta_1 \\ \gamma_1 & \delta_1 \end{pmatrix}.$$

Since $N(\gamma_1) < N(\gamma)$, the element $H_1 := U^{-q_1}T^{-p_1}A$ has the claimed properties.

We loop the above construction in order to generate a sequence of matrices of the form $MH_1 \cdots H_\ell$. Let γ_ℓ be the bottom-left entry of matrix $MH_1 \cdots H_\ell$. We repeat this process until we obtain a matrix $MH_1 \cdots H_k$ that satisfies the condition $\gamma_k = 0$.

The following observations guarantee that the above process both terminates and does so correctly. First, $N(\gamma_\ell) = |\gamma_\ell|^2 \in \mathbb{Z}_{\geq 0}$ for each ℓ since $\gamma_\ell \in \mathbb{Z}[\mathrm{i}]$. Second, the sequence of these norms is strictly decreasing (and so $N(\gamma_{\ell+1}) \leq N(\gamma_\ell) - 1$ for $1 \leq \ell \leq k-1$). Hence there exists $k \in \mathbb{N}$ such that $N(\gamma_k) = 0$, from which we deduce that $\gamma_k = 0$. Thus we have constructed an element $H_1 \cdots H_k =: H \in \mathrm{PSL}(2, \mathbb{Z}[\mathrm{i}])$ such that MH is upper-triangular. For our second step, consider

$$MH = \begin{pmatrix} \alpha_k & \beta_k \\ 0 & \delta_k \end{pmatrix} =: \begin{pmatrix} \rho & \sigma \\ 0 & \tau \end{pmatrix}$$

with $\rho, \sigma, \tau \in \mathbb{Z}[\mathrm{i}]$. Since $1 = \det(MH) = \rho\tau$, we deduce that $\rho, \tau \in \mathbb{Z}[\mathrm{i}]^\times = \{\pm 1, \pm\mathrm{i}\}$ and, moreover, $\bar{\tau} = \tau^{-1} = \rho$. The possible pairs (ρ, τ) ensure that

$$MH = \begin{pmatrix} \mathrm{i} & 0 \\ 0 & -\mathrm{i} \end{pmatrix}^\epsilon \begin{pmatrix} 1 & \sigma' \\ 0 & 1 \end{pmatrix} = \begin{pmatrix} \mathrm{i} & 0 \\ 0 & -\mathrm{i} \end{pmatrix}^\epsilon \begin{pmatrix} 1 & 1 \\ 0 & 1 \end{pmatrix}^{p_0} \begin{pmatrix} 1 & \mathrm{i} \\ 0 & 1 \end{pmatrix}^{q_0} = L^\epsilon T^{p_0} U^{q_0}$$

for some $\epsilon \in \{0, 1\}$ and $\sigma' = p_0 + q_0\mathrm{i} \in \mathbb{Z}[\mathrm{i}]$ (an associate of σ).

Taken together, these two procedural steps construct representations for elements $H \in \mathrm{PSL}(2, \mathbb{Z}[\mathrm{i}])$ and $L^\epsilon T^{p_0} U^{q_0}$ for which $M = L^\epsilon T^{p_0} U^{q_0} H^{-1}$. Using these products, we can output a word representation of the desired form for a given M.

Polynomial Runtime. It is useful to introduce the following notations. For $1 \leq \ell \leq k$, we shall write

$$M_\ell := MH_1 \cdots H_\ell = \begin{pmatrix} \theta_\ell \alpha_{\ell-1} + \beta_{\ell-1} & -\alpha_{\ell-1} \\ \theta_\ell \gamma_{\ell-1} + \delta_{\ell-1} & -\gamma_{\ell-1} \end{pmatrix} =: \begin{pmatrix} \alpha_\ell & \beta_\ell \\ \gamma_\ell & \delta_\ell \end{pmatrix}. \tag{1}$$

The update matrix M_ℓ is obtained by performing a Euclidean division in the Gaussian integers $\mathcal{O}_1 = \mathbb{Z}[\mathrm{i}]$. Recall that a division $a = qb + r$ in the Gaussian integers satisfies $N(r) < N(b)$ where N is the associated field norm. In fact, working in $\mathbb{Z}[\mathrm{i}]$ we have the tighter upper bound $N(r) < \frac{1}{2}N(b)$ since $\frac{1}{2}$ is the Euclidean minimum of $\mathbb{Z}[\mathrm{i}]$. Further background on Euclidean minima can be found in [2, Appendix A].

For $\ell < k$, repeated application of the Euclidean minimum gives

$$1 \leq N(\gamma_\ell) < \tfrac{1}{2}N(\gamma_{\ell-1}) < \cdots < \tfrac{1}{2^{\ell-1}}N(\gamma_1) < \tfrac{1}{2^\ell}N(\gamma) = \tfrac{1}{2^\ell}|\gamma|^2 \leq \tfrac{1}{2^\ell}\|M\|.$$

Taking logarithms, we obtain the upper bound $\log_2 \|M\| > \ell$, from which the desired inequality $1 + \log_2 \|M\| > k$ follows.

We now exhibit bounds on the matrix norms $\langle \|M_\ell\|\rangle_{\ell=1}^k$. The determinant condition on M_ℓ tells us that

$$\alpha_\ell\delta_\ell - \beta_\ell\gamma_\ell = -(\theta_\ell\alpha_{\ell-1} + \beta_{\ell-1})\gamma_{\ell-1} + (\theta_\ell\gamma_{\ell-1} + \delta_{\ell-1})\alpha_{\ell-1}$$
$$= -(\theta_\ell\alpha_{\ell-1} + \beta_{\ell-1})\gamma_{\ell-1} + \gamma_\ell\alpha_{\ell-1} = 1.$$

For $\ell - 1 < k$, we have that $\gamma_{\ell-1} \neq 0$ and so

$$|\alpha_\ell|^2 = N(\alpha_\ell) = N(\theta_\ell\alpha_{\ell-1} + \beta_\ell) = N\left(\frac{\gamma_\ell\alpha_{\ell-1} - 1}{\gamma_{\ell-1}}\right) \leq$$
$$\frac{N(\gamma_\ell)N(\alpha_{\ell-1})}{N(\gamma_{\ell-1})} + \frac{1}{N(\gamma_{\ell-1})} \leq \frac{1}{2}N(\alpha_{\ell-1}) + 1. \quad (2)$$

In 2, the rightmost inequality follows from the division $-\delta_{\ell-1} = \theta_\ell\gamma_{\ell-1} + \gamma_\ell$.
We make the following claim.

Claim 1. *For $M_{\ell-1}$ and M_ℓ in $\mathrm{PSL}(2, \mathbb{Z}[i])$ as above, we necessarily have that $\|M_\ell\| \leq \|M_{\ell-1}\|$.*

All that remains is to bound the sizes of the integer exponents in the word representation. For $\ell \leq k$, we bound the quotients $\theta_\ell := -p_\ell - q_\ell i$ as follows,

$$|p_\ell|^2 + |q_\ell|^2 = N(\theta_\ell) = N\left(\frac{\delta_{\ell-1} + \gamma_\ell}{\gamma_{\ell-1}}\right)$$
$$\leq N(\delta_{l-1}) + \frac{N(\gamma_\ell)}{N(\gamma_{\ell-1})} \leq \|M_{\ell-1}\| + \tfrac{1}{2} \leq \|M\| + \tfrac{1}{2}.$$

In the above line, the rightmost inequality follows from Claim 1. Since $N(\theta_\ell), \|M\| \in \mathbb{Z}$, we have $N(\theta_\ell) \leq \|M\|$ and so $|p_\ell|^2, |q_\ell|^2 \leq \|M\|$. We consider the matrix MH in order to bound p_0 and q_0. Once again, it is clear that

$$|p_0|^2, |q_0|^2 \leq |\sigma|^2 \leq \max_{1 \leq i,j \leq 2} |(MH)_{ij}|^2 = \|M_k\| \leq \|M\|.$$

The above observations bound the number of iterations in the initial looping procedure as well as the sizes of the exponents in the word representation. Taken together with standard results on the division algorithm, we obtain the stated polynomial runtime. $\qquad\square$

Proof (Proof of Claim 1). By 1, we have $|\beta_\ell|^2, |\delta_\ell|^2, |\gamma_\ell|^2 \leq \|M_{\ell-1}\|$. Thus $\|M_\ell\| \leq \|M_{\ell-1}\|$ unless $|\alpha_\ell|^2 > \|M_{\ell-1}\|$. Let us assume, for a contradiction, that $|\alpha_\ell|^2 > \|M_{\ell-1}\|$. By 2, it follows that

$$\|M_{\ell-1}\| < \|M_\ell\| = N(\alpha_\ell) \leq \frac{1}{2}N(\alpha_{\ell-1}) + 1 < \frac{1}{2}\|M_{\ell-1}\| + 1,$$

and so we deduce that $\|M_{\ell-1}\| < 2$. Thus $(M_{\ell-1})_{ij} \in \{0, \pm 1, \pm i\}$ for each (i,j).
We first argue that it is not possible that each entry of $M_{\ell-1}$ is non-zero; for otherwise, each entry is a unit and the determinant condition $\alpha_{\ell-1}\delta_{\ell-1} -$

$\beta_{\ell-1}\gamma_{\ell-1} = 1$ is not satisfied by any tuple of units in $\mathbb{Z}[i]^{\times}$. The determinant condition also ensures that $M_{\ell-1} \neq 0_{2\times 2}$. Thus $M_{\ell-1}$ takes one of the following forms

$$\begin{pmatrix} 0 & \beta_{\ell-1} \\ \gamma_{\ell-1} & \delta_{\ell-1} \end{pmatrix}, \begin{pmatrix} 0 & \beta_{\ell-1} \\ \gamma_{\ell-1} & 0 \end{pmatrix}, \begin{pmatrix} \alpha_{\ell-1} & 0 \\ \gamma_{\ell-1} & \delta_{\ell-1} \end{pmatrix}, \text{ or } \begin{pmatrix} \alpha_{\ell-1} & \beta_{\ell-1} \\ \gamma_{\ell-1} & 0 \end{pmatrix}$$

where an entry not denoted by 0 is a unit (i.e., in $\{\pm 1, \pm i\} = \mathbb{Z}[i]^{\times}$). In the first two forms, where $\alpha_{\ell-1} = 0$, it is clear that the update 1 ensures that $|\alpha_\ell|^2 = |\beta_{\ell-1}|^2 > 0$ and so the required inequality trivially holds. Likewise, the inequality holds for the third form since, by the update 1, $|\alpha_\ell|^2 = |\alpha_{\ell-1}|^2$. All that remains is to treat the fourth form. The update to the fourth form gives $\begin{pmatrix} \alpha_\ell & \beta_\ell \\ \gamma_\ell & \delta_\ell \end{pmatrix} = \begin{pmatrix} \beta_{\ell-1} & -\alpha_{\ell-1} \\ 0 & -\gamma_{\ell-1} \end{pmatrix}$ because $0 = \theta_\ell \gamma_{\ell-1} + \gamma_\ell = \theta_\ell \gamma_{\ell-1}$ implies that $\theta_\ell = 0$. Thus $|\alpha_\ell|^2 = |\beta_{\ell-1}|^2 = 1 = |\alpha_{\ell-1}|^2$. Thus the required inequality holds. □

As seen in [1], a procedure that generates a word representation of an element in $\mathrm{PSL}(2, \mathbb{Z})$ implies a procedures that generates the word representation of an element in $\mathrm{SL}(2, \mathbb{Z})$. In a similar vein, we have the following corollary.

Corollary 4. *For each of the groups* $\mathrm{SL}(2, \mathcal{O}_d)$ *with* $d \in \{1, 2, 3, 7, 11\}$, *there is a procedure that, given an element* $M \in \mathrm{SL}(2, \mathcal{O}_d)$ *outputs a word representation for* M *in terms of the generators of* $\mathrm{SL}(2, \mathcal{O}_d)$.

4 Word Embeddings Into Matrix (Semi)Groups

Let us first recall two known group embeddings from the literature (cf. [4, 8]).

Proposition 5. *Let* $\Sigma_2 = \{a, b\}$. *Then* $\varphi : \mathrm{F}(\Sigma_2) \to \mathbb{Z}^{2\times 2}$ *defined by*

$$\varphi(a) = \begin{pmatrix} 1 & 2 \\ 0 & 1 \end{pmatrix}, \quad \varphi(b) = \begin{pmatrix} 1 & 0 \\ 2 & 1 \end{pmatrix}$$

is an embedding.

Proposition 6. *Let* $\Sigma_2 = \{a, b\}$. *Then* $\varphi : \mathrm{F}(\Sigma_2) \to \mathbb{C}^{2\times 2}$ *defined by*

$$\varphi(a) = \begin{pmatrix} \frac{3}{5} + \frac{4}{5}i & 0 \\ 0 & \frac{3}{5} - \frac{4}{5}i \end{pmatrix}, \quad \varphi(b) = \begin{pmatrix} \frac{3}{5} & \frac{4}{5} \\ -\frac{4}{5} & \frac{3}{5} \end{pmatrix}$$

is an embedding.

For both of the above embeddings, all four sets $\{x \in \mathrm{F}(\Sigma_2) : (\varphi(x))_{ij} \neq 0\}$ (where $1 \leq i, j \leq 2$) are non-empty, i.e., the codomains of both embeddings contain matrices with non-zero values in each of the four entries. For comparison, let ψ denote Paterson's classical embedding for pairs of binary semigroup words into $\mathbb{N}^{3\times 3}$ [22]. Then $\{x \in \mathrm{S}(\Sigma_2) \times \mathrm{S}(\Sigma_2) : (\psi(x))_{21} \neq 0\}$ is empty, i.e., Paterson's embedding uses only the elements on and above the main diagonal. Further, it is known that one cannot embed $\mathrm{F}(\Sigma_2)$ into the upper triangular matrices of any dimension. That is to say,

Proposition 7. *There is no embedding from* $F(\Sigma_2)$ *into* $U(n, \mathbb{C})$ *for any* n. *Here* $U(\mathbb{C}, n)$ *is the group of* $n \times n$ *upper triangular matrices with complex entries.*

The result in Proposition 7 (and the generalisation that one cannot embed $F(\Sigma_2)$ into any solvable group) follows from standard results in group theory involving normal series [24, Chapter 5].

We briefly sketch the proof of Proposition 7. Recall that $U(n, \mathbb{C})$ is solvable and that every subgroup of a solvable group is itself solvable [24, Theorem 5.15]. Further, no subgroup of a solvable group is isomorphic to $F(\Sigma_2)$ (cf. [24, Chapter 5.3]). The existence of an embedding $\varphi \colon F(\Sigma_2) \to U(n, \mathbb{C})$ contradicts the above observation. that $F(\Sigma_2)$ is isomorphic to some subgroup of $U(n, \mathbb{C})$.

The analogous result (to Theorem 7) for semigroups is as follows: one cannot embed $S(\Sigma_2)$ into any nilpotent group. This follows from the observation that the number of words in $S(\Sigma_2)$ with length at most n grows exponentially, whilst nilpotent groups exhibit polynomial growth [20].

Motivated by the above existence and non-existence results for embeddings, it is interesting to investigate the (algebraic) complexity of the word structures one can embed into restricted groups and semigroups of low-dimensional matrices. The next two propositions provide a summary of our results. For the rest of the section, we assume that $\Sigma_2 = \{a, b\}$ and $\Sigma_1 = \{c\}$.

First, we consider some existence results. The details of the embeddings build on the well-known embeddings from the literature and can be found in [2, Appendix C].

Proposition 8.   – *There exists an embedding from* $F(\Sigma_1) \times F(\Sigma_1)$ *into* $U(2, \mathbb{N})$.
  – *There exists an embedding from* $F(\Sigma_2) \times F(\Sigma_1)$ *into* $\mathbb{Q}^{2 \times 2}$.
  – *There exists an embedding from* $S(\Sigma_2) \times F(\Sigma_1)$ *into* $\mathbb{Z}^{3 \times 3}$.
  – *There exists an embedding from* $F(\Sigma_2) \times S(\Sigma_1)$ *into* $\mathbb{Z}^{2 \times 2}$.

We now turn our attention to non-existence results in similar settings.

Proposition 9. *Let* $d \in \{1, 2, 3, 7, 11\}$.

  – *There is no embedding from* $S(\Sigma_2) \times F(\Sigma_1)$ *into* $\mathcal{O}_d^{2 \times 2}$.
  – *There is no embedding from* $F(\Sigma_2) \times F(\Sigma_1)$ *into* $\mathcal{O}_d^{2 \times 2}$.
  – *There is no embedding from* $F(\Sigma_2) \times S(\Sigma_1)$ *into* $SL(2, \mathcal{O}_d)$.
  – *There is no embedding from* $S(\Sigma_2) \times S(\Sigma_2)$ *into* $SL(3, \mathcal{O}_d)$.
  – *There is no embedding from* $F(\Sigma_2) \times S(\Sigma_2)$ *into* $\mathcal{O}_d^{3 \times 3}$.

Proof. Let us prove the first and second statements. Proofs of the other three statements use similar approaches akin to techniques found in [8, 18]. Additional details can be found in [2, Appendix C].

Let $\Sigma_2 = \{a, b\}$ and $\Sigma_1 = \{c\}$. Assume, for a contradiction, that there exists an embedding $\varphi : S(\Sigma_2) \times F(\Sigma_1) \to \mathcal{O}_d^{2 \times 2}$. Embedding φ maps the generators as follows:

$$(a, \varepsilon) \mapsto \begin{pmatrix} a_1 & a_2 \\ a_3 & a_4 \end{pmatrix}, \quad (b, \varepsilon) \mapsto \begin{pmatrix} b_1 & b_2 \\ b_3 & b_4 \end{pmatrix},$$

$$(\varepsilon, c) \mapsto \begin{pmatrix} c_1 & c_2 \\ c_3 & c_4 \end{pmatrix}, \quad (\varepsilon, c^{-1}) \mapsto \frac{1}{c_1 c_4 - c_2 c_3} \begin{pmatrix} c_4 & -c_2 \\ -c_3 & c_1 \end{pmatrix}.$$

The last mapping implies that $c_1 c_4 - c_2 c_3 \in \mathcal{O}_d^\times$; that is, $c_1 c_4 - c_2 c_3$ is a unit. Furthermore, we have the following relations:

$$\begin{aligned} \varphi((a, \varepsilon)(b, \varepsilon)) &\neq \varphi((b, \varepsilon)(a, \varepsilon)), \\ \varphi((a, \varepsilon)(\varepsilon, c)) &= \varphi((\varepsilon, c)(a, \varepsilon)), \\ \varphi((b, \varepsilon)(\varepsilon, c)) &= \varphi((\varepsilon, c)(b, \varepsilon)). \end{aligned} \tag{3}$$

Let us consider the products $\varphi((a, \varepsilon)(\varepsilon, c))$ and $\varphi((\varepsilon, c)(a, \varepsilon))$:

$$\varphi((a, \varepsilon)(\varepsilon, c)) = \begin{pmatrix} a_1 c_1 + a_2 c_3 & a_1 c_2 + a_2 c_4 \\ a_3 c_1 + a_4 c_3 & a_3 c_2 + a_4 c_4 \end{pmatrix}, \tag{4}$$

$$\varphi((\varepsilon, c)(a, \varepsilon)) = \begin{pmatrix} a_1 c_1 + a_3 c_2 & a_2 c_1 + a_4 c_2 \\ a_1 c_3 + a_3 c_4 & a_2 c_3 + a_4 c_4 \end{pmatrix}. \tag{5}$$

In order to derive a contradiction, we first assume that $c_2 = 0$. Under this assumption, the top-right corner elements in the matrices $\varphi((a, \varepsilon)(\varepsilon, c))$ and $\varphi((\varepsilon, c)(a, \varepsilon))$ are $a_2 c_4$ and $a_2 c_1$, respectively. By (3), we know these two matrices are equal and so infer that $c_1 = c_4$. Analogously reasoning about the bottom-left corner entries, we have that $a_1 = a_4$. From the third relation in (3), we similarly deduce that $b_1 = b_4$.

We now turn our attention to the bottom-right corner entries in (4) and (5), which, from the preceding work, are $a_1 c_1$ and $a_2 c_3 + a_1 c_1$, respectively. By (3), these entries are equal. Thus we have that either $a_2 = 0$ (and analogously $b_2 = 0$) or $c_3 = 0$. In the latter case, $\varphi(\varepsilon, c) = \begin{pmatrix} c_1 & 0 \\ 0 & c_1 \end{pmatrix}$ and $\varphi(\varepsilon, c^{-1}) = \frac{1}{c_1^2}\begin{pmatrix} c_1 & 0 \\ 0 & c_1 \end{pmatrix}$. This implies that $c_1 \in \mathcal{O}_d^\times$. For each d, the elements of $\mathcal{O}_d^\times$ are listed in [2, Lemma 11]. It is straightforward to see that $(\varphi(\varepsilon, c))^k = \mathrm{Id}_2$ for some $k \geq 1$, which contradicts our assumption that φ is an embedding. As an example, consider $d = 3$ and the unit $\omega = \frac{1}{2} + \frac{\sqrt{-3}}{2}$. Then $\varphi(\varepsilon, c)^6 = \begin{pmatrix} \omega & 0 \\ 0 & \omega \end{pmatrix}^6 = \mathrm{Id}_2$. We deduce that $c_3 \neq 0$.

Let us turn to the second subcase, where we assume that $a_2 = b_2 = 0$. Under this assumption, we obtain the following equality

$$\varphi((a, \varepsilon)(b, \varepsilon)) = \begin{pmatrix} a_1 b_1 & 0 \\ a_3 b_1 + a_1 b_3 & a_1 b_1 \end{pmatrix} = \varphi((b, \varepsilon)(a, \varepsilon)),$$

which contradicts the inequality in (3). We derive an analogous contradiction if we instead work under the assumption that $c_3 = 0$ (rather than $c_2 = 0$).

Observe that in Eqs. (4) and (5), the top-left corner elements are equal if and only if $a_2c_3 = a_3c_2$. The preceding work showed that both c_2 and c_3 are non-zero. We note that a_2 and a_3 (and b_2 and b_3) are also non-zero. If this was not the case, then

$$\varphi((a,\varepsilon)(b,\varepsilon)) = \begin{pmatrix} a_1b_1 & a_1b_2 \\ a_4b_3 & a_4b_4 \end{pmatrix} \quad \text{and} \quad \varphi((b,\varepsilon)(a,\varepsilon)) = \begin{pmatrix} a_1b_1 & a_4b_2 \\ a_1b_3 & a_4b_4 \end{pmatrix} \tag{6}$$

as well as

$$\varphi((a,\varepsilon)(\varepsilon,c)) = \begin{pmatrix} a_1c_1 & a_1c_2 \\ a_4c_3 & a_4c_4 \end{pmatrix} \quad \text{and} \quad \varphi((\varepsilon,c)(a,\varepsilon)) = \begin{pmatrix} a_1c_1 & a_4c_2 \\ a_1c_3 & a_4c_4 \end{pmatrix}. \tag{7}$$

By (3), the two matrices at (6) are not equal, whilst the two matrices at (7) are equal. For both of these statements to be true, both $a_1 \neq a_4$ and $a_1 = a_4$ have to hold simultaneously, a contradiction.

Finally, observe that the top-left corners of the products imply that $a_2c_3 = a_3c_2$, $b_2c_3 = b_3c_2$, and $a_2b_3 \neq a_3b_2$. Since a_2, a_3, b_2, b_3, c_2 and c_3 are all non-zero, we have $\frac{a_2}{a_3} = \frac{c_2}{c_3} = \frac{b_2}{b_3}$ which contradicts the inequality $a_2b_3 \neq a_3b_2$. Thus there is no embedding from $S(\Sigma_2) \times F(\Sigma_1)$ into $\mathcal{O}_d^{2\times 2}$.

The second statement follows directly: if there is no embedding from a semigroup alphabet, then there is also no embedding from a group alphabet. $\square$

Remark 1. *In the proof of Proposition 9, we use the fact that the unit group $\mathcal{O}_d^\times$ is finite. Mutatis mutandis, the same non-existence results hold for any integer ring with a torsion unit group. By Dirichlet's unit theorem (cf. [21, Chapter 1, Theorem 7.4]), a number field $\mathbb{K}$ has a torsion unit group (i.e., every unit has finite order) if and only if $\mathbb{K} = \mathbb{Q}$ or $\mathbb{K} = \mathbb{Q}(\sqrt{-d})$ for square-free $d \geq 1$. Thus the non-existence statements in Proposition 9 for embeddings into $\mathcal{O}_d^{2\times 2}$ with $d \in \{1,2,3,7,11\}$ also hold for the codomains $\mathbb{Z}^{2\times 2}$ and $\mathcal{O}_d^{2\times 2}$ for square-free $d \geq 1$.*

5   Conclusion and Future Directions

In Sect. 3, we presented the first step towards showing the decidability of the identity problem for matrices over $\mathrm{SL}(2, \mathcal{O}_d)$. Indeed, in [1], the word representation algorithm for $\mathrm{SL}(2, \mathbb{Z})$ was used to construct finite petal graphs where the existence of a short path can be checked in NP. The construction utilised a complete, confluent, and monadic term rewriting system for the generators of $\mathrm{SL}(2, \mathbb{Z})$. No such system is known for Bianchi groups. One can obtain either a confluent or a monadic rewriting system, but this is not sufficient for the construction as the finiteness of the petal graph is no longer guaranteed. A pertinent question is whether there exists a confluent and monadic rewriting system for $\mathrm{SL}(2, \mathcal{O}_d)$ or whether one can ensure that the resulting petal graph is finite.

In Sect. 4, we considered various matrix semigroups and showed that there is no way to embed pairs of words over different alphabets. It is interesting to ask, how narrow is the boundary between existence and non-existence? In particular, when considering $\mathbb{K}^{n \times n}$, what is the smallest n and the largest $\mathbb{K}$ such that the embedding no longer exists. It is also worth noting that we considered embeddings of pairs of words. One can extend the problem to the k-fold products, i.e., $\mathrm{S}(\Sigma)^k$ and $\mathrm{F}(\Sigma)^k$; see [7].

Acknowledgements. We are grateful to the anonymous reviewers' careful readings and helpful suggestions that led to many improvements in this manuscript. George Kenison was partially supported by the FWO grants G0F5921N (Odysseus) and G023721N, and by the KU Leuven grant iBOF/23/064. Igor Potapov was supported by the APEX Awards 2024 - APX\R1\241186.

References

1. Bell, P.C., Hirvensalo, M. and Potapov, I.: The identity problem for matrix semigroups in $\mathrm{SL}_2(\mathbb{Z})$ is NP-complete. In: Klein, P.N. (ed.) Proceedings of the Twenty-Eighth Annual ACM-SIAM Symposium on Discrete Algorithms, SODA 2017, Barcelona, Spain, Hotel Porta Fira, January 16–19, pp. 187–206. SIAM (2017)

2. Bell, P.C., Kenison, G., Niskanen, R., Potapov, I., Semukhin, P.: On word representations and embeddings in complex matrices. CoRR arXiv (2026)

3. Bell, P.C., Niskanen, R., Potapov, I., Semukhin, P.: On the identity and group problems for complex heisenberg matrices. In: Bournez, O., Formenti, E., Potapov, I. (eds.) Reachability Problems — 17th International Conference, RP 2023, Nice, France, 11–13 October 2023, Proceedings. LNCS,vol. 14235, pp. 42–55. Springer (2023). https://doi.org/10.1007/978-3-031-45286-44

4. Bell, P.C., Potapov, I.: Reachability problems in quaternion matrix and rotation semigroups. Inf. Comput. **206**(11), 1353–1361 (2008)

5. Bianchi, L.: Sui gruppi di sostituzioni lineari con coefficienti appartenenti a corpi quadratici immaginarî. Math. Ann. **40**(3), 332–412 (1892)

6. Button, J.O.: Groups and Embeddings in SL(2;C). Commun. Algebra **44**(1), 265278 (2016). https://doi.org/10.1080/00927872.2014.975347

7. Campagnolo, C., Kammeyer, H.: Products of free groups in Lie groups. J. Algebra **579**, 237–255 (2021)

8. Cassaigne, J., Harju, T., Karhumäki, J.: On the undecidability of freeness of matrix semigroups. Int. J. Algebra Comput **9**(03n04), 295–305 (1999)

9. Cassaigne, J., Nicolas, F.: On the decidability of semigroup freeness. RAIRO Theor. Inf. Appl. **46**(3), 355–399 (2012)

10. Cohen, H.: A course in computational algebraic number theory, vol. 138, Springer (2013)

11. Dong, R.: Semigroup algorithmic problems in metabelian groups. In: Mohar, B., Shinkar, I., O'Donnell, R. (eds.) Proceedings of the 56th Annual ACM Symposium on Theory of Computing, STOC 2024, Vancouver, BC, Canada, 24–28 June 2024, pp. 884–891. ACM (2024)

12. Dong, R.: Semigroup intersection problems in the Heisenberg groups. SIAM J. Discret. Math. **38**(4), 3176–3197 (2024)

13. Dubois, D.W., Steger, A.: A note on division algorithms in imaginary quadratric number fields. Canadian J. Math. **10**, 285–286 (1958)

14. Fine, B.: The Picard group and the modular group, pp. 150–163. London Mathematical Society Lecture Note Series, Cambridge University Press (1987)
15. Fine, B.: Algebraic theory of the Bianchi groups. Monographs and Textbooks in Pure and Applied Mathematics, vol. 129. Marcel Dekker Inc, New York (1989)
16. Hardy, G.H., Wright, E.M.: An introduction to the theory of numbers. Oxford University Press, Oxford, 6th edn: Revised Heath-Brown, D.R., Silverman, J.H. With a foreword by Andrew Wiles (2008)
17. Hatcher, A.: Topology of Numbers. American Mathematical Society, Providence, RI (2022)
18. Ko, S.K., Niskanen, R., Potapov, I.: On the identity problem for the special linear group and the Heisenberg group. In: Proceedings of ICALP 2018, vol. 107 of LIPIcs, pp. 132:1–132:15 (2018)
19. Magnus, W., Karrass, A., Solitar, D.: Combinatorial Group Theory: Presentations of Groups in Terms of Generators and Relations. Interscience Publishers, Wiley, New York-London-Sydney (1966)
20. Milnor, J.: Growth of finitely generated solvable groups. J. Differ. Geom. **2**(4), (1968)
21. Neukirch, J.: Algebraic number theory. Grundlehren der mathematischen Wissenschaften, vol. 322, Springer (1999). Fundamental Principles of Mathematical Sciences
22. Paterson, M.S.: Unsolvability in 3×3 matrices. Stud. Appl. Math. **49**(1), 105 (1970)
23. Potapov, I., Semukhin, P.: Vector and scalar reachability problems in $SL(2,\mathbb{Z})$. J. Comput. Syst. Sci. **100**, 30–43 (2019)
24. Rotman, J.J.: An Introduction to the Theory of Groups. Springer, New York (1995)
25. Stewart, I., Tall, D.: Algebraic Number Theory and Fermat's Last Theorem, 4th edn. CRC Press, Boca Raton, FL (2016)

Generation and Enumeration of Floorplans Determined by HV-Matrices

Andrea Frosini[1]($\boxtimes$) , Shin-Ichi Nakano[2] , and Simone Rinaldi[3]

[1] Department of Mathematics and Informatics, University of Florence,
Florence, Italy
andrea.frosini@unifi.it
[2] Department of Computer Science, Gunma University, Maebashi City, Japan
[3] Department of Information Engineering and Mathematics, University of Siena,
Siena, Italy

Abstract. We study rectangulations of an axis-aligned rectangle in the integer lattice and the associated notions of weak and strong equivalence, which give rise to mosaic floorplans and floorplans. Using the representation via HV-matrices, we describe how such matrices generate rectangulations and investigate the correspondence between matrices, associated HV-words, and floorplans.

Our main objectives are twofold. First, given a diagonal HV-matrix M, we provide a constructive procedure to generate all floorplans related to M and determine their cardinality. Second, given a family $\mathcal{M}$ of diagonal HV-matrices, we enumerate the floorplans generated by its elements. To obtain these results, we develop suitable generating trees that describe recursively the structure of floorplans induced by diagonal HV-matrices and some of their subfamilies. This approach leads to explicit formulas for their enumeration.

Keywords: floorplans · permutations · pattern avoidance · enumerative combinatorics

1 Introduction

Let R be an axes-aligned rectangle in the integer lattice $\mathbb{Z} \times \mathbb{Z}$ and n a positive integer number. We call a *rectangulation of R* a partition $\mathfrak{R}$ of R into finitely many interior-disjoint rectangles such that no four rectangles share a common vertex. $\mathfrak{R}$ has *size n* when it consists in exactly n interior rectangles. We will often refer to the interior-disjoint rectangles within R as *rooms*, and the line segments that form the boundaries of each room as *walls*.

We introduce the *leftright* and *abovebelow* order relations between two rooms r and $\tilde{r}$ of a rectangulation $\mathfrak{R}$. We say that:

M.-P. Béal and P. Caron (Eds.): DLT 2026, LNCS 16578, pp. 335–348, 2026.
https://doi.org/10.1007/978-3-032-28404-4_25

- r is *on the left* $\tilde{r}$ (equivalently, $\tilde{r}$ is *on the right* of r) if there is a sequence of rooms of $\mathfrak{R}$, $r = r_1, r_2, \ldots, r_k = \tilde{r}$ such that the right side of r_i and the left side of r_{i+1} intersect, for $i = 1, 2, \ldots, k - 1$;
- r is *above* $\tilde{r}$ (equivalently, $\tilde{r}$ is *below* r) if there is a sequence of rooms of $\mathfrak{R}$, $r = r_1, r_2, \ldots, r_k = \tilde{r}$ such that the upper side of r_i and the bottom side of r_{i+1} intersect, for $i = 1, 2, \ldots, k - 1$.

So, given a room r, the outer rectangle R can be partitioned into four regions (possibly empty), that group the rooms on the left, on the right, above or below r. Two rectangulations $\mathfrak{R}_1$ and $\mathfrak{R}_2$ for a rectangle R are said to be *weakly equivalent* if they induce the same left–right and above–below order relations among their rooms (see Fig. 1). Two rectangulations $\mathfrak{R}_1$ and $\mathfrak{R}_2$ are *strongly equivalent* if they are weakly equivalent and preserve all adjacency relations between rooms.

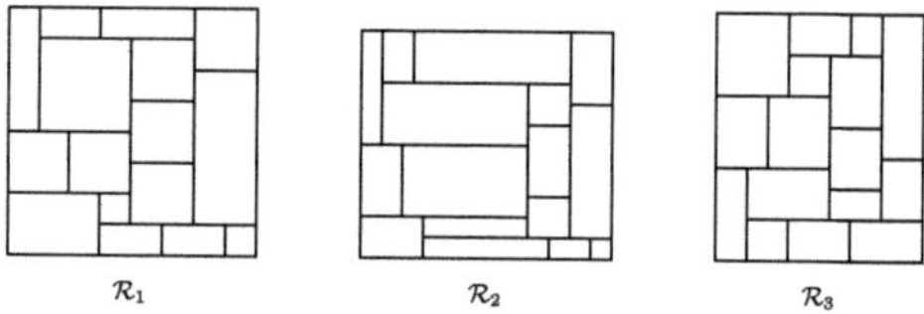

Fig. 1. Three weakly equivalent rectangulations; $\mathcal{R}_1$ and $\mathcal{R}_2$ are strong equivalent.

A *floorplan* (resp. *mosaic floorplan*) is an equivalence class under the strong (resp. weak) equivalence relation, and is typically represented by one of the rectangulations within that class.

Several research fields benefit from studies on floorplans: mainly, but not exclusively the wide area of computer science, from VLSI to data base design, and graph representation (see [9] for a review).

Following [12,13], let $M = (m_{ij})$ be an $n \times n$ *HV–matrix*, that is, a matrix with entries in $\{H, V, 0\}$ such that each row and each column contains exactly one nonzero entry. To each nonzero entry m_{ij} we associate a rectilinear segment in the square $[0, n+1] \times [0, n+1]$ as follows: if $m_{ij} = H$, we associate a horizontal segment lying in row i and intersecting column j, while if $m_{ij} = V$, we associate a vertical segment lying in column j and intersecting row i.

The endpoints of each segment are determined by extending it maximally until it meets either the boundary of the rectangle or another segment of the opposite orientation.

The collection of all such segments induces subdivisions of the rectangle into $n + 1$ interior-disjoint axis-aligned rectangles. This subdivision is a rectangulation, and the strong equivalence classes of these rectangulations are the *floorplan generated by* M. We indicate by $F(M)$ the set of rectangulations (up to strong equivalence) generated by M (Fig. 2).

In this paper we investigate some purely combinatorial problems that are particularly noteworthy, concerning HV-matrices, associated binary words, and floorplans, precisely the following:

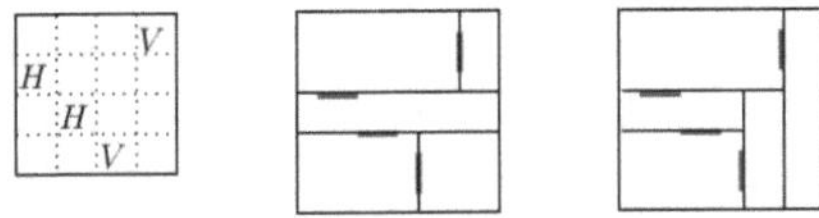

Fig. 2. Two floorplans generated by the HV-matrix on the left.

1. Given a HV-matrix M, we want to provide a generating procedure to compute $F(M)$ and to determine its cardinality for any n.
2. Given a family of HV-matrices $\mathcal{M}$, how many different floorplans of size n are generated by the elements of $\mathcal{M}$? We denote $F(\mathcal{M})$ the set of floorplans generated by any $M \in \mathcal{M}$.

Precisely, we will provide a characterization of floorplans generated by some subfamilies of diagonal HV-matrices (defined in the next section), and provide exact formulas for the number of different floorplans generated by a diagonal HV-matrix. The following researches were inspired by the study of F. Sciammacca, whose master thesis [15] provided some interesting hints for the development of the involved combinatorial structures.

2 Preliminaries

In this section we recall some notions and results which will be useful in the following.

Mosaic floorplans and diagonal rectangulations. Let $\mathcal{R}$ be a rectangulation of size n. The union of the transitive relations *left* and *below* provides a total order on the rooms. Hence, we can label them by the numbers from 1 to n, according to this order, say *SWNE labeling*. The rectangle with label j ($1 \leq j \leq n$) will be denoted by r_j.

A *diagonal rectangulation* of size n is a rectangulation $\mathcal{D}$ of size n, drawn on a rectangle R of size n, such that every rectangle of $\mathcal{D}$ intersects the SW–NE diagonal of S. Diagonal rectangulations have the following properties [1]:

1. Every rectangulation $\mathcal{R}$ is weakly equivalent to a unique diagonal rectangulation $\mathcal{D}$ (up to strong equivalence).
2. In a diagonal rectangulation we have the following: for every horizontal segment s, all the vertical segments above s occur to the right of all the vertical segments below s; and for every vertical segment t, all the horizontal segments on the left of t occur below all the horizontal segments on the right of t.
3. The order in which the SW–NE diagonal of R meets the rectangles of a diagonal rectangulation is the SW–NE order.

According to the statement above, every mosaic floorplan can be represented by a unique diagonal rectangulation (up to strong equivalence), called the *diagonal representative* of $\mathcal{R}$.

Baxter numbers. Yao et al. [16] proved that weak rectangulations (mosaic floor-plans) are enumerated by Baxter numbers B_n that are identified as sequence A001181 in [14]. These numbers are classically related to *Baxter permutations*, which can be defined as permutations avoiding the two vincular patterns $2-41-3$ and $3-14-2$. Since [8], an explicit formula for $B(n)$, with $n \geq 1$, is known. In [5], M. Bousquet-Mélou provides a recursive method for the enumeration of Baxter permutations via a two-labels generating tree.

Generating trees. Let $\mathcal{C}$ be a combinatorial class, i.e. a set of discrete objects endowed with a notion of size such that the number of objects of size n is finite for every n, and assume that $\mathcal{C}$ contains a unique object of size 1. A *generating tree* for $\mathcal{C}$ is an infinite rooted tree whose vertices are the objects of $\mathcal{C}$ (each appearing exactly once), and such that objects of size n lie at level n (the root being at level 1) (see [2,3]). The children of an object $c \in \mathcal{C}$ are obtained by adding an *atom* (i.e. a piece increasing the size by 1) according to prescribed rules ensuring the unique appearance property. As a classical example, consider Dyck paths. A Dyck path of semi-length n is a lattice path from $(0,0)$ to $(2n,0)$ with steps $U = (1,1)$ and $D = (1,-1)$, staying weakly above the x-axis. The atoms are UD factors (peaks), inserted only along the *last descent*, i.e. the longest suffix consisting of D steps. If $w = u\,UD^k$, then its children are

$$\phi_{Cat}(w) = u\,UUDD^k,\ u\,UDUDD^{k-1}, \ldots,\ u\,UD^{k-1}UDD,\ u\,UD^kUD.$$

When the growth is sufficiently regular, it can be encoded by a *succession rule*, consisting of an axiom (the label of the root) and productions describing how labels propagate. For Dyck paths, the controlling statistic is the length of the last descent, the axiom is (1) and the production is $(k) \rightarrow (1), (2), \ldots, (k+1)$

where the $\rightarrow$ notation indicates that an object of label k produces $k+1$ objects of immediately greater size and labels $(1), (2), \ldots, (k+1)$. Dyck paths are counted by the Catalan numbers (sequence A000108 in [14]).

3   Diagonal HV-Matrices and Their Floorplans

Given two HV-matrices A and B we say that they are *f-equivalent* if they generate all the same floorplans. For instance, the reader can check that the two patterns $\begin{pmatrix} & V & \\ H & & \\ & H & \end{pmatrix}$ and $\begin{pmatrix} & V & \\ H & & \\ H & & \end{pmatrix}$ are *f*-equivalent. It is natural to seek conditions that guarantee two sets of patterns are *f*-equivalent. Let A and B be two HV-matrices, $n \times n$ and $m \times m$, respectively. The *direct sum* $A \oplus B$ is the $(n+m) \times (n+m)$ block matrix $A \oplus B = \begin{pmatrix} 0_{n\times m} & B \\ A & 0_{m\times n} \end{pmatrix}$, where $0_{n\times m}$ denotes the $n \times m$ matrix with all entries equal to zero. We can easily prove that:

Proposition 1. *Let A and A' be two f-equivalent $n \times n$ HV-matrices, and B a $m \times m$ HV-matrix, then $A \oplus B$ is f-equivalent to $A' \oplus B$.*

The previous result can be easily extended to the case of HV-matrices which are direct sum of HV-matrices, each containing only one entry. Therefore, we start our investigation focusing on the family of *diagonal HV-matrices*, denoted by $\mathcal{D}$, i.e., those HV-matrices where non zero entries lie only on the main diagonal. This family is trivially f-equivalent to the larger family of all HV-matrices M obtained as the direct sum:

$$M = W_1 \oplus W_2 \oplus \ldots \oplus W_k,$$

where the non zero entries of each matrix W_i are either only H, say H-matrix, or only V, say V-matrix.

The aim of the paper is the characterization of the floorplans generated by the diagonal HV-matrices and some of their subfamilies. In particular, we first consider the class of diagonal HV-matrices.

Theorem 1. *Let $\mathfrak{R}_1$ and $\mathfrak{R}_2$ be two rectangulations generated by the same diagonal HV-matrix $M \in \mathcal{D}$. If $\mathfrak{R}_1$ and $\mathfrak{R}_2$ are weakly equivalent, then they are also strongly equivalent.*

Proof (Sketch of proof). Since M is diagonal, every segment introduced in the construction is uniquely associated with a specific row or column, and the induced walls form a staircase structure with no crossings.

The weak equivalence assumption implies that all left–right and above–below relations among rooms coincide in $\mathfrak{R}_1$ and $\mathfrak{R}_2$. For diagonal matrices, these order relations uniquely determine adjacency relations between rooms: indeed, whenever two rooms are consecutive in the partial order, they must share a wall.

Hence all adjacencies are preserved, which implies strong equivalence.   □

Theorem 2. *Every mosaic floorplan is generated by exactly one diagonal HV-matrix.*

Proof (Sketch of proof). Let $\mathcal{F}$ be a mosaic floorplan, i.e. a weak equivalence class of rectangulations. By a standard normalization procedure, $\mathcal{F}$ admits a unique representative $\mathfrak{R}$ called its *diagonal rectangulation*, where all interior walls are arranged in monotone staircase form. From this diagonal rectangulation, we construct an HV-matrix M as follows: for each horizontal wall crossing the i-th row, set $m_{ii} = H$, while for each vertical wall crossing the i-th column, set $m_{ii} = V$. This procedure is algorithmic and invertible. Therefore, M is uniquely determined and generates exactly the given mosaic floorplan.   □

Theorem 3. *The set of floorplans generated by diagonal HV-matrices coincides with the set of mosaic floorplans.*

A generating tree to generate mosaic floorplans associated with a HV-matrix. Let $F(\mathcal{D})$ denote the set of floorplans generated by diagonal HV-matrices, which, by Theorem 3, coincides with the family of mosaic floorplans. We now describe a recursive method for generating mosaic floorplans via a generating tree, which keeps track of the HV-matrix associated with each generated floorplan.

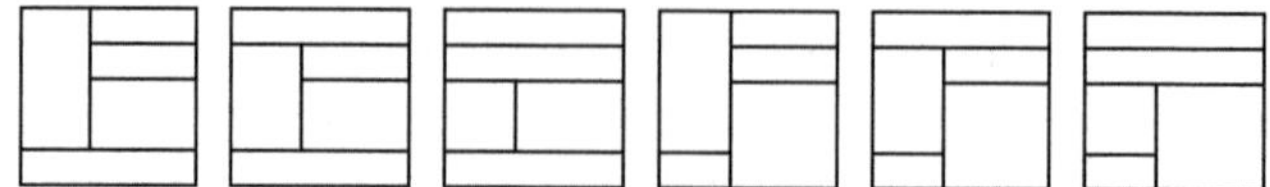

Fig. 3. The six mosaic floorplans with HV-word $w(M) = HVHH$.

For convenience, we represent a diagonal HV-matrix M of size n as a binary HV-word, called the *HV-word* $w = w(M) \in \{H, V\}^n$ (see Fig. 3).

Let $\mathcal{F} \in F(\mathcal{D})$ be a floorplan of size n, with HV-word $w = w_1 \ldots w_n$. We assign to $\mathcal{F}$ a label that determines how $\mathcal{F}$ grows in the generating tree. Specifically, the label of $\mathcal{F}$ is $(h, v)_{w_n}$, where v (resp. h) denotes the number of vertical (resp. horizontal) segments incident to the top *horizontal* (resp. right *vertical*) segments of $\mathcal{F}$, and $w_n \in \{H, V\}$ is the last entry of w.

The generating tree of $F(\mathcal{D})$ is constructed as follows. The root of the tree is the unit rectangle, considered as a special case with label $(0, 0)_\epsilon$ and HV-word equal to the empty word ϵ.

Assume $\mathcal{F}$ to be a mosaic floorplan having label $(h, v)_{w_n}$, HV-word w, and $P_1, \ldots, P_v$ (resp. $Q_1, \ldots, Q_h$) to be the sequence of points where the internal vertical (resp. horizontal) segments of F meet the top (resp. right) segments of F, listed from right to left (resp. from top to bottom).

We define two production operators acting on $\mathcal{F}$:

φ_H: **(i)** It produces h floorplans by inserting a rectangle at the top $\mathcal{F}$ (called an *h-rectangle*) that spans from the north-east corner of $\mathcal{F}$ to the point P_i. The labels of the resulting floorplans are: $(1, v + 1)_H, \ldots, (h, v + 1)_H$.
  (ii) It produces one additional floorplan by inserting an h-rectangle at the top of $\mathcal{F}$, spanning completely from right to left, with label $(0, v + 1)_H$. The $h + 1$ mosaic floorplans obtained from φ_H have HV-word $w' = wH$.

φ_V: **(i)** It produces v floorplans by inserting a rectangle at the right of $\mathcal{F}$ (called a *v-rectangle*) that spans from the north-east corner of $\mathcal{F}$ to the point Q_i. The labels of the resulting floorplans are: $(h+1, 1)_V, \ldots, (h+1, v)_V$.
  (ii) It produces one additional floorplan by inserting a v-rectangle at the right of $\mathcal{F}$, spanning completely from top to bottom, with label $(h + 1, 0)_V$. The $v + 1$ mosaic floorplans obtained from φ_V have HV-word $w' = wV$.

Thus, the children of $\mathcal{F}$ are the $h + v + 2$ floorplans obtained by applying φ_H and φ_V to $\mathcal{F}$. The succession rule associated with the generating tree of mosaic floorplans can be written as:

$$
\Omega_{Mos} \begin{cases}
(0, 0)_\epsilon \\
(0, 0)_\epsilon \rightarrow (0, 1)_H \, (1, 0)_V \\
(h, v)_H \rightarrow (0, v + 1)_H, (1, v + 1)_H, \ldots, (h, v + 1)_H, \\
\qquad (h + 1, 0)_V, (h + 1, 1)_V, \ldots, (h + 1, v)_V \\
(h, v)_V \rightarrow (h + 1, 0)_V, (h + 1, 1)_V, \ldots, (h + 1, v)_V, \\
\qquad (0, v + 1)_H, (1, v + 1)_H, \ldots, (h, v + 1)_H.
\end{cases}
$$

The first levels of the generating tree are shown in Fig. 4.

It is worthwhile noting that, if we forget about the labels H and V, we get the known rule for Baxter permutations [5].

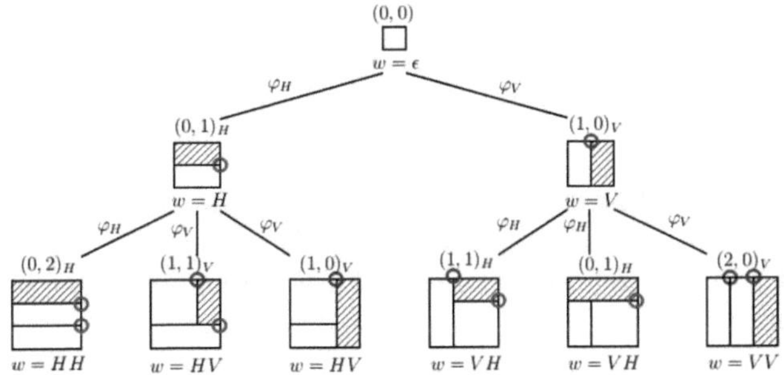

Fig. 4. The first three levels of the generating tree of mosaic floorplans with the corresponding HV-words.

For a word $w \in \{H, V\}^n$, say HV-word, let us call the *complexity* of w the number $f(w)$ of mosaic floorplans at level n of the generating tree of Ω_{Mos} with HV-word equal to w. Our goal is to derive a formula for the complexity $f(w)$ of $w \in \{H, V\}^n$, for specific classes of diagonal HV-matrices (HV-word). So, let $w = w_1 w_2 \ldots w_n \in \{H, V\}^n$. By symmetry reasons we will assume wlog that the first entry w_1 of w is H.

Exact Formulas for Two and Three Blocks. We start by considering the simplest cases of $w = H^{h_1} V^{v_1}$ and $w = H^{h_1} V^{v_1} H^{h_2}$, $h_1, h_2, v_1 \geq 0$.

Proposition 2. *Let $h_1, v_1 \geq 1$. For $w = H^{h_1} V^{v_1}$, we have:*

$$f(H^{h_1} V^{v_1}) = \binom{h_1 + v_1}{v_1}.$$

Proof. Start at the root which is the unique node $(0, 1)$. After reading the initial block H^{h_1} we are at a unique H-node whose second coordinate equals h_1 and whose first coordinate is 0. Each subsequent single V-step replaces a node (h, v) by the V-nodes

$$(h + 1, 0)_V, (h + 1, 1)_V, \ldots, (h + 1, v)_V.$$

Thus, after v_1 consecutive V-steps every final V-node is determined by a nondecreasing sequence:

$$0 \leq i_1 \leq i_2 \leq \cdots \leq i_{v_1} \leq h_1,$$

where i_t indicates which index among $0, \ldots, h_1$ was picked at the t-th V-step. The number of such nondecreasing sequences of length v_1 with values in $\{0, \ldots, h_1\}$ is the stars-and-bars count

$$\binom{(h_1 + 1) + v_1 - 1}{v_1} = \binom{h_1 + v_1}{v_1}.$$

Hence the claimed formula. $\qquad\square$

The argument used above can be further generalized to the case three-blocks.

Proposition 3. *Let $h_1, v_1, h_2 \geq 1$. For $w = H^{h_1} V^{v_1} H^{h_2}$, the number of nodes with label sequence w is*

$$f(H^{h_1} V^{v_1} H^{h_2}) = \binom{h_1 + v_1}{v_1} \binom{v_1 + h_2}{h_2}.$$

The proof is omitted here for brevity, as it relies on combinatorial arguments analogous to those used in the previous proof. However, this technique does not readily extend to more general settings, and different methods are required to handle the remaining cases.

Alternating HV-Matrices. A special subclass of diagonal HV-matrices is given by the *alternating HV-matrices*, namely those diagonal HV-matrices whose diagonal entries alternate between H and V. For the sake of simplicity we include the empty matrix. We denote by $\mathcal{A}$ the set of alternating HV-matrices. The characterization of the corresponding floorplans is rather straightforward.

Proposition 4. *A floorplan is generated by an alternating HV-matrix if and only if it contains neither two adjacent h-rectangles nor two adjacent v-rectangles.*

As in the previous case, we use the *alternating word* representation for an alternating HV-matrix, and we may assume without loss of generality that the first entry is H. Thus, the construction presented for mosaic floorplans can be restricted to the case of alternating words. Here, we benefit from the fact that for any $n \geq 0$ there is a unique (empty if $n = 0$) alternating word $w = w(n) = HVHVHV \cdots$ of length n, so the number $f(w)$ simply corresponds to the number of nodes in the generating tree obtained by restricting the generating tree for mosaic floorplans to allow, at each level, only one of the two operators φ_H or φ_V. More precisely, assuming that the root is at level 1, we apply φ_H at odd levels and φ_V at even levels (see Fig. 5). This recursive construction yields the following generating tree:

$$\Omega_A \begin{cases} (0,0)_\epsilon \\ (0,0)_\epsilon \to (0,1)_H \\ (h,v)_H \to (h+1,0)_V, \ (h+1,1)_V, \ \ldots \ (h+1,v)_V \\ (h,v)_V \to (0,v+1)_H, \ (1,v+1)_H, \ \ldots \ (h,v+1)_H. \end{cases}$$

For $n \geq 1$, let $a(n)$ be the number of nodes at level n of the generating tree, i.e. the number of floorplans with n rooms and alternating HV-word $w(n-1)$, i.e. $a(n) = f(w(n-1))$.

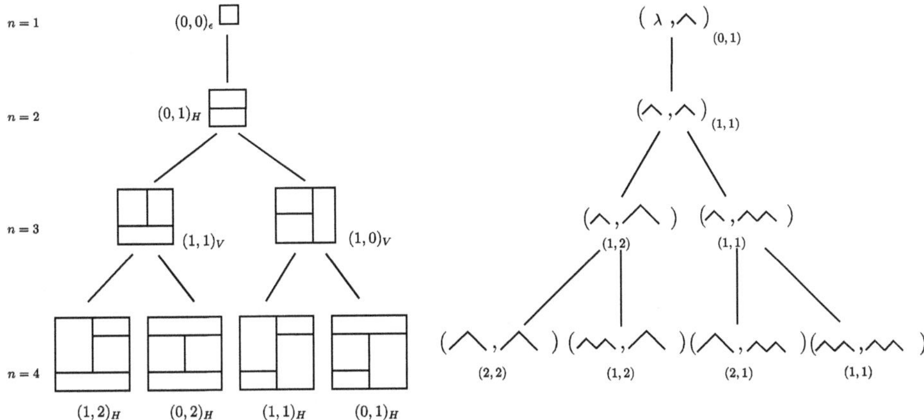

Fig. 5. Left: The first few levels of the generating tree of floorplans generated by alternating words; Right: The isomorphic generating tree of pairs of Dyck paths.

Theorem 4. *For all integers $m \geq 0$,*

$$a(2\,m) = C(m)^2 \qquad and \qquad a(2\,m+1) = C(m)\,C_(m+1),$$

where $C(m)$ denotes the m-th Catalan number, $C(m) = \dfrac{1}{m+1}\dbinom{2\,m}{m}$. Equivalently, $a(n) = C(\lfloor n/2\rfloor)\,C(\lceil n/2\rceil)$ (sequence A005817 in [14])

Proof. Let $\mathcal{D}_r$ denote the set of Dyck paths of semilength r. Recall from the introductory section the construction of Dyck paths via the operator ϕ_{Cat} and its associated generating tree Ω_{Cat}, and recall that $|\mathcal{D}_r| = C_r$. For $n \geq 1$, let $\mathcal{DD}_n$ be the family of ordered pairs (D_1, D_2) of Dyck paths such that

$$\begin{cases} D_1, D_2 \in \mathcal{D}_{n/2}, & \text{if } n \text{ is even} \\ D_1 \in \mathcal{D}_{\lfloor n/2\rfloor}, \ D_2 \in \mathcal{D}_{\lceil n/2\rceil}, & \text{if } n \text{ is odd.} \end{cases}$$

Let $\mathcal{DD} = \bigcup_{n\geq 1} \mathcal{DD}_n$. We construct a bijection between floorplans of size n generated by alternating HV-matrices of size $n-1$ and pairs $(D_1, D_2) \in \mathcal{DD}_n$. The construction is recursive and proceeds in two steps: first, we define a generating tree $\Omega_{\mathcal{DD}}$ for $\mathcal{DD}$; then, we show that it is isomorphic to the generating tree Ω_A of floorplans generated by alternating matrices.

Given a pair $(D_1, D_2) \in \mathcal{DD}_n$, we associate to it the label (r, s), where r (resp. s) denotes the length of the last descent of D_1 (resp. D_2). We define the operator $\phi_{\mathcal{DD}}$ by

$$\phi_{\mathcal{DD}}(D_1, D_2) = \begin{cases} \{(D_1, D_2') : D_2' \in \phi_{\mathrm{Cat}}(D_2)\}, & \text{if } n \text{ is even} \\ \{(D_1', D_2) : D_1' \in \phi_{\mathrm{Cat}}(D_1)\}, & \text{if } n \text{ is odd.} \end{cases}$$

Thus, at even levels the second component is grown according to the usual Dyck path rule, while at odd levels the first component is grown. In particular, at even levels the two Dyck paths have the same semilength $n/2$, whereas at odd levels the semilength of the second path exceeds that of the first by one. The root of the generating tree $\Omega_{\mathcal{DD}}$ is placed at level 0 and consists of the pair (λ, λ) of empty Dyck paths, labelled $(0,0)$. It produces at level 1 the pair (λ, UD), labelled $(0,1)$. Using the standard succession rule for Dyck paths, we obtain the following succession rule for $\Omega_{\mathcal{DD}}$:

$$\Omega_{\mathcal{DD}} \begin{cases} (0,0) \\ (r,s) \to (r,1), \ldots, (r,s+1), \text{ if } (r,s) \text{ is at even level} \\ (r,s) \to (1,s), \ldots, (r+1,s), \text{ if } (r,s) \text{ is at odd level.} \end{cases}$$

We now consider the generating tree $\Omega^+_{\mathcal{DD}}$ obtained from $\Omega_{\mathcal{DD}}$ by taking (λ, UD), labelled $(0,1)$, as the root. Equivalently, $\Omega^+_{\mathcal{DD}}$ is the subtree of $\Omega_{\mathcal{DD}}$ starting from level 1. This normalization aligns the construction with the generating tree Ω_A of floorplans (see Fig. 5).

We claim that $\Omega^+_{\mathcal{DD}}$ is isomorphic to Ω_A. Recall that in Ω_A the root is the node $(0,0)_\varepsilon$ at level 1, which produces the node $(0,1)_H$ at level 2. Levels then alternate: even levels are of type H, odd levels of type V, and the production rules are

$$(h,v)_H \to (h+1,0)_V, \ (h+1,1)_V, \ \ldots, \ (h+1,v)_V$$
$$(h,v)_V \to (0,v+1)_H, \ (1,v+1)_H, \ \ldots, \ (h,v+1)_H.$$

Let (D_1, D_2) be a node of $\Omega^+_{\mathcal{DD}}$ at level n with label (r,s). We distinguish two cases.

Case 1: n even. Then (r,s) produces the nodes $(r,s) \to (r,1), \ldots, (r,s+1)$. The corresponding node in Ω_A has label $(r-1,s)_H$ and produces $(r-1,s)_H \to (r,0)_V, \ (r,1)_V, \ \ldots, \ (r,s)_V$.

Case 2: n odd. Then (r,s) produces $(r,s) \to (1,s), \ldots, (r+1,s)$. The corresponding node in Ω_A has label $(r,s-1)_V$ and produces $(r,s-1)_V \to (0,s)_H, \ \ldots, \ (r,s)_H$. In both cases, the number of children coincides, and the correspondence between labels is given by

$$(r,s) \in \Omega^+_{\mathcal{DD}} \longleftrightarrow (r-1,s)_H \in \Omega_A, \text{ if } n \text{ is even,}$$
$$(r,s) \in \Omega^+_{\mathcal{DD}} \longleftrightarrow (r,s-1)_V \in \Omega_A, \text{ if } n \text{ is odd.}$$

This establishes an isomorphism between the two generating trees, so a bijection between floorplans of size n and pairs of Dyck paths in $\mathcal{DD}_n$. Consequently, $a(n) = |\mathcal{DD}_n| = C(\lfloor n/2 \rfloor) \, C(\lceil n/2 \rceil)$. $\qquad\square$

4   An Effective Method to Compute the Complexity of a Diagonal HV-Matrix

Let us go back to consider the general problem of determining $f(w)$ for a generic $w \in \{H, V\}^n$, using the generating tree Ω_{Bax} of diagonal mosaic floorplans. We proved that the number of nodes at level n of the generating tree is the nth Baxter number $B(n)$. For $n \geq 2$, it holds:

$$B(n) = 2 \cdot \sum_{\substack{w \in \{H,V\}^{n-1} \\ w_1 = H}} f(w) \tag{1}$$

where $f(w)$ is the number of nodes with HV-word w, as defined previously. To provide a concrete way to compute the numbers $f(w)$ for a generic w, we need to recall some important facts about Baxter numbers, and in particular Baxter permutations.

The Generating Tree for Baxter Permutations. Let us first consider Baxter permutations. To that effect, recall that a LTR (left-to-right) maximum of a permutation π is an element π_i such that $\pi_i > \pi_j$ for all $j < i$. Similarly, a RTL maximum (resp. RTL minimum) of π is an element π_i such that $\pi_i > \pi_j$ (resp. $\pi_i < \pi_j$) for all $j > i$. Following [5], we can make Baxter permutations grow by adding new maximal elements to them, which may be inserted either immediately before a LTR maximum or immediately after a RTL maximum. Giving to any Baxter permutation the label (h, k) where h (resp. k) is the number of its RTL (resp. LTR) maxima, this gives the most classical succession rule for Baxter numbers.

Proposition 5 ([5]). *The growth of Baxter permutations by insertion of a maximal element is encoded by the succession rule:*

$$\Omega_{Bax} = \begin{cases} (1,1) \\ (h,k) \to (1, k+1), \ldots, (h, k+1) \, (h+1, 1), \ldots, (h+1, k), \end{cases}$$

where h (resp. k) is the number of RTL (resp. LTR) maxima.

We point out that, if we increase by 1 the two entries of the label of a mosaic floorplan for uniformity reasons, then mosaic floorplans and Baxter permutations both grow according to the generating tree Ω_{Bax}. This gives rise to a recursive bijection between the two families. Observe that this bijection maps each rectangle lying on the top (resp. right, bottom, left) boundary of the mosaic floorplan to a left-to-right maximum (resp. right-to-left maximum, right-to-left minimum, left-to-right minimum) of the corresponding Baxter permutation. In the next section, we will investigate how the HV-word of a mosaic floorplan can be read from the corresponding Baxter permutation.

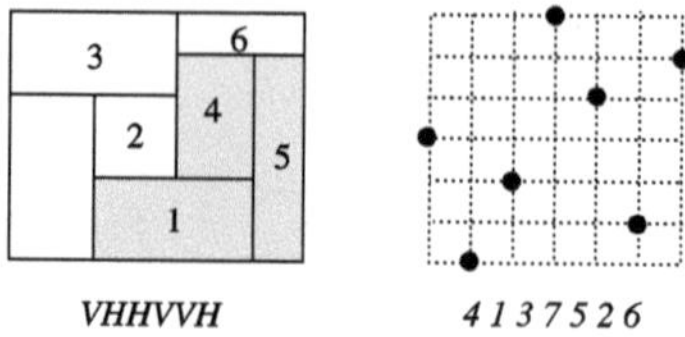

Fig. 6. Left: A mosaic floorplan F of size 7, label $(1,1)$, HV-word $w(F)$; the rectangles added with V operation are in grey color. Right: the corresponding Baxter permutation, with label $(2,2)$ and positional HV-word $w(F)$. (Color figure online)

Baxter Numbers by Ascent–Descent Words. The *positional word* of a Baxter permutation π is a compact way of recording the sequence of operations performed to obtain π from the root of the generating tree. We write H whenever an insertion is made among the left-to-right minima, and V otherwise. Formally, define the *positional HV-word* $P(\pi) = p_1 \cdots p_{n-1} \in \{H, V\}^{n-1}$ by comparing the positions of consecutive values:

$$p_k = \begin{cases} H & \text{if } \pi^{-1}(k+1) < \pi^{-1}(k) \quad \text{(value } k+1 \text{ lies to the left of value } k \text{ in } \pi\text{),} \\ V & \text{if } \pi^{-1}(k+1) > \pi^{-1}(k) \quad \text{(value } k+1 \text{ lies to the right of value } k \text{ in } \pi\text{).} \end{cases}$$

By construction, it is trivial that our recursive bijection maps a floorplan with HV-word w in a Baxter permutation with positional word w. For example, referring to the permutation in Fig. 6, the positional word is $VHHVVH$.

Let $n \geq 1$. For a permutation $\pi = \pi_1 \cdots \pi_n \in S_n$ define its *ascent–descent word*

$$\mathrm{ad}(\pi) = w_1 w_2 \cdots w_{n-1} \in \{a, d\}^{n-1}, \qquad w_i = \begin{cases} a & \text{if } \pi_i < \pi_{i+1}, \\ d & \text{if } \pi_i > \pi_{i+1}. \end{cases}$$

Let $\mathcal{B}_n$ be the set of Baxter permutations of size n. So, for instance, $\pi = 41352 \in \mathcal{B}_5$ has $\mathrm{ad}(\pi) = daad$. For a binary word $w \in \{a, d\}^{n-1}$, we set $\ell(w) := \#\{\pi \in \mathcal{B}_n : \mathrm{ad}(\pi) = w\}$. Let us investigate some basic properties of ascent–descent words.

Proposition 6 *If $\overline{w}$ denotes the bit-flip of w (swap $a \leftrightarrow d$), then $\ell(w) = \ell(\overline{w})$, and*

$$B(n) = \sum_{\substack{w \in \{a,d\}^{n-1} \\ w \leq_{\mathrm{lex}} \overline{w}}} \ell(w) = \tfrac{1}{2} \sum_{w \in \{a,d\}^{n-1}} \ell(w).$$

Hence B_n admits a refinement as a sum of exactly 2^{n-1} terms (one representative per $\{w, \overline{w}\}$-pair).

This is the same expression as in (1), and in the following we will explain it in a simple way.

Ascent–Descent Words Versus Positional Words for Baxter Permutations. We show that the family of words $AD(n) = \{\mathrm{ad}(\pi) : \pi \in \mathcal{B}_n\}$ and the family $\{P(\pi) : \pi \in \mathcal{B}_n\}$ coincide up to a fixed relabelling of the alphabet.

Theorem 5. *For every $n \geq 1$, the map $\pi \longmapsto \pi^{-1}$ induces a bijection*

$$\{\pi \in \mathcal{B}_n : \mathrm{ad}(\pi) = w\} \longleftrightarrow \{\sigma \in \mathcal{B}_n : P(\sigma) = \tau(w)\},$$

where $\tau : \{a, d\}^{n-1} \to \{H, V\}^{n-1}$ is the letter-to-letter map defined by $\tau(a) = V, \tau(d) = H$.

In particular, for each fixed word $w \in \{a, d\}^{n-1}$, the number of Baxter permutations with ascent–descent word w equals the number of Baxter permutations with positional word $\tau(w)$.

Proof. Let $\sigma = \pi^{-1}$. For each $k \in \{1, \ldots, n-1\}$, by definition of the positional word,

$$P(\sigma)_k = \begin{cases} H & \text{if } \sigma^{-1}(k+1) < \sigma^{-1}(k), \\ V & \text{if } \sigma^{-1}(k+1) > \sigma^{-1}(k), \end{cases} \quad \text{that is,} \quad P(\sigma)_k = \begin{cases} H & \text{if } \pi_{k+1} < \pi_k, \\ V & \text{if } \pi_{k+1} > \pi_k. \end{cases}$$

On the other hand, by definition it holds $\mathrm{ad}(\pi)_k = \begin{cases} a & \text{if } \pi_k < \pi_{k+1}, \\ d & \text{if } \pi_k > \pi_{k+1}. \end{cases}$

Hence, for every k, $P(\pi^{-1})_k = H$ iff $\mathrm{ad}(\pi)_k = d$, and $P(\pi^{-1})_k = V$ iff $\mathrm{ad}(\pi)_k = a$, Therefore $P(\pi^{-1}) = \tau(\mathrm{ad}(\pi))$, where $\tau(a) = V$ and $\tau(d) = H$. Since the class of Baxter permutations is closed under inversion, the map $\pi \mapsto \pi^{-1}$ is a bijection of $\mathcal{B}_n$ onto itself. $\square$

Example 1. Let $\pi = 4137526 \in \mathcal{B}_7$. The ascent–descent word of π is $\mathrm{ad}(\pi) = daadda$. The inverse permutation is $\pi^{-1} = 2631574$, and its ascent–descent word is $\mathrm{ad}(\pi^{-1}) = addaad$. Applying the relabelling $\tau(a) = V$, $\tau(d) = H$ to $\mathrm{ad}(\pi^{-1})$, we obtain $\tau(\mathrm{ad}(\pi^{-1})) = VHHVVH$. This is precisely the positional word of π, that is, $P(\pi) = VHHVVH$.

Counting Baxter Permutations with a Fixed Ascent–Descent Word. The most effective general formula for $\ell(w)$ (and hence for $f(w)$) arises from the classical bijections between Baxter permutations, plane bipolar orientations, and triples of non-intersecting lattice paths (see [10] and related literature).

Under the orientation/tree interpretation, fixing the ascent–descent word w prescribes the horizontal and vertical lengths of the segments on the outer face of the orientation, which correspond exactly to the run lengths of w. This information determines the starting and ending points of a triple of non-intersecting lattice paths (NILPs) with steps $E = (1, 0)$ and $N = (0, 1)$. Via the Lindström–Gessel–Viennot (LGV) lemma, we reduce the problem to the computation of a 3×3 determinant whose entries are binomial coefficients.

References

1. Asinowski, A., Cardinal, J., Felsner, S., Fusy, E.: Combinatorics of rectangulations: old and new bijections (2024). https://doi.org/10.48550/arxiv.2402.01483
2. Banderier, C., Bousquet-Mélou, M., Denise, A., Flajolet, P., Gardy, D., Gouyou-Beauchamps, D.: Generating functions for generating trees. Disc. Math. **246**, 29–55 (2002)
3. Barcucci, E., Del Lungo, A., Pergola, E., Pinzani, R.: ECO: a methodology for the enumeration of combinatorial objects. J. Diff. Eq. App. **5**, 435–490 (1999)
4. Beaton, N.R., Bouvel, M., Guerrini, V., Rinaldi, S.: Slicings of Parallelogram Polyominoes: Catalan, Schröder, baxter, and other sequences. The Electronic J. Combinatorics **26**(3), P3.13 (2019)
5. Bousquet-Mélou, M.: Four classes of pattern-avoiding permutations under one roof: generating trees with two labels. Electron. J. Combin. **9**(2), R19 (2003)
6. Brooks, R.L., Schmidt, C.A.B., Stone, A.H., Tutte, W.T.: The dissection of rectangles into squares. Duke Math. J. **7**, 312–340 (1940)
7. Cardinal, J., Sacristán, V., Silveira, R.I.: A note on flips in diagonal rectangulations. Discret. Math. Theor. Comput. Sci. **20**(2) (2018)
8. Chung, F.R.K., Graham, R., Hoggatt, V., Kleiman, M.: The number of Baxter permutations. J. Comb. Theory A **24**(3), 382–394 (1978)
9. Di Battista, G., Eades, E., Tamassia, R., Tollis, I.G.: Algorithms for drawing graphs: an annotated bibliography. Comput. Geom. Theory Appl. **4**(5), 235–282 (1994)
10. Felsner, S., Fusy, E., Noy, M., Orden, D.: Bijections for Baxter families and related objects. J. Comb. Theory Ser. A **118**(3), 993–1020 (2011)
11. Kimura, K., Haramiishi, T., Amano, K., Nakano, S.: Escape from the room. IEICE Trans. Inf. E108-D(3), 186–191 (2025)
12. Nakano, S., Yamanaka, K.: Enumerating floorplans with walls. Discret. Appl. Math. **342**, 1–11 (2024)
13. Nakano, S., Yamanaka, K., Rahman, M.S.: Floorplans with columns. Lect. Notes Comput. Sci., 33–40 (2017)
14. OEIS Foundation Inc., The On-line Encyclopedia of Integer Sequences (2011). http://oeis.org
15. Sciammacca, F.: Toward Floorplan Enumeration through Wall and Pattern Avoidance Analysis: A Focus on Guillotine and Diagonal Floorplans, Florence IT (2024). Master Thesis
16. Yao, B., Chen, H., Cheng, C.K., Graham, R.L.: Floorplan representations: complexity and connections. ACM Trans. Design Autom. Electron. Syst. **8**, 55–80 (2003)

Author Index